Oceanography: A Study of the Oceans

Oceanography: A Study of the Oceans

Editor: Suzy Bullock

R CALLISTO REFERENCE

www.callistoreference.com

Callisto Reference,
118-35 Queens Blvd., Suite 400,
Forest Hills, NY 11375, USA

Visit us on the World Wide Web at:
www.callistoreference.com

ISBN: 978-1-63239-990-8 (Hardback)

Cataloging-in-Publication Data

Oceanography : a study of the oceans / edited by Suzy Bullock.
   p. cm.
Includes bibliographical references and index.
ISBN 978-1-63239-990-8
1. Oceanography. 2. Ocean. I. Bullock, Suzy.
GC11.2 .O24 2018
551.46--dc23

Table of Contents

Preface...IX

Chapter 1 **Turnover time of fluorescent dissolved organic matter in the dark global ocean**...................1
Teresa S. Catalá, Isabel Reche, Antonio Fuentes-Lema, Cristina Romera-Castillo,
Mar Nieto-Cid, Eva Ortega-Retuerta, Eva Calvo, Marta Álvarez, Cèlia Marrasé,
Colin A. Stedmon and X. Antón Álvarez-Salgado

Chapter 2 **The timescales of global surface-ocean connectivity**...9
Bror F. Jönsson and James R. Watson

Chapter 3 **Catch reconstructions reveal that global marine fisheries catches are higher
than reported and declining**...15
Daniel Pauly and Dirk Zeller

Chapter 4 **Selective silicate-directed motility in diatoms**..24
Karen Grace V. Bondoc, Jan Heuschele, Jeroen Gillard, Wim Vyverman
and Georg Pohnert

Chapter 5 **Near-island biological hotspots in barren ocean basins**..30
Jamison M. Gove, Margaret A. McManus, Anna B. Neuheimer, Jeffrey J. Polovina,
Jeffrey C. Drazen, Craig R. Smith, Mark A. Merrifield, Alan M. Friedlander,
Julia S. Ehses, Charles W. Young, Amanda K. Dillon and Gareth J. Williams

Chapter 6 **An extreme event of sea-level rise along the Northeast coast of North America
in 2009–2010**..38
Paul B. Goddard, Jianjun Yin, Stephen M. Griffies and Shaoqing Zhang

Chapter 7 **Biological and physical controls in the Southern Ocean on past millennial-scale
atmospheric CO_2 changes**..47
Julia Gottschalk, Luke C. Skinner, JörgLippold, Hendrik Vogel, Norbert Frank,
Samuel L. Jaccard and Claire Waelbroeck

Chapter 8 **A role for diatom-like silicon transporters in calcifying coccolithophores**........................58
Grażyna M. Durak, Alison R. Taylor, Charlotte E. Walker, Ian Probert,
Colomban de Vargas, Stephane Audic, Declan Schroeder, Colin Brownlee
and Glen L. Wheeler

Chapter 9 **Deep-reaching thermocline mixing in the equatorial pacific cold tongue**........................70
Chuanyu Liu, Armin Köhl, Zhiyu Liu, Fan Wang and Detlef Stammer

Chapter 10 **A better-ventilated ocean triggered by Late Cretaceous changes in continental
configuration**...81
Yannick Donnadieu, Emmanuelle Pucéat, Mathieu Moiroud, François Guillocheau
and Jean-François Deconinck

Chapter 11   **Tropical Atlantic temperature seasonality at the end of the last interglacial**.................................**93**
Thomas Felis, Cyril Giry, Denis Scholz, Gerrit Lohmann, Madlene Pfeiffer,
Jürgen Pätzold, Martin Kölling and Sander R. Scheffers

Chapter 12   **Initialized decadal prediction for transition to positive phase of the Interdecadal
Pacific Oscillation**.................................**101**
Gerald A. Meehl, Aixue Hu and Haiyan Teng

Chapter 13   **Global marine protected areas do not secure the evolutionary history of tropical
corals and fishes**.................................**108**
D. Mouillot, V. Parravicini, D. R. Bellwood, F. Leprieur, D. Huang, P. F. Cowman,
C. Albouy, T. P. Hughes, W. Thuiller and F. Guilhaumon

Chapter 14   **Ocean currents generate large footprints in marine palaeoclimate proxies**.................................**116**
Erik van Sebille, Paolo Scussolini, Jonathan V. Durgadoo, Frank J. C. Peeters,
Arne Biastoch, Wilbert Weijer, Chris Turney, Claire B. Paris and Rainer Zahn

Chapter 15   **The exposure of the Great Barrier Reef to ocean acidification**.................................**124**
Mathieu Mongin, Mark E. Baird, Bronte Tilbrook, Richard J. Matear, Andrew Lenton,
Mike Herzfeld, Karen Wild-Allen, Jenny Skerratt, Nugzar Margvelashvili,
Barbara J. Robson, Carlos M. Duarte, Malin S. M. Gustafsson, Peter J. Ralph
and Andrew D. L. Steven

Chapter 16   **Morphological plasticity of the coral skeleton under CO_2-driven
seawater acidification**.................................**132**
E. Tambutté, A. A. Venn, M. Holcomb, N. Segonds, N. Techer, D. Zoccola,
D. Allemand and S. Tambutté

Chapter 17   **Bidecadal North Atlantic ocean circulation variability controlled by timing of
volcanic eruptions**.................................**141**
Didier Swingedouw, Pablo Ortega, Juliette Mignot, Eric Guilyardi,
Valérie Masson-Delmotte, Paul G. Butler, Myriam Khodri and Roland Séférian

Chapter 18   **Both respiration and photosynthesis determine the scaling of plankton
metabolism in the oligotrophic ocean**.................................**153**
Pablo Serret, Carol Robinson, María Aranguren-Gassis, Enma Elena García-Martín,
Niki Gist, Vassilis Kitidis, José Lozano, John Stephens, Carolyn Harris and
Rob Thomas

Chapter 19   **Glacial ice and atmospheric forcing on the Mertz Glacier Polynya over the past
250 years**.................................**162**
P. Campagne, Xavier Crosta, M. N. Houssais, D. Swingedouw, S. Schmidt, A. Martin,
E. Devred, S. Capo, V. Marieu, I. Closset and G. Massé

Chapter 20   **Global pulses of organic carbon burial in deep-sea sediments during
glacial maxima**.................................**171**
Olivier Cartapanis, Daniele Bianchi, Samuel L. Jaccard and Eric D. Galbraith

Chapter 21   **The absence of an Atlantic imprint on the multidecadal variability of wintertime
European temperature**.................................**178**
Ayako Yamamoto and Jaime B. Palter

Chapter 22   **Evidence for an ice shelf covering the central Arctic Ocean during the penultimate glaciation**..**186**
Martin Jakobsson, Johan Nilsson, Leif Anderson, Jan Backman, Göran Björk,
Thomas M. Cronin, Nina Kirchner, Andrey Koshurnikov, Larry Mayer,
Riko Noormets, Matthew O'Regan, Christian Stranne, Roman Ananiev,
Natalia Barrientos Macho, Denis Cherniykh, Helen Coxall, Björn Eriksson,
Tom Flodén, Laura Gemery, Örjan Gustafsson, Kevin Jerram, Carina Johansson,
Alexey Khortov, Rezwan Mohammad and Igor Semiletov

Chapter 23   **Recent increases in Arctic freshwater flux affects Labrador Sea convection and Atlantic overturning circulation**..**196**
Qian Yang, Timothy H. Dixon, Paul G. Myers, Jennifer Bonin, Don Chambers
and M. R. van den Broeke

Chapter 24   **Evidence for link between modelled trends in Antarctic sea ice and underestimated westerly wind changes**..**203**
Ariaan Purich, Wenju Cai, Matthew H. England and Tim Cowan

Chapter 25   **Oxygen depletion recorded in upper waters of the glacial Southern Ocean**...**212**
Zunli Lu, Babette A. A. Hoogakker, Claus-Dieter Hillenbrand, Xiaoli Zhou,
Ellen Thomas, Kristina M. Gutchess, Wanyi Lu, Luke Jones
and Rosalind E. M. Rickaby

Permissions

List of Contributors

Index

Preface

This book was inspired by the evolution of our times; to answer the curiosity of inquisitive minds. Many developments have occurred across the globe in the recent past which has transformed the progress in the field.

Oceanography focuses on both the physical aspects and living organisms of the ocean. The oceans are an important part of our ecology. Oceanography helps in comprehending issues such as global warming. Plate tectonics, ocean currents, geology of the sea floor, etc. are some of the widely studied aspects of oceanography. This book includes some of the vital pieces of work being conducted across the world, on various topics related to oceanography. This book is a complete source of knowledge on the present status of this important field.

This book was developed from a mere concept to drafts to chapters and finally compiled together as a complete text to benefit the readers across all nations. To ensure the quality of the content we instilled two significant steps in our procedure. The first was to appoint an editorial team that would verify the data and statistics provided in the book and also select the most appropriate and valuable contributions from the plentiful contributions we received from authors worldwide. The next step was to appoint an expert of the topic as the Editor-in-Chief, who would head the project and finally make the necessary amendments and modifications to make the text reader-friendly. I was then commissioned to examine all the material to present the topics in the most comprehensible and productive format.

I would like to take this opportunity to thank all the contributing authors who were supportive enough to contribute their time and knowledge to this project. I also wish to convey my regards to my family who have been extremely supportive during the entire project.

Editor

Turnover time of fluorescent dissolved organic matter in the dark global ocean

Teresa S. Catalá[1], Isabel Reche[1], Antonio Fuentes-Lema[2], Cristina Romera-Castillo[3,4], Mar Nieto-Cid[3], Eva Ortega-Retuerta[4], Eva Calvo[4], Marta Álvarez[5], Cèlia Marrasé[4], Colin A. Stedmon[6] & X. Antón Álvarez-Salgado[3]

Marine dissolved organic matter (DOM) is one of the largest reservoirs of reduced carbon on Earth. In the dark ocean ($>200\,m$), most of this carbon is refractory DOM. This refractory DOM, largely produced during microbial mineralization of organic matter, includes humic-like substances generated *in situ* and detectable by fluorescence spectroscopy. Here we show two ubiquitous humic-like fluorophores with turnover times of 435 ± 41 and 610 ± 55 years, which persist significantly longer than the ~ 350 years that the dark global ocean takes to renew. In parallel, decay of a tyrosine-like fluorophore with a turnover time of 379 ± 103 years is also detected. We propose the use of DOM fluorescence to study the cycling of resistant DOM that is preserved at centennial timescales and could represent a mechanism of carbon sequestration (humic-like fraction) and the decaying DOM injected into the dark global ocean, where it decreases at centennial timescales (tyrosine-like fraction).

[1] Departamento de Ecología and Instituto del Agua, Universidad de Granada, 18071 Granada, Spain. [2] Departamento de Ecoloxía e Bioloxía Animal, Universidade de Vigo, 36208 Vigo, Spain. [3] CSIC Instituto de Investigacións Mariñas, 36208 Vigo, Spain. [4] CSIC Institut de Ciencies del Mar, 08003 Barcelona, Spain. [5] IEO Centro Oceanográfico de A Coruña, 15006 A Coruña, Spain. [6] National Institute of Aquatic Resources, Technical University of Denmark, 2920 Charlottenlund, Denmark. Correspondence and requests for materials should be addressed to I.R. (email: ireche@ugr.es).

The biological pump has long been recognized as a mechanism to remove CO_2 from the atmosphere, through photosynthesis and export of particulate and dissolved organic matter (DOM) to the deep ocean[1]. More recently, the transformation of biologically labile organic matter into refractory (long lifetime in the dark ocean)[2] compounds by prokaryotic activity has been termed the 'microbial carbon pump' and may also constitute an effective mechanism to accumulate reduced carbon in the dark ocean[3,4]. Given the large pool of refractory DOM (RDOM) in the oceans (~ 656 pg C)[2], understanding its generation, transformation and role in carbon sequestration is crucial in regards to understanding present and future CO_2 emission scenarios.

Some components of the marine RDOM pool absorb light and a fraction of them also emit fluorescence (fluorescent DOM, FDOM)[5,6]. These optical properties are used as tracers for circulation and biogeochemical processes in the dark ocean[7].

Here we obtain the distribution of FDOM in the ocean interior by measuring fluorescence intensities scanned over a range of excitation/emission wavelengths during the Spanish Malaspina 2010 circumnavigation of the globe (Fig. 1). Discrete fluorescent fractions can be discriminated from the measured spectra by applying parallel factor analysis (PARAFAC)[8] (see Methods). Four fluorescent components are isolated from the whole data set and appear to be ubiquitous and common in the dark global ocean[5,6]. Two components (C1 and C2) have a broad excitation and emission spectra, with excitation and emission maxima in the ultraviolet and visible region, respectively. These are traditionally referred to as humic-like components, which accumulate and have turnover times of 435 ± 41 and 610 ± 55 years, respectively. These turnover times are longer than the ~ 350 years that the dark (>200 m) global ocean takes to renew[9]. The two other components (C3 and C4) have narrower spectra with excitation and emission maxima below 350 nm and are similar to the spectra of tryptophan and tyrosine, respectively[10]. The dark global ocean appears to be a sink for fluorescent tyrosine-like (C4) component and has a turnover time of 379 ± 103 years that is comparable to turnover time of DOC pool (estimated in 370 years[11]). Thus, we propose that the fluorescent fractions of DOM are suitable proxies for determining the cycling of the RDOM that produces (humic-like) and decays (tyrosine-like) in the dark ocean at centennial timescales.

Results

Water masses across the circumnavigation. The water mass composition of each water sample was described through the mixing of prescribed water types (WT) with a multi-parameter analysis (see Methods). We identified 22 WT (see Table 1) with 12 of them representing 90% of the total volume of water samples collected during the global cruise (Fig. 2). Circumpolar Deep

Water (CDW) was widely distributed in the Indian and Pacific Oceans accounting for up to 25.6% of the total volume sampled. North Atlantic Deep Water of 2.0 °C ($NADW_{2.0}$) and 4.6 °C ($NADW_{4.6}$) accounted 21.4% of the total volume and mostly located in the Atlantic basin; Antarctic Intermediate Water of 3.1 °C ($AAIW_{3.1}$) and 5.0 °C ($AAIW_{5.0}$) accounted for 7.6% of the total volume and spread out through the South Atlantic and Indian basins; and North Pacific Intermediate Water accounted for 5.7% of the total volume sampled.

Global distribution of fluorescent components. The maximum fluorescence intensity (Fmax) of the two humic-like components (C1 and C2) obtained during the Malaspina circumnavigation showed a global distribution similar to the apparent oxygen utilization (AOU; Fig. 3a–c), reaching their maxima in the Eastern North Pacific central and intermediate waters and their minima in the Indian Ocean central waters. In contrast, this global trend with AOU was not evident for the two amino-acid-like components (C3 and C4; Fig. 3d,e).

Relationships between the archetypal AOU and fluorophores. Combining the outputs from the water mass, AOU and fluorescence analyses, we calculated WT proportion-weighted average values, hereafter referred to as archetypal, for the AOU and the fluorescence intensities of the four components (see Methods). Archetypal values retain the variability associated with the initial concentrations at the site where each WT is defined and its transformation by basin-scale mineralization processes up to the study site[12]. Archetypal concentrations explained 81, 77, 75, 24 and 26% of the total variability of AOU, C1, C2, C3 and C4, respectively (Table 1).

In Fig. 4, we show the measured maximum fluorescence intensity (grey dots), archetypal values for each WT (white dots) and archetypal values for each sample (black dots) for the components C1, C2, C3 and C4 (Fig. 4a–d, respectively). We obtained direct relationships between the archetypal humic-like components (C1 and C2) and archetypal AOU (Fig. 4a,b) suggesting a net production of these components in parallel with the water mass ageing. The relatively high archetypal fluorescence of C1 (blue dots in Fig. 4a) for North Atlantic Deep Water (NADW) is related to the high load of terrestrial fluorescent materials transported by the Arctic[6,13,14] rivers, whereas the cause of the high value for the Mediterranean water (MW) is due to the low proportion of this water mass ($9 \pm 14\%$) compared to the proportion of NADW ($76 \pm 33\%$) in the same sample. Therefore, it is expectable that the archetypal concentration of C1 that our data set produces for the MW should be close to the NADW archetype. For C2, the North Pacific Subtropical Mode Water ($STMW_{NP}$; green dot in Fig. 4b) also departed from the general archetypal C2–AOU trend. Archetypal C1 and C2 can be modelled with archetypal AOU values using power functions (Fig. 4a,b). Given that the high fluorescence of NADW and MW in C1 and of $STMW_{NP}$ in C2 is not related to ageing, these water masses were excluded from their corresponding regression models. It is noticeable that the power factor for C1 (0.51 ± 0.04) is almost twice than for C2 (0.31 ± 0.04), indicating a higher C1 production rate per unit of consumed oxygen. Furthermore, in hypoxic waters (dissolved oxygen concentrations $<60 \mu M$ (ref. 15); orange dots in Fig. 4a,b), C1 and C2 also behave differently; whereas C1 production was enhanced, C2 did not change substantially.

In contrast, we do not observe a significant relationship between the archetypal values of the tryptophan-like fluorescence component (C3) and AOU ($R^2 = 0.001$, $n = 22$, $P = 0.88$; Fig. 4c). This lack of correlation is caused by the low archetypal

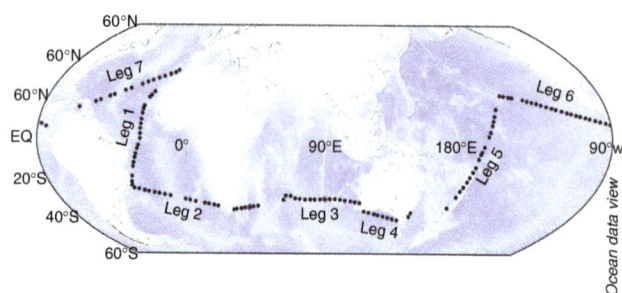

Figure 1 | Cruise track of the Malaspina 2010 circumnavigation. The cruise departed from Cartagena (Spain) on 17 December 2010 and returned to the same port on 15 July 2011, and was divided into seven legs.

Table 1 | Archetypal values of the depth (Z_i), AOU (AOU_i) and the four fluorescent components ($Fmax1_i$, $Fmax2_i$, $Fmax3_i$ and $Fmax4_i$) for the different WTs intercepted during the Malaspina 2010 circumnavigation.

Source water type	Acronym	VOL_i (%)	Z_i (m)	AOU_i ($\mu mol\,kg^{-1}$)	$Fmax1_i$ ($\times 10^{-3}$ RU)	$Fmax2_i$ ($\times 10^{-3}$ RU)	$Fmax3_i$ ($\times 10^{-3}$ RU)	$Fmax4_i$ ($\times 10^{-3}$ RU)
Eighteen Degrees Water	EDW	0.7	264±20	46±8	8.2±0.6	6.9±0.7	6.0±1.4	6.4±1.9
Eastern North Atlantic Central Water (12 °C)	$ENACW_{12}$	3.7	641±40	114±7	11.6±0.2	9.3±0.1	5.3±0.6	5.0±0.5
Eastern North Atlantic Central Water (15 °C)	$ENACW_{15}$	1.8	327±25	63±8	9.1±0.4	8.0±0.4	5.8±0.9	6.6±0.9
Equatorial Atlantic Central Water (13 °C)	13EqAtl	1.6	427±37	61±7	7.4±0.3	6.9±0.3	4.9±0.8	6.8±0.9
South Atlantic Central Water (12 °C)	$SACW_{12}$	2.6	303±26	110±16	9.3±0.6	8.4±0.5	5.9±0.8	7.8±0.7
South Atlantic Central Water (18 °C)	$SACW_{18}$	1.5	211±11	38±12	6.6±0.5	6.6±0.6	5.7±1.2	8.3±0.9
Indian Subtropical Mode Water	$STMW_I$	0.8	259±35	26±3	5.4±0.2	5.7±0.2	3.5±0.4	7.3±0.8
Indian Central Water (13 °C)	ICW_{13}	4.3	395±28	32±2	5.5±0.1	5.4±0.1	3.2±0.2	5.8±0.4
South Pacific Subtropical Mode Water	$STMW_{SP}$	0.2	277±84	49±11	6.2±0.4	6.2±0.3	2.8±0.1	6.0±1.1
South Pacific Central Water (20 °C)	$SPCW_{20}$	0.5	269±26	70±15	6.5±0.5	7.8±0.5	2.9±0.3	5.3±0.8
Equatorial Pacific Central Water (13 °C)	13EqPac	5.7	483±35	231±10	14.5±0.6	10.8±0.3	4.5±0.3	4.8±0.4
North Pacific Central Mode Water (12 °C)	CMW_{NP}	3.3	253±13	234±10	14.1±0.5	12.0±0.2	5.3±0.3	5.9±0.5
North Pacific Subtropical Mode Water (16 °C)	$STMW_{NP}$	0.2	207±36	111±6	9.0±0.4	11.8±1.7	6.3±1.6	6.8±0.1
Mediterranean Water	MW	0.4	1276±354	84±9	11.8±0.8	9.4±0.3	4.5±0.9	4.3±1.5
Sub-Antarctic Mode Water	SAMW	7.9	719±42	72±6	6.9±0.3	5.9±0.2	2.7±0.1	4.4±0.2
Antarctic Intermediate Water (3.1 °C)	$AAIW_{3.1}$	4.4	1317±108	134±5	10.5±0.4	7.8±0.2	3.1±0.3	3.8±0.4
Antarctic Intermediate Water (5.0 °C)	$AAIW_{5.0}$	3.2	677±36	128±8	10.1±0.4	8.1±0.3	5.2±0.8	6.1±0.6
North Pacific Intermediate Water	NPIW	5.7	671±65	255±6	16.4±0.4	11.4±0.2	4.3±0.2	4.0±0.3
Circumpolar Deep Water	$CDW_{1.6}$	25.6	2412±76	183±4	14.6±0.2	9.9±0.1	3.6±0.1	3.4±0.2
North Atlantic Deep Water (2 °C)	$NADW_{2.0}$	13.6	3279±66	88±2	13.3±0.1	10.1±0.1	6.2±0.4	6.1±0.4
North Atlantic Deep Water (4.6 °C)	$NADW_{4.6}$	7.8	1582±99	103±4	12.5±0.2	9.6±0.2	6.2±0.5	6.3±0.4
Antarctic Bottom Water	AABW	4.4	3780±64	149±6	13.4±0.2	9.6±0.1	3.7±0.3	3.9±0.4
R^2 (N_i versus $<N_i>$)				0.81	0.77	0.75	0.24	0.26
s.d. of the estimate				34	1.8	1.0	2.4	2.6
Analytical error				1.0	0.09	0.06	0.11	0.17

AOU, apparent oxygen utilization; WT, water types.
%VOL_i is the contribution of each WT to the total volume of water collected along the cruise and R^2 is the determination coefficient.

fluorescence ($< 4 \times 10^{-3}$ RU) of the relatively young ($AOU < 75\,\mu mol\,kg^{-1}$) central waters of the Indian and South Pacific oceans (ICW_{13}, $STMW_I$, SAMW, $SPCW_{20}$ and $STMW_{SP}$; Table 1; purple dots in Fig. 4c). However, such low archetypal values are not observed in the tyrosine-like component (C4), leading to a weak but significant inverse power relationship with AOU (Fig. 4d). The archetypal tyrosine-like fluorescence of the aged Central waters ($AOU > 200\,\mu mol\,kg^{-1}$) of the Equatorial (13EqPac) and Central North Pacific (CMW_{NP}) exceeded the expected value from their AOU (cyan dots in Fig. 4d). These two WT were excluded from the regression model.

Net FDOM production and turnover times. On the basis of the relationships observed between the archetypal fluorescence intensity of three out of four fluorescence components (C1, C2 and C4) and AOU, we calculated the net production rate of each component, termed net FDOM production (NFP). A positive value of NFP indicates net production, as for the case of the humic-like components C1 and C2, and a negative value indicates

net decay, as for the case of the amino-acid-like C4. The NFP of each component was calculated by multiplying the WT proportion-weighted average fluorescence production values per unit of AOU by the oxygen consumption rate (OCR) for the dark ocean (see Methods). Here we have used a conservative OCR estimate of 0.827 pmol O_2 per year[16]. The net humic-like fluorescence production obtained was $2.8 \pm 0.2 \times 10^{-5}$ Raman units per year (RU per year) for C1 and $1.5 \pm 0.1 \times 10^{-6}$ RU per year for C2, whereas C4 was consumed at a net rate of $-1.3 \pm 0.2 \times 10^{-5}$ RU per year in the dark global ocean.

Turnover times of components C1, C2 and C4 were calculated dividing the WT weighted-average fluorescence values of the dark global ocean by its corresponding NFP rate (see Methods). These values represent the time required to produce (C1, C2) or consume (C4) a fluorescence signal of the same intensity than the actual fluorescence of the dark ocean. The resulting turnover of C2, 610 ± 55 years, was significantly longer than the turnover of C1, 435 ± 41 years, and both exceeded the turnover times of the bulk DOC pool—estimated in 370 years[11]—and the terrestrial

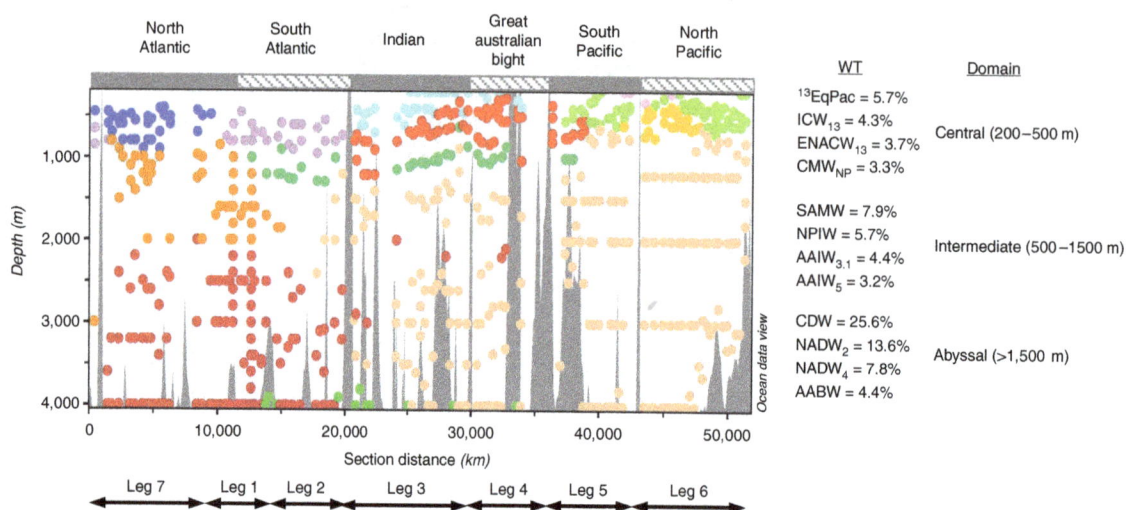

Figure 2 | Proportions of the main WT intercepted during Malaspina 2010. The percentages represent the proportion of the total volume of water sampled during the cruise that corresponds to each WT. Here the 12 most abundant WT are presented. Note that the depth range starts at 200 m. For more details see Table 1.

DOC in the open ocean—estimated in <100 years[17]—as well as the renewal time of the dark ocean (water depths >200 m)—estimated in 345 years[9]. Conversely, the turnover time of the tyrosine-like component C4 in 379 ± 103 years was compatible with the turnover time of the bulk DOC.

Discussion

The global pattern in the dark ocean of an increase in the humic-like components concomitant to water mass ageing (high AOU values) has been previously reported[5,6,18,19]. Although it has been recently hypothesized that this relationship could also be caused by further transformations of terrestrial humic-like materials in the open ocean[20], culture experiments have unequivocally demonstrated that these materials can be produced *in situ* in the oceans[21]. In fact, we observe positive and significant relationships between the archetypes of both humic-like components (C1 and C2) and the AOU (Fig. 4a,b). The higher C1 production rate per unit of consumed oxygen in comparison with C2 could be related to different mechanisms of production[6] that might be linked to the phylogenetic nature of producers (bacteria, archaea or eukarya)[22] and/or the sensitivity to environmental oxygen concentration.

The particularly high archetypal fluorescence of C1 for North Atlantic Deep Water (NADW) has been previously described and appears to be clearly related to the high load of FDOM of terrestrial origin transported from the Arctic rivers to the North Atlantic Ocean[6,13,14] with a relevant proportion of unaltered high molecular weight lignin[23]. However, the cause of the high C2 signature for the North Pacific Subtropical Mode Water has not been previously reported. We hypothesize that it could be related to intense rainfall south of the Kuroshio extension where these water mass is formed[24], since it is known that rainwater is particularly enriched in these fluorescent compounds[25,26] and this WT is very shallow (archetypal depth $= 277 \pm 84$ m; Table 1), which means that rainwater would dilute in a few tenths of metres during formation of that warm water mass. Indeed, lignin-derived phenols, highly modified by photo-oxidation, have been found in dissolved and submicron particles suspended in the North Pacific Subtropical Mode Water, suggesting an aerosol source for these fluorescent materials[27]. The high archetypal tyrosine-like fluorescence (C4) of the aged Central waters (AOU > 200 μmol kg^{-1}) of the Equatorial (13EqPac) and Central North Pacific (CMW$_{NP}$) might be due also to both WT

that occupy shallow layers (archetypal depths 253 ± 13 m for CMW$_{NP}$ and 483 ± 35 m for 13EqPac; Table 1), where protein-like fluorescence is higher because of the proximity to the epipelagic waters, where these materials are usually produced[6].

The turnover times of the fluorescent materials (timescale of centuries) are of the same order of magnitude of the turnover time of the bulk DOC[11], but an order of magnitude faster than the apparent age of the ocean DOC as derived from ^{14}C measurements (timescale of millennia)[2]. However, it should be noted that mean age, derived from ^{14}C involved by nuclear tests, is not homologous with turnover time (mean transit time), derived from total reservoir and fluxes entering/leaving the reservoir.

It is remarkable that the observed decrease in the tyrosine-like fluorescence in the dark ocean is at centennial timescales. It has been reported that the turnover of these fluorophores in the surface ocean is on a timescale of days[6,28], but this long-term decline in tyrosine-like fluorescence in the dark global ocean, coupled to water mass ageing, has never been reported. We can hypothesize that a minor fraction of the tyrosine-like fluorescence is processed on the scale of centuries, whereas the bulk of the signal has a turnover time on the order of days to weeks. Furthermore, this apparent discrepancy could also be related to the different turnover of the set of compounds that are represented by this fluorescence signature[29].

We conclude that humic-like fluorescence (C1 and C2) reveals a suitable marker of the production of optically active RDOM with turnover times of 400–600 years. Using the oceans as an incubator, our measurements indicate that the *in situ* microbial production of fluorescent humic-like materials in the dark global ocean is a sink of reduced carbon in the timescale of hundreds of years. Conversely, the turnover time of the tyrosine-like component (C4) was compatible with the turnover time of the bulk DOC and both decline with water mass ageing. This coincidence between the turnover times of the bulk DOC pool and the tyrosine-like component could also have applications for tracing long-term DOC reactivity in the dark ocean.

Methods

Sample collection. A total of 147 stations were sampled from 40°S to 34°N in the Atlantic, Indian and Pacific Oceans (Malaspina 2010 Expedition, from December 2010 to July 2011). Vertical profiles of salinity (S), potential temperature (θ) and dissolved oxygen (O$_2$ μmol kg^{-1}) were recorded continuously with the

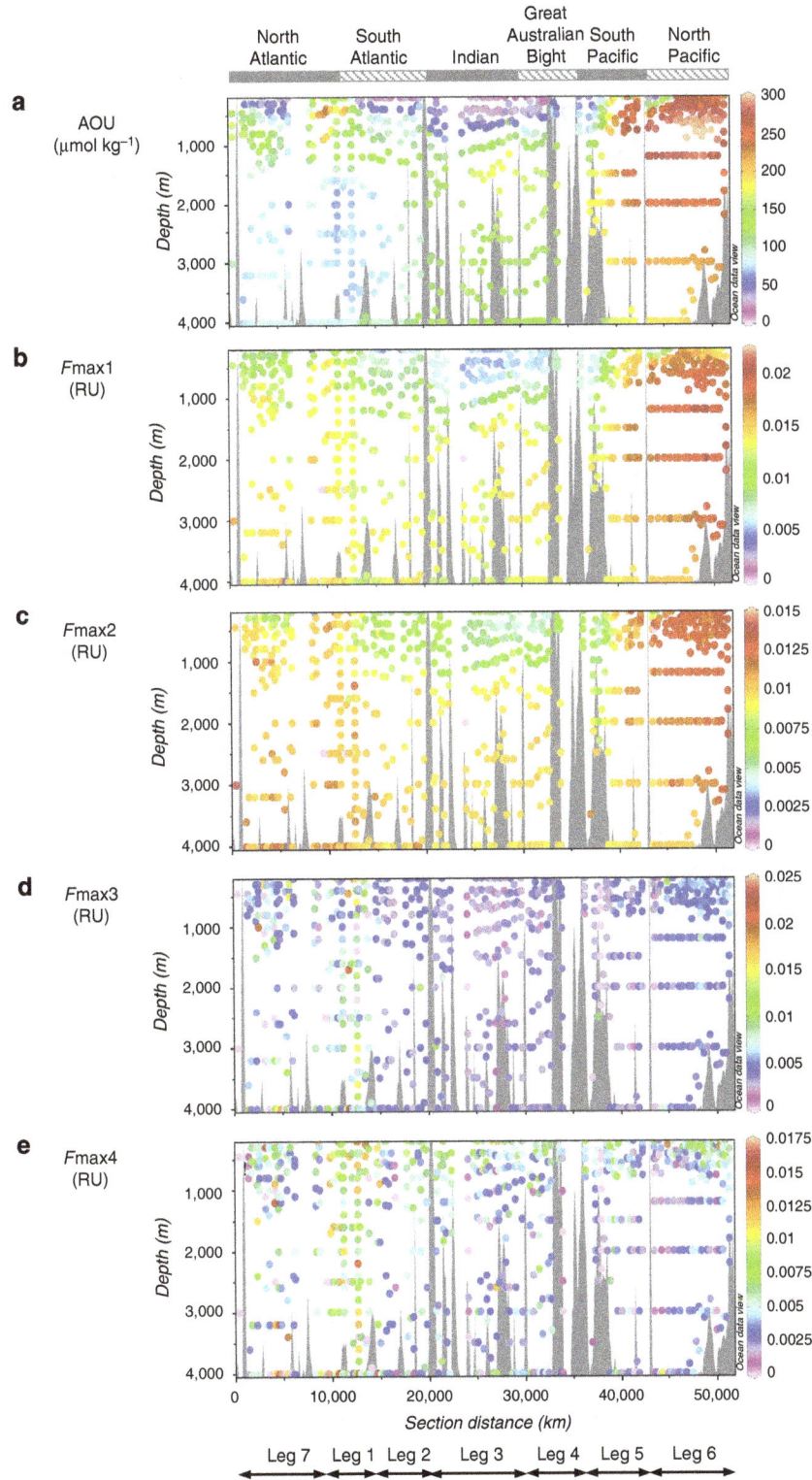

Figure 3 | Global distribution of the AOU and fluorescence components. The AOU (**a**) and the fluorescence intensity at the excitation–emission maxima of each component (Fmax) of the four components (**b-e**) discriminated by the PARAFAC analysis in the global ocean data set of the Malaspina 2010 Expedition are plotted. Note that the depth range starts at 200 m. See Methods and Supplementary Fig. 2 for a detailed description of the four fluorescence components.

conductivity–temperature–depth (CTD) and oxygen sensors installed in the rosette sampler. Salinity and dissolved oxygen were calibrated against bottle samples determined on board with a salinometer and the Winkler method, respectively. The AOU was calculated as the difference between the saturation and measured dissolved oxygen. Oxygen saturation was calculated from salinity and potential temperature[30]. Bottle depths were chosen on the basis of the CTD-O_2 profiles to cover as much water masses as possible of the dark global ocean. Seawater samples for fluorescence measurements, collected in 12 l Niskin bottles, were immediately poured into glass bottles and stored in dark conditions until measurement within 6 h from collection. We collected 800 water samples from 200 to 4,000 m depth.

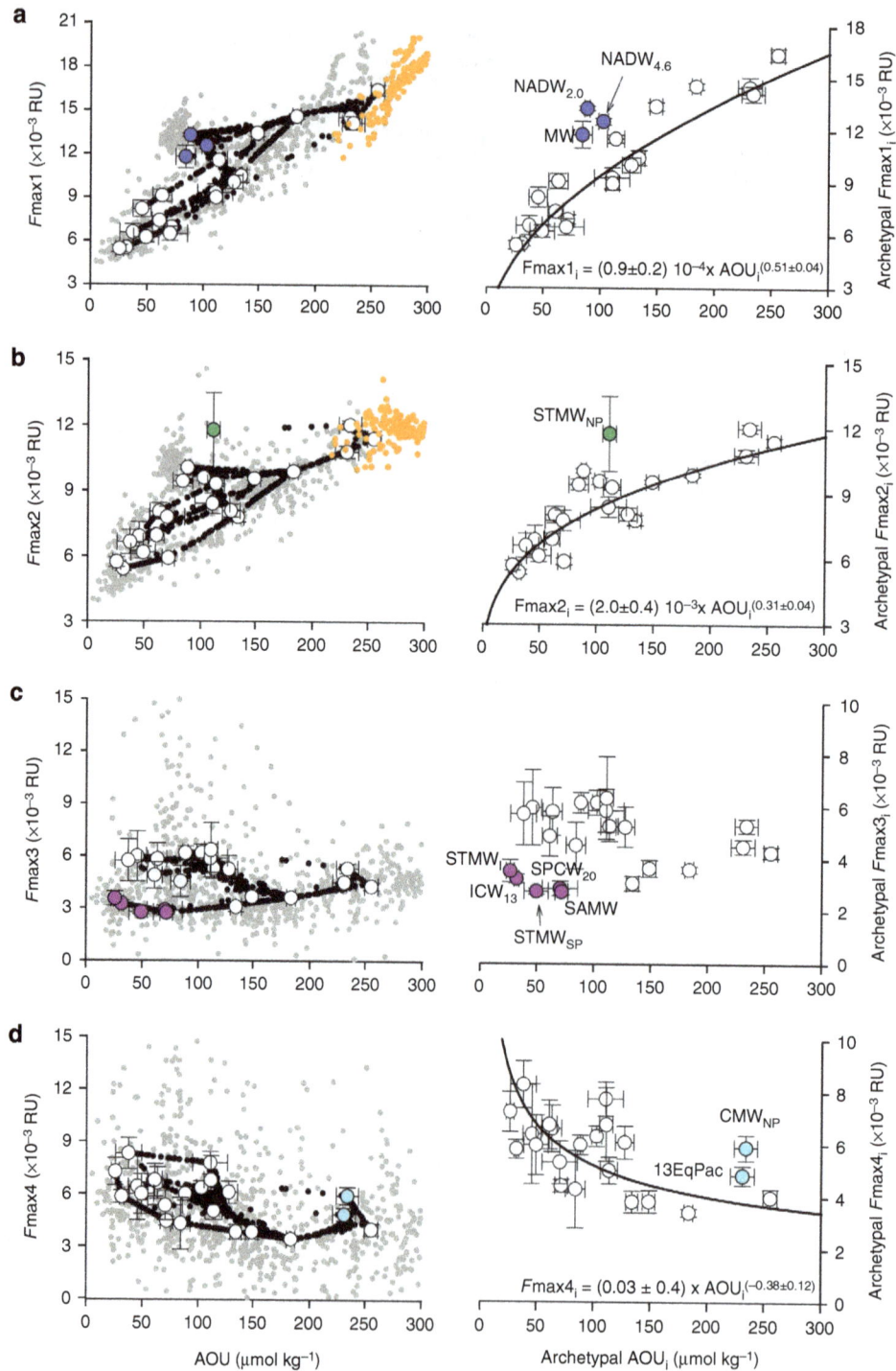

Figure 4 | Relationships between the fluorescence components and the AOU. Measured concentrations (grey dots), archetypal concentrations for each WT (white dots) and archetypal concentrations for each sample (black dots) for the components C1 (**a**), C2 (**b**), C3 (**c**) and C4 (**d**). Orange dots represent hypoxic samples. In the right column are presented the regression models and fitting equations for the archetypal values of C1 ($R^2 = 0.90$, $P < 0.001$, $n = 19$), C2 ($R^2 = 0.79$, $P < 0.001$, $n = 21$) and C4 ($R^2 = 0.31$, $P < 0.05$, $n = 20$) with AOUi. WT shown in blue, green and cyan dots were excluded from their respective regression models. Error bars represent the s.d. of the estimated archetypal values of AOU and the four fluorescent components for each WT (see Methods; equation (6)).

Fluorescence spectral acquisition. When the coloured fraction of marine DOM (CDOM) is irradiated with ultraviolet light, it emits a fluorescence signal characteristic of both amino-acid- and humic-like compounds, which is collectively termed FDOM[10]. Fluorescence excitation–emission matrices (EEMs) were collected with a JY-Horiba Spex Fluoromax-4 spectrofluorometer at room temperature (around 20 °C) using 5 nm excitation and emission slit widths, an integration time of 0.25 s, an excitation range of 240–450 nm at 10 nm increments and an emission range of 300–560 nm at 2 nm increments. To correct for lamp

spectral properties and to compare results with those reported in other studies, spectra were collected in signal-to-reference (S:R) mode with instrument-specific excitation and emission corrections applied during collection, and EEMs were normalized to the Raman area (RA). In our case, the RA and its baseline correction were performed with the emission scan at 350 nm of the Milli-Q water blanks and the area was calculated following the trapezoidal rule of integration[29].

To track the variability of the instrument in the Raman, protein- and humic-like regions of the spectrum during the 147 working days of the expedition and assess gradual spectral bias, three standards were run daily: (1) a P-terphenyl block (Stranna) that fluoresces in the protein region, between 310 and 600 nm exciting at 295 nm; (2) a tetraphenyl butadiene block (Stranna) that fluoresces in the humic region, between 365 and 600 nm exciting at 348 nm; and (3) a sealed Milli-Q cuvette (Perkin Elmer) scanned between 365 and 450 nm exciting at 350 nm. Supplementary Figure 1a shows that the temporal evolution of the RA of the Milli-Q water used on board and the reference P-terphenyl and tetraphenyl butadiene materials were parallel, which confirms that the Raman normalization was successful in both the protein- and the humic-region of the EEMs. Therefore, no additional drift corrections were necessary. The comparison between the reference sealed Milli-Q (sMQ) and the daily Milli-Q water allowed us to demonstrate that the Milli-Q water used on board was of a good quality (Supplementary Fig. 1b). The average coefficient of variation between the sMQ and Milli-Q water throughout the circumnavigation was only 0.81%. Similarly, two scans of the reference sealed Milli-Q were measured at the beginning (sMQ$_1$) and at the end (sMQ$_2$) of each session, which reveals a slight shift of the fluorescence intensities along each working day (Supplementary Fig. 1c). The initial and final sMQ spectra were separated by about 10 h of continuous work of the spectrofluorometer. We found that the average coefficient of variation between sMQ$_1$ and sMQ$_2$ was 1.62%. Therefore, the daily instrument shift was low and about twice the long-term variability throughout the circumnavigation.

Inner-filter correction was not applied due to the low absorption coefficient of CDOM of the samples collected during the circumnavigation: $1.01 \pm 0.04\,m^{-1}$ (average ± s.d.) at 250 nm, that is, much lower than the threshold of $10\,m^{-1}$ above which this correction is required[8]. Raman-normalized Milli-Q blanks were subtracted to remove the Raman scattering signal[31,32]. RA normalization, blank subtraction and generation of EEMs were performed using MATLAB (version R2008a).

Global PARAFAC modelling. PARAFAC was used to identify the fluorescent components that comprise the EEMs in the global ocean. PARAFAC was performed using the DOMFluor 1_7 Toolbox[8]. Before the analysis, Rayleigh scatter bands (first order at each wavelength pair where Ex = Em ± bandwidth; second order at each wavelength pair where Em = 2 × Ex ± (2 × bandwidth)) were trimmed. The global PARAFAC model was derived based on 1,574 corrected EEMs and was validated using split-half validation and random initialization[8]. A four-component model was obtained (Supplementary Fig. 2), two of them humic-like nature, peak A/C (at Ex/Em < 270–370/470 nm) and peak M (at Ex/Em 320/400 nm), and two of amino-acid-like nature, attributed to tryptophan and tyrosine at 290/340 and 270/310 nm[8], respectively. Here we report the maximum fluorescence (Fmax) in Raman units (RU) (refs. 31,32).

Multi-parameter water mass analysis. The dark ocean (from 200 m to the bottom) can be described by the mixing of prescribed WT, characterized by a unique combination of thermohaline and chemical property values. Water mass analysis quantifies the proportions of the WT that contribute to a given water sample. In our case, we have characterized the WT on the basis of its salinity (S) and potential temperature (θ), which are assumed to be conservative parameters and, therefore, they do not change from the area where the WT are defined to the study area.

The equations to be solved for a water sample j are:

$$100 = \sum_i x_{ij} \tag{1}$$

$$\theta_j = \sum_i x_{ij} \cdot \theta_i \tag{2}$$

$$S_j = \sum_i x_{ij} \cdot S_i, \tag{3}$$

where x_{ij} is the proportion of WT i in sample j; θ_j and S_j are the thermohaline characteristics of sample j; and θ_i and S_i are the prescribed thermohaline characteristics of WT i in the area where it is defined. Furthermore, the solution of the multi-parameter water mass analysis includes an additional constraint, that is, all contributions (x_{ij}) have to be non-negative.

We identified 18 water masses and 22 WT on the route sampled and their characteristics are summarized in Supplementary Table 1. They were divided into three domains according to their depth: central, intermediate and abyssal waters. In the central domain, there are Eighteen Degrees Water (EDW), Eastern North Atlantic Central Water (ENACW), defined by two WT of 12 and 15 °C, Equatorial Atlantic Central Water (13 °C; 13EqAtl), South Atlantic Central Water (SACW), defined by two WT of 12 and 18 °C, Indian Subtropical Mode Water (STMW$_I$), Indian Central Water (13 °C; ICW$_{13}$), South Pacific Subtropical Mode Water (STMW$_{SP}$), South Pacific Central Water (20 °C; SPCW$_{20}$), Equatorial Pacific Central Water (13 °C; 13EqPac), North Pacific Subtropical Mode Water (STMW$_{NP}$) and North Pacific Central Mode Water (12 °C; CMW$_{NP}$). In the intermediate domain, we found Mediterranean Water (MW), Sub-Antarctic Mode Water (SAMW), Antarctic Intermediate Water (AAIW), defined by two WT of 3.1 and 5.0 °C, and North Pacific Intermediate Water (NPIW). In the abyssal domain, there are Circumpolar Deep

Water (1.6 °C; CDW), North Atlantic Deep Water (NADW), defined by two types of 2 and 4.6 °C, and Antarctic Bottom Water (AABW).

Equations (1)–(3) can be solved for a maximum of three WT simultaneously. Given that we have identified 22 WT, we have grouped the WT in the triads presented in Supplementary Table 2 on the basis of reasonable vertical and geographical constraints to the water mass mixing usually applied in the analysis of water masses. Concerning the vertical constraints, for a given region of the ocean, every WT will mix only with the WT situated immediately above and below according to their density. Concerning the geographical constraints, every WT will mix preferentially with WT in their surroundings.

The multi-parameter water mass analysis was applied to the 800 samples from the dark ocean ($\theta < 18$ °C, AOU > 0) where corresponding measurements of fluorescence (Fmax1, Fmax2, Fmax3 and Fmax4) and AOU were obtained.

Once the WT proportions (x_{ij}) are known, the proportion of the total volume of water sampled that corresponds to WT i (%VOL$_i$) can be calculated as:

$$\% \, VOL_i = \frac{\sum_j x_{ij}}{n}, \tag{4}$$

Where $n = 800$ is the number of deep samples.

Archetypal values of Fmax and AOU for each WT. Using the measured values of fluorescence and AOU (N) and the proportions of the 22 WT identified in this study (x_{ij}), the water mass proportion-weighted average concentration of N in each WT, N_i, termed archetypal value of N, can be calculated as[12]:

$$N_i = \frac{\sum_j x_{ij} \cdot N_j}{\sum_j x_{ij}} \tag{5}$$

where N_j is the concentration of N in sample j.

The s.d. of the estimated archetypal value of N_i was obtained by:

$$s.d. \, (N_i) = \frac{\sqrt{\sum_j x_{ij} \cdot \left(N_j - N_i\right)^2}}{\sum_j x_{ij}}. \tag{6}$$

Similarly, the archetypal value of variable N in every sample (N_j) was calculated as follows:

$$N_j = \frac{\sum_j x_{ij} \times N_i}{100} \tag{7}$$

The determination coefficient (R^2) and the s.d. of the residual (s.d. res) of the linear regressions of N_j (measured values) versus <N_j> (archetypal values) allows assessing the degree of dependence of variable N on WT mixing in the dark ocean.

Application of equations (5) and (6) to the collection depth of the samples (Z_i) allows obtaining the archetypal depth of each WT (Z_i) and its corresponding s.d. (Z_i).

NFP and turnover times of the components. The tight relationship between fluorescence intensity and AOU for components C1, C2 and C4 allows estimating the rate of change of Fmax per AOU unit (∂Fmax/∂AOU). It was calculated as the first derivative of the power functions fitting Fmax$_i$ and AOU$_i$ (Fig. 4). Then, the global net production of each fluorescence component, NFP (in RU per year), was calculated as:

$$NFP = \frac{\sum_j VOL_i \cdot \left(\frac{\partial Fmax}{\partial AOU}\right)_i}{100} \cdot OCR \tag{8}$$

where $\sum VOL_i \cdot (\partial Fmax/\partial AOU)_i/100$ is the WT proportion-weighted fluorescence change rate per AOU unit of the global dark ocean. The OCR was calculated by dividing 0.827 pmol O$_2$ per year, a conservative estimate of the global oxygen demand of the dark ocean based on organic carbon sedimentation fluxes[16], by 1.38×10^{21} kg, which is the mass of the ocean at a depth > 200 m.

Once the NFP of the three components was obtained, we calculated their respective turnover time (τ):

$$\tau = \frac{\sum_i VOL_i \cdot Fmax_i}{100 \cdot NEP} \tag{9}$$

Where $\sum VOL_i \cdot Fmax_i/100$ is the WT proportion-weighted average fluorescence of the dark global ocean.

References

1. Ducklow, H. W., Steinberg, D. K. & Buesseler, K. O. Upper ocean carbon export and the biological pump. *Oceanography* **14,** 50–58 (2001).
2. Hansell, D. A. Recalcitrant dissolved organic carbon fractions. *Annu. Rev. Mar. Sci.* **5,** 421–445 (2013).

3. Ogawa, H., Amagai, Y., Koike, I., Kaiser, K. & Benner, R. Production of refractory dissolved organic matter by bacteria. *Science* **292**, 917–920 (2001).

4. Jiao, N. *et al.* Microbial production of recalcitrant dissolved organic matter: long-term carbon storage in the global ocean. *Nat. Rev. Microbiol.* **8**, 593–599 (2010).

5. Yamashita, Y. & Tanoue, E. Production of bio-refractory fluorescent dissolved organic matter in the ocean interior. *Nat. Geosci.* **1**, 579–582 (2008).

6. Jørgensen, L. *et al.* Global trends in the fluorescence characteristics and distribution of marine dissolved organic matter. *Mar. Chem.* **126**, 139–148 (2011).

7. Nelson, N. B. & Siegel, D. A. The global distribution and dynamics of chromophoric dissolved organic matter. *Annu. Rev. Mar. Sci.* **5**, 447–476 (2013).

8. Stedmon, C. A. & Bro, R. Characterizing dissolved organic matter fluorescence with parallel factor analysis: a tutorial. *Limnol. Oceanogr. Methods* **6**, 572–579 (2008).

9. Laruelle, G.G. *et al.* Anthropogenic perturbations of the silicon cycle at the global scale: key role of the land-ocean transition. *Glob. Biogeochem. Cycles* **23**, GB4031 (2009).

10. Coble, P. G. Characterization of marine and terrestrial DOM in seawater using excitation–emission matrix spectroscopy. *Mar. Chem.* **51**, 325–346 (1996).

11. Hansell, D. A., Carlson, C. A., Repeta, D. J. & Schlitzer, R. Dissolved organic matter in the ocean. A controversy stimulates new insights. *Oceanography* **22**, 202–211 (2009).

12. Álvarez-Salgado, X. A. *et al.* New insights on the mineralization of dissolved organic matter in central, intermediate, and deep water masses of the northeast North Atlantic. *Limnol. Oceanogr.* **58**, 681–696 (2013).

13. Amon, R. M. W., Budéus, G. & Meon, B. Dissolved organic carbon distribution and origin in the Nordic Seas: exchange with the Arctic Ocean and the North Atlantic. *J. Geophys. Res.* **108**, 3221 (2003).

14. Benner, R., Louchouarn, P. & Amon, R. M. W. Terrigenous dissolved organic matter in the Arctic Ocean and its transport to surface and deep waters of the North Atlantic. *Glob. Biogeochem. Cycles* **19**, GB2025 (2005).

15. Naqvi, S. W. A. *et al.* Marine hypoxia/anoxia as a source of CH_4 and N_2O. *Biogeosciences* **7**, 2159–2190 (2010).

16. Andersson, J. H. *et al.* Respiration patterns in the deep ocean. *Geophys. Res. Lett.* **31**, L03304 (2004).

17. Hernes, P. J. & Benner, R. Photochemical and microbial degradation of dissolved lignin phenols: implications for the fate of terrigenous dissolved organic matter in marine environments. *J. Geophys. Res.* **108**, 3291–3299 (2003).

18. Chen, R. F. & Bada, J. L. The fluorescence of dissolved organic matter in seawater. *Mar. Chem.* **37**, 191–221 (1992).

19. Hayase, K. & Shinozuka, N. Vertical distribution of fluorescent organic matter along with AOU and nutrients in the equatorial Central Pacific. *Mar. Chem.* **48**, 283–290 (1995).

20. Andrew, A. A., Del Vecchio, R., Subramaniam, R. A. & Blough, N. V. Chromophoric dissolved organic matter (CDOM) in the Equatorial Atlantic Ocean: optical properties and their relation to CDOM structure and source. *Mar. Chem.* **148**, 33–43 (2013).

21. Jørgensen, L., Stedmon, C. A., Granskog, M. A. & Middelboe, M. Tracing the long-term microbial production of recalcitrant fluorescent dissolved organic matter in seawater. *Geophys. Res. Lett.* **41**, 2481–2488 (2014).

22. Romera-Castillo, C., Sarmento, H., Álvarez-Salgado, X. A., Gasol, J. M. & Marrasé, C. Net production and consumption of fluorescent colored dissolved organic matter by natural bacterial assemblages growing on marine phytoplankton exudates. *Appl. Environ. Microbiol.* **77**, 7490–7498 (2011).

23. Hernes, P. J. & Benner, R. Terrigenous organic matter sources and reactivity in the North Atlantic Ocean and a comparison to the Arctic and Pacific oceans. *Mar. Chem.* **100**, 66–79 (2006).

24. Qiu, B. The Kuroshio Extension system: its large-scale variability and role in the mid latitude ocean-atmosphere interaction. *J. Oceanogr.* **58**, 57–75 (2002).

25. Kieber, R. J., Whitehead, R. F., Reid, S. N., Willey, J. D. & Seation, P. J. Chromophoric dissolved organic matter (CDOM) in rainwater, Southeastern North Carolina, USA. *J. Atmos. Chem.* **54**, 21–41 (2006).

26. Santos, S. M. P., Otero, M., Duarte, R. M. B. O. & Duarte, A. C. Spectroscopic characterization of dissolved organic matter isolated from rainwater. *Chemosphere* **74**, 1053–1061 (2009).

27. Hernes, P. J. & Benner, R. Transport and diagenesis of dissolved and particulate terrigenous organic matter in the North Pacific Ocean. *Deep-Sea Res. I* **49**, 2119–2132 (2002).

28. Lønborg, C. *et al.* Assessing the microbial bioavailability and degradation rate constants of dissolved organic matter by fluorescence spectroscopy in the coastal upwelling system of the Ría de Vigo. *Mar. Chem.* **119**, 121–129 (2010).

29. Stubbins, A. *et al.* What's in an EEM? Molecular signatures associated with dissolved organic fluorescence in boreal Canada. *Environ. Sci. Technol.* **48**, 10598–10606 (2014).

30. Benson, B. B. & Krauss, Jr D. The concentration and isotopic fractionation of oxygen dissolved in freshwater and seawater in equilibrium with the atmosphere. *Limnol. Oceanogr.* **29**, 620–632 (1984).

31. Murphy, K. R. *et al.* Measurement of dissolved organic matter fluorescence in aquatic environments: an interlaboratory comparison. *Environ. Sci. Technol.* **44**, 9405–9412 (2010).

32. Stedmon, C. A., Markager, S. & Bro, R. Tracing dissolved organic matter in aquatic environments using a new approach to fluorescence spectroscopy. *Mar. Chem.* **82**, 239–254 (2003).

Acknowledgements

We thank C.M. Duarte for the coordination of the Malaspina expedition; the members of the physical oceanography party for collecting, calibrating and processing the CTD data; the chief scientists of the seven legs, the staff of the Marine Technology Unit (CSIC-UTM) and the Captain and crew of R/V Hespérides for their outright support during the circumnavigation; N. Mladenov for the initial protocols; F. Iuculano for fluorescence measurements; and Cintia L. Ramón for support with Matlab software. This study was financed by the Malaspina 2010 circumnavigation expedition (grant number CSD2008-00077). C.R.-C. acknowledges funding through a Beatriu de Pinos postdoctoral fellowship from the Generalitat de Catalunya. M.N.-C. was funded by the CSIC Program 'Junta para la Ampliación de Estudios' cofinanced by the ESF.

Author contributions

T.S.C., I.R. and X.A.A.-S. contributed equally to this work. T.S.C., A.F.-L., C.R.-C., M.N.-C., E.O.-R., C.M., E.C. and X.A.A.-S. contributed to data acquisition during the circumnavigation. T.S.C., M.A. and X.A.A.-S. carried out the post-cruise data preparation and run the water mass analyses. T.S.C. and C.A.S. executed the PARAFAC modelling. All authors contributed to the interpretation of the data and the discussion of the results presented in the manuscript.

Additional information

2

The timescales of global surface-ocean connectivity

Bror F. Jönsson[1] & James R. Watson[2,3]

Planktonic communities are shaped through a balance of local evolutionary adaptation and ecological succession driven in large part by migration. The timescales over which these processes operate are still largely unresolved. Here we use Lagrangian particle tracking and network theory to quantify the timescale over which surface currents connect different regions of the global ocean. We find that the fastest path between two patches—each randomly located anywhere in the surface ocean—is, on average, less than a decade. These results suggest that marine planktonic communities may keep pace with climate change—increasing temperatures, ocean acidification and changes in stratification over decadal timescales—through the advection of resilient types.

[1] Department of Geosciences, Princeton University, Princeton, New Jersey 08544, USA. [2] College of Earth, Ocean and Atmospheric Sciences, Oregon State University, Corvallis, Oregon 97331-5503, USA. [3] The Stockholm Resilience Centre, Stockholm University, 118 14 Stockholm, Sweden. Correspondence and requests for materials should be addressed to B.F.J. (email: bjonsson@princeton.edu) or to J.R.W. (email: james.watson@su.se).

Two different paradigms are used to explain the structuring of planktonic communities (bacteria, phytoplankton and zooplankton) in ocean ecosystems. One fundamental idea is that 'everything is everywhere but the environment selects'[1,2]. That is, different regions of the ocean are connected by ocean currents, resulting in potentially panmictic planktonic communities[3]. It is then the differential response of species to environmental conditions that leads to community structure[4,5]. The alternative is that regions of the oceans are not so well connected, and that this isolation leads to divergent evolution and hence differences in which species are where[6,7]. Recent work has shown that neither concept is wholly accurate, with a number of examples showing slight spatial structure on global and regional scales in marine microbial communities[8–10] and sometimes strong genetic differentiation at small spatial scales[11]. Such examples suggest that dispersal limitation is important in specific areas of the ocean. However, most studies have focused on patterns of community composition, or genetic differences of individuals within a species. The key mechanism itself—dispersal by surface ocean currents—has rarely[12] been explored and quantified.

One common method for investigating the timescales of dispersal for marine organisms is to calculate oceanographic distances by tracking virtual particles in modeled velocity fields[13,14]. Particle tracking has been used in a variety of ways to explain Lagrangian processes in the ocean, such as biological dispersal on regional scales[14–17], connectivity between different coastal habitats like coral reefs[18,19] and deep-water transport pathways[20].

Normally, oceanographical distances are defined as the expected connectivity times, or the mean time it takes for particles to travel from one location to another[15,21]. Another option is to instead use minimum connectivity times (Min-T), or distances based on the fastest times that particles can travel from one location to another. Minimum connection times have major advantages over expected connection times for problems concerning global plankton communities. First, the minimum connection time is a more appropriate metric for phytoplankton and bacterial connectivity since asexually reproducing organisms have high reproductive output that attenuates low dispersal probabilities, and only a few individuals are required to 'connect' two places, especially in terms of population genetics[22]. It is therefore possible for such organisms to exploit dispersal routes where the probability to reach a given destination is very low. Previous empirical work have also shown that minimum connection times provide better correspondence with the genetic similarity of groups[13] and that minimum connection times are more relevant to community similarity than the mean. Second, mean or median transit times in the global ocean are not well defined, as water can recirculate eternally and, hence, every particle seeded in a given patch eventually will reach all other patches if enough time is provided. Constraining the particles to a maximum advection time or using a lower percentile of connections (for example, the fastest 20%) would create results that mainly depend on on those arbitrary cutoffs. It is challenging to identify one physically motivated timescale (loop time of the the subtropical gyres, typical time spent in the gulf stream, time to circumfer the Antarctic Circumpolar Current and so on) that is generally applicable to all regions in the ocean. The estimates of minimum connection times for the global surface ocean are stored in the form of a matrix (the Min-T connectivity matrix) where each i, j element represents the shortest transit time between a given source patch i and destination patch j.

Connectivity matrices produced from Lagrangian particle tracking tend to be highly sparse with most pairs of patches unconnected. Indeed, to estimate connection times between all pairs of patches globally, an infeasible number of Lagrangian particles[23] would be needed. To circumvent this obstacle, a shortest path algorithm can be used to calculate missing connections. Here the network is the global ocean, with patches in the ocean as nodes and minimum connection times as edges connecting the nodes. Applied to this network, shortest path algorithms identify the shortest path between every global ocean patch pair, accounting for all possible multistep connections. For example, if there is no direct connection between nodes A and D, then these algorithms identify the multistep connection from $A \rightarrow B \rightarrow C \rightarrow D$. We use Dijkstra's algorithm[24], which is one of the most commonly used shortest path algorithms, and which fits our specific application.

Each step along these minimum-time routes may be unlikely, making the conditional multistep probability (of going from A to B to C to D...) very low as well. However, one can assume that the effect of these low probabilities is attenuated by the large reproductive output of microorganisms drifting with ocean currents. Over monthly to annual timescales, microorganisms moving with water masses can grow by the million[25]. Hence, there will still be planktonic organisms traveling along the potentially low probability paths identified here. Indeed, if one considers the dispersal of genetic material, then there need only be a small number of individuals traveling along these Min-T routes to make them evolutionarily relevant[17,22,25].

The outcome of our study when applying Dijkstra's algorithm to a raw Min-T matrix is a full matrix that contains estimates of minimum connection times between every region of the world's surface ocean. Both the raw and full Min-T matrices are rich with spatial information, but most importantly are the distribution of minimum connection times themselves. We find that different regions in the global surface ocean are connected on very short timescales, within ~10 years. This is in contrast to deep-water circulation, where water is thought to recirculate around the globe in roughly 1,000 years. These short surface-connection times are relevant to anyone studying dispersion in the surface ocean beyond planktonic species, including radioactive materials, plastics and other forms of pollution.

Results

Particle advection. We seeded particles in near-surface velocity fields from the ECCO2 $1/4° \times 1/4°$ state estimate[26] over 9 years and advected them for 100 years by looping fields for the years 2000–2010. The resulting paths were used to estimate the shortest time taken for water to travel from one patch in the surface ocean to another. Minimum connectivity times were then calculated by aggregating the ECCO2 grid cells (Supplementary Fig. 1) into 8×8 patches, each approximately $2° \times 2°$ in size (11,116 patches in total: Supplementary Fig. 2). On average, particles seeded in any given source patch reached 1,150 destination patches after 100 years of advection by ocean currents.

Connectivity matrices. The raw Min-T connectivity matrix, produced from the 2D Lagrangian particle simulations, is highly sparse (Fig. 1a, grey areas). Connections are made primarily within each ocean basin, reflecting the computational limits of the simulation integration period (see Methods), with values hugging the diagonal. Some cross-basin connections are made, and these typically take much longer, on the order of 20–30 years. In contrast, the Min-T matrix, modified by applying Dijkstra's algorithm (Fig. 1b), is full with minimum connection times for every ocean–patch pair. Short values still hug the diagonal, but now the cross-basin connection times are shorter, on the order of 10–20 years. For example, the largest connection time values in the full Min-T matrix occur between the Arctic and Southern

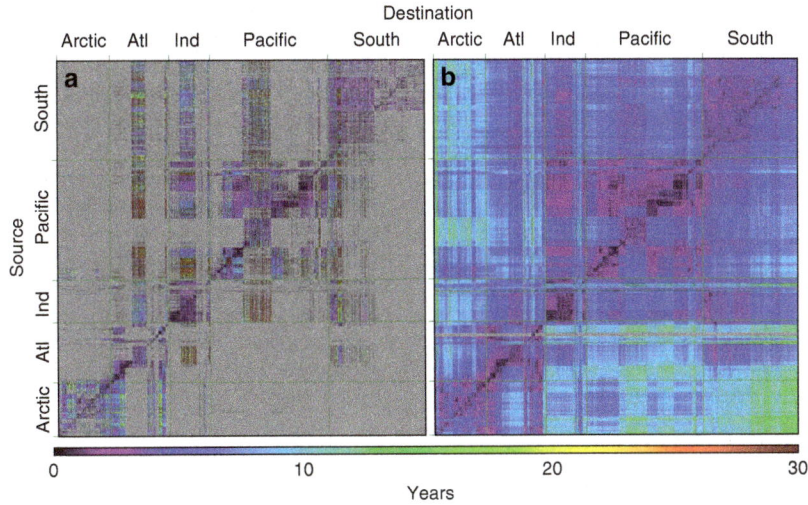

Figure 1 | Connectivity matrices. The raw minimum time connectivity matrix (**a**) and after Dijkstra's algorithm was applied (**b**). Major oceans are delimited by green lines. The number of patches (and hence the number of rows and columns) is 11,116.

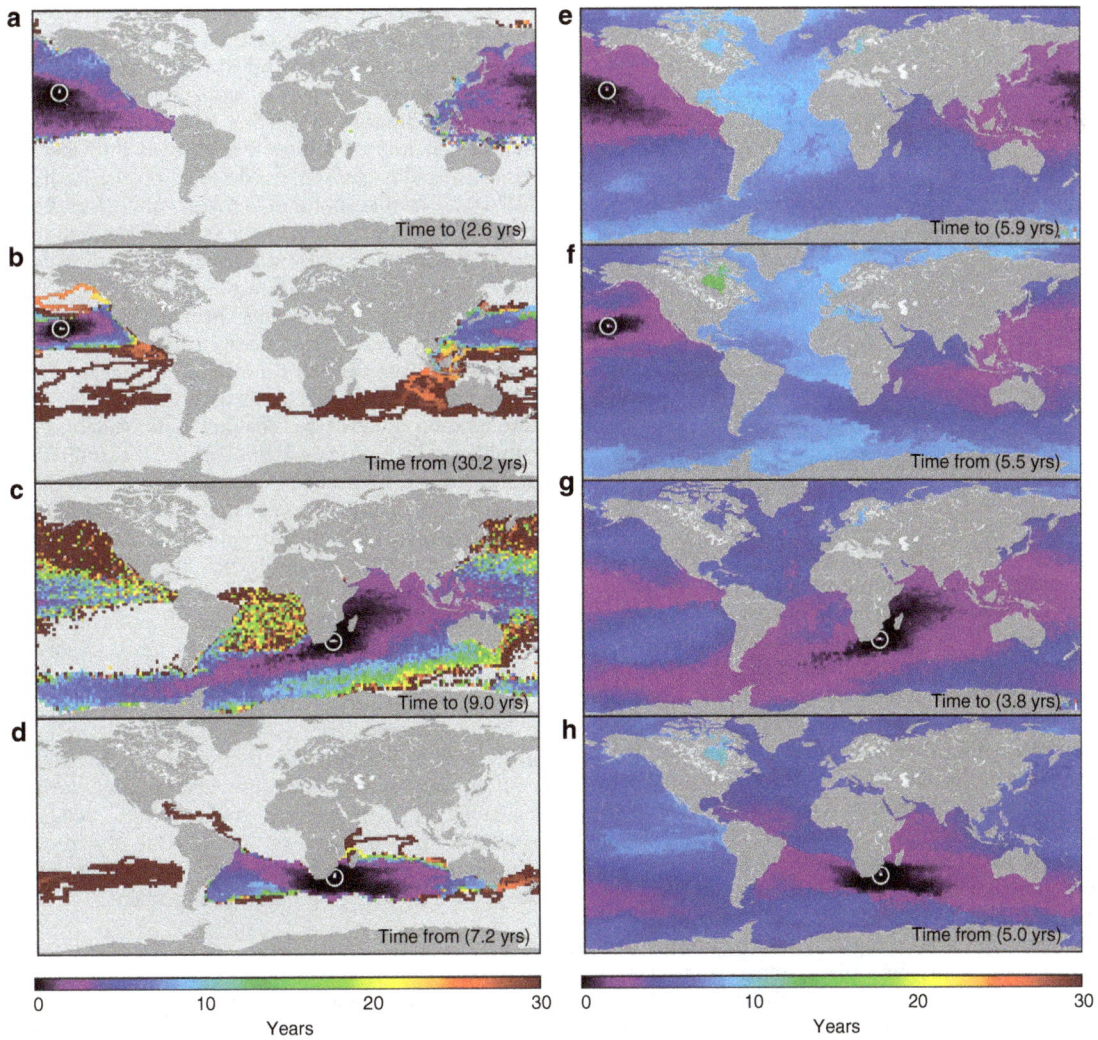

Figure 2 | Connectivity examples. Examples of minimum connection times (Min-T) to and from two locations identified by white circle-dots: off Hawaii (**a,b,e,f**) and off South Africa (**c,d,g,h**). Times 'to' are the shortest times taken for water from other patches to arrive at these locations. Times 'from' are the shortest times taken for water from these locations to go to all others. The left column shows raw minimum connection times, with the large number of no-connections noted in grey, and median Min-T in parentheses. The right column panels show Min-T values generated using Dijkstra's algorithm. Here connections occur between all areas of the ocean and median values are much lower on average than those of the raw minimum connection times.

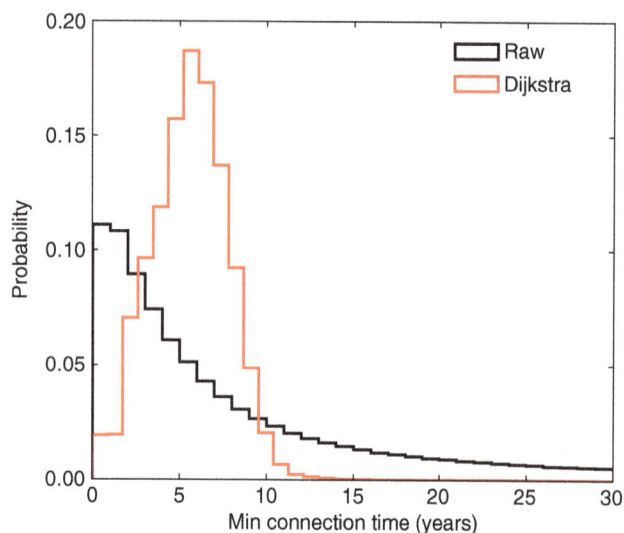

Figure 3 | Global connectivity distributions. Probability distributions of raw minimum connections times (blue) and those produced from Dijkstra's algorithm (red). Median minimum connection times (identified by the dashed vertical lines) are 6.13 years for the raw matrix, and 6.11 years for the modified. Note that connection times shorter than 1 year for Dijkstra are in fact raw connection times.

Table 1 | Median minimum connectivity time.

Source	Destination				
	Arctic	**Atlantic**	**Indian**	**Pacific**	**Southern**
Arctic	3.7 (1.9)	5.8 (2.6)	8.6 (2.0)	8.8 (1.9)	11.2 (1.6)
Atlantic	6.4 (2.2)	3.5 (2.3)	7.7 (2.6)	9.2 (1.8)	7.9 (3.0)
Indian	8.3 (2.1)	5.5 (1.9)	2.3 (3.1)	5.2 (2.0)	5.0 (2.5)
Pacific	8.1 (2.3)	7.2 (1.7)	3.9 (2.1)	3.4 (1.7)	5.9 (1.7)
Southern	9.1 (2.0)	6.4 (2.2)	5.0 (2.2)	6.2 (1.8)	4.1 (1.8)

Median minimum connectivity time between ocean basins in years. S.d. in parentheses.

Oceans. Asymmetry is present, too, revealing that there are differences in the time taken to go to, and come from, two places.

Spatial properties of connectivity. Rows of the raw and full Min-T matrices describe the minimum connection times from particular patches to all other patches in the global surface ocean. Similarly, the columns of the Min-T matrices describe the minimum time it takes for water to go from all patches to a given patch. This information is shown in Fig. 2 for two locations: Hawaii and a coastal location off of South Africa. From the raw Min-T information (Fig. 2a), the limitations of the particle tracking are evident in the large number of locations that are not connected by any particle trajectories (Fig. 2, ocean areas in grey). Of those that are connected, patches near the release point have low Min-T values relative to those locations farther away, with median connection times varying from location to location. In contrast, the full Min-T values (Fig. 2b) have connections everywhere, as expected from using Dijkstra's algorithm. Spatial structure is still seen, with some places more connected than others, but long connection times are now absent, and all median values have changed relative to their raw Min-T counterparts (Fig. 2, values in parentheses).

Timescales of global surface ocean connectivity. The most notable result from our analysis is the distribution of Min-T values themselves. The distribution of raw Min-T values (Fig. 3,

blue distribution) is roughly log-normal, with a median value of 6.13 years, and a long tail extending towards 100 years when the simulations were stopped. After being modified by Dijkstra's algorithm, the global distribution of minimum connection times is changed (Fig. 3, red distribution) with a median minimum connection time of 5.61 years and with the bulk of the distribution now below 15 years, showing that the global ocean can be connected over timescales of a decade. The maximum full Min-T value is still about 100 years, relating to water traveling from the Weddell Sea to the California coast. In scaling up, the average Min-T values between different ocean basins are shown in Table 1. These aggregated metrics again highlight the short connection times between ocean regions, but also show physical consistency (that is, on average basins farther away take longer time to reach).

Discussion

Our results mirror the data from unintended, often tragic, natural experiments that quantify analogous connectivity in the Pacific Ocean. Large quantities of shoes[27] and toys[28] washed overboard from container ships en route from Asia to North America have been useful in estimating connectivity times by acting as drifters. Such results show timescales that are similar to, or often shorter than, our findings in the North Pacific. These kinds of drifters are, however, susceptible to wind-drift and could record faster times than our models. A more comparable experiment is the 2011 Fukushima disaster, in which a Japanese nuclear reactor released a large quantity of radioactive isotopes into the Pacific Ocean. Traces of radioactivity were detected on the Pacific Coast of the U.S. in November of 2014—3.6 years later (Ken Buessler WHOI, personal communication). Our estimated minimum connectivity time between the Fukushima release site and its detection site of the U.S. west coast is 3.5 years.

In summary, our results provide evidence for a highly connected global surface ocean with all regions connected to each other over decadal timescales. This suggests that plankton communities may keep pace with climate change through the immigration of new types that are better suited in changing local conditions[5,29,30]. Beyond this result, the utility of calculating global surface connectivity extends to its spatial information. For example, in many regional studies it is common to identify connectivity modules or subpopulations[31] and also the location of key stepping-stone patches, which are central to maintaining the overall connectance of the system[32,33]. These network theoretic analyses have an applied nature, such as in the design of spatial management units[34]; but they are also important for basic research, for example in generating hypotheses about genetic or taxonomic similarity across the ocean[7], or when testing models of community assembly[35]. Finally, it is important to note that we have only estimated the timescales of physical connectivity without addressing environmental factors such as nutrient availability or temperature gradients[36]. Gauging the effect of environmental barriers on global-scale dispersal will further contribute to our understanding of how marine communities adapt to their changing ocean environment.

Methods

Lagrangian particle tracking. Two-dimensional Lagrangian particle tracking was used to make our connectivity calculations[15,21,33,37]. We used velocity fields from ECCO2 (http://ecco2.org), a high-resolution (1/4) global ocean model that assimilates available satellite and the *in situ* data[26], to advect particles in the surface ocean (Supplementary Fig. 1). ECCO2 is based on a global full-depth ocean and sea-ice configuration of the Massachusetts Institute of Technology general circulation model (MITgcm) and applies an ad-joint approach to generate the physically consistent data assimilations. ECCO2's resolution is high enough to permit the formation of eddies and other narrow current systems within the ocean.

Particles were advected in the surface ocean using TRACMASS (http://tracmass.org), an off-line particle tracking code that calculates trajectories using Eulerian velocity fields. TRACMASS estimates the trajectory path through each grid cell of every Lagrangian particle, using an analytical solution to a differential equation that depends on the velocities on the grid-box walls. The scheme was originally developed for stationary velocity fields[20,38], and thereafter extended for time-dependent fields by solving a linear interpolation of the velocity field both in time and in space over each grid box[39]. This differs from the Runge-Kutta method, where trajectories are iterated forward in time with short time steps.

Particle seeding and connectivity patches. We seeded six particles in the second depth layer of each ECCO2 grid cell (a total of 4 million particles at each seeding time, or 36 million particles in total over all seeding times). When calculating connectivity, we aggregated the model's $1/4° \times 1/4°$ grid cells to 11,116 discrete $2° \times 2°$ patches. The size of these connectivity patches was selected as a balance of computational feasibility and biogeographic detail. Each connectivity patch is therefore seeded with 384 particles at each seeding event (9 in total). The second depth layer is between 5 and 20 m depth and was used to avoid potential numerical problems due to how ECCO2 implement a varying sea surface, precipitation, and evaporation. See Supplementary Fig. 2 for the spatial distribution of connectivity patches. Particles were seeded at 9 points in time: 1 January 2001, 1 February 2002, 1 March 2003, 1 April 2004, 1 May 2005, 1 June 2006; 1 July 2007, 1 August 2008; and 1 September 2009 in model years. As a consequence of the multiple seeding times, a total of 3,456 particles were used per patch to estimate connectivity. Particles were then advected using horizontal velocity fields from the second depth layer in ECCO2 so that they were locked in the surface ocean. We looped velocity fields for the years 2000–2010 continuously and advected the particles for 100 years in total. Particle positions were saved every 3 days and used to calculate minimum connection times. No extra diffusivity was added to the movement of the particles. Supplementary Figure 6 shows the relationship between advection time and number of other patches reached. It is clear from this figure that the number of connectivity patches reached saturates after about 12 years. In other words, like the number of particles released, there are diminishing returns to running simulations for longer integration times.

Estimating the timescales of connectivity. The resulting Lagrangian particle trajectories were used to estimate the shortest time taken for water to travel from one patch in the surface ocean to another. This minimum connection time is a variant on the standard measure of ocean distance, which is the expected transit time for water to travel from one patch to another[13,14]. We use the minimum and not the expected connection time for two reasons. First, the minimum connection time is a more appropriate metric for phytoplankton and bacterial connectivity since asexually reproducing organisms have high reproductive output that attenuates low dispersal probabilities, and only a few individuals are required to 'connect' two places, especially in terms of population genetics[22]. It is therefore possible for such organisms to exploit dispersal routes where the probability to reach a given destination is very low. Second, expected transit times in the global ocean are not properly defined, as water can recirculate for an infinitely long time. There is no limit, therefore, to the distribution of connectivity times over which to calculate expected connection times. Thus, the minimum connection time is a preferable alternative measure of ocean distance for this global application.

Minimum connection times for the global surface ocean are called Min-T, and they are stored in the form of a matrix—the Min-T connectivity matrix, where each i, j element represents the shortest transit time between a given source patch i and destination patch j (Fig. 1a). The *raw* Min-T matrix, produced from the Lagrangian particle tracking, is highly sparse with most pairs of patches being unconnected.

Network analysis of shortest/quickest paths. Estimation of connection times between all pairs of patches globally using Lagrangian particle simulations alone would require a currently infeasible number of particles[23] (see particle density sensitivity test described below). To circumvent this obstacle, a shortest-path algorithm was used to calculate missing values in the raw Min-T connectivity matrix. Here the network is the global ocean, with patches in the ocean as nodes, and minimum connection times as edges connecting the nodes. Applied to this network, shortest path algorithms identify the shortest path between every global ocean patch-pair, accounting for all possible multistep connections (see Supplementary Online Material for details). For example, if there is no direct connection between nodes A and D, then these algorithms identify the multistep connection from $A \to B \to C \to D$. We use Dijkstra's algorithm[24], which is one of the most commonly used shortest path algorithms, and which fits our specific application. The end result is a modified Min-T connectivity matrix (Fig. 1b), where all possible minimum connection times between patches are calculated.

Each step along these minimum-time routes may be unlikely, and so the conditional multistep probability (of going from A to B to C to D...) can have a very low probability as well. However, we assume that the effect of these low probabilities is attenuated by the large reproductive output of microorganisms drifting with ocean currents. Over the timescales that we are considering, microorganisms moving with water masses can grow by the million[25]. Hence, there will still be planktonic organisms traveling along the potentially low probability

paths identified here. Indeed, if one considers the dispersal of genetic material, then there need only be a small number of individuals traveling along these Min-T routes, to make them evolutionarily relevant[17,22,25].

While nodes in a network are usually defined as singular nodes with well-defined distances between them, our ocean patches have relatively large areas and are continuously adjacent to one another. This difference creates a problem when using Dijkstra's algorithm since a particle seeded next to the boundary of its initial patch can rapidly move to an adjacent patch. However, shortest path algorithms assumes that the travel time across each intermediate node is zero, or at least included in the edge distances. (This phenomenon is also a problem when analysing the speed of tracer transport in General Circulation Models[40].) By removing all calculated connectivity times shorter than 1 year before applying the shortest-path algorithm, we limit the effect of not including within-patch crossing times. The 365-day cutoff is based on calculated typical residence times in the patches, which are on the order of weeks. All initial minimum connectivity times are based on travel distances at least an order of magnitude longer than typical patch crossing distances.

The removed connectivity times were added back to the final connectivity matrix, allowing for connection times shorter than 365 days, as shown in Figs 2 and 3. It should be noted that the absolute number of connectivity times shorter than 1 year in Fig. 3 are identical for the Raw and Dijkstra cases. The lack of discontinuities between sub-annual and longer connection times in Fig. 2 (and all other cases we have explored) give us confidence that the resulting connectivity matrix is reasonable and that our approach works.

After applying Dijkstra's algorithm, we find that the resulting minimum connection time matrices are all connected. However, we do find some areas that are only connected in one direction (that is, there are connection time to, but not from, particular regions). These areas are mainly inland seas—the Baltic and Mediterranean, for example. However, they only account for a small fraction (2%) of the modified Min-T matrix and, consequently, do not impact the general result of the timescales of global surface ocean connectivity.

Particle seeding sensitivity test. Since the number of particles seeded per grid-cell and the seeding times are limited, we have not accounted for all possible Min-T pathways. As a result, our estimates of the timescales of global surface connectivity are conservative, since adding more particles and seeding dates could only lead to shorter Min-T pathways (that is, we look for the shortest connection times over all possibilities including seeding times). Thus, the few seeding dates—although arguably numerically incomplete—strengthen our conclusion that the global surface ocean is well connected over a few decades.

To examine the effect of particle seeding density, we performed a particle sensitivity test. Minimum connection times from a patch in the north pacific to all others were estimated using simulations with increasing numbers of seeded particles. Supplementary Figure 3 shows the results of these simulations. It is clear that a larger oceanic extent is reached as the number of particles released increases. However, when we examine only those patches that were reached in all seeding experiments (Supplementary Fig. 4), we can see that increasing the number of particles serves only to decrease the minimum connection times in these patches (Supplementary Fig. 4: with 84 particles some areas are reached after 100 years—the patches in gold, in contrast with 16,660 particles, these same patches are reached after around 20 years—patches now in light red).

Finally, we show the aggregated results of the sensitivity test in Supplementary Fig. 5, where we plot the fraction of patches reached (over the whole ocean) and the median minimum connection time from this study patch. The fraction of patches reached saturates at around 30%, which means that there are diminishing returns (in terms of estimating minimum connection times to new patches) to adding more particles. It also indicates that, to release enough particles to estimate minimum connection times to all patches globally, a currently impossible number of Lagrangian particles would be required. Similarly, the median minimum connection time from this patch saturates at around 8,000 particles released. In our simulations we use 3,456 particles per patch as this achieved a balance of connectivity sampling power and computational efficiency.

References

1. Baas-Becking, L. *Geobiologie of Inleiding tot de Milieukunde* (ed Van Stockum, W. P. & Zoon) (The Hague, 1934).
2. Fenchel, T. & Finlay, B. J. The ubiquity of small species: patterns of local and global diversity. *Bioscience* **54**, 777–784 (2004).
3. De Wit, R. & Bouvier, T. 'Everything is everywhere, but, the environment selects'; what did Baas Becking and Beijerinck really say? *Environ. Microbiol.* **8**, 755–758 (2006).
4. McGillicuddy, D. J. *et al.* Eddy/wind interactions stimulate extraordinary mid-ocean plankton blooms. *Science* **316**, 1021–1026 (2007).
5. Thomas, M. K., Kremer, C. T., Klausmeier, C. A. & Litchman, E. A global pattern of thermal adaptation in marine phytoplankton. *Science* **338**, 1085–1088 (2012).
6. Martiny, J. *et al.* Microbial biogeography: putting microorganisms on the map. *Nat. Rev. Microbiol.* **4**, 102–112 (2006).

7. Casteleyn, G. *et al.* Limits to gene flow in a cosmopolitan marine planktonic diatom. *Proc. Natl Acad. Sci. USA* **107**, 12952–12957 (2010).

8. Saez, A. G. *et al.* Pseudo-cryptic speciation in coccolithophores. *Proc. Natl Acad. Sci. USA* **100**, 7163–7168 (2003).

9. Rynearson, T. A. & Virginia Armbrust, E. Genetic differentiation among populations of the planktonic marine diatom Ditylum brightwellii (Bacillariophyceae). *J. Phycol.* **40**, 34–43 (2004).

10. Sul, W. J., Oliver, T. A., Ducklow, H. W., Amaral-Zettler, L. A. & Sogin, M. L. Marine bacteria exhibit a bipolar distribution. *Proc. Natl Acad. Sci. USA* **110**, 2342–2347 (2013).

11. Godhe, A. *et al.* Seascape analysis reveals regional gene flow patterns among populations of a marine planktonic diatom. *Proc. R. Soc. B Biol. Sci.* **280**, 1773 (2013).

12. Froyland, G., Stuart, R. M. & van Sebille, E. How well-connected is the surface of the global ocean? *Chaos* **24**, 3 (2014).

13. Alberton, F. *et al.* Isolation by oceanographic distance explains genetic structure for Macrocystis pyrifera in the Santa Barbara Channel. *Mol. Ecol.* **20**, 2543–2554 (2011).

14. Watson, J. R. *et al.* Currents connecting communities: nearshore community similarity and ocean circulation. *Ecology* **92**, 1193–1200 (2011).

15. Mitarai, S., Siegel, D. & Winters, K. A numerical study of stochastic larval settlement in the California Current system. *J. Mar. Syst.* **69**, 295–309 (2008).

16. Cowen, R., Paris, C. & Srinivasan, A. Scaling of connectivity in marine populations. *Science* **311**, 522–527 (2006).

17. Kool, J. T., Paris, C. B. & Andre, S. Complex migration and the development of genetic structure in subdivided populations: an example from Caribbean coral reef ecosystems. *Evolution, September* **2009**, 1–10 (2010).

18. Mora, C. *et al.* High connectivity among habitats precludes the relationship between dispersal and range size in tropical reef fishes. *Ecography* **35**, 89–96 (2012).

19. Wood, S., Paris, C. B., Ridgwell, A. & Hendy, E. J. Modelling dispersal and connectivity of broadcast spawning corals at the global scale. *Global Ecol. Biogeogr.* **23**, 1–11 (2014).

20. Döös, K. Interocean exchange of water masses. *J. Geophys. Res.* **100**, 13499–13514 (1995).

21. Cowen, R., Gawarkiewicz, G., Pineda, J., Thorrold, S. & Werner, F. E. Population connectivity in marine systems. *Oceanography* **20**, 14–20 (2007).

22. Hedgecock, D., Barber, P. H. & Edmands, S. Genetic approaches to measuring connectivity. *Oceanography* **20**, 70–79 (2007).

23. Simons, R. D., Siegel, D. A. & Brown, K. S. Model sensitivity and robustness in the estimation of larval transport: A study of particle tracking parameters. *J. Mar. Syst.* **119-120**, 19–29 (2013).

24. Dijkstra, E. W. A note on two problems in connexion with graphs. *Numerische Mathematik* **1**, 269–271 (1959).

25. Falkowski, P. G. *et al.* The evolution of modern eukaryotic phytoplankton. *Science* **305**, 354–360 (2004).

26. Wunsch, C., Heimbach, P., Ponte, R. M. & Fukumori, I. The ECCO-GODAE Consortium Members. The global general circulation of the ocean estimated by the ECCO-consortium. *Oceanography* **22**, 88–103 (2009).

27. Ebbesmeyer, C. C. & Ingraham, W. J. Shoe spill in the North Pacific. *Eos Trans. Am. Geophys. Union* **73**, 361–365 (1992).

28. Ebbesmeyer, C. C. & Ingraham, W. J. Pacific toy spill fuels ocean current pathways research. *Eos Trans. Am. Geophys. Union* **75**, 425–430 (1994).

29. Schaum, E., Rost, B., Millar, A. J. & Collins, S. Variation in plastic responses of a globally distributed picoplankton species to ocean acidification. *Nat. Clim. Change* **3**, 298–302 (2012).

30. Barton, A. D., Dutkiewicz, S., Flierl, G., Bragg, J. & Follows, M. J. Patterns of diversity in marine phytoplankton. *Science* **327**, 1509–1511 (2010).

31. Jacobi, M. N., André, C., Döös, K. & Jonsson, P. R. Identification of subpopulations from connectivity matrices. *Ecography* **35**, 1004–1016 (2012).

32. Jacobi, M. N. & Jonsson, P. R. Optimal networks of nature reserves can be found through eigenvalue perturbation theory of the connectivity matrix. *Ecol. Appl.* **21**, 1861–1870 (2011).

33. Watson, J. R. *et al.* Identifying critical regions in small-world marine metapopulations. *Proc. Natl Acad. Sci. USA* **108**, 907–913 (2011).

34. Treml, E., Halpin, P., Urban, D. & Pratson, L. Modeling population connectivity by ocean currents, a graph-theoretic approach for marine conservation. *Landscape Ecol.* **23**, 19–36 (2008).

35. Chust, G., Irigoien, X., Chave, J. & Harris, R. P. Latitudinal phytoplankton distribution and the neutral theory of biodiversity. *Global Ecol. Biogeogr.* **22**, 531–543 (2013).

36. Sarmiento, J. L. *et al.* Response of ocean ecosystems to climate warming. *Global. Biogeochem. Cycles.* **18**, GB3003 (2004).

37. Treml, E. A. & Halpin, P. N. Marine population connectivity identifies ecological neighbors for conservation planning in the Coral Triangle. *Conserv. Lett.* **5**, 441–449 (2012).

38. Blanke, B. & Raynaud, S. Kinematics of the pacific equatorial undercurrent: An Eulerian and Lagrangian approach from GCM results. *J. Phys. Oceanogr.* **27**, 1038–1053 (1997).

39. de Vries, P. & Döös, K. Calculating Lagrangian trajectories using time-dependent velocity fields. *J. Atmos. Oceanic Technol.* **18**, 1092–1101 (2001).

40. Griffies, S. M. Elements of the Modular Ocean Model (MOM): 2012 release. Geophysical Fluid Dynamics Laboratory (GFDL) Ocean Group Technical Report No. 7, 1–631 (2012).

Acknowledgements

We thank Dave Siegel and Debora Iglesias-Rodriguez for comments and discussions and Amy Ehntholt for help with the manuscript. The project was in part funded by the NSF Coupled Natural-Human Systems grant GEO-1211972, NASA ROSES NNX13AC52G and the Nippon Foundation Nereus Program.

Author contributions

The work was initiated by J.R.W. and B.F.J. equally. B.F.J. conducted the particle tracking runs and the data analysis. Figures were produced by B.F.J. with support from J.R.W. The text was written by B.F.J. and J.R.W. equally.

Additional information

3

Catch reconstructions reveal that global marine fisheries catches are higher than reported and declining

Daniel Pauly[1] & Dirk Zeller[1]

Fisheries data assembled by the Food and Agriculture Organization (FAO) suggest that global marine fisheries catches increased to 86 million tonnes in 1996, then slightly declined. Here, using a decade-long multinational 'catch reconstruction' project covering the Exclusive Economic Zones of the world's maritime countries and the High Seas from 1950 to 2010, and accounting for all fisheries, we identify catch trajectories differing considerably from the national data submitted to the FAO. We suggest that catch actually peaked at 130 million tonnes, and has been declining much more strongly since. This decline in reconstructed catches reflects declines in industrial catches and to a smaller extent declining discards, despite industrial fishing having expanded from industrialized countries to the waters of developing countries. The differing trajectories documented here suggest a need for improved monitoring of all fisheries, including often neglected small-scale fisheries, and illegal and other problematic fisheries, as well as discarded bycatch.

[1] Sea Around Us, Global Fisheries Cluster, University of British Columbia, 2202 Main Mall, Vancouver, British Columbia, Canada V6T 1Z4. Correspondence and requests for materials should be addressed to D.P. (email: d.pauly@oceans.ubc.ca) or to D.Z. (email: d.zeller@oceans.ubc.ca).

Marine fisheries are the chief contributors of wholesome seafood (finfish and marine invertebrates; here 'fish'). In many developing countries (and likely also in many 'transition' countries), fish is the major animal protein source that rural people can access or afford[1]; and they are also an important source of micronutrients essential to people with otherwise deficient nutrition[2]. However, the growing popularity of fish in countries with developed or rapidly developing economies creates a demand that cannot be met by fish stocks in their own waters (for example, the EU, the USA, China and Japan). These markets are increasingly supplied by fish imported from developing countries, or caught in the waters of developing countries by various distant-water fleets[3-5], with the consequences that:

(a) Foreign and/or export-oriented domestic industrial fleets are increasingly fishing in the waters of developing countries[5,6],
(b) Industrially caught fish has become a globalized commodity that is mostly traded between continents rather than consumed in the countries where it was caught[7], and
(c) The small-scale fisheries that traditionally supplied seafood to coastal rural communities and the interior of developing countries (notably in Africa)[8] are forced to compete with the export-oriented industrial fleets without much support from their governments.

The lack of attention that small-scale fisheries suffer in most parts of the world[9] manifests itself in potentially misleading statistics that are submitted annually by many member countries of the Food and Agriculture Organization of the United Nations (FAO), which may omit or substantially underreport small-scale fisheries data[10]. FAO harmonizes the data submitted by its members, which then becomes the only global data set of fisheries statistics in the world, widely used by policy makers and scholars[11].

This data set, however, may not only underestimate artisanal (that is, small scale, commercial) and subsistence fisheries[10], but also generally omit the catch of recreational fisheries, discarded bycatch[12] and illegal and otherwise unreported catch, even when some estimates are available[13]. Thus, except for a few obvious cases of over-reporting[14], the landings data updated and disseminated annually by the FAO on behalf of member countries may considerably underestimate actual fisheries catch. While this underestimation is widely known among many fisheries scientists working with FAO catch data, and is freely acknowledged by FAO, its global magnitude has not been explicitly presented until now.

Here we present the results of an approach called 'catch reconstruction'[15,16] that utilizes a wide variety of data and information sources to derive estimates for all fisheries components missing from the official reported data. We find that reconstructed global catches between 1950 and 2010 were 50% higher than data reported to FAO suggest, and are declining more strongly since catches peaked in the 1990s. These findings and the country-specific technical work underlying these results will hopefully contribute to member countries submitting more accurate fisheries statistics to FAO. Such improved and more comprehensive data contribute a foundation that can facilitate the implementation of ecosystem-based fisheries management[17], which is a component of the 'FAO Code of Conduct for Responsible Fisheries'[18].

Results

Global pattern. The sum of the reconstructed catches of all sectors in all Exclusive Economic Zones (EEZs) of the world, plus the catch of tuna and other large pelagic fishes in the High Seas leads to two major observations (Fig. 1; Supplementary Table 1).

First, the trajectory of reconstructed catches differs substantially from those reported by FAO on behalf of its member countries. The FAO statistics suggest that, starting in 1950, the world catch (actually 'landings', as discarded catches are explicitly excluded from the global FAO data set) increased fairly steadily to 86 million tonnes (mt) in 1996, stagnated and then slowly declined to around 77 mt by 2010 (Fig. 1). In contrast, the reconstructed catch peaked at 130 mt in 1996 and declined more strongly since. Thus, the reconstructed catches are overall 53% higher than the reported data.

Furthermore, since the year of peak catches in 1996, the reconstructed catch declined strongly at a mean rate of -1.22 mt · per year, whereas FAO, at least until 2010, described the reported catch cautiously as characterized by 'stability'[19,20], though it exhibited a gradual decline (-0.38 mt · per year). The reconstructed total catches therefore represent a decline of over three times that of the reported data as presented by FAO on behalf of countries. A segmented regression[21] identifies two breakpoints in the catch time series (that is, change in trend) of the reconstructed total catches as well as the reported catches. These are in 1967 as a result of a changing slope of the catch time series from a stronger increase prior to 1967 (reconstructed catches $=2.82$ mt · per year; reported catches $=1.88$ mt · per year) to a slower increase after 1967 (reconstructed catches $=1.86$ mt · per year; reported catches $=1.30$ mt · per year). The second breakpoint is in 1996 (the year of peak catch), with a subsequently decreasing trend (that is, slope) of -1.22 mt · per year for reconstructed catches and -0.38 mt per · year for reported catches, as also presented for the simple regression above (Fig. 1; see also Supplementary Table 2).

Note that the recent, stronger decline in reconstructed total catches is not due to some countries reducing catch quotas so that stocks can rebuild. For example, a similar decline (-1.01 mt · per year) in reconstructed catches is obtained when the catch from the Unites States, Northwestern Europe, Australia and New Zealand (that is, countries where quota management predominates) is excluded (Fig. 2; Supplementary Table 3).

Spatial pattern. Closer examination of the reconstructed versus reported catches in each of the 19 maritime FAO statistical areas suggests that some of the areas where industrial fishing originated, such as the Northwest Atlantic (FAO area 21), are the

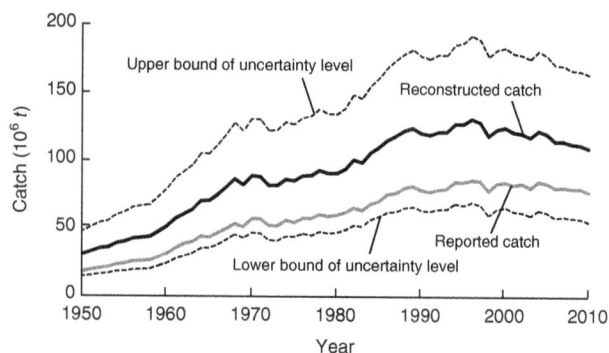

Figure 1 | Trajectories of reported and reconstructed marine fisheries catches 1950-2010. Contrast between the world's marine fisheries catches, assembled by FAO from voluntary submissions of its member countries ('reported') and that of the catch 'reconstructed' to include all fisheries known to exist, in all countries and in the High Sea ('reconstructed' = 'reported' + estimates of 'unreported'). The mean weighted percentage uncertainty of the reconstructed total catches (over all countries and fisheries sectors) based on the quality scores attributed to each sector in each country and territory (dashed line) is also shown.

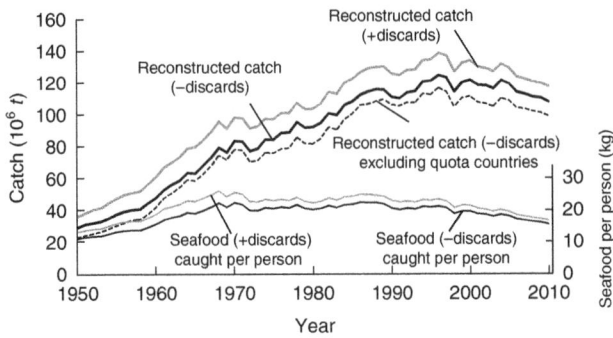

Figure 2 | Trajectories of marine fisheries catches 1950-2010. Effects of removing discards on estimates of seafood caught *per capita*, and of removing the catches of the major countries using quota management (that is, USA, New Zealand, Australia and Western Europe) on reconstructed total catches.

first regions of the world to demonstrate declining catches (Fig. 3). In contrast, lower-latitude areas demonstrate declines later, or still appear to have increasing catches, for example, the Indian and Western Central Pacific Oceans still showing generally increasing trends in reported catches (Fig. 3).

Catches by fishing sector. We present, for the first time, global reconstructed marine fisheries catches by fisheries sectors (Fig. 4; Supplementary Table 4). They are dominated by industrial fisheries, which contribute 73 mt of landings in 2010, down from 87 mt in 2000 (Fig. 4). At the global scale it is a declining industrial catch (combined with the smaller contribution of gradually reduced levels of discarding)[12] that leads to declining global catches since 1996, while the artisanal sector, which generates a catch increasing from about 8 mt · per year in the early 1950s to 22 mt · per year in 2010, continues to show gradual growth in catches at the global scale (Fig. 4).

Also noticeable is that the inter-annual variations (small peaks and troughs) in both reconstructed catches and reported catches (Fig. 1) are mainly driven by industrial data, which are relatively well documented and reported in time series, while the small-scale sector data are smoother over time (Fig. 4), and more strongly influenced by continuity assumptions over time as part of the national reconstructions.

While some countries increasingly include subsets of artisanal catches in official catch statistics provided to FAO, subsistence fisheries catches (Fig. 4) rarely are[10]. Worldwide, subsistence fisheries caught an estimated 3.8 mt · per year between 2000 and 2010 (Fig. 4; Supplementary Table 4). The current global estimate of just under 1 mt · per year of recreational catches is rather imprecise, and recreational fishing is declining in developed, but increasing in developing countries.

Discarded bycatch, generated mainly by industrial fishing, notably shrimp trawling[22], was estimated at 27 mt · per year (± 10 mt) and 7 mt · per year (± 0.7 mt) in global studies conducted for FAO in the early 1990s and 2000s, respectively[23,24]. However, these point estimates were not incorporated into FAO's global 'capture' database, which thus consists only of landings. Here, these studies are used, along with numerous other sources, to generate time series of discards (Fig. 4). Discards, after peaking in the late 1980s, have declined, and during 2000–2010, an average of 10.3 mt · per year of fish were discarded.

Discussion
Our reconstructed catch data, which combines the data reported to FAO with estimates of unreported catches (that is,

reconstructed data are 'reported FAO data + unreported catches') include estimates of uncertainty (Fig. 1) associated with each national reconstruction. Note that many reconstructions are associated with high uncertainty, especially for earlier decades, for sectors such as subsistence which receive less data collection attention by governments, and for small countries or territories (Fig. 1; Supplementary Table 5)[10]. We include uncertainty estimates here, despite the fact that reconstructions address an inherent negative bias in global catch data (that is, address the 'accuracy' of data) and not the replicability of catch data collection (that is, the statistical 'precision' of such estimates), which is what 'uncertainty' estimates (for example, confidence limits) generally are used for. We do recognize that any estimates of unreported catches implies a certain degree of uncertainty, but so do officially reported data. Most countries in the world use sampling schemes, estimations and raising factors to derive their national catch data they officially report domestically and internationally, all without including estimates of the uncertainty inherent in the numbers being reported as official national catches.

Our comparison of the reconstructed versus reported catches in each of the 19 maritime FAO statistical areas suggests that some of the lower-latitude areas still appear to have increasing reported catches. This generally increasing trend is most pronounced in the Indian and Western Central Pacific Oceans (Fig. 3), where the reconstructed catches are most uncertain, as the statistics of various countries could only partially correct a regional tendency to exaggerate reported catches[5]. FAO's Indian and Western Central Pacific Oceans areas are also the only ones with an increasing FAO reported catch, which, when added to that of other FAO areas, makes the FAO reported world catch appear more stable than it is based on our global reconstructions.

Our data and analyses show that, at the global scale, it is a declining industrial catch (plus a smaller contribution of gradually declining discards)[12] that provide for the declining global catches, while artisanal fishing continues to show slight growth in catches (Fig. 4). Thus, the gradually increasing incorporation of artisanal and other small-scale catches in the officially reported data presented by FAO on behalf of countries is partly masking the decline in industrial catches at the global level. Since officially reported data are not (at the international level) separated into large-scale versus small-scale sectors[25], this trend could not be easily documented until now. Obviously, these patterns may vary between countries. Furthermore, while parts of artisanal catches are increasingly included in official catch statistics by some countries, non-commercial subsistence fisheries catches, a substantial fraction of it through gleaning by women in coastal ecosystems such as coral reef flats and estuaries[26] are generally neglected. The importance of subsistence fishing for the food security of developing countries, particularly in the tropical Indo-Pacific, cannot be overemphasized[10,27].

Our preliminary and somewhat imprecise reconstruction of recreational catches indicates that this sector is largely missing from official reported data, despite FAO's annual data requests explicitly allowing inclusion of recreational catch data. This activity, however, generates an estimated 40 billion USD · per year of global benefits, involves between 55 and 60 million persons, and generates about one million jobs worldwide[28].

Finally, our country-by-country reconstructed data supports previous studies illustrating that global discards have decreased[12,24]. Discarded catches should therefore be included in catch databases, if only to allow for correct inferences on the state of the fisheries involved in this problematic practice.

The reconstructed catch data presented here for the first time for all countries in the world can contribute to formulating better policies for governing the world's marine fisheries, with a first

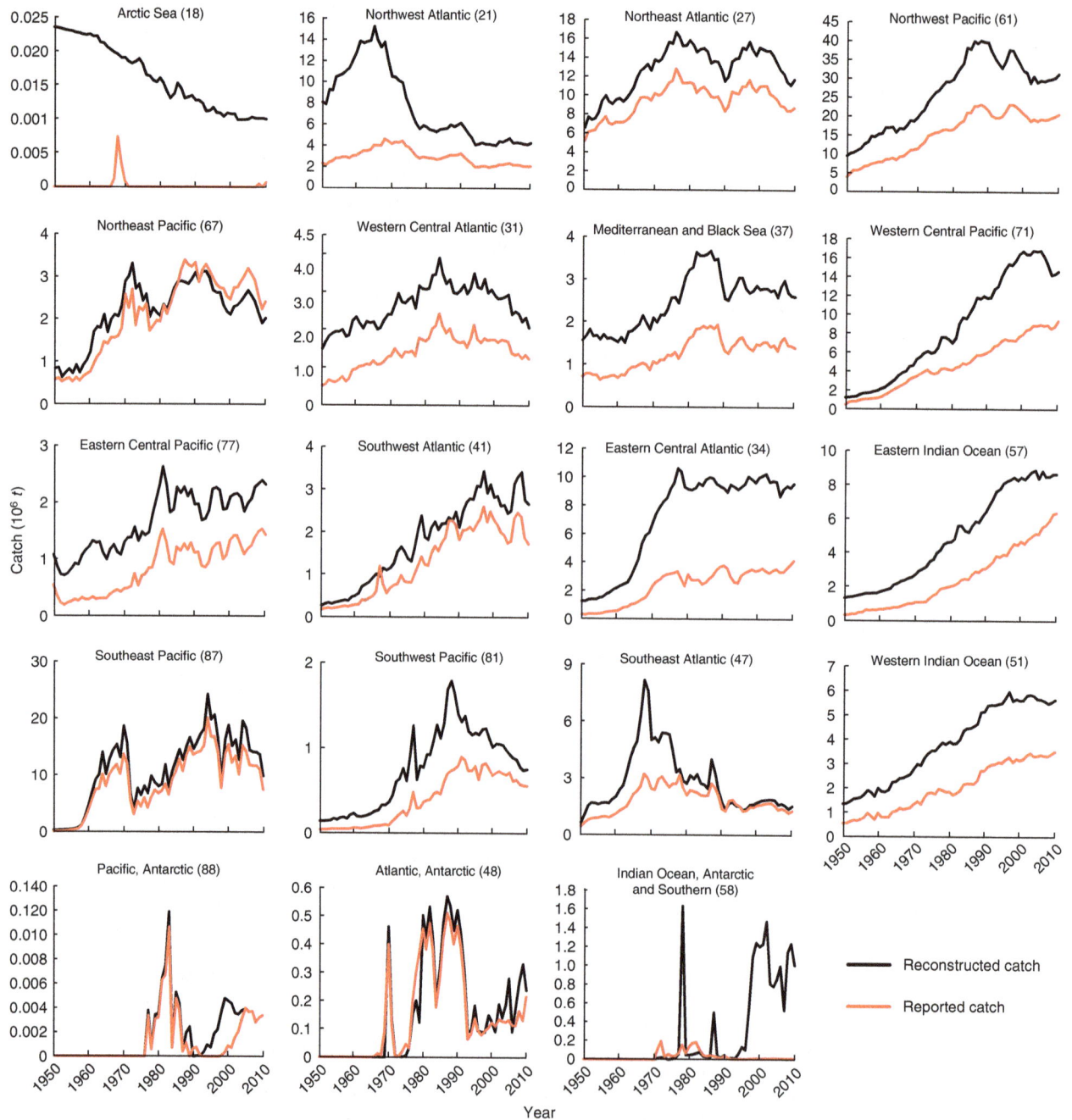

Figure 3 | Reconstructed and reported catches by FAO areas. Contrasting reconstructed and reported catches in the 19 maritime 'Statistical Areas' which FAO uses to roughly spatialize the world catch. Note that for Area 18 (Arctic), the reported catch by the U.S. and Canada was zero, while only Russia (former-USSR) reported a small catch in the late 1960s, even though the coastal fishes of the high Arctic are exploited by Inuit and others.

step being the recognition in national policies of the likely magnitude of fisheries not properly captured in the official national data collection systems. This recognition will hopefully contribute to improvements in national data collection systems, an aspiration that we share with FAO. For example, in Mauritania and Guinea Bissau, which, in large part as a result of the reconstructions[29,30] and our ongoing direct engagement with these countries, are now initiating national data collection systems for recreational fisheries (a growth industry in both countries and missing from current data systems). It is hoped that this type of data, and other missing data (for example, subsistence catches)[10], will be included in future national data reports to FAO, as is the case for some other countries such as Finland[31].

The taxonomic composition of this reconstructed catch (not presented here but available from the *Sea Around Us* and through the individual catch reconstruction reports, see Supplementary Table 5) can also contribute to the development of more useful first-order indicators of fisheries status[32–34] than has been possible previously, especially in the absence of comprehensive stock assessments for all taxa targeted.

A policy change that would be straightforward for FAO to coordinate and implement with all countries around the world is to request countries to submit their annual catch statistics separately for large-scale and small-scale fisheries[25], which would be an excellent contribution towards the implementation of the '*Voluntary Guidelines for Securing Sustainable Small-scale*

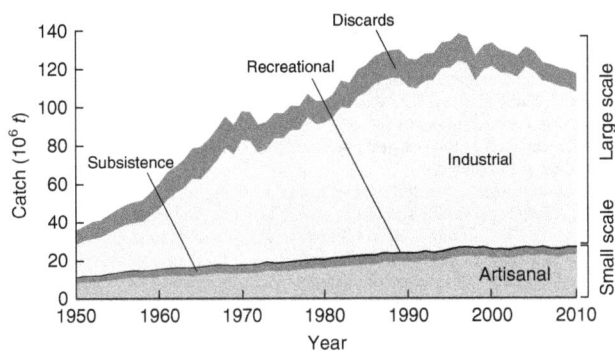

Figure 4 | Reconstructed global catch by fisheries sectors. Reconstructed catches for all countries in the world, plus High Seas, by large-scale (industrial) and small-scale sectors (artisanal, subsistence, recreational), with discards (overwhelmingly from industrial fisheries) presented separately.

Fisheries in the Context of Food Security and Poverty Eradication' recently adopted and endorsed at the thirty-first Session of the FAO Committee on Fisheries and Aquaculture (COFI) in June 2014 (ref. 35). While we have found that many countries already have such data or data structure at hand, until all countries can implement such a data-change request, FAO could incorporate such a split into their internal data harmonization procedures, based, for example, on the same or similar information sources as used by the reconstructions.

The very high catches that were achieved globally in the 1990s were probably not sustainable. However, they do suggest that stock rebuilding, as successfully achieved in many Australian and US fisheries, and beginning to be applied in some European fisheries, is a policy that needs wider implementation, and which would generate even higher sustained benefits than previously estimated from reported catches[36]. On the other hand, the recent catch decline documented here is of considerable concern in its implication for food security, as evidenced by the decline in *per capita* seafood availability (Fig. 2). Note that the recent, strong decline in reconstructed total catches is also evident if catches in countries with well-established quota management systems (United States, Northwestern Europe, Australia and New Zealand) are excluded (Fig. 2). Low quotas are generally not imposed when a stock is abundant; rather low and reduced quotas in fully developed fisheries are generally a management intervention to reduce fishing pressure as a result of past overfishing. Similarly, it has been proposed that strongly declining catches in unmanaged, heavily exploited fisheries are likely a sign of overfishing[32-34]. The often raised suggestion that aquaculture production can replace or compensate for the shortfall in wild capture seafood availability, while being questionable for various reasons[37], is not addressed here.

The last policy relevant point to be made here transcends fisheries in that it deals with the accuracy of the data used by the international community for its decision making, and the generation of factual knowledge that this requires. After the creation of the United Nations and its technical organizations, including the FAO, a major project of 'quantifying the world'[38] began to provide data for national and international agencies on which they could base their policies. As a result, large databases, for example on agricultural crops and forest cover, were created whose accuracy is becoming increasingly important given the expanding exploitation of our natural ecosystems[39].

Periodic validation of these databases should therefore be a priority to ensure they avoid producing 'poor numbers'[40]. For example, reports of member countries to FAO about their forest cover, when aggregated at the global level, suggest that the annual rate of forest loss between 1990/2000 and 2001/2005 was nearly halved, while the actual loss rate doubled when assessed by remote sensing and rigorous sampling[41]. Similarly, here we show that the main trend of the world marine fisheries catches is not one of 'stability' as cautiously suggested earlier by FAO[42], but one of decline. Moreover, this decline, which began in the mid-1990s, started from a considerably higher peak catch than suggested by the aggregate statistics supplied by FAO members, implying that we have more to lose if this decline continues. Thankfully, this also means that there may be more to gain by rebuilding stocks.

For the global community, a solution could therefore be to provide the FAO the required funds to more intensively assist member countries in submitting better and more comprehensive fishery statistics, especially statistics that cover all fisheries components, and report data by sector[25]. Such improved statistics can then lead to better-informed policy changes for rebuilding stocks and maintaining (sea)food security. Alternatively, or in addition, FAO could team up with other groups (as was done for forestry statistics) to improve the fisheries statistics of member countries that often have fisheries departments with very limited human and financial resources.

Ultimately, the only database of international fisheries statistics that the world has (through FAO) can be improved. The more rapid decline of fisheries catches documented here is a good reason for this.

Methods

Catch reconstruction principles. The catch reconstruction approach rests on two basic principles[16]:

- When 'no data are available' on a fishery that is known to exist, it is not appropriate to enter 'NA' or 'no data' into the database. Such entries will later be turned into a zero, which is a bad estimate of the catch of an existing fishery. This concern about the problematic 'elegance of the number zero' is also something that affects other scientific activities, such as climate modelling[43];
- Rather, a best estimate should be inserted in all such cases, based on the fact that fishing is a social activity that is bound to throw a 'shadow' on the society in which it is embedded, and from which an approximate and conservative (but better than zero) estimate of catch can be derived if fishing of this type is known to occur (for example, from the seafood or the fuel consumed locally, or the number of vessels engaged in fisheries and the average catch rate of vessels of this type and so on).

This approach addresses an inherent negative bias in national and, by extension, global catch data, although considerable uncertainty in catch data is likely to remain.

Notably, when doing reconstructions, it became apparent that the perception of 'no data' being available was not always correct: the 'social shadow' yields hundreds of articles in the peer-reviewed and report literature with catch data, or data from which catch rates could be inferred, even for remote islands[10]. Also, countries may sometimes send to FAO a stripped down version of the national catch data their fisheries research institutes actually possess, and may even publish on their websites.

What is covered here are both 'coastal' waters, defined as the waters within the EEZ (Supplementary Fig. 1) that countries have claimed since this was allowed under the United Nations Convention on the Law of the Sea (UNCLOS), or which they could claim under UNCLOS rules, but have not (such as many countries around the Mediterranean), and the open oceans, or High Seas, that is, the waters beyond national jurisdiction (that is, beyond the EEZs). The delineations provided by the Flanders' Marine Institute (VLIZ, see www.vliz.be) are used for our definitions of EEZs. Countries that have not formally claimed an EEZ are assigned areas equivalent to EEZs based on the basic principles of EEZs as outlined in UNCLOS (that is, 200 nm and/or mid-line rules). Note that we (a) include territorial waters within our EEZs; and (b) treat disputed zones (that is, EEZ areas claimed by more than one country) as being 'owned' by each claimant with respect to their fisheries catches. We treat EEZ areas prior to each country's year of EEZ declaration as 'EEZ-equivalent waters' (with open access to all fishing countries during that time). If the year of EEZ declaration could not be determined (and for 'EEZs' that were derived by us for non-claimant countries), we assign the year 1982 as declaration year, that is, the year of conclusion of UNCLOS.

We use different catch reconstruction approaches for EEZs (40% of the global ocean), and High Seas (60%), where the catches are mainly large pelagic fishes (notably tuna). Note that we also exclude the Caspian Sea from all considerations.

Domestic catch reconstruction method. Reconstructing time series of fisheries catch for all countries of the world from 1950 (the first year that FAO published its 'Yearbook' of global fisheries statistics) to 2010 was undertaken by fisheries 'sectors'. However, because a standardized global definition of fishing sectors based on vessel size does not exist (for example, a vessel considered large-scale (industrial) in a developing country may be considered small-scale (artisanal) in developed countries), reconstructions utilize each country's individual definitions for sectors, or a regional equivalent. These are described in each country reconstruction publication underlying this work. We consider four sectors:

  i Industrial: large-scale fisheries (using trawlers, purse-seiners, longliners) with high capital input into vessel construction, maintenance and operation, and which may move fishing gear across the seafloor or through the water column using engine power (for example, demersal and pelagic trawlers), irrespective of vessel size. This corresponds to the 'commercial' sectors of countries such as the USA;

  ii Artisanal: small-scale fisheries whose catch is predominantly sold (hence they are also 'commercial fisheries'), and which often use a large variety of generally static or stationary (passive) gears. Our definition of artisanal fisheries relies also on adjacency: they are assumed to operate only in domestic waters (that is, in their country's EEZ). Within their EEZ, they are further limited to a coastal area to a maximum of 50 km from the coast or to 200 m depth, whichever comes first. This area is defined as the Inshore Fishing Area (IFA)[44]. Note that the definition of an IFA assumes the existence of a small-scale fishery, and thus unpopulated islands, although they may have fisheries in their EEZ (which by our definition are industrial, whatever the gear used), have no IFA;

  iii Subsistence: small-scale non-commercial fisheries whose catch is predominantly consumed by the persons fishing it, and their families (this may also include the 'take-home' fraction of the catch of commercial fishers, which usually by-passes reporting systems); and

  iv Recreational: small-scale non-commercial fisheries whose major purpose is enjoyment.

In addition to the reconstructions by sector, we also assign catches to either 'landings' (that is, retained and landed catch) or 'discards' (that is, discarded catch), and label all catches as either 'reported' or 'unreported' with regards to national and FAO data. Thus, reconstructions present 'catch' as the sum of 'landings' plus 'discards'.

Discarded fish and invertebrates are generally assumed to be dead, except for the US fisheries where the fraction of fish and invertebrates reported to survive is generally available on a per species basis[45]. Due to a distinct lack of global coverage of information, we do not account for so-called under-water discards, or net-mortality of fishing gears[46]. We also do not address mortality caused by ghost-fishing of abandoned or lost fishing gear[47].

For commercially caught jellyfishes (particularly Rhizostomeae, but also other taxa), it has been shown that over 2.5 time more are caught than reported to FAO (mostly as 'Rhizostoma spp.')[48]. This factor is used to estimate missing catches of unidentified jellyfish. However, this additional catch is, pending further study, not allocated to any specific country or FAO area, and is thus counted only in the world's total catch.

We exclude from consideration all catches of marine mammals, reptiles, corals, sponges and marine plants (the bulk of the plant material is not primarily used for human consumption, but for cosmetic or pharmaceutical use). In addition, we do not estimate catches made for the aquarium trade, which can be substantial in some areas in terms of number of individuals, but relatively small in overall tonnage, as most aquarium fish are small or juvenile specimens[49].

Most catch reconstructions consist of six steps[15]:

(1) Identification, sourcing and comparison of baseline catch times series, that is, (a) FAO reported landings data by FAO statistical areas, taxon and year; and (b) national or regional data series by area, taxon and year. Implicit in this first step is that the spatial entity be identified and named that is to be reported on (for example, EEZ of Germany in the Baltic Sea), something that is not always obvious, and which poses problems to some of our external collaborators, notably those in countries with a claimed EEZ overlapping with that of their neighbour.

For most countries, the baseline data are the statistics reported by member countries to FAO. We treat all countries recognized in 2010 (or acting like independent countries with regards to fisheries) by the international community as having existed from 1950 to 2010. This is necessary, given our emphasis on 'places', that is, on time-series of catches taken from specific ecosystems. This also applies to islands and other territories, many of which were colonies, and which have changed status and borders since 1950.

For several countries, the baseline data are provided by international bodies. In the case of EU countries, the baseline data originate from the International Council for the Exploration of the Sea (ICES), which maintains fisheries statistics by smaller statistical areas, as required given the Common Fisheries Policy of the EU. A similar area is the Antarctic waters and surrounding islands, whose fisheries are managed by the Commission for the Conservation of Antarctic Marine Living Resources (CCAMLR), where catch data are available by relatively small statistical areas[50].

When FAO data are used, care is taken to maintain their assignment to different FAO statistical areas for each country (Supplementary Fig. 1), as they often distinguish between strongly different ecosystems. For example, the Caribbean Sea versus the coast of the Eastern Central Pacific in the case of Panama, Costa Rica, Nicaragua, Honduras and Guatemala. For each maritime country, the area covered extends from the coastline to the edge of the EEZ, including any major coastal lagoons connected to the sea, and the mouths of rivers, that is, estuaries. However, freshwaters are excluded.

(2) Identification of sectors (for example, subsistence, recreational), time periods, species, gears and so on, not covered by (1), that is, missing data components. This is conducted via literature searches and consultations with local experts. This step is one where the contribution of local co-authors and experts is crucial. Potentially, all four sectors defined by us can occur in the marine fisheries of a given coastal country, with the distinction between large-scale and small-scale being the most important[25]. For any entity, we check whether catches originating from the four sectors were included in the reported baseline of catch data, notably by examining their taxonomic composition, and any metadata, which were particularly detailed in the early decades of the FAO 'Yearbooks'[51].

The absence of a taxon known to be caught in a country or territory from the baseline data (for example, cockles gleaned by women on the shore of an estuary)[26] can also be used to identify a fishery that has been overlooked in the official data collection scheme, as can the absence of reef fishes in the coastal data of a Pacific Island state[10]. To avoid double counting, tuna and other large pelagic fishes, unless known to be caught by a local small-scale fishery (and thus in the past not likely reported to a Regional Fisheries Management Organization or RFMO), are not included in this reconstruction step (see below under 'High Seas and other catches of large pelagic fishes').

Finally, if gears are identified in national data, but a gear known to exist in a given country is not included, then it can be assumed that its catch has been missed, as documented for weirs (hadrah) in the Persian Gulf[52].

(3) Sourcing of alternative information sources on missing sectors identified in (2), via literature searches (peer-reviewed and grey) and consultations with local experts. Information sources include social science studies (anthropology, economics and so on), reports, data sets and expert knowledge. The major initial source of information for catch reconstructions is governments' websites and publications (specifically their Department of Fisheries or equivalent agency), both online and in hard copies. Contrary to what could be expected, it is often not the agency responsible for fisheries research and initial data collection that supplies the catch statistics to FAO, but other agencies, for example, statistical office or agency. As a result, much of the granularity of the original data (that is, catch by sector, by species or by gear) may be lost even before data are prepared for submission to FAO. Furthermore, the data request form sent by FAO each year to each country does not encourage improvements or changes in taxonomic composition, as the form that requests the most recent year's data contains the country's previous years' data in the same composition as submitted in earlier years. This encourages the pooling of detailed data at the national level into the taxonomic categories inherited through earlier (often decades old) FAO reporting schemes, as was discovered, for example, for Bermuda in the early 2000s (ref. 53). Thus, by getting back to the original data, much of the original granularity can be regained during reconstructions.

Additional sources of information on national catches are international organizations such as FAO, ICES or SPC (Secretariat of the Pacific Community), or a Regional Fisheries Management Organization (RFMO) such as NAFO (Northwest Atlantic Fisheries Organization), or CCAMLR[54], or current or past regional fisheries development and/or management projects (many of them launched and supported by FAO), such as the Bay of Bengal Large Marine Ecosystem project (BOBLME). All these organizations and projects issue reports and publications describing—sometimes in considerable details—the fisheries of their member countries. Another source of information is the academic literature, now widely accessible through Google Scholar.

A good source of information for the earlier decades (especially the 1950s and 1960s) for countries that were part of former colonial empires (especially British or French) are the colonial archives in London (British Colonial Office) and the 'Archives Nationales d'Outre-Mer', in Aix-en-Provence, and the publications of ORSTOM (Office de la recherche scientifique et technique d'outre-mer), for former French colonies. A further source of information and data are non-fisheries sources, including household and/or nutritional surveys, which are occasionally used for estimating unreported subsistence catches. Our global network of local collaborators is also crucial in this respect, as they have access to key data sets, publications and local knowledge not available elsewhere, often in languages other than English.

Supplementary Figure 2 shows a plot of the publications used for slightly over 110 reconstructions against their date of publication. Although, recent publications predominate, older publications firmly anchor the 1950s catch estimates of many reconstructions. On average, around 35 unique publications were used per reconstruction (not counting online sources and personal communications).

Potential language bias is taken seriously in the *Sea Around Us*, to ensure that data are collated in languages other than English. Besides team members who read Chinese, others speak Arabic, Danish, Filipino/Tagalog, French, German, Hindi, Japanese, Portuguese, Russian, Spanish, Swedish and Turkish. To deal with other languages, research assistants are hired who speak, for example, Korean or

Table 1 | Scoring system for deriving uncertainty bands for the quality of time series data of reconstructed catches.

Score		± %[*]	Corresponding IPCC criteria[†]
4	Very high	10	High agreement and robust evidence
3	High	20	High agreement and medium evidence **or** medium agreement and robust evidence
2	Low	30	High agreement and limited evidence **or** medium agreement and medium evidence **or** low agreement and robust evidence.
1	Very low	50	Less than high agreement and less than robust evidence

[*]Percentage uncertainty derived from Monte-Carlo simulations[66,67].
[†]'Confidence increases' (and hence percentage ranges are reduced) 'when there are multiple, consistent independent lines of high-quality evidence'[61].

Malay/Indonesian. We also rely on our multilingual network of colleagues and friends throughout the world, for example, for Greek or Thai. While it is true that English has now become the undisputed language of science[55], other languages are used by billions of people, and assembling knowledge about the fisheries of the world is not possible without the capacity to explore the literature in languages other than English.

(4) Development of data 'anchor points' in time for each missing data item, and expansion of anchor point data to country-wide catch estimates. 'Anchor' points are catch estimates usually pertaining to a single year and sector, and often to an area not exactly matching the limits of the EEZ or IFA in question. Thus, an anchor point pertaining to a fraction of the coastline of a given country may need to be expanded to the country as a whole. For expansion, we use fisher or population density, or relative IFA or shelf area as raising factor, as appropriate given the local condition. In all cases, we consider that case studies underlying or providing the anchor point data may had a case-selection bias (for example, representing an exceptionally good area or community for study, compared with other areas in the same country), and thus use raising factors very conservatively.

(5) Interpolation for time periods between data anchor points, either linearly or assumption based for commercial fisheries, and generally via per capita (or per fisher) catch rates for non-commercial sectors. Fisheries are often difficult to govern, as they are social activities involving multiple actors. In particular, fishing effort is often difficult to reduce, at least in the short term. Thus, if anchor points are available for years separated by multi-year intervals, it usually will be more reasonable to assume that the underlying fishing activity continues in the intervening years with no data. We tread this 'continuity' assumption as a default proposition. Exceptions to such continuity assumptions are major environmental impacts such a hurricanes or tsunamis[56], or major socio-political disturbances, such as military conflicts or civil wars[57], which we explicitly consider with regards to the use of raising factors and the structure of time series estimates. In such cases, our reconstructions mark the event through a temporary change (for example, decline) in the catch time series, which is documented in the text of each catch reconstruction. At the very least, this provides pointers for future research on the relationship between fishery catches and natural catastrophes or conflicts. We note that the absence of such signals (such as a reduction in catch for a year or two) in the officially reported catch statistics for countries having experienced a major natural or socio-political disturbance can be a sign that their official catch data may not accurately reflect what occurs on the ground. This contributes to the emergence of 'poor numbers'[40]. Overall, our reconstructions assume—when no information to the contrary is available—that commercial catches (that is, industrial and artisanal) can be linearly interpolated between anchor points, while non-commercial catches (that is, subsistence and recreational) can generally be interpolated between anchor points using non-linear trends in human population numbers or number of fishers over time (via per capita rates).

Radical and rapid effort reductions as a result of an intentional policy decision and implementation do not occur widely. One example we are aware of is the trawl ban of 1980 in Western Indonesia[58]. The ban had little or no impact on official Indonesian fisheries statistics for Western Indonesia, another indication that these statistics may have little to do with the realities on the ground. FAO hints at this being widespread in the Western Central Pacific and the Eastern Indian Ocean (the only FAO areas where reported catches appear to be increasing) when they note that 'while some countries (i.e., the Russian Federation, India and Malaysia) have reported decreases in some years, marine catches submitted to FAO by Myanmar, Vietnam, Indonesia and China show continuous growth, i.e., in some cases resulting in an astonishing decadal increase (e.g., Myanmar up 121 percent, and Vietnam up 47 percent)'.[42]

(6) Estimation of total catch times series. A reconstruction is completed when the estimated catch time series derived through steps 2–5 are combined and harmonized with the reported catch of step 1. Generally, this results in an increase of the overall catch, but several cases exist where the reconstructed total catch is lower than the reported catch. The best documented case of this is that of mainland China[14], whose over-reported catches for local waters in the Northwest Pacific are compensated for by under-reported catches taken by Chinese distant water fleets fishing elsewhere. In the 2000s, Chinese distant water fleets operated in the EEZs of over 90 countries, that is, in most parts of the world's oceans[5]. Harmonizing reconstructed catches with the reported baselines goes hand-in-hand with documenting the entire reconstruction procedure. Thus, every reconstruction is documented and published, either in the peer-reviewed scientific literature, or as detailed technical reports in the publicly accessible and indexed Fisheries Centre

Research Reports series or the Fisheries Centre Working Paper series, or other regional organization reports (Supplementary Table 5).

Several reconstructions were conducted in the mid- to late 2000s, when official reported data (that is, FAO statistics or national data) were not available to 2010 (refs 15,59). All these cases are updated to 2010, in line with each country's individual reconstruction approach to estimating missing catch data. Thus, all reconstructions are brought to 2010 to ensure identical time coverage (Supplementary Table 5).

Since these six points were originally proposed, a seventh point has come to the fore that cannot be ignored[10]:

(7) Quantifying the uncertainty associated with each reconstruction. In fisheries research, catch data are rarely associated with a measure of uncertainty, at least not in the form resembling confidence intervals. This may reflect the fact that the issue with catch data is not a lack of precision (that is, whether we could expect to produce similar results upon re-estimation), but about accuracy, that is, attempting to eliminate a systematic bias, a type of error which statistical theory does not really address.

We deal with this issue through a procedure related to 'pedigrees'[60] and the approach used by the Intergovernmental Panel on Climate Change to quantify the uncertainty in its assessments[61]. The authors of the reconstructions are asked to attribute a 'score' expressing their evaluation of the quality of the time series data to each fisheries sector (industrial, artisanal and so on) for each of the three time periods (1950–1969, 1970–1989 and 1990–2010). These 'scores' are (1) 'very low', (2) 'low', (3) 'high' and (4) 'very high' (Table 1). There is a deliberate absence of an uninformative 'medium' score, to avoid the effective 'non-choice' that this option would represent. Each of these scores is assigned a percentage uncertainty range (Table 1). Thereafter, the overall mean weighted percentage uncertainty (over all countries and sectors) was computed (Fig. 1).

Foreign catches. We define foreign catches as taken by vessels of a maritime state in the EEZ, or EEZ-equivalent waters of another coastal state. Based on our definition of sectors, all foreign fishing in the waters of another country is deemed to be industrial in nature. As the High Seas legally belong to no one (or to everyone), there can be no 'foreign' catches in the High Seas. Prior to UNCLOS, and the declaration of EEZs by maritime countries, foreign catches were illegal only if conducted without explicit permission within the territorial waters of such countries (generally 12 nautical miles). Since the declarations of EEZs by the overwhelming majority of maritime countries, foreign catches are considered illegal if conducted within the EEZ but without access being granted by the coastal state. A distinct exception is the EU, whose waters are managed by a 'Common Fisheries Policy', which implies a multilateral 'access agreement'.

Access permission can be tacit and based on historic rights ('observed' access), or more commonly in the form of explicit access agreements and involving compensatory payment for the coastal state. The Sea Around Us, building on previous work by FAO[62], has created a database of such access and agreements, which is used to allocate the catches of distant-water fleets to the waters where they were taken.

This information is then harmonized with the catches reported by FAO for countries fishing outside their country's 'home' FAO areas, which always identifies this catch as distant-water industrial catch (see below for tuna catches reported to RFMOs).

In line with INTERPOL and others[63], we define illegal fishing as foreign fishing within the EEZ waters of another country without a permission to access these resources. We do not treat domestic fisheries' violations of 'fishing regulations' as 'illegal'. In general, our reconstruction method cannot readily distinguish between legal and illegal foreign fishing, as we do not necessarily know about all access agreements[5,6]. Thus, our data only pertain to 'reported' versus 'unreported' status, irrespective of legality of foreign fleets in a host country[5]. However, for around two dozen countries (mainly in West Africa) where the number of illegally operating vessels could be inferred, the fleet size can be multiplied by appropriate catch per unit of effort rates, leading to an estimate of illegal catch in these EEZs.

Industrial catches of large pelagic fishes. Nominal landings data. To date, there is no single, publicly available data set presenting industrial landings of tuna and large pelagic fishes for the entire world that is separate from the amalgamated FAO statistics, despite these fisheries being among the most valuable in the world[64].

Here, we first compile nominal industrial landings of tuna and other large pelagic fish caught either in the High Seas or within EEZs by fishing gear, taxon, countries and statistical reporting areas from data published by Regional Fisheries Management Organizations. Second, we use partially spatialized landings data provided by staff of the French 'Institut de recherche pour le développement' to spatially pre-assign the nominal landings data derived from RFMOs (Supplementary Table 6).

For each ocean, the nominal landings data are spatialized according to reported proportions in the previously spatialized data (Supplementary Table 6). For example, if the nominal data reports France catching 100 tonnes of yellowfin tuna (*Thunnus albacares*) in 1983 using longlines, but the spatial data only present 85 tonnes of yellowfin tuna reported in 1983 by France using longlines in four separate statistical cells, the nominal 100 tonnes for France are split into these four spatial cells according to their reported proportion of catch in the spatial dataset. This matching of the nominal and spatial records is done over a series of successive refinements, with the first being the best-case scenario, in which there are matching records for year, country, gear and taxon. The last refinement is the worst-case scenario, in which there are no matching records except for the year of catch. For example, if Sri Lanka reports 100 tonnes of yellowfin tuna caught in 1983 using longlines, but there are no spatial records for any country catching yellowfin tuna in 1983, the nominal 100 tonnes for Sri Lanka are split into spatial cells according to their reported proportions of total catch of any species and gear in 1983. The end result is a baseline landings database containing all matched and spatialized catch records, which sum to the original nominal catch tonnages.

Discards. A review of the literature for each ocean provided limited country- and fleet-specific discard data. Therefore, we average the discard rates across the entire time period and apply these to the region of origin of the fleet (for example, East Asia or Western Europe), rather than the actual country of origin of the fleet. Discards were spatialized in conjunction with nominal landings data.

Assembly of total catches. Ultimately, the total catch extracted from a given area, such as a given EEZ or EEZ-equivalent waters, or high seas waters within a given FAO area is computed as the sum of three data layers: (1) the reconstructed domestic catches within home EEZs ('Layer 1' data); (2) the derived catch by foreign fleets ('Layer 2' data); and (3) the tuna and other large pelagic fishes caught in the High Seas and in EEZs ('Layer 3' data).

Documentation of the catch reconstructions. The references and web-links of the contributions documenting the catch reconstructions that went into the re-estimation of the global catch of marine fisheries are documented in Supplementary Table 5. Altogether, 273 EEZs (or EEZ 'components') were covered in 247 catch reconstructions, which had 103 unique first authors and 279 unique co-authors in over 50 countries.

Analyses. To examine if significant breakpoints exist in the catch data time series of both reconstructed total catches and reported catches that may illustrate a change in trends of catches over time (that is, a change in the slope), we analyse the time series trajectories using segmented regression[21]. For both the reconstructed as well as reported time series, we identify two breakpoints, being 1967 and 1996, respectively (Supplementary Table 2). These breakpoints suggest a change in regression slope, with the second breakpoint suggesting a trend reversal. This was validated by testing for a significant difference-in-slope parameter using the Davies test[65], which tests for a non-zero difference-in-slope of a segmented regression relationship.

References

1. Mohan Dey, M. *et al.* Fish consumption and food security: a disaggregated analysis by types of fish and classes of consumers in selected Asian countries. *Aquacult. Econ. Manage* **9**, 89–111 (2005).
2. Kawarazuka, N. & Béné, C. The potential role of small fish species in improving micronutrient deficiencies in developing countries: building evidence. *Public Health Nutr* **14**, 1927–1938 (2011).
3. Swartz, W., Sala, E., Tracey, S., Watson, R. & Pauly, D. The spatial expansion and ecological footprint of fisheries (1950 to present). *PLoS ONE* **5**, e15143 (2010).
4. Swartz, W., Sumaila, U. R., Watson, R. & Pauly, D. Sourcing seafood for the three major markets: the EU, Japan and the USA. *Mar. Policy* **34**, 1366–1373 (2010).
5. Pauly, D. *et al.* China's distant water fisheries in the 21st century. *Fish Fish.* **15**, 474–488 (2014).
6. Le Manach, F. *et al.* European Union's public fishing access agreements in developing countries. *PLoS ONE* **8**, e79899 (2013).
7. Alder, J. & Sumaila, U. R. Western Africa: the fish basket of Europe past and present. *J. Environ. Dev.* **13**, 156–178 (2004).
8. Belhabib, D., Sumaila, U. R. & Pauly, D. Feeding the poor: contribution of West African fisheries to employment and food security. *Ocean Coast. Manage.* **111**, 72–81 (2015).
9. Pauly, D. Major trends in small-scale marine fisheries, with emphasis on developing countries, and some implications for the social sciences. *Marit. Studies* **4**, 7–22 (2006).
10. Zeller, D., Harper, S., Zylich, K. & Pauly, D. Synthesis of under-reported small-scale fisheries catch in Pacific-island waters. *Coral Reefs* **34**, 25–39 (2015).
11. Garibaldi, L. The FAO global capture production database: a six-decade effort to catch the trend. *Mar. Policy* **36**, 760–768 (2012).
12. Zeller, D. & Pauly, D. Good news, bad news: global fisheries discards are declining, but so are total catches. *Fish Fish.* **6**, 156–159 (2005).
13. Zeller, D., Booth, S., Pakhomov, E., Swartz, W. & Pauly, D. Arctic fisheries catches in Russia, USA and Canada: Baselines for neglected ecosystems. *Polar Biol.* **34**, 955–973 (2011).
14. Watson, R. & Pauly, D. Systematic distortions in world fisheries catch trends. *Nature* **414**, 534–536 (2001).
15. Zeller, D., Booth, S., Davis, G. & Pauly, D. Re-estimation of small-scale fishery catches for U.S. flag-associated island areas in the western Pacific: the last 50 years. *Fish. Bull.* **105**, 266–277 (2007).
16. Pauly, D. Rationale for reconstructing catch time series. *EC Fish. Coop. Bull.* **11**, 4–10 (1998).
17. Pikitch, E. K. *et al.* Ecosystem-based Fishery Management. *Science* **305**, 346–347 (2004).
18. FAO. in *Code of Conduct for Responsible Fisheries* (Food and Agriculture Organization of the United Nations (FAO), 1995).
19. FAO. in *The State of World Fisheries and Aquaculture (SOFIA) 2010* 197 (Food and Agriculture Organization, 2011).
20. Pauly, D. & Froese, R. Comments on FAO's State of Fisheries and Aquaculture, or 'Sofia 2010'. *Mar. Policy* **36**, 746–752 (2012).
21. Oosterbaan, R. J. in *Drainage Principles and Applications* Publication 16 (ed Ritzema, H. P.) 175–224 (International Institute for Land Reclamation and Improvement (ILRI), 1994).
22. Andrew, N. L. & Pepperell, J. G. The by-catch of shrimp trawl fisheries. *Oceanogr. Mar. Biol. Annu. Rev* **30**, 527–565 (1992).
23. Alverson, D. L., Freeberg, M. H., Pope, J. G. & Murawski, S. A. *A Global Assessment of Fisheries by-Catch And Discards* 233 (FAO Fisheries Technical Papers T339, 1994).
24. Kelleher, K. *Discards in the World's Marine Fisheries. An Update.* 131 (FAO Fisheries Technical Paper 470, Food and Agriculture Organization, 2005).
25. Pauly, D. & Charles, T. Counting on small-scale fisheries. *Science* **347**, 242–243 (2015).
26. Harper, S., Zeller, D., Hauzer, M., Sumaila, U. R. & Pauly, D. Women and fisheries: contribution to food security and local economies. *Mar. Policy* **39**, 56–63 (2013).
27. Chapman, M. D. Women's fishing in Oceania. *Hum. Ecol.* **15**, 267–288 (1987).
28. Cisneros-Montemayor, A. M. & Sumaila, U. R. A global estimate of benefits from ecosystem-based marine recreation: potential impacts and implications for management. *J. Bioecon* **12**, 245–268 (2010).
29. Belhabib, D. *et al.* in *Marine Fisheries Catches in West Africa, 1950-2010, part I.* Fisheries Centre Research Reports 20(3) (eds Belhabib, D., Zeller, D., Harper, S. & Pauly, D.) 61–78 (Fisheries Centre, University of British Columbia, 2012).
30. Belhabib, D., Nahada, V. A., Blade, D. & Pauly, D. *Fisheries in Troubled Waters: A Catch Reconstruction for Guinea-Bissau, 1950-2010.* Working Paper #2015-72 21 (Fisheries Centre, University of British Columbia, 2015).
31. Zeller, D. *et al.* The Baltic Sea: estimates of total fisheries removals 1950-2007. *Fish. Res.* **108**, 356–363 (2011).
32. Kleisner, K., Zeller, D., Froese, R. & Pauly, D. Using global catch data for inferences on the world's marine fisheries. *Fish Fish.* **14**, 293–311 (2013).
33. Froese, R., Zeller, D., Kleisner, K. & Pauly, D. Worrisome trends in global stock status continue unabated: a response to a comment by R.M. Cook on 'What catch data can tell us about the status of global fisheries'. *Mar. Biol. (Berlin)* **160**, 2531–2533 (2013).
34. Froese, R., Zeller, D., Kleisner, K. & Pauly, D. What catch data can tell us about the status of global fisheries. *Mar. Biol. (Berlin)* **159**, 1283–1292 (2012).
35. FAO. in *Voluntary Guidelines for Securing Sustainable Small-Scale Fisheries in the Context of Food Security and Poverty Eradication* xii + 18 (Food and Agriculture Organization of the United Nations, 2015).
36. Sumaila, U. R. *et al.* Benefits of rebuilding global marine fisheries outweigh costs. *PLoS ONE* **7**, e40542 (2012).
37. Cao, L. *et al.* China's aquaculture and the world's wild fisheries. *Science* **347**, 133–135 (2015).
38. Ward, M. in *Quantifying the World: UN Ideas and Statistics* (Indiana University Press, 2004).
39. Rockström, J. *et al.* Planetary boundaries: exploring the safe operating space for humanity. *Nature* **461**, 472–475 (2009).
40. Jerven, M. in *Poor Numbers: How We Are Misled by African Development Statistics and What to Do About It* (Cornell University Press, 2013).

41. Lindquist, E. J. *et al.* in *FAO/JRC Global Forest Land-Use Change from 1990 to 2005*. FAO Forestry Paper 169 xi + 40 (Food and Agriculture Organization of the United Nations and European Commission Joint Research Center, 2012).

42. FAO. in *The State of World Fisheries and Aquaculture (SOFIA)* 223 (Food and Agriculture Organization, 2014).

43. Covey, C. Beware the elegance of the number zero. *Clim. Change* **44**, 409–411 (2000).

44. Chuenpagdee, R., Liguori, L., Palomares, M. D. & Pauly, D. in *Bottom-up, Global Estimates of Small-Scale Marine Fisheries Catches. 112 (Fisheries Centre Research Reports 14(8)* (University of British Columbia, 2006).

45. McCrea-Strub, A. in *Reconstruction of Total Catch by U.S. Fisheries in the Atlantic and Gulf of Mexico: 1950-2010*. Working Paper #2015-79 46 (Fisheries Centre, University of British Columbia, 2015).

46. Rahikainen, M., Peltonen, H. & Poenni, J. Unaccounted mortality in northern Baltic Sea herring fishery—magnitude and effects on estimates of stock dynamics. *Fish. Res.* **67**, 111–127 (2004).

47. Bullimore, B. A., Newman, P. B., Kaiser, M. J., Gilbert, S. E. & Lock, K. M. A study of catches in a fleet of 'ghost-fishing' pots. *Fish. Bull.* **99**, 247–253 (2001).

48. Brotz, l. in *So Long, and Thanks For the All Fish: the Sea Around Us, 1999-2014—A Fifteen-Year Retrospective.* (eds Pauly, D. & Zeller, D.) 81–85 (A Sea Around Us Report to The Pew Charitable Trusts, University of British Columbia, 2014).

49. Rhyne, A. L. *et al.* Revealing the appetite of the marine aquarium fish trade: the volume and biodiversity of fish imported into the United States. *PloS ONE* **7**, e35808 (2012).

50. Ainley, D. & Pauly, D. Fishing down the food web of the Antarctic continental shelf and slope. *Polar Rec.* **50**, 92–107 (2013).

51. FAO. in *Catches and landings (1977)* Vol. **44**, 343 (Yearbook of Fishery Statistics, Food and Agriculture Organization, 1978).

52. Al-Abdulrazzak, D. & Pauly, D. Managing fisheries from space: Google Earth improves estimates of distant fish catches. *ICES J. Mar. Sci.* **71**, 450–455 (2014).

53. Luckhurst, B., Booth, S. & Zeller, D. in *From Mexico to Brazil: Central Atlantic Fisheries Catch Trends and Ecosystem Models* (eds Zeller, D., Booth, S., Mohammed, E. & Pauly, D.) 163–169 (Fisheries Centre Research Reports 11(6), University of British Columbia, 2003).

54. Cullis-Suzuki, S. & Pauly, D. Failing the high seas: a global evaluation of regional fisheries management organizations. *Mar. Policy* **34**, 1036–1042 (2010).

55. Ammon, U. in *The Dominance of English as a Language of Science: Effects on Other Languages and Language Communities* (Walter de Gruyter, 2001).

56. Ramdeen, R., Ponteen, A., Harper, S. & Zeller, D. in *Fisheries Catch Reconstructions: Islands, Part III* (eds Harper, S. *et al.*) 69–76 (Fisheries Centre Research Reports 20(5), University of British Columbia, 2012).

57. Belhabib, D. *et al.* in *When 'Reality Leaves A Lot To the Imagination': Liberian Fisheries from 1950 to 2010* 18 (Fisheries Centre Working Paper #2013-06, University of British Columbia, 2013).

58. Pauly, D. & Budimartono, V. in *Marine Fisheries Catches of Western, Central and Eastern Indonesia, 1950-2010.* 51 (Fisheries Centre Working Paper #2015-61, University of British Columbia, 2015).

59. Zeller, D., Booth, S., Craig, P. & Pauly, D. Reconstruction of coral reef fisheries catches in American Samoa, 1950-2002. *Coral Reefs* **25**, 144–152 (2006).

60. Funtowicz, S. O. & Ravetz, J. R. *Uncertainty and Quality of Science for Policy* XI + 231 (Springer, 1990).

61. Mastrandrea, M. D. *et al.* in *Guidance Note for Lead Authors of the IPCC Fifth Assessment Report on Consistent Treatment of Uncertainties* (Intergovernmental Panel on Climate Change (IPCC), 2010).

62. FAO. in *FAO's Fisheries Agreement Register (FARISIS)* 4 (Committee on Fisheries, 23rd Session, 15-19 February 1999, Food and Agriculture Organization, COFI/99/Inf9E, 1998).

63. UNODC. in *Transnational Organized Crime in the Fishing Industry* 140 (United Nations Office on Drug and Crime, 2011).

64. FAO. in *The State of World Fisheries and Aquaculture* 209 (Food and Agriculture Organization of the United Nations (FAO), 2012).

65. Davies, R. B. Hypothesis testing when a nuisance parameter is present only under the alternative. *Biometrika* **74**, 33–43 (1987).

66. Ainsworth, C. H. & Pitcher, T. J. Estimating illegal, unreported and unregulated catch in British Columbia's marine fisheries. *Fish. Res.* **75**, 40–55 (2005).

67. Tesfamichael, D. & Pitcher, T. J. Estimating the unreported catch of Eritrean Red Sea fisheries. *Afr. J. Mar. Sci.* **29**, 55–63 (2007).

Acknowledgements

The Pew Charitable Trusts, Philadelphia funded the *Sea Around Us* from 1999 to 2014, during which the bulk of the catch reconstruction work was performed. Since mid-2014, the *Sea Around Us* has been funded mainly by The Paul G. Allen Family Foundation and assisted by the staff of Vulcan, Inc., with additional funding from the Rockefeller, MAVA, and Prince Albert II Foundations. We thank our many collaborators, as listed in the Supplementary Acknowledgements, and also numerous additional colleagues who assisted in various aspects of this work over the last 15 years.

Author contributions

D.P. conceptualized the rationale for catch reconstructions, and as Principal Investigator of the *Sea Around Us*, leads the project since 1999. D.Z. operationalized the methods employed for reconstructions, and as Senior Scientist coordinated and/or led many of the reconstructions contributing to this paper. D.P. and D.Z. co-wrote the paper. A very large number of colleagues have contributed to individual catch reconstructions and our core database work over the last 15 years, and are acknowledged and listed in the Supplementary Acknowledgements.

Additional information

Competing financial interests: The authors declare no competing financial interests.

4

Selective silicate-directed motility in diatoms

Karen Grace V. Bondoc[1,2], Jan Heuschele[3,4], Jeroen Gillard[5,6], Wim Vyverman[5] & Georg Pohnert[1,2]

Diatoms are highly abundant unicellular algae that often dominate pelagic as well as benthic primary production in the oceans and inland waters. Being strictly dependent on silica to build their biomineralized cell walls, marine diatoms precipitate 240×10^{12} mol Si per year, which makes them the major sink in the global Si cycle. Dissolved silicic acid (dSi) availability frequently limits diatom productivity and influences species composition of communities. We show that benthic diatoms selectively perceive and behaviourally react to gradients of dSi. Cell speed increases under dSi-limited conditions in a chemokinetic response and, if gradients of this resource are present, increased directionality of cell movement promotes chemotaxis. The ability to exploit local and short-lived dSi hotspots using a specific search behaviour likely contributes to micro-scale patch dynamics in biofilm communities. On a global scale this behaviour might affect sediment–water dSi fluxes and biogeochemical cycling.

[1] Institute for Inorganic and Analytical Chemistry, Bioorganic Analytics, Department of Chemistry and Earth Sciences, Friedrich-Schiller-Universität Jena, Lessingstrasse 8, D-07743 Jena, Germany. [2] Max Planck Institute for Chemical Ecology, Max Planck Fellow, Hans-Knöll-Str. 8, D-07745 Jena, Germany. [3] Centre for Ocean Life, National Institute of Aquatic Resources, Technical University of Denmark, Charlottenlund Slot, Jægersborg Allé, DK-2920 Charlottenlund, Denmark. [4] Department of Biology, Aquatic Ecology Unit, Lund University, SE-22362 Lund, Sweden. [5] Laboratory of Protistology and Aquatic Ecology, Department of Biology, University Gent, Krijgslaan 281, S8, 9000 Gent, Belgium. [6] California State University, Department of Biology, 9001 Stockdale Hwy, Bakersfield, California 93311, USA. Correspondence and requests for materials should be addressed to W.V. (email: Wim.Vyverman@ugent.be) or to G.P. (email: Georg.Pohnert@uni-jena.de).

Diatoms contribute about 20% to the global primary production and are key players in marine and freshwater benthic and planktonic communities[1]. A hallmark of diatom physiology is their biomineralized cell wall that is formed by template-catalysed precipitation of silicic acid[2]. Given their vast abundance, diatoms are thereby driving the silicate cycle[3,4]. Dissolved silicic acid (dSi) availability is often the limiting factor controlling diatom growth and thus also shaping species composition in marine communities[5]. While the pelagic zone is often dSi-limited below 1 µM, the benthic zone typically shows strong and steep gradients of this resource with higher dSi concentrations (around 150 µM) in the sediment due to the continuous dissolution of deposited minerals[6]. Because of their high productivity and biomineralization activity, benthic diatom biofilms can influence sediment properties[7] and alter dSi fluxes within the sediment–water interface, thus regulating dSi concentrations in the oceans[8]. Such processes have implications for the transfer of energy to higher trophic levels, bentho-pelagic coupling and hence population and ecosystem productivity[4]. Most benthic diatoms belong to the pennates, comprising the youngest (90 Myr old) yet most species-rich clade within the diatoms[9,10]. Many species evolved a strong capacity for vertical migration in sediments under the control of photoperiod and/or tidal cycles[11]. However, these processes alone are not fully explaining observed spatiotemporal dynamics of microbial biofilms, and since many years other factors including the direct and indirect influence of herbivory and microbe–microbe interactions are assumed to guide diatom movement[11-13]. Here we identify an additional guiding factor by showing that diatoms detect and actively move towards dSi sources.

We used the pennate diatom Seminavis robusta to explore cell movement and aggregation in response to dSi. Like many other pennate diatoms, this biofilm-forming species moves by gliding through the excretion of extracellular polymeric substances from its raphe, an elongate slit in the cell wall[14]. This allows pennate diatoms to move back and forth. Observed turning movements were suggested to result from the action of extracellular polymeric substance-derived pseudopods or stalks. When a pseudopod or stalk is adhering to the substratum resulting torque supports the whole-cell rotation[15]. In this contribution, we describe three sets of experiments where we first look at the general influence of dSi concentration on diatom motility, then we observe and analyse diatom behaviour in a dSi gradient and last, we test the specificity of the response by comparing the reaction towards dSi and dGe gradients. These experiments clearly demonstrate that diatoms have means to selectively perceive and orient towards the essential resource dSi. A search behaviour in form of increased cell motility and cell speed is observed when the nutrient dSi is depleted. The unicellular algae are also capable of directional movement towards the sources of dSi gradients, a behaviour that supports foraging in the patchy natural environment of benthic diatoms. The fact that structurally closely related dissolved germanium dioxide (Ge(OH)$_4$, dGe) sources are not eliciting attraction suggests a specific receptor-mediated response.

Results

dSi-dependent diatom motility and speed.
To determine if dSi availability affects cell behaviour, we counted motile cells in conjunction with dSi depletion in batch cultures. The proportion of motile cells steeply increased along with decreasing dSi availability as cells entered the stationary growth phase (Fig. 1a). Addition of dSi (106 µM) to stationary-phase cultures elicited within 1 h a marked drop in the proportion of motile cells, indicating a reversible reaction controlled by dSi (Fig. 1b). In addition to motility, cell speed is also dependent on dSi

concentration. When stationary-phase S. robusta cultures were transferred to artificial sea water without added dSi (low-dSi medium) and were further starved for 3 days, cell movement was more than twice as fast than that of cells transferred to dSi-rich control medium (Fig. 1c). Moreover, cell speed decreased after 1 h of dSi addition while blank addition did not affect the speed. The observed increased proportion of motile cells and higher speed under limiting conditions is a chemokinetic response, that is, a motile response to chemicals. Since dSi is not required for movement, speeding up is an effective mechanism for fast location of this limiting resource during starvation[16,17].

Directed movement towards dSi sources.
Since steep vertical and horizontal dSi gradients prevail in benthic environments, an additional strategy to exploit this resource would be a directed movement within dSi gradients. Requirements for such a behaviour are the cells' ability to perceive the resource in a quantitative manner, as described above, and a directed movement towards it. Track analysis revealed that S. robusta moves in a back and forth manner that enables cells to reverse direction after each stop (Supplementary Movie 1). This behaviour found in many raphid diatoms and in certain bacteria allows for orientation towards higher concentrations in chemical gradients[18,19]. S. robusta thus fulfils both above-mentioned requirements for a directed movement. We therefore verified if it indeed has the capability of chemotaxis that would lead to the accumulation of Si-starved cells at local dSi hotspots. Gradients of dSi were generated from micrometer-sized point sources in form of aluminium oxide (alox) beads that were loaded with dSi in different concentrations. To analyse movement along the dSi concentration gradient, the microscopic observation area was divided into three bins (A–C) of equidistant concentric rings around each observed bead covering a radius of 336 µm (Fig. 2a). Since a directed orientation would likely be most relevant under low dSi concentrations found at the water–sediment interface, we aimed to adjust the local concentration gradient in this range. If 1.4 nmol dSi per microscopic bead were applied, ca. 5% diffused out within the 600 s assay period. In a matter of ~460 s a concentration gradient is established with ~100 µM dSi at the surface of the bead decreasing to ~5 µM dSi at the edges of the microscopic observation field (Supplementary Fig. 1). This concentration gradient mimics conditions at the transition of the pelagic and benthic zones[6]. Si-starved S. robusta responded to such dSi gradients with an accumulation of cells around the beads while control beads were not attractive (Fig. 2). This behavioural response is observed in different S. robusta isolates (Supplementary Movies 2–5). If the observation period is extended to 1 h, continuous movement of cells towards the bead is observed until ~25 min. After that, chemoattraction becomes less obvious, presumably due to diffusion of dSi. Control beads are not active through the entire assay period (Supplementary Movies 2 and 4). Response to dSi is not limited to S. robusta, since Navicula sp., another benthic diatom, also accumulated around dSi sources (Supplementary Movie 6). The administered dSi concentration is in the optimum range to elicit a response. Attraction became less pronounced if lower concentrations of dSi were administered, higher concentrations still resulted in a substantial accumulation around the beads but data became noisier, indicating a more erratic search (Supplementary Fig. 2). Such a concentration-dependent response is typical for receptor-mediated interactions, since higher concentrations of dSi on the beads might cause receptor saturation, whereas lower concentrations fall below the detection limit[20]. The finding capability is thus most efficient in an environmentally relevant concentration range.

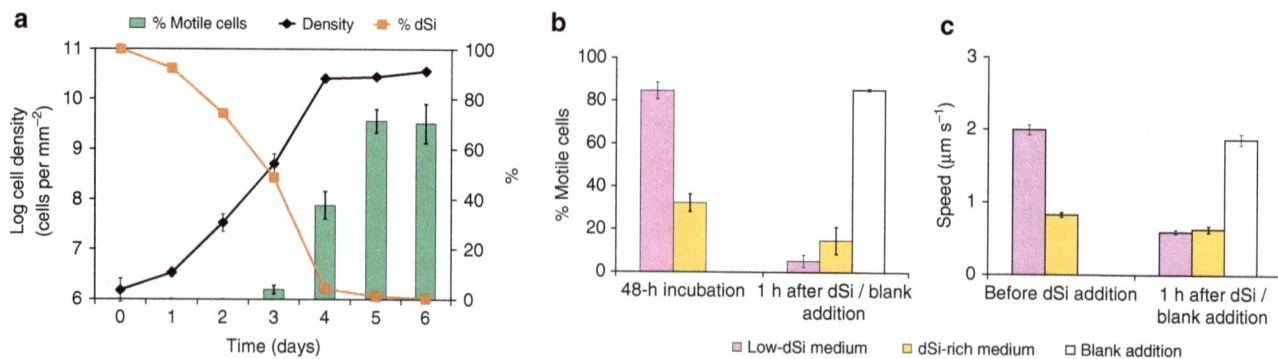

Figure 1 | Seminavis robusta cell motility is modulated by environmental silicic acid concentrations. (a) During silicon (Si)-limited batch culture growth, cell motility increases while dSi (100% dSi = 106 μM) is being depleted and the culture enters stationary phase. Error bars show the s.e.m. of five replicates ($n = 300$). **(b)** There is a higher percentage of motile cells in low-Si medium than Si-rich medium (one-way analysis of variance (ANOVA), $n = 5$, $P < 0.001$). After addition of dSi (106 μM) to 48-h silicon-starved cultures, cell motility drops within 1 h (one-way ANOVA, $n = 3$, $P < 0.0001$). This is not the case after a blank addition of sea water to such silicon-starved cultures ($P = 1.00$). Error bars show s.e.m. of three replicates. **(c)** The mean cell speed is significantly higher (LME with Tukey's honest significance difference (HSD) test, $n = 70$-200 cells per movie, three movies analysed, $P < 0.0001$) in cultures grown in low-dSi medium compared with dSi-rich medium (dSi = 246 μM). One hour after the addition of dSi to starved cultures, cell speed significantly dropped (LME with Tukey's HSD, $n = 70$-200 cells per movie, three movies analysed, $P < 0.0001$) while blank addition did not induce any change on cell behaviour (LME with Tukey's HSD, $n = 70$-200 cells per movie, three movies analysed, $P = 0.665$,). Error bars show s.e.m. of all tracked cells from three 60-s movies.

The fact that starved cells accumulate in the immediate vicinity of the bead suggests that they sense local dSi concentrations and direct their movement up the gradient and ultimately to the source. Surprisingly, the mean swimming speed of cells exposed to a dSi gradient was higher in the close proximity of the beads (Fig. 3a and Supplementary Fig. 3). This only holds true under the influence of the steep dSi gradient in the immediate proximity of the beads and is already not manifested any more in the more distant bin C (Supplementary Fig. 3). While cells generally move faster when starved of dSi (Fig. 1c), it is apparent that an additional response to a steep gradient of dSi is observed. The defined dSi source in an otherwise limited environment causes locally increased speed, which could be the cell's mechanism to avoid diffusion limitation during Si uptake around a dSi hotspot.

Analysis of movement. The mode of orientation within the dSi gradient was verified in detail by fitting the Taylor's equation[21] to analyse the swimming characteristics based on tracking individual cell behaviour in response to dSi-loaded and control beads (details in Methods section). *S. robusta* cells have higher directional persistence when perceiving an ascending dSi gradient, as exemplified by longer correlation length and timescales (λ and τ) (Table 1). Cells within the dSi gradient also had on average a 57% higher diffusivity (D) and 81% higher encounter kernel (β) for the dSi beads, showing that they maximize their encounter rate to find dSi beads. Analysis of the angular orientation by monitoring the angle of the vectorized tracks relative to the bead centre did not reveal any differences between control and Si-loaded beads, thereby excluding a directed orientation during reversing events (Fig. 3b). However, cells persistently migrate towards the dSi source as indicated by the change in the sum of distances between bead and all cells over time in the treatment compared with the control (Fig. 3c). The observed behaviour can be explained by a preferential forward movement in an ascending gradient of dSi. This directionality during the chemokinetic response thus promotes chemotaxis. The persistent orientation implies a biased random walk, wherein cells adapt their movement patterns to find a dSi source.

Selectivity of the directed response. To learn more about the selectivity of the response to dissolved minerals we determined

how Si-starved cells react to dGe sources. dGe and dSi share very similar chemical properties. Ge uptake and incorporation instead of Si in diatom frustules inhibits growth and causes morphological aberrations and toxicity[22,23]. When we administered dGe-loaded beads (1.4 nmol per bead) to dSi-starved cultures, a negative response was elicited as cells in average moved away from the dGe sources. In contrast, dSi-loaded beads were attractive, and near constant cell densities were observed around control treatments (Fig. 4 and Supplementary Fig. 4). *S. robusta* thus discriminates between the two very similar inorganic resources. This remarkably specific behavioural response combined with discrimination of the elements during uptake[24] represents an efficient mechanism to protect the cells against Ge toxicity.

Discussion
Our results clearly indicate the specific modulation of foraging behaviour of benthic diatoms in response to silicate. A detailed analysis of movement indicates a chemokinetic response since cell speed changes in dependence of dSi concentrations. In addition the observed attraction of cells within a gradient of dSi indicates a chemotactic search capability. Interestingly this search behaviour is not regulated by directed turns of the cells towards the dSi source but rather by a longer directional persistence within an ascending dSi gradient. This finding mechanism thus differs from the chemoattraction mechanism in brown algae where the pheromone-directed movement is mediated by signal molecule-induced turning events of gametes[25]. Such orientation towards dSi or any other dissolved mineral has to our knowledge not been observed before in diatoms. The specificity of the attraction is demonstrated by a selective movement towards dSi while the structurally closely related dGe does not stimulate an attraction response. These observations might be explained by a receptor-mediated process[20], but until now no candidate receptor mediating a specific recognition of dissolved minerals is known. Since different isolates of *S. robusta* as well as another tested diatom species *Navicula sp.* exhibit this search behaviour, the response might be general for benthic diatoms. The observed dSi-directed movement might thus help to explain the often patchy species composition and structure of marine biofilms.

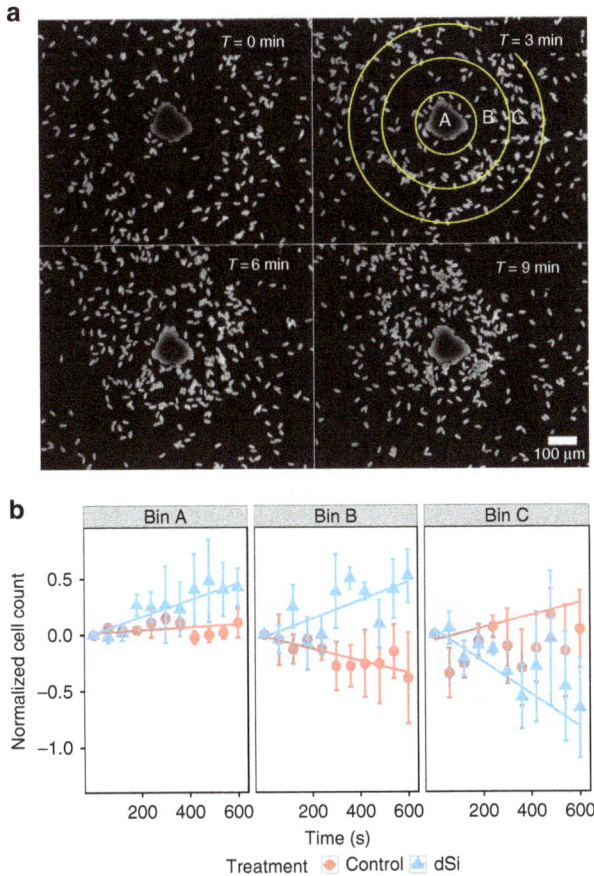

Figure 2 | Chemoattraction of dSi-starved cells to dSi-loaded alox beads. (a) Light microscopic observation of the cell accumulation (scale bar, 100 μm). The observation area with the alox bead as centre was divided into three concentric rings (Δradius of 112 μm) enclosing bins A-C. (b) Plot of the mean normalized cell counts every 60 s, and overlaid shaded area indicating the LME model fit with s.e.m. for each bin ($n = 70$-200 cells per movie, three movies analysed). The observed increased cell density in bin B (LME, $P = 0.0034$) and decreased cell density in bin C over time (LME, $P = 0.0079$) in the dSi treatment reflects the movement of the cells towards the dSi bead.

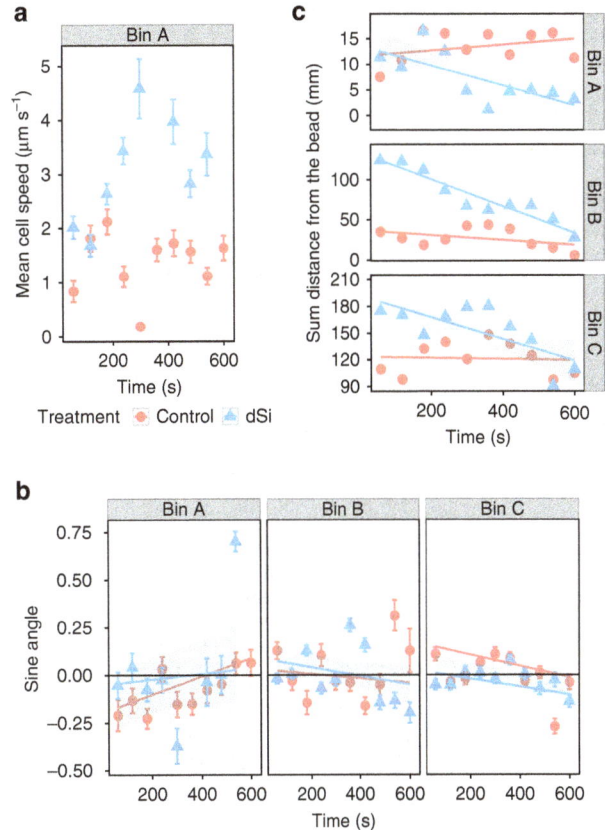

Figure 3 | Analysis of track data from dSi and control movies. (a) Cell speed increases in bin A (blue triangles) under the influence of the steep dSi gradient. The log + 1-transformed mean speed of cells was fitted using generalized additive mixed modelling (GAMM) for each bin. Cells move faster as they approach the bead in bin A (GAMM, $n_{(dSi)} = 12$, $n_{(Control)} = 7$, $p_{(Si)} = 8.27^{-8}$, $p_{(control)} = 0.176$). (b) Plot of mean sine angle every 60 s, with overlaid shared area showing LME model fit with s.e.m. Non-directional movement of cells is observed as dSi had no effect on the angular orientation (LME, $n_{(dSi)} = 34$, $n_{(Control)} = 28$, $P > 0.05$). (c) Plot of the sum distance of each cell from the bead every 60 s, with overlaid shaded area showing linear model fit with s.e.m. The decreasing sum distance over time is indicative for the preferential migration of the starved cells towards the dSi-loaded beads in all bins (linear model, $n_{(dSi)} = 34$, $n_{(Control)} = 28$, $P = < 0.05$).

Tactic behaviour towards nutrient sources such as phosphate and different sources of nitrogen has been demonstrated for other algae and bacteria[26,27]. These responses can provide important adaptive advantage for the organisms due to increased acquisition capability for the resources[16,17]. The ability of motile diatoms to trace dSi gradients and exploit micro-scale hotspots by changing their foraging behaviour enables them to thrive and dominate phototrophic biofilm communities such as the intertidal microphytobenthos. More general, this capability might be a key factor explaining their explosive radiation in marine and freshwater benthos.

Microbial activity is known to greatly affect global biogeo-chemical processes involved in the cycling of elements[28–30]. Several mechanisms have been suggested on how microbial behavioural responses to patchy resources can influence ocean biogeochemistry[31]. The ability of diatoms to track dSi availability in the environment has thus implications on a global scale by affecting dSi fluxes and on a micro-scale by shaping biofilm communities.

Methods

Cultures. We used the *S. robusta* strains F2-31B and P36 MT$^+$ maintained cryopreserved in the BCCM/DCG diatom culture collection at Ghent University

(http://bccm.belspo.be/about-us/bccm-dcg)[20]. *Navicula sp.* was isolated from a mudflat at Solana Beach, California 32° 58′ 37.5″ N 117° 16′ 08.8″ W. For both species, cells were grown in batch culture[20] either with natural sea water and F/2 medium[32] or artificial, buffered sea water (ASW) prepared as described by Maier & Calenberg[33] to avoid overlaying effects of pH changes due to the treatments. In low-dSi treatments no dSi was added while high-dSi treatments were supplemented with 106 μM dSi for F/2 medium and 246 μM dSi for ASW. Experimental cultures were prepared by 10-fold dilution of aliquot of stock cultures using fresh culture medium and then grown in tissue culture flasks with standard caps, Petri or well plates (Greiner Bio-One, Frickenhausen, Germany). Observations, cell culture photography and video recording were done on an Leica DM IL LED inverted light microscope with a Leica DFC 280 camera system (Heerbrugg, Switzerland).

Preparation of dSi- or dGe-loaded aluminium oxide particles. Aluminium oxide (100 mg alox, Merck, Darmstadt, Germany; 90 active neutral; 0.063–0.200 mm particle diameter) was used to adsorb silicate by fully evaporating (overnight at 50–90 °C) 800 μl freshly prepared sodium silicate solution (440 mM Na$_2$SiO$_3$ · 9H$_2$O; Sigma-Aldrich, Deisenhofen, Germany) or germanium dioxide solution (440 mM GeO$_2$; Alfa Aesar, Karlsruhe, Germany). To determine the most effective concentration of dSi, 50, 800 and 1,270 μl silicate stock were added to 100 mg alox and evaporated, resulting in ∼0.088, 1.40 and 2.23 nmol dSi per particle, respectively (Supplementary Fig. 2 and Supplementary Table 7). For all succeeding experiments, the concentration 1.40 nmol per particle was used for both

Table 1 | Motility parameters of cells exposed to control and dSi beads.

Parameter	Control	Si
N	28	34
τ (s)	4.68	14.85
λ (µm)	10.70	23.86
D (µm^2 s^{-1})	12.22	19.17
β (µm^3 s^{-1})	8,200	14,900

Calculated using Taylor's equation, where N is number of tracks, τ is decorrelation timescale, λ is decorrelation length scale, D is diffusivity and β is encounter kernel (observation area: bins A–C; observation time: 600 s).

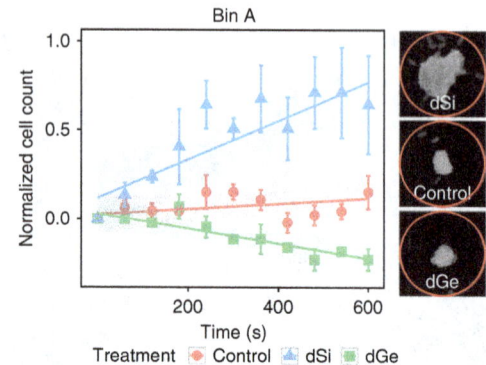

Figure 4 | Substrate-specific response of Si-starved cells to dSi- and dGe-loaded (1.4 nmol per bead), and control beads. Plot of mean normalized cell counts with overlaid LME model demonstrates avoidance of starved cells to dGe (LME, $n = 70$–200 cells per movie, three movies analysed, $P = 0.0047$) and attraction to dSi (LME, $n = 70$–200 cells per movie, three movies analysed, $P = 0.0088$). Right panels show cells in bin A in selected treatments 600 s after addition of beads (the red circle has a radius of 112 µm).

dSi and dGe treatments. Blank alox particles for control treatments were identically prepared by evaporation of bi-distilled water. The amount of alox particles per unit weight was determined by counting the number of particles within 10 mg alox beads on tissue culture plates. We counted three randomly chosen microscopic fields (area of each field = 5.6 mm^2) from microscopic photographs. On average 2,640 (± 660) particles were present per mg alox.

Determination of dSi diffusing from the bead. To determine the total flux of dSi diffusing from the bead (i), dSi was quantified in water exposed to Si-loaded or control beads after 600 s by standard colorimetric methods[34]. The steady-state concentration of dSi was calculated from the initial concentration on the particle using the formula[18] $C_{(r)} = i/4\pi\sqrt{r}D$, where i is the total diffusive flux of dSi, r is the radius where dSi diffused, and D is the diffusivity constant for dSi (10^{-5} cm^2 s^{-1})[35]. Time to steady state was determined as the time $\geq d^2/D$ where d is the diameter of the whole observation area (672 µm). The $\sqrt{r}$ was used to correct the shape of a gradient in a flat chamber[36]. The i ($1.21^{-13} \pm 1.01^{-14}$ mol s^{-1}) was substituted in the equation and the steady-state C was calculated based on the distance (radius) from the bead r. A plot showing steady-state concentration of dSi against distance can be found in Supplementary Fig. 1.

Data processing. The open-source software Fiji (http://fiji.sc/Fiji)[37] with the plug-ins Cell Counter and TrackMate (http://fiji.sc/TrackMate) was used for cell counting and tracking, respectively. All data analyses were done using the open-source statistical and graphic software R version 3.0.3 (http://www.R-project.org/)[38].

Motility and speed of S. robusta controlled by dSi. For the experiment in Fig. 1a, replicate cultures were grown in Petri dishes (Greiner Bio-One; 60 × 15 mm) in F/2 medium as described above. Five cultures were used for density and proportion of motile cells determinations, performed ∼9 h after the onset of light. In addition, the medium was collected from one more replicate culture each day, by filtration over a 0.2-µm pore size filter and frozen until analysis for silicate content. Dissolved silicate was measured on a spectrophotometer using the molybdosilicate method[39]. For Fig 1b,c, cultures were grown on tissue culture flasks (75 cm^2 growth surface; filter caps) in 20 ml growth medium with three replicates for each condition. Cultures were grown for 3 days in F/2 medium (106 µM dSi, Fig. 1b) and 7 days in dSi-enriched ASW (246 µM dSi, Fig. 1c). For Fig. 1b, at the beginning of the light period on day 4, the supernatant growth medium was replaced by careful aspiration with a Pasteur pipette attached to a water pump, immediately followed by the addition of ASW with 246 µM dSi or without dSi enrichment. Further culturing was performed for 48 h in continuous light to avoid interference of light–dark alterations on cell motility. Proportion of motile cells was assessed by overlaying two photographs taken at a 15-s interval from the same observation field. Cells ($n > 300$) located at the exact same position in both photographs were counted as immotile; others were counted as motile. Cell densities were microscopically estimated by counting cells in at least 35 observation fields with an area of 0.993 mm^2. The percentage of motile cells was compared before and after addition of dSi (106 µM) or addition of blank artificial sea water without added dSi to cultures grown for 48 h in low-dSi or dSi-rich medium (one-way analysis of variance, Fig. 1b). For Fig. 1c, cell speeds of starved and non-starved cells with addition of bulk dSi (246 µM) or blank addition of artificial sea water without added dSi were assessed by tracking cells from 60-s movies. Differences on treatments were determined by fitting the log + 1-transformed speed to a linear mixed effects (LME) model with unique track ID as random factor and a constant variance function structure (varIdent) for treatment. Multiple pairwise comparisons were done through Tukey's honest significance difference (HSD) test (outcome of statistical analysis is given in Supplementary Table 1).

Movement of Si-starved cells in response to dSi gradients. For all succeeding experiments, S. robusta cultures were grown in tissue culture flasks in artificial, buffered sea water with dSi[33] until they reached stationary phase. On the seventh

day, 1 ml of cell suspension was transferred to 12-well tissue culture plates supplemented with 2 ml low-Si medium. Normal light–dark cycle was followed for the 3-day incubation period. Alox particles were carefully administered to each well using a spatula, ensuring that the total number of beads per well does not exceed 30. For obtaining cell count data, photos were taken after exposure to the beads every 60 s for 600 s. Movies for tracking were also recorded for 600 s (1 frame per s). Cell accumulation around the particles was determined from microscopically acquired photographs by counting the number of cells within a circle having an area of 0.300 mm^2 (for Supplementary Fig. 2) or 0.355 mm^2 (Figs 2b and 3). For Figs 2b and 4, the observation area was divided into three concentric rings, called bins (bins A–C), having a Δradius of 112 µm with the alox bead as the central point (Fig. 2a).

Modelling. A representative movie from dSi and control treatments was chosen and cells were randomly selected to be tracked ($n = 29$ for control and 34 for dSi) for 600 s. To analyse the track data, cell density (n), speed (µm s^{-1}), angular orientation (sine angle) and distance (µm) of cells relative to the bead were taken as parameters. Mixed models were used to analyse and fit the data to be able to account for the nested and longitudinal design of the study. Linear modelling of sum distance and LME modelling of cell count and angular orientation were done using the R package nlme[40] while general additive mixed modelling of cell speed was done using the R package mgcv[41]. To correct correlated data between independent variables, a correlation structure, autoregressive order 1 (AR-1) was used. A constant variance function structure (varIdent) was also added to the model for correcting residual spreads. Individual models for each bin were chosen based on the Akaike information criterion. For each model, a Wald test was performed to determine the significance of the fitted estimates on each term. All results are shown in Supplementary Tables 2–6.

Cell counts. Cell counts were standardized according to standard Z-score calculation per treatment: standard score $Z = (X - \mu)/\sigma$, where μ is mean, X is score and σ is s.d. For the starting point to be normalized to 0, we subtracted the standardized cell count on each time point to the value at $T = 0$ s. A value of 0 indicates that the cell density is equal to the mean. Positive values indicate a cell density higher than the mean and a negative value the opposite. To compare control and dSi treatments (Fig. 2b and Supplementary Table 2), a model for each bin was fitted using the interaction between treatment and time as independent variables and replicate ID as a random factor. An AR-1 correlation structure for successive measurements within the replicates and a varIdent variance structure for treatment on bins A and B and replicates for bin C were added to the model. For the comparison of substrate specificity (Fig. 4, Supplementary Fig. 4 and Supplementary Table 6), the model for each bin was fitted the same way as described above and a varIdent variance structure for treatment was added for all the bins.

Cell speed. A log + 1 transformation was used to normalize cell speed. The mean speed of the cells every 30 s for each bin was fitted using general additive mixed model (Fig. 3a, Supplementary Table 3 and Supplementary Fig. 3). Data for a time point were excluded when only a single-track data contributed to the mean. The independent variable was fitted with a simple factor smooth and penalized with cubic regression splines of time on each treatment, and track ID (that is, unique cell ID) was assigned as a random factor. An AR-1 correlation structure between treatment and track ID was added to the model. In addition, a varIdent structure on treatment was used in the model to decrease the Akaike information criterion significantly and for a better fit.

Angular orientation. To determine angular orientation, the sine angle of each coordinate position of the cell relative to the coordinate position of the bead was calculated[42]. A cell is moving towards the bead if it has a positive value and away from it in case of a negative value. Average of the sine angles was determined per bin every 60 s (Fig. 3b and Supplementary Table 4). Data for a time point were excluded when only a single-track data contribute to the mean. Each bin data were fitted via LME using the interaction between treatment and time as independent variables and replicate ID as a random factor. An AR-1 correlation structure was used between treatment and replicate ID as well as a varIdent variance structure for treatment.

Sum distance. The sum of distance from each cell's coordinate position relative to the bead centre was used to determine the migration pattern of cells. Each bin was fitted using a simple linear model (Fig. 3c and Supplementary Table 5) with sum distance as the response variable and the interaction of time and treatment as explanatory variable.

Motility parameters. To determine whether the motility characteristics between the control and dSi treatment were different, we analysed track data from two representative movies. Motility parameters were computed by fitting the root mean square of the net distance as a function of time to Taylor's equation using nonlinear least-squares estimation[21]: root mean square $= [2v^2\tau\,(t - \tau\,(1 - e^{-t/\tau})]^{0.5}$ where v is effective swimming speed, τ is the decorrelation timescale, t is the time and λ is decorrelation length scale: $\lambda = v\tau$.

The decorrelation length (λ) and timescale (τ) give the distance and time, respectively, wherein there is directional persistence in motility over a period of 600 s.

We also calculated the effective diffusivity of motility ($D = v^2\tau/n$) and encounter kernel ($\beta = 4\pi RD$), where n is number of dimensions and R is radius of the bead. D describes the spread of the cell tracks and β determines the water volume screened by cells within the observation time[21].

References

1. Field, C. B. Primary production of the biosphere: integrating terrestrial and oceanic components. *Science* **281**, 237–240 (1998).
2. Kroger, N., Lorenz, S., Brunner, E. & Sumper, M. Self-assembly of highly phosphorylated silaffins and their function in biosilica morphogenesis. *Science* **298**, 584–586 (2002).
3. Tréguer, P. J. & De La Rocha, C. L. The world ocean silica cycle. *Annu. Rev. Mar. Sci.* **5**, 477–501 (2013).
4. Yool, A. & Tyrrell, T. Role of diatoms in regulating the ocean's silicon cycle. *Global Biogeochem. Cycles* **17**, 1103 (2003).
5. Martin-Jezequel, V., Hildebrand, M. & Brzezinski, M. A. Silicon metabolism in diatoms: implications for growth. *J. Phycol.* **36**, 821–840 (2000).
6. Leynaert, A., Longphuirt, S. N., Claquin, P., Chauvaud, L. & Ragueneau, O. No limit? The multiphasic uptake of silicic acid by benthic diatoms. *Limnol. Oceanogr.* **54**, 571–576 (2009).
7. Ziervogel, K. & Forster, S. Do benthic diatoms influence erosion thresholds of coastal subtidal sediments? *J. Sea Res.* **55**, 43–53 (2006).
8. Sigmon, D. E. & Cahoon, L. B. Comparative effects of benthic microalgae and phytoplankton on dissolved silica fluxes. *Aquat. Microb. Ecol.* **13**, 275–284 (1997).
9. Bowler, C. *et al.* The *Phaeodactylum* genome reveals the evolutionary history of diatom genomes. *Nature* **456**, 239–244 (2008).
10. Cahoon, L. B. in *Oceanography and Marine Biology: An Annual Review* Vol. 37 (eds Ansell, A. D., Gibson, R. N. & Barnes, M.) 47–86 (University College London Press, 1999).
11. Consalvey, M., Paterson, D. M. & Underwood, G. J. C. The ups and downs of life in a benthic biofilm: migration of benthic diatoms. *Diatom Res.* **19**, 181–202 (2004).
12. Van Colen, C., Underwood, G. J. C., Serôdio, J. & Paterson, D. M. Ecology of intertidal microbial biofilms: mechanisms, patterns and future research needs. *J. Sea Res.* **92**, 2–5 (2014).
13. Agogué, H., Mallet, C., Orvain, F., De Crignis, M., Mornet, F. & Dupuy, C. Bacterial dynamics in a microphytobenthic biofilm: a tidal mesocosm approach. *J. Sea Res.* **92**, 36–45 (2014).
14. Molino, P. J. & Wetherbee, R. The biology of biofouling diatoms and their role in the development of microbial slimes. *Biofouling* **24**, 365–379 (2008).
15. Wang, J., Cao, S., Du, C. & Chen, D. Underwater locomotion strategy by a benthic penate diatom *Navicula* sp. *Protoplasm* **250**, 1203–1212 (2013).
16. Amsler, C. D. & Iken, K. B. in *Marine Chemical Ecology* (eds McClintock, J. B. & Baker, B. J.) 413–430 (CRC, 2001).
17. Hutz, A., Schubert, K. & Overmann, J. *Thalassospira sp* isolated from the oligotrophic eastern mediterranean sea exhibits chemotaxis toward inorganic phosphate during starvation. *Appl. Environ. Microbiol.* **77**, 4412–4421 (2011).
18. Barbara, G. M. & Mitchell, J. G. Marine bacterial organisation around point-like sources of amino acids. *FEMS Microbiol. Ecol.* **43**, 99–109 (2003).
19. Witkowski, A., Brehm, U., Palinska, K. A. & Rhiel, E. Swarm-like migratory behaviour in the laboratory of a pennate diatom isolated from North Sea sediments. *Diatom Res.* **27**, 95–100 (2012).
20. Gillard, J. *et al.* Metabolomics enables the structure elucidation of a diatom sex pheromone. *Angew. Chem. Int. Ed.* **52**, 854–857 (2013).
21. Visser, A. W. & Kiorboe, T. Plankton motility patterns and encounter rates. *Oecologia* **148**, 538–546 (2006).
22. Froelich, P. N. & Andreae, M. O. The marine geochemistry of germanium—ekasilicon. *Science* **213**, 205–207 (1981).
23. Azam, F., Hemmingsen, B. B. & Voleani, B. E. Germanium incorporation into the silica of diatom cell walls. *Arch. Microbiol.* **92**, 11–20 (1973).
24. Sutton, J., Ellwood, M. J., Maher, W. A. & Croot, P. L. Oceanic distribution of inorganic germanium relative to silicon: germanium discrimination by diatoms. *Global Biogeochem. Cycles* **24**, GB2017 (2010).
25. Pohnert, G. & Boland, W. The oxylipin chemistry of attraction and defense in brown algae and diatoms. *Nat. Prod. Rep.* **19**, 108–122 (2002).
26. Ikegami, S., Imai, I., Kato, J. & Ohtake, H. Chemotaxis toward inorganic phosphate in the red alga *Chattonella antiqua*. *J. Plankton Res.* **17**, 1587–1591 (1995).
27. Ermilova, E. V., Nikitin, M. M. & Fernandez, E. Chemotaxis to ammonium/methylammonium in *Chlamydomonas reinhardtii*: the role of transport systems for ammonium/methylammonium. *Planta* **226**, 1323–1332 (2007).
28. Azam, F. Microbial control of oceanic carbon flux: the plot thickens. *Science* **280**, 694–696 (1998).
29. Seymour, J. R., Marcos & Stocker, R. Resource patch formation and exploitation throughout the marine microbial food web. *Am. Nat.* **173**, E15–E29 (2009).
30. Seymour, J. R., Simo, R., Ahmed, T. & Stocker, R. Chemoattraction to dimethylsulfoniopropionate throughout the marine microbial food web. *Science* **329**, 342–345 (2010).
31. Stocker, R. Marine microbes see a sea of gradients. *Science* **338**, 628–633 (2012).
32. Guillard, R. R. L. in *Culture of Marine Invertebrate Animals*. (eds Smith, L. H. & Chanley, M. H.) 26–60 (Plenum, 1975).
33. Maier, I. & Calenberg, M. Effect of extracellular Ca^{2+} and Ca^{2+}-antagonists on the movement and chemoorientation of male gametes of *Ectocarpus siliculosus* (Phaeophyceae). *Bot. Acta* **107**, 451–460 (1994).
34. Strickland, J. D. H. & Parson, T. R. *A Practical Handbook of Seawater Analysis* 2nd edn (Fisheries Research Board of Canada, 1978).
35. Wollast, R. & Garrels, R. M. Diffusion coefficient of silica in seawater. *Nat. Phys. Sci.* **229**, 94 (1971).
36. Blackburn, N., Fenchel, T. & Mitchell, J. Microscale nutrient patches in planktonic habitats shown by chemotactic bacteria. *Science* **282**, 2254–2256 (1998).
37. Schindelin, J. *et al.* Fiji: an open-source platform for biological-image analysis. *Nat. Methods* **9**, 676–682 (2012).
38. Team, R. C. R Foundation for Statistical Compouting www.R-project.org/ (2013).
39. Eaton, A. D., Clesceri, L. S., Greenberg, A. E. & Franson, M. A. H. *Standard Methods for the Examination of Water and Wastewater* (American Public Healtch Association, 1998)).
40. Pinheiro, J., Bates, D., DebRoy, S., Sarkar, D. & Team, R. C. nlme: Linear and Nonlinear Mixed Effects Models. R package version 3.1–120 http://CRAN.R-project.org/package=nlme (2015).
41. Wood, S. N. Fast stable restricted maximum likelihood and marginal likelihood estimation of semiparametric generalized linear models. *J. R. Stat. Soc. Ser. B* **73**, 3–36 (2011).
42. Fenchel, T. Orientation in two dimensions: chemosensory motile behaviour of Euplotes vannus. *Eur. J. Protistol.* **40**, 49–55 (2004).

Acknowledgements

This work was supported by the Volkswagen Foundation, the IMPRS Exploration of Ecological Interactions with Molecular and Chemical Techniques, the Flemish Research foundation project TG.0374.11N and the Ugent research grants 01/04611 and BOF15/GOA/17.

Author contributions

G.P. and W.V. designed the project; K.G.V.B. and J.G. discussed and performed the experiments; K.G.V.B. performed cell tracking; K.G.V.B. and J.H. discussed and performed the modelling; all authors contributed to writing the manuscript.

Additional information

Competing financial interests: The authors declare no competing financial interests.

5

Near-island biological hotspots in barren ocean basins

Jamison M. Gove[1], Margaret A. McManus[2], Anna B. Neuheimer[2], Jeffrey J. Polovina[1], Jeffrey C. Drazen[2], Craig R. Smith[2], Mark A. Merrifield[2,3], Alan M. Friedlander[4,5], Julia S. Ehses[3,6], Charles W. Young[3,6], Amanda K. Dillon[3,6] & Gareth J. Williams[7,8]

Phytoplankton production drives marine ecosystem trophic-structure and global fisheries yields. Phytoplankton biomass is particularly influential near coral reef islands and atolls that span the oligotrophic tropical oceans. The paradoxical enhancement in phytoplankton near an island-reef ecosystem—Island Mass Effect (IME)—was first documented 60 years ago, yet much remains unknown about the prevalence and drivers of this ecologically important phenomenon. Here we provide the first basin-scale investigation of IME. We show that IME is a near-ubiquitous feature among a majority (91%) of coral reef ecosystems surveyed, creating near-island 'hotspots' of phytoplankton biomass throughout the upper water column. Variations in IME strength are governed by geomorphic type (atoll vs island), bathymetric slope, reef area and local human impacts (for example, human-derived nutrient input). These ocean oases increase nearshore phytoplankton biomass by up to 86% over oceanic conditions, providing basal energetic resources to higher trophic levels that support subsistence-based human populations.

[1] Ecosystems and Oceanography Program, Pacific Islands Fisheries Science Center, 1845 Wasp Blvd Building 176, Honolulu, 96818 Hawaii, USA. [2] Department of Oceanography, University of Hawai'i at Mānoa, 1000 Pope Road, Marine Sciences Building, Honolulu, 96822 Hawaii, USA. [3] Joint Institute for Marine and Atmospheric Research, University of Hawai'i at Mānoa, 1000 Pope Road, Marine Sciences Building, Honolulu, 96822 Hawaii, USA. [4] Fisheries Ecology Research Laboratory, Department of Biology, University of Hawai'i at Mānoa, 2538 McCarthy Mall, Honolulu, 96822 Hawaii, USA. [5] Pristine Seas, National Geographic Society, 1145 17th St NW, Washington, DC 20036, USA. [6] Coral Reef Ecosystem Program, Pacific Islands Fisheries Science Center, 1845 Wasp Blvd Building 176, Honolulu, 96818 Hawaii, USA. [7] Center for Marine Biodiversity and Conservation, Scripps Institution of Oceanography, UC San Diego, 9500 Gilman Drive, La Jolla, 92093 California, USA. [8] School of Ocean Sciences, Bangor University, Menai Bridge, LL59 5AB Anglesey, UK. Correspondence and requests for materials should be addressed to J.M.G. (email: jamison.gove@noaa.gov).

Phytoplankton production is an essential source of energy in the marine environment[1]. The extent and availability of phytoplankton biomass drives the trophic-structure of entire marine ecosystems[2], dictating the distribution and production of the world's fisheries[3]. The ecological impacts of enhanced phytoplankton biomass are especially acute near tropical coral reef islands and atolls as these ecosystems predominantly reside in nutrient impoverished waters that lack new production[4]. For example, across the central and western Pacific, islands and atolls exposed to elevated levels of nearshore phytoplankton support higher fish biomass and a greater abundance of reef-building organisms than those found in more oligotrophic waters[5,6]. Hence, mechanisms that act to promote nearshore phytoplankton biomass are critical for coral reef ecosystem development and persistence[5].

The increase in phytoplankton biomass proximate to island-reef ecosystems—'Island Mass Effect' (IME)—was first documented over a half century ago[7] (Fig. 1). Much of our current knowledge of the IME, however, stems from studies in a small number of locations in geographically confined areas[8–10]. Thus, whether or not the IME is a pervasive phenomenon across broad gradients in oceanic conditions has historically remained unknown. Furthermore, the relative influence of natural vs anthropogenic drivers of variations in the magnitude of the IME has remained a mystery until now.

Here we present a basin-scale investigation of the 60-year-old IME hypothesis. Using 35 coral reef islands and atolls spanning 43° of latitude and 60° of longitude (Fig. 2 and Supplementary Table 1) that cross multiple gradients in oceanic forcing[11],

geophysical attributes[11], reef-community composition[5,12] and local human impacts[12], we quantify the prevalence of the IME across the tropical Pacific. We use long-term satellite-derived observations of chlorophyll-a (a proxy for phytoplankton biomass) to show the IME is a near-ubiquitous feature and identify key biogeophysical drivers of variations in the magnitude of the IME among Pacific island- and atoll-reef ecosystems. We also incorporate ship-based surveys at 29 of these locations to confirm that the nearshore enhancement in chlorophyll-a occurs over the full euphotic depth range. Finally, we show that the IME increases the nearshore standing stock of phytoplankton biomass by up to 85.6% over background oceanic conditions, providing basal energy sources for higher trophic levels across an otherwise barren ocean landscape.

Results

Nearshore phytoplankton enhancement. We found that 91% ($n = 32$) of island- and atoll-reef ecosystems displayed localized nearshore enhancement in long-term chlorophyll-a associated with the IME. The magnitude of the IME varied among locations (evidenced by differences in the linear slope of log–log transformed data, $F_{1,32} = 22.24$, $P < 0.0001$, Fig. 3a,b, see Methods section). To identify the proximate drivers of the IME, we quantified a suite of biogeophysical predictor variables for each island- and atoll-reef ecosystem (Supplementary Table 2), namely latitude, land area, reef area, bathymetric slope, ocean currents, precipitation, sea-surface temperature, geomorphic type (atoll vs island) and human population status (unpopulated vs populated).

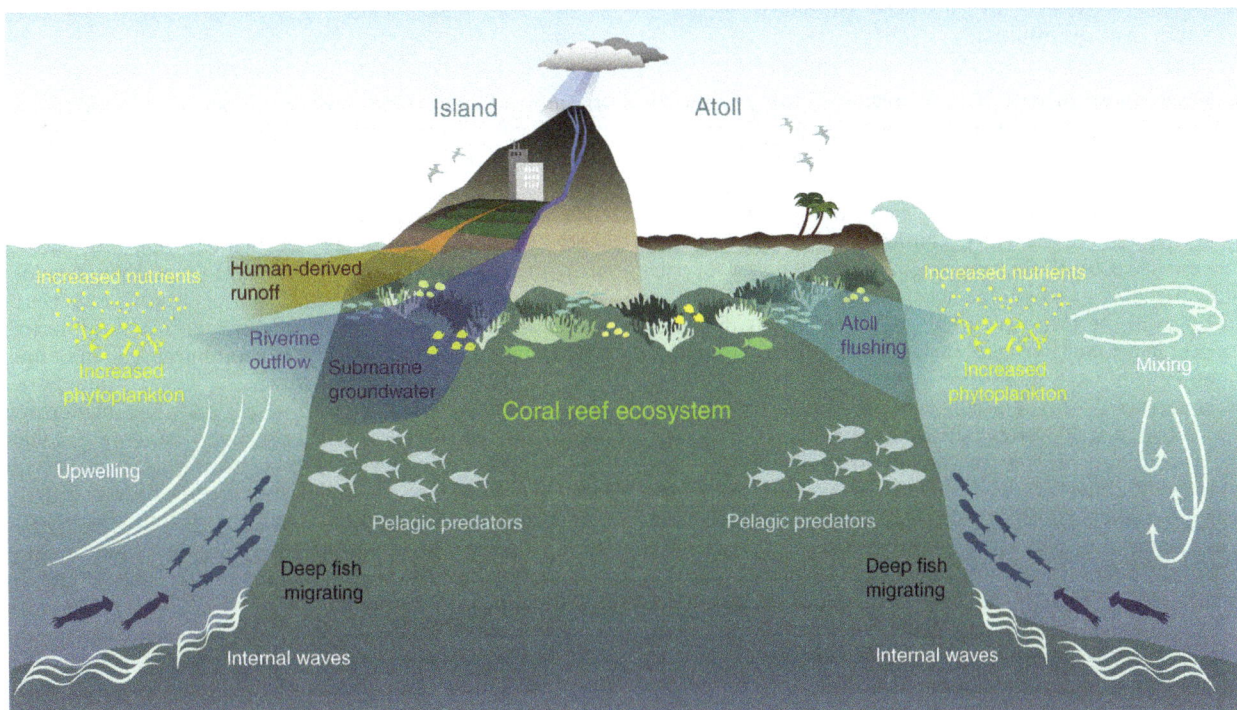

Figure 1 | The Island Mass Effect. Localized increases in phytoplankton biomass near island- and atoll-reef ecosystems—Island Mass Effect—may be the result of several causative mechanisms that enhance nearshore nutrient concentrations, including coral reef ecosystem processes, such as nitrogen fixation or decomposition, and animal waste products, such as reef-associated fishes; current-bathymetric interactions that can drive vertical transport of water masses via upwelling, downstream mixing and eddies, and internal waves; island-associated inputs, such as submarine groundwater discharge and outflow from rivers, which can mobilize and transport sediment and other terrigenous material laden with nutrients; the flushing and associated outflow of lagoonal waters from atoll environments; human-derived runoff of agricultural production, urban development and wastewater input. Enhanced nearshore phytoplankton can influence food-web dynamics and elicit a biological response in higher trophic groups, for example: horizontal and vertical migration patterns in squids, fishes and other micronekton (collectively referred to as the 'mesopelagic boundary layer community') that move nearshore at night to feed on increased food resources; inshore migration of pelagic predators, such as tuna, to feed on the island-associated micronekton community; greater reef fish biomass and increased cover of calcifying benthic organisms in coral reef ecosystems.

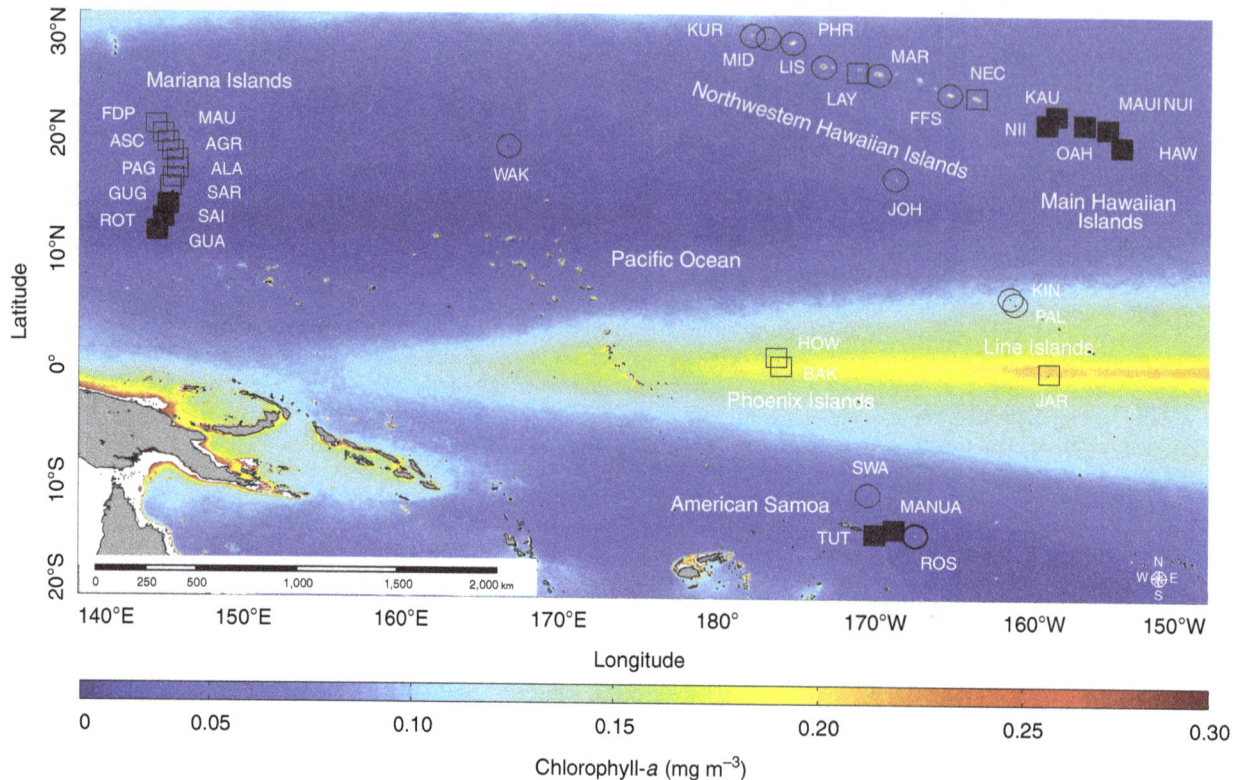

Figure 2 | Map of the Pacific highlighting the coral reef islands and atolls used in this study. Long-term mean (10 year) chlorophyll-*a* with coral reef islands (squares) and atolls (circles) that either have local human populations (filled) or are unpopulated (open). Please see Supplementary Table 1 for study location name designations.

Generalized linear models (GLMs) revealed that geomorphic type, bathymetric slope, reef area and population status were the primary drivers of spatial variations in the IME among Pacific island- and atoll-reef ecosystems, together explaining 78% of the variation observed ($n = 28$, only locations with significant increases in nearshore chlorophyll-*a* were modelled, P range = <0.0001–0.029, r^2 range = 0.59–0.99, see Supplementary Tables 3,4 and see Methods section).

Variation in the IME across our study region was driven by differences in geomorphological make-up; nearshore chlorophyll-*a* enhancements were more pronounced at atolls than islands (34% of explained variation, Fig. 3a,b, Supplementary Table 4). Atolls, unlike islands, have partially enclosed interior lagoons often containing thriving ecosystems (Fig. 1). Wave- and tidal-driven flushing of these lagoons to surrounding waters may export nutrients fuelling enhanced nearshore phytoplankton biomass. Across our study system, the IME was most pronounced in the Northwestern Hawaiian Islands (Supplementary Table 1) at semi-enclosed atolls with naturally occurring channels to the open ocean. Large ocean swells generated from North Pacific storms and northeast trade-winds pump considerable amounts of water over the emergent barrier reef that then flow through the entire atoll system and eventually exit the channel[13]. Wave forcing is a highly efficient atoll flushing mechanism, advecting detritus and other sources of nutrients generated via coral reef ecosystem processes out of the atoll[14]. This rapid mobilization of material can exceed the assimilation ability of the benthic community[15], thereby providing increased nutrients that drive nearshore phytoplankton biomass enhancement.

Along with geomorphic type, island- and atoll-reef ecosystems with more gradual sloping bathymetry exhibited a stronger IME (28% of explained variation, Fig. 3a, Supplementary Table 4). Bathymetric influences on ocean currents can force vertical

transport of subsurface nutrient-rich waters that fuel nearshore productivity (Fig. 1). For example, vertical transport can be driven by current impingement that uplifts isotherms on the upstream side of an island[4,16] or through turbulent mixing, lee eddy and wake effects on the downstream side[17–19]. Internal waves, generated from tidal currents interacting with underlying bathymetry, can also drive vertical perturbations in the background stratification that deliver cooler, nutrient-rich waters to the near-surface[20] resulting in increased nearshore phytoplankton biomass[21]. The shoreward propagation of internal waves is directly related to bathymetric slope[22]; internal waves more readily reach shallower waters and fuel phytoplankton production where the underlying slope is more gradual. In contrast, internal waves are reflected offshore at steeper sloped locations. Across our study region, we found the IME to be particularly pronounced at locations within the Hawaiian Archipelago, a region characterized by islands and atolls with gradual sloping bathymetry (Supplementary Table 1) and highly active internal wave generation[23].

Pacific island- and atoll-reef ecosystems with greater reef area exhibited increased nearshore chlorophyll-*a* enhancements and thus had a more pronounced IME (26% of explained variation, Fig. 3b, Supplementary Table 4). The mechanisms underlying this relationship include autochthonous nutrient sources in coral reef ecosystems such as nitrogen fixation, regeneration (either through decomposition of primary producers or from sediment deposition), and recycling from other biota[15,24]. Animal waste products, such as those derived from sea-bird guano[25], reef-associated fishes[26] and mobile marine invertebrates[27] also enhance nutrient concentrations in coral reef ecosystems. The total reef-derived nutrients available to phytoplankton are likely variable, dependent upon biogeochemical processes within coral reef ecosystems that are influenced by physical factors such as water

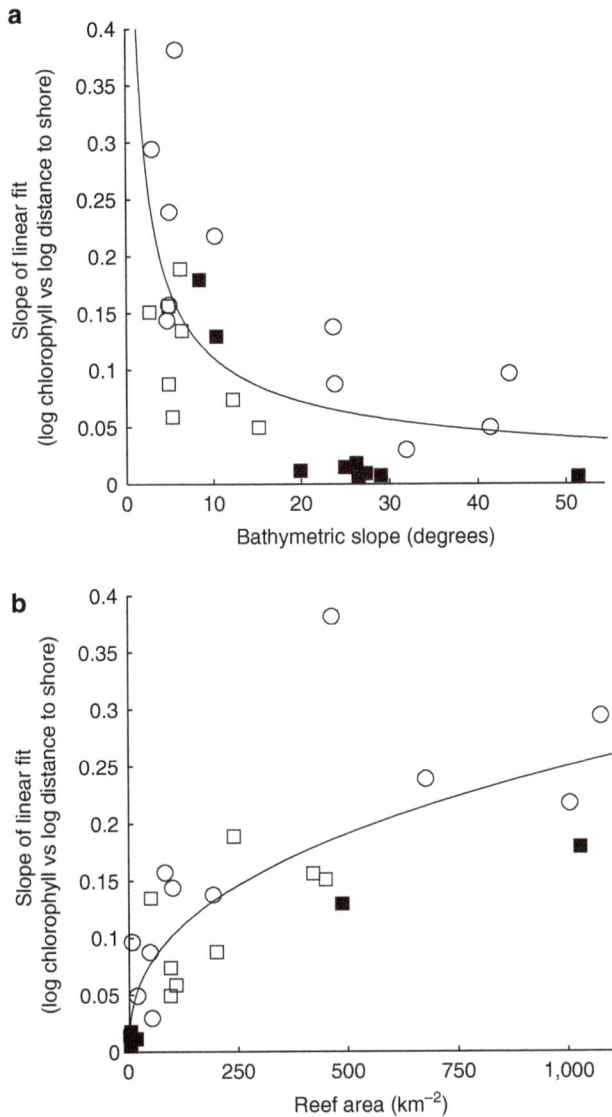

Figure 3 | Biogeophysical drivers of variations in Island Mass Effect strength. Relationship between study locations that had significant phytoplankton biomass enhancement (y-axis; increasing y represents a greater rate of increase in chlorophyll-a towards shore) and significant drivers identified from model results; bathymetric slope (**a**) and reef area (**b**) with geomorphic type (atolls vs islands, represented as circles vs squares, respectively) and population status (populated vs unpopulated, represented as filled vs open icons, respectively). Geomorphic type and population status are identified the same as in Fig. 2. Nonlinear regressions ($P<0.05$) with r^2 values of 0.45 (**a**) and 0.68 (**b**).

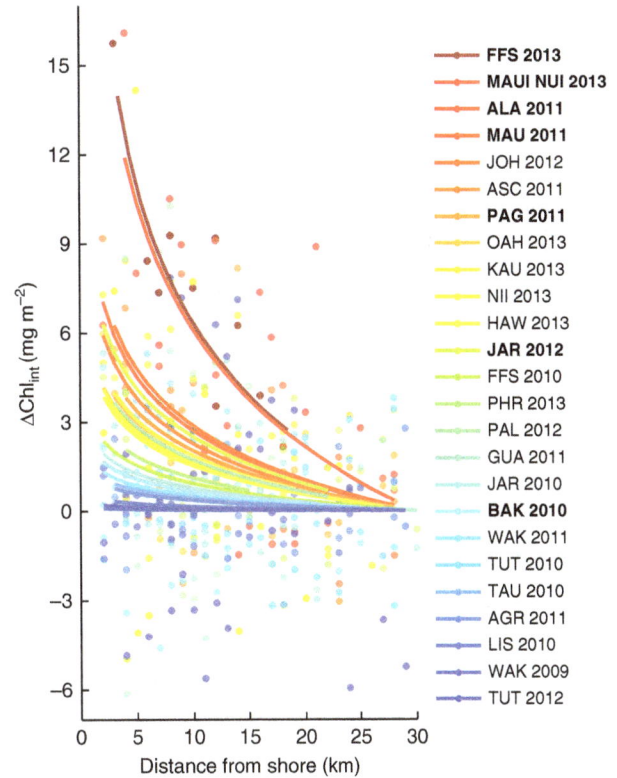

Figure 4 | Phytoplankton enhancements below the surface. Relationship between changes in depth-integrated chlorophyll-a (ΔChl_{int}) and distance from shore for 25 *in situ* surveys across 21 Pacific coral reef islands and atolls. Nonlinear regression fits are colour-coded based on the rate of increase in ΔChl_{int}; red (blue) fits have a stronger (weaker) rate of increase in ΔChl_{int} towards shore across all surveys. Location and survey year are shown (right). Bold indicates significant relationships ($P<0.05$). All information is centred to have a value of zero Chl_{int} at 30 km, the spatial extent of satellite observations presented herein. Please see Fig. 2 for island and atoll geographic locations.

residence times and incoming light energy[15,24]. Nevertheless, total ecosystem processes presumably scales with total reef area, thereby driving increased phytoplankton biomass and an increased IME at larger island- and atoll-reef ecosystems.

A further 7% of overall variation was explained through an interaction between reef area and geomorphic type; chlorophyll-a enhancement was greater with increased reef area at islands vs atolls (Fig. 3a,b, Supplementary Table 4). Near island-reef ecosystems, phytoplankton biomass can be influenced by a variety of sources that increase ambient nutrient concentrations. For example, riverine outflow can export large amounts of nutrient-laden terrigenous material to the nearshore[28], while submarine groundwater discharge can also drive considerable,

albeit highly variable, increases in nearshore nutrient concentrations[29]. This significant interaction effect among drivers adds to a growing body of evidence that multiple and simultaneously changing biogeophysical drivers shape ecological communities in the marine realm.

The presence of local anthropogenic impacts also influenced variations in the IME; nearshore chlorophyll-a enhancements were greater at populated ($n=7$) than unpopulated ($n=21$) island- and atoll-reef ecosystems (12% of explained variation, Fig. 3a,b, Supplementary Table 4). Human activities can increase nearshore nutrient concentrations well beyond natural levels, artificially elevating planktonic production in coastal marine ecosystems[30] (Fig. 1). This occurs through a variety of mechanisms including runoff from urban development and agricultural land use[28]. Wastewater effluent can also increase nearshore nutrient concentrations[31], particularly in areas where treatment occurs on-site (for example, cesspools), a common waste disposal practice across the Pacific, including the heavily populated Main Hawaiian Islands[32].

Phytoplankton enhancement below the surface. Remotely sensed chlorophyll-a provides an estimate of phytoplankton biomass in the upper 10s of metres in the ocean[33]. However, phytoplankton biomass often increases deeper in the water

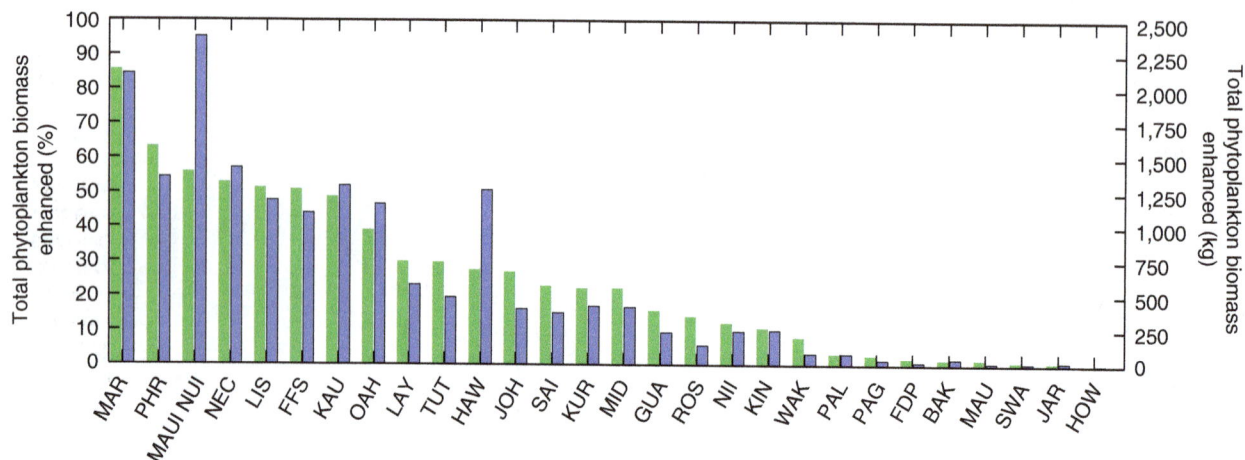

Figure 5 | Total phytoplankton biomass enhanced near Pacific island- and atoll-reef ecosystems. Bars represent the total increase in long-term phytoplankton biomass at each location over offshore, oceanic phytoplankton biomass. Values are shown in percentage (green bars; left y-axis) and kg (blue bars; right y-axis), oriented in decreasing percent biomass from left to right. All locations had significant linear fits ($P < 0.05$).

column, reaching a subsurface maximum that can be far greater than that observed in surface waters[34]. Using ship-based surveys, we examined nearshore enhancements in phytoplankton biomass down through the upper water column (5–300 m, see Methods section). Across the 29 island- and atoll-reef ecosystems surveyed, depth-integrated chlorophyll-*a* levels increased towards shore at 73% (21 of 29) of locations (Fig. 4). This is clear evidence that the IME propagates well below surface waters and occurs over the full euphotic depth range.

While *in situ* surveys provided direct observations of nearshore phytoplankton enhancements in the upper water column, they also demonstrated clear spatiotemporal variability exists in the IME. Five locations surveyed (JAR, TUT, PHR, FFS and WAK; Fig. 4) exhibited temporal variation in the strength of nearshore subsurface phytoplankton enhancement. In contrast, three locations exhibited opposing spatial trends; phytoplankton biomass both increased and decreased towards shore within the same location during different survey years (data not shown). Temporal differences observed between surveys at individual island- and atoll-reef ecosystems likely reflect the variable nature of biogeophysical processes that drive increased nearshore phytoplankton biomass (Fig. 1). Current-topographic interactions and wave-driven lagoonal flushing, for example, exhibit spatiotemporal variability that can drive nutrient supply fluctuations resulting in a locally variable phytoplankton response[35]. Therefore, the extent of nearshore phytoplankton biomass enhancement may not be fully captured during our brief 1–3-day ship-based sampling efforts at a given island or atoll in a given year. Nevertheless, our *in situ* observations provide important verification that the observed surface phytoplankton biomass enhancements are indeed reflective of subsurface phytoplankton gradients near Pacific island- and atoll-reef ecosystems.

Total increase in marine food resources. We examined the ecological implications of the IME by quantifying the total increase in standing stock of phytoplankton, and thus the increase in basal resources available to higher trophic levels, driven by the presence of each island- and atoll-reef ecosystem. The IME resulted in a long-term averaged (10 years) combined increase in nearshore phytoplankton biomass of 703.6% across our study locations ($n = 28$; see Methods section). This represented a total increase of 17.21 metric tonnes over the background oceanic standing stock of phytoplankton. On a per island basis, we saw a range of 0.2–85.6% increase in total phytoplankton biomass (Fig. 5). The Hawaiian Archipelago harboured the top nine reef

ecosystems with the greatest long-term increases in total phytoplankton biomass relative to background oceanic conditions (range; 29.9–85.6%). Other locations exhibited more modest enhancements in phytoplankton biomass. The equatorial islands of Howland, Baker and Jarvis, for example, showed a combined long-term enhancement in phytoplankton biomass of only 2.6% over background oceanic conditions (Fig. 5). This result was unexpected considering *in situ* observations of strong localized upwelling[16], high cover of reef-building organisms[5] and high planktivorous and predatory fish biomass[12] at these locations. However, these islands are uniquely situated at the western edge of the equatorial cold tongue; a geographic area in the Pacific that experiences consistent trade wind-driven equatorial upwelling and high chlorophyll-*a* concentrations (Fig. 2). Therefore, a more muted biological response associated with the IME was observed at these island-reef ecosystems owing to the already elevated background phytoplankton levels.

Discussion

The consequences of increased phytoplankton biomass span multiple trophic groups within coral reef island and atoll marine food-webs. Calcium carbonate-forming benthic organisms, namely hard (scleractinian) corals and crustose coralline algae, show increased abundance across the central and western Pacific Ocean when exposed to increased phytoplankton biomass[5]. Similarly, the biomass of planktivorous and piscivorous fishes and baseline estimates of Pacific reef sharks are far greater at locations with greater mean phytoplankton biomass[6,36]. In addition, a distinct community of squids, fishes and other deep-water associated micronekton are found in high densities near island-reef ecosystems relative to offshore waters[37]. This community—the mesopelagic boundary layer community—exhibits strong diel migration in association with subsurface island topography, transiting long distances (> 5 km) towards shore at night and peaking in abundance and density in waters with increased phytoplankton biomass[38]. Pelagic predators, such as dolphins[39] and tuna[40] cue in on the shoreward migration of organisms, exploiting the island-associated micronekton community as a food resource (Fig. 1). Moreover, inter-island migratory patterns of marine apex predators, such as tiger sharks (*Galeocerdo cuvier*), appear to be driven by variations in phytoplankton biomass, presumably owing to net energetic gain associated with bottom-up driven increases in prey abundance[41].

Phytoplankton biomass and the underlying drivers that enhance it may not always act to bolster ecological communities. For example, sewage pollution can increase nutrient levels that drive entire regime shifts in marine ecosystems[42] while the associated phytoplankton enhancement can drive coastal eutrophication and toxic algal blooms, resulting in mass mortalities of fishes, marine mammals and seabirds[28]. In addition, cold ocean temperatures in chronic upwelling environments may limit coral growth and suppress reef-building processes that are important for coral reef persistence[43]. Similarly, upwelled nutrients can significantly exceed background concentrations and enhance fleshy (non reef-building) algal growth[20], potentially disrupting benthic competitive interactions and reshaping coral reef communities. Because ecological communities can exhibit threshold-type responses, or tipping points, to variations in human[44] and natural physical drivers[45], similar relationships potentially exist between island- and atoll-reef ecosystems and the processes that enhance nearshore phytoplankton biomass. Future research is needed, however, to properly identify and better understand the existence of non-linearities in marine ecosystems associated with the IME.

Global shifts in biogeochemical cycling and ocean mixing associated with climate change are projected to decrease nutrient availability and primary production in the tropics and subtropics[46], impacting fisheries and in turn compromising human food supplies[47]. However, state-of-the art climate models used in these projections do not resolve complex biophysical interactions, such as the IME, that occur at the island-reef ecosystem scale. The projected strengthening of the Pacific equatorial undercurrent[48], for example, may increase vertical transport of nutrient-rich waters that drive phytoplankton biomass enhancements near equatorial island- and atoll-reef ecosystems despite projected regional declines in marine primary production. The number of island- and atoll-reef ecosystems that stand to benefit from future equatorial undercurrent strengthening is but a fraction of reef ecosystems in the Pacific. Nevertheless, these systems represent potentially important refugia for coral reef ecosystems and the food-webs they support in a rapidly changing climate. Biogeophysical drivers of the IME may also serve to bolster coral reef ecosystem resiliency to future thermal stress events and associated coral bleaching. The delivery of particle-laden deep ocean waters via upwelling and internal waves, for example, could provide important energetic subsidies[49] and a thermal reprieve[50] for corals during prolonged periods of anomalously warm temperatures.

Our basin-scale investigation of the IME demonstrates that nearshore phytoplankton enhancement is a long-term, near-ubiquitous feature among Pacific coral reef islands and atolls. Moreover, we found the magnitude of nearshore phytoplankton enhancement differed among island- and atoll-reef ecosystems owing to variations in key biogeophysical drivers, namely geomorphic type (atoll vs island), bathymetric slope, reef area and factors associated with the presence of local human populations. Individual coral reef island and atolls were capable of increasing the nearshore standing stock of phyto-plankton biomass by up to 86% over background oceanic conditions, forming biological hotspots across an otherwise barren ocean landscape. Ecosystem services vital to human populations, such as fisheries production and coastal protection, are intrinsically linked to the nearshore phytoplankton enhance-ment associated with the IME. Such ecosystem-scale biophysical phenomenon must therefore be incorporated into future model-ling efforts if we are to accurately predict the trajectory of marine ecosystems and the millions of people they support in this era of rapid change.

Methods

Quantifying chlorophyll-_a_ enhancements. Spatial gradients in chlorophyll-_a_ were quantified following Gove et al.[11]. In brief, long-term mean (July 2002 to June 2012) chlorophyll-_a_ was calculated from 8-day, 0.0417° Moderate Resolution Imaging Spectroradiometer (MODIS) data. Eight sectors of ~ 3.27 km width (0.0295°; ½ the diagonal of a satellite pixel) that were perpendicular to the 30 m isobath were quantified at each location (for example, Supplementary Fig.1). Sectors extended ~ 3.27–6.54 km to ~ 26.16–29.43 km (0.0295–0.0590° to 0.2360–0.2655°) offshore from the 30 m isobath. Data within ~ 3.27 km of the 30 m isobath were removed to avoid optically shallow waters and errors induced by terrigenous input, re-suspended material or bottom substrate properties[51] (Supplementary Fig. 2). Where island or atoll proximity resulted in sectors from different locations containing common data pixels, pixels were identified and removed before pixel averaging to avoid potential biases associated with the IME signal from one location influencing another's. However, when proximity resulted in a large proportion of pixels shared between locations (>50%), bathymetry was combined and sectors recalculated for the larger formed island-complex. Specifically, Molokai, Maui, Lanai and Kahoolawe were combined to form MAUI NUI; Saipan, Tinian and Agujian were combined to form SAI; Ofu, Olosega and Tau were combined to form MANUA (Fig. 2 and Supplementary Table 1).

We found the relationship between long-term chlorophyll-_a_ and distance to shore was best described by a power function at all locations (for example, Supplementary Fig.1b). Given that a power function is equivalent to a linear fit on log–log transformed data, seamless transition can be made between nonlinear and linear fits and associated analyses. To test for differences in the relationship between long-term chlorophyll-_a_ and distance to shore between locations, an analysis of covariance was performed on log–log transformed data. Two tests were performed; one test that included all islands and atolls ($n = 35$) and another that only included locations that showed a significant increase in chlorophyll-_a_ with decrease distance to shore ($n = 28$; $P < 0.05$).

Biogeophysical drivers. We incorporated a series of biogeophysical parameters that are potential drivers of increased phytoplankton biomass near oceanic island- and atoll-reef ecosystems. The following were quantified for each location: latitude, geomorphic type (atoll vs island), reef area, land area, bathymetric slope, elevation, human population status (unpopulated vs populated) and the long-term mean and standard deviation for: sea-surface temperature, precipitation and ocean currents (Supplementary Table 2).

Latitude (°) represented the centre point of each location. Geomorphic type was either 'atoll' or 'island'. Reef area (km^2) was calculated from the shore-line to the 30-m isobath and land area (km^2) was calculated for all emergent land. Bathymetric slope (°) was derived from bathymetric grids in ArcGIS v10.1 using the Spatial Analyst 'slope' function, calculated between 30–300 m depth and then averaged across the entire location. A detailed description of these factors (latitude, geomorphic type, reef area and land area) can be obtained in Gove et al.[11].

Elevation (m) was obtained from a variety of sources, including the U.S. Central Intelligence Agency (https://www.cia.gov), NOAA's Coral Reef Information System (http://www.coris.noaa.gov) and the U.S. Fish and Wildlife Service (www.fws.gov). Population status was either 'unpopulated' or 'populated' following Williams et al..[12] Locations were considered populated with a human habitation of >160 people.

Island- and atoll-scale SST (°C) was obtained following Gove et al.[11]. In brief, SST was quantified using 0.0439°, 7-day information from the Pathfinder v5.0 data set (http://pathfinder.nodc.noaa.gov). Data were excluded if deemed of poor quality (quality value < 4 ref. 52) or if individual pixels were masked as land. Island- and atoll-specific SST data were produced by spatially averaging the individual pixels that were intersected by or contained within the 30 m isobath for each location.

Precipitation data was obtained from the Global Precipitation Climatology Project v2.2 (http://www.esrl.noaa.gov); a global, 2.5° spatial resolution, monthly data set that merges remotely sensed (microwave and infrared) and surface rain gauge observations.

Ocean current data were obtained from NOAA's OSCAR (http://www.oscar. noaa.gov/); a global, 1° spatial resolution, monthly ocean current data set derived from satellite altimetry (sea-level) and scatterometer (wind). The magnitude of current was calculated from the zonal (u) and meridional (v) components of flow for each time step. Grid cells for precipitation and ocean currents were chosen based on the centre point of each island or atoll.

The long-term mean and the standard deviation were calculated for SST, precipitation and ocean currents for each location over the 10-year time period concurrent with the long-term chlorophyll-_a_ values (July 2002 to June 2012). Where locations were combined to form a larger island-complex (that is, MAUI NUI, MANUA and SAI), time series data among islands were averaged for each time step before long-term mean and standard deviation calculations while remaining biogeophysical metrics were summed (land area, reef area), averaged (slope, latitude) or the maximum value was obtained (elevation).

Underlying bathymetry data for all locations were provided by the Pacific Islands Benthic Habitat Mapping Center (PIBHMC), Hawaii Mapping Research Group (HMRG), National Geophysical Data Center (NGDC) and satellite-derived global topography.

Statistical analysis and model selection. Our research hypothesis was that the IME varied with island type (atoll vs island), reef area, land area, bathymetric slope, elevation, population status (populated vs unpopulated), SST (mean and standard deviation), precipitation (mean and standard deviation) and current speed (mean and standard deviation). We used slope ('b') from the regression output of log–log transformed data (chlorophyll-a vs distance to shore) to represent the IME (that is, the response variable), enabling a standardized comparison in chlorophyll-a gradients among study locations. This hypothesis was tested by initially fitting a GLM. While also assuming linear relationships among response and predictor variables (that is, like traditional linear regression), GLMs relax the requirement of a normal error distribution, 'generalizing' the model to other distributions (within the exponential family) by relating the response and predictor variables through a 'link' function. Only locations with significant negative relationships (that is, increased chlorophyll-a with decreased distance to shore) were used in the model ($n = 28$; $P < 0.05$). Initial examination of response vs predictor(s) appeared to show a combination of linear and non-linear relationships were possible. A GLM was used as the starting model and the assumption of linearity (that is, use of GLM) was tested (as were all other assumptions). An error distribution of gamma was assumed (requiring a positive transformation of slope values before model input) and a log link was used.

Collinearity among predictors was tested both by calculating Pearson's correlation values and variance inflation factors (VIF, for multicollinearity; for example[53] via car package[54]). Predictors with the highest VIF value were inspected with respect to their linear correlation with other predictors (via Pearson's correlation), as well as the mechanistic underpinnings of the research hypothesis. Following removal of the predictor representing the most concern (for example, highest collinearity), the model was refit and remaining predictors were assessed until all VIFs were < 3 ref. 53. Following this iterative process, elevation, latitude, SST (mean and standard deviation) and the standard deviations of current speed and precipitation were removed. Remaining predictors were geomorphic type (atoll vs island), reef area, land area, bathymetric slope, population status (populated vs unpopulated), mean precipitation and mean current speed. Model variants representing all possible combinations of these predictors (main effects, see interactions below) were tested and the models were ranked according to Akaike Information Criteria corrected (AICc) for small sample size. Two candidate models were chosen based on ΔAICc ≤ 2 (Supplementary Table 3). On the basis of these results (Supplementary Table 3), models representing all possible combinations of main effects and two-way interactions were tested with geomorphic type, reef area, bathymetric slope, population status and mean current speed as predictors (MuMIn package[55]). The interaction between geomorphic type and population status was removed as no atolls were populated. Two models were chosen based on ΔAICc ≤ 2 (Supplementary Table 3). The highest AICc weight occurred for the simpler model and further analysis indicated mean current speed did not significantly improve model explanatory power (based on analysis of deviance via χ^2-test, $P = 0.219$). In contrast, the interaction term between geomorphic type (island) and reef area was significant (based on analysis of deviance via χ^2-test, $P = 0.001$), increasing the overall explained deviance of the model from 78 to 85%. The resulting best-specified model was:

(1) abs(b) $\sim$ Geomorphic Type + Reef Area + Bathymetric Slope + Population Status + Reef Area:Geomorphic Type

Next, the assumption of independence of response estimates was tested by fitting a generalized linear mixed model (via the lme4 package[56]) and including region (see Supplementary Table 1) as a random effect on the intercept as:

(2) abs(b) $\sim$ Geomorphic Type + Reef Area + Bathymetric Slope + Population Status + Reef Area:Geomorphic Type + (1|Region)

There was no significant difference found when comparing the model with and without this random effect (based on analysis of deviance via χ^2-test, $P = 0.5$) and the fixed effects model was chosen (model 1).

The assumption of linearity of relationships among predictors and response was then tested by refitting model 1 as a Generalized Additive Model (GAM; mgcv package[57]) which included smoothing terms on the continuous variables of slope and reef area (no interaction terms possible):

(3) Geomorphic Type + s(Reef Area) + s(Bathymetric Slope) + Population Status

There was a significant difference between the GLM (model 1) and GAM (model 3) with $P = 0.012$ (analysis of deviance via χ^2-test). The GLM is simpler and with a lower AICc (GLM AICc: -105 vs GAM AICc: -95) and thus the GLM (model 1) was chosen as our best-fit model.

The resulting model exhibited well-behaved uniform residuals with no significant outliers. Residuals deviated somewhat from an ideal normal distribution when inspected graphically but this deviation was not significant (that is, residual distribution did not differ significantly from normal; Shapiro–Wilks $P = 0.12$). Models were refit with alternate link functions (for example, 'inverse') but models fit with a log link function provided the most well-behaved residuals. The model was significant with $P < 0.0001$ and explains 85% of the overall deviance. Predictor coefficients and significance are shown in Supplementary Table 4.

Chlorophyll enhancement increases (that is, slope b is more negative) at atolls (vs islands, Wald's test $P < 0.0001$), decreases in bathymetric slope (Wald's test $P = 0.004$) and at populated locations (vs unpopulated; Wald's test $P = 0.001$). While reef area on its own was not significant (Wald's test $P = 0.27$), there is a significant interaction term between reef area and geomorphic type that indicates that chlorophyll

enhancement increases (b is more negative) with increased reef area at islands more than atolls (Wald's test $P = 0.002$). In all cases, the magnitude of enhancement will be relative to the surrounding waters (vs absolute chlorophyll abundance).

The relative importance of each predictor in explaining the deviance was determined by hierarchical partitioning, which examines the effect of removing each predictor from models representing all possible orders of variables (Supplementary Table 4). In this way, the average independent contribution to explained deviance of each predictor is obtained[58]. Hierarchical partitioning was performed using the hier.part package in R[59] (note that interaction terms are not included in this analysis) and indicated that relative importance of geomorphic type, bathymetric slope, reef area and population status in explaining the overall deviance of 34, 28, 26 and 12%, respectively. As above, these predictors explained 78% of the overall variability in the data, with the interaction term of reef area and geomorphic type increasing the explained deviance to 85%. All data manipulation and statistical analyses were performed using Matlab v2013b and R (R Core Team 2013 and related packages) unless otherwise specified.

Total phytoplankton biomass. The total long-term mean (July 2002 to June 2012) in nearshore standing stock phytoplankton biomass enhanced by each island and atoll over background oceanic phytoplankton was calculated using remotely sensed observations. We first calculated the long-term mean in depth-integrated chlorophyll-a (ΣChl; mg m^{-2}) by multiplying the long-term mean in chlorophyll-a (see above for details) by the long-term mean in the depth of light penetration for each satellite pixel. Depth of light penetration, or the depth at which the satellite can 'see' into the water column, was calculated by taking the reciprocal of the long-term mean in 8 day, 0.0417° MODIS k490 ref. 33. All long-term mean ΣChl values within each sector (see Supplementary Fig.1) were then averaged for each island and atoll. The relationship between ΣChl and distance to shore was then quantified by applying a linear fit on log–log transformed data. Only locations with significant ($P < 0.05$) negative relationships (that is, increased ΣChl with decreased distance to shore) were used ($n = 28$; same locations used in modelling efforts). Using the furthest sector from each location (sector 8, for example see Supplementary Fig.1) to represent offshore, oceanic conditions in phytoplankton, we calculated the change (Δ) in ΣChl with each subsequent, more proximate sector. We then multiplied $\Delta\Sigma$Chl for each location by the respective sector area (m^2; the change in latitude was accounted for in longitude to distance conversions for all locations) to calculate the phytoplankton enhanced (kg and percentage) for each sector. Summing over all sectors (1–7) enabled a quantitative estimate of the long-term phytoplankton biomass enhanced by the presence of each island and atoll in our study.

Ship-based phytoplankton. Vertical chlorophyll-a profiles were obtained using a profiling fluorometer during 37 ship-based surveys of 29 individual islands and atolls in our study region. Surveys consisted of horizontal transects, starting ~ 2–4 km from shore and extending 20–30 km offshore in one or more cardinal directions over 1–3 days. Depth-integrated *in situ* chlorophyll-a (mg m^{-2}) was calculated over the upper water column (5–300 m depth). Nonlinear least squares regression fits were applied over the spatial distance covered within each survey.

References

1. Duarte, C. & Cebrian, J. The fate of marine autotrophic production. *Limnol. Oceanogr.* **41,** 1758–1766 (1996).
2. Iverson, R. L. Control of marine fish production. *Limnol. Oceanogr.* **35,** 1593–1604 (1990).
3. Chassot, E. *et al.* Global marine primary production constrains fisheries catches. *Ecol. Lett.* **13,** 495–505 (2010).
4. Hamner, W. M. & Hauri, I. R. Effects of island mass: water flow and plankton pattern around a reef in the Great Barrier Reef Lagoon, Australia. *Limnol. Oceanogr.* **26,** 1084–1102 (1981).
5. Williams, G. J., Gove, J. M., Eynaud, Y., Zgliczynski, B. J. & Sandin, S. A. Local human impacts decouple natural biophysical relationships on Pacific coral reefs. *Ecography* **38,** 751–761 (2015).
6. Williams, I. D. *et al.* Human, oceanographic and habitat drivers of central and western Pacific coral reef fish assemblages. *Plos ONE* **10,** e0120516 (2015).
7. Doty, M. S. & Oguri, M. The island mass effect. *J. Cons.* **22,** 33–37 (1956).
8. Signorini, S. R., McClain, C. R. & Dandonneau, Y. Mixing and phytoplankton bloom in the wake of the Marquesas Islands. *Geophys. Res. Lett.* **26,** 3121–3124 (1999).
9. Palacios, D. M. Factors influencing the island-mass effect of the Galapagos Archipelago. *Geophys. Res. Lett.* **29,** 2134 (2002).
10. Andrade, I., Sangrà, P., Hormazabal, S. & Correa-Ramirez, M. Island mass effect in the Juan Fernández Archipelago (33°S), Southeastern Pacific. *Deep Sea Res. Part 1 Oceanogr. Res. Pap.* **84,** 86–99 (2014).
11. Gove, J. M. *et al.* Quantifying climatological ranges and anomalies for Pacific coral reef ecosystems. *Plos ONE* **8,** e61974 (2013).
12. Williams, I. D. *et al.* Differences in Reef Fish Assemblages between Populated and Remote Reefs Spanning Multiple Archipelagos Across the Central and Western Pacific. *Journal of Marine Biology* **2011,** 14 (2011).

13. Aucan, J., Hoeke, R. & Merrifield, M. A. Wave-driven sea level anomalies at the Midway tide gauge as an index of North Pacific storminess over the past 60 years. *Geophys. Res. Lett.* **39**, L17603 (2012).

14. Callaghan, D. P., Nielsen, P., Cartwright, N., Gourlay, M. R. & Baldock, T. E. Atoll lagoon flushing forced by waves. *Coast Eng.* **53**, 691–704 (2006).

15. Atkinson, M. in *Coral Reefs: An Ecosystem in Transition* (eds Dubinsky, Z. & Stambler, N.) 199–206 (Springer, 2011).

16. Gove, J. M., Merrifield, M. A. & Brainard, R. E. Temporal variability of current-driven upwelling at Jarvis Island. *J. Geophys. Res. Oceans* **111**, C12011 (2006).

17. Coutis, P. F. & Middleton, J. H. Flow-topography interaction in the vicinity of an isolated, deep ocean island. *Deep Sea Res. Part 1 Oceanogr. Res. Pap.* **46**, 1633–1652 (1999).

18. Hernández-León, S. Accumulation of mesozooplankton in a wake area as a causative mechanism of the 'island-mass effect'. *Mar Biol* **109**, 141–147 (1991).

19. Heywood, K. J., Barton, E. D. & Simpson, J. H. The effects of flow disturbance by an oceanic island. *J. Mar. Res.* **48**, 55–73 (1990).

20. Leichter, J. J., Stewart, H. L. & Miller, S. L. Episodic nutrient transport to Florida coral reefs. *Limnol. Oceanogr.* 1394–1407 (2003).

21. Leichter, J. J., Shellenbarger, G., Genovese, S. J. & Wing, S. R. Breaking internal waves on a Florida (USA) coral reef: a plankton pump at work? *Mar. Ecol. Prog. Ser.* **166**, 83–97 (1998).

22. Carter, G. S., Gregg, M. C. & Merrifield, M. A. Flow and Mixing around a Small Seamount on Kaena Ridge, Hawaii. *J. Phys. Oceanogr.* **36**, 1036–1052 (2006).

23. Merrifield, M. A. & Holloway, P. E. Model estimates of M2 internal tide energetics at the Hawaiian Ridge. *J. Geophys. Res. Oceans* **107**, 5-1–5-12 (2002).

24. Suzuki, Y. & Casareto, B. E. The role of dissolved organic nitrogen (DON) in coral biology and reef ecology. In: *Coral Reefs: An Ecosystem in Transition.* (ed Dubinsky, Z. & Stambler, N.) (Springer, 2011).

25. McCauley, D. J. *et al.* From wing to wing: the persistence of long ecological interaction chains in less-disturbed ecosystems. *Sci. Rep.* **2**, 409 (2012).

26. Burkepile, D. E. *et al.* Nutrient supply from fishes facilitates macroalgae and suppresses corals in a Caribbean coral reef ecosystem. *Sci. Rep.* **3**, 1493 (2013).

27. Williams, S. L. & Carpenter, R. C. Nitrogen-limited primary productivity of coral reef algal turfs: potential contribution of ammonium excreted by *Diadema antillarum*. *Mar. Ecol. Prog. Ser.* **47**, 145–152 (1988).

28. Anderson, D. M., Glibert, P. M. & Burkholder, J. M. Harmful algal blooms and eutrophication: nutrient sources, composition, and consequences. *Estuaries* **25**, 704–726 (2002).

29. Street, J. H., Knee, K. L., Grossman, E. E. & Paytan, A. Submarine groundwater discharge and nutrient addition to the coastal zone and coral reefs of leeward Hawai'i. *Mar. Chem.* **109**, 355–376 (2008).

30. Vitousek, P. M., Mooney, H. A., Lubchenco, J. & Melillo, J. M. Human domination of Earth's ecosystems. *Science* **277**, 494–499 (1997).

31. Smith, V. H., Tilman, G. D. & Nekola, J. C. Eutrophication: impacts of excess nutrient inputs on freshwater, marine, and terrestrial ecosystems. *Environ. Pol.* **100**, 179–196 (1999).

32. Whittier, R. B. & El-Kadi, A. I. *Human Health and Environmental Risk of Onsite Sewage Disposal Systems for The Hawaiian Islands of Kauai, Maui, Molokai, and Hawaii* (Hawaii's Department of Health, 2014).

33. Gordon, H. R. & McCluney, W. R. Estimation of the depth of sunlight penetration in the sea for remote sensing. *Appl. Optics* **14**, 413–416 (1975).

34. Furuya, K. Subsurface chlorophyll maximum in the tropical and subtropical western Pacific Ocean: Vertical profiles of phytoplankton biomass and its relationship with chlorophylla and particulate organic carbon. *Mar. Biol.* **107**, 529–539 (1990).

35. Hasegawa, D., Lewis, M. R. & Gangopadhyay, A. How islands cause phytoplankton to bloom in their wakes. *Geophys. Res. Lett.* **36**, L20605 (2009).

36. Nadon, M. O. *et al.* Re-creating missing population baselines for Pacific reef sharks. *Conserv. Biol.* **26**, 493–503 (2012).

37. Reid, S. B., Hirota, J., Young, R. E. & Hallacher, L. E. Mesopelagic-boundary community in Hawaii: micronekton at the interface between neritic and oceanic ecosystems. *Mar. Biol.* **109**, 427–440 (1991).

38. McManus, M. A., Benoit-Bird, K. J. & Brock Woodson, C. Behavior exceeds physical forcing in the diel horizontal migration of the midwater sound-scattering layer in Hawaiian waters. *Mar. Ecol. Prog. Ser.* **365**, 91–101 (2008).

39. Benoit-Bird, K. J. & Au, W. W. L. Prey dynamics affect foraging by a pelagic predator (Stenella longirostris) over a range of spatial and temporal scales. *Behav. Ecol. Sociobiol.* **53**, 364–373 (2003).

40. Musyl, M. K. *et al.* Vertical movements of bigeye tuna (*Thunnus obesus*) associated with islands, buoys, and seamounts near the main Hawaiian Islands from archival tagging data. *Fish. Oceanogr.* **12**, 152–169 (2003).

41. Papastamatiou, Y. P. *et al.* Telemetry and random-walk models reveal complex patterns of partial migration in a large marine predator. *Ecology* **94**, 2595–2606 (2013).

42. Pastorok, R. A. & Bilyard, G. R. Effects of sewage pollution on coral-reef communities. *Mar. Ecol. Prog. Ser.* **21**, 175–189 (1985).

43. Benzoni, F., Bianchi, C. N. & Morri, C. Coral communities of the northwestern Gulf of Aden (Yemen): variation in framework building related to environmental factors and biotic conditions. *Coral Reefs* **22**, 475–484 (2003).

44. McClanahan, T. R. *et al.* Critical thresholds and tangible targets for ecosystem-based management of coral reef fisheries. *Proc. Natl Acad. Sci. USA* **108**, 17230–17233 (2011).

45. Gove, J. M. *et al.* Coral reef benthic regimes exhibit non-linear threshold responses to natural physical drivers. *Mar. Ecol. Prog. Ser.* **522**, 33–48 (2015).

46. Doney, S. C. The growing human footprint on coastal and open-ocean biogeochemistry. *Science* **328**, 1512–1516 (2010).

47. Blanchard, J. L. *et al.* Potential consequences of climate change for primary production and fish production in large marine ecosystems. *Philos. Trans. R Soc. Lond. B Biol. Sci.* **367**, 2979–2989 (2012).

48. Karnauskas, K. B. & Cohen, A. L. Equatorial refuge amid tropical warming. *Nat. Clim. Change* **2**, 530–534 (2012).

49. Grottoli, A. G., Rodrigues, L. J. & Palardy, J. E. Heterotrophic plasticity and resilience in bleached corals. *Nature* **440**, 1186–1189 (2006).

50. Riegl, B. & Piller, W. E. Possible refugia for reefs in times of environmental stress. *Int. J. Earth Sci.* **92**, 520–531 (2003).

51. Boss, E. & Zaneveld, J. R. V. The effect of bottom substrate on inherent optical properties: evidence of biogeochemical processes. *Limnol. Oceanogr.* **48**, 346–354 (2003).

52. Kilpatrick, K. A., Podesta, G. P. & Evans, R. Overview of the NOAA/NASA advanced very high resolution radiometer Pathfinder algorithm for sea surface temperature and associated matchup database. *J. Geophys. Res. Oceans* **106**, 9179–9197 (2001).

53. Zuur, A. F., Ieno, E. N. & Elphick, C. S. A protocol for data exploration to avoid common statistical problems. *Methods Ecol. Evol.* **1**, 3–14 (2010).

54. Hayden, R. W. A Review of: 'An R Companion to Applied Regression, Second Edition, by J. Fox and S. Weisberg'. *J Biopharmaceut. Stat.* **22**, 418–419 (2012).

55. Barton, K. MuLIn: Multi-Model Inference. R package version 1.13.4. (2015).

56. Bates, D., Mächler, M., Bolker, B. & Walker, S. Fitting linear mixed-effects models using lme4. *J. Stat. Software* **67**, 1–48 (2015).

57. Wood, S. N. Fast stable restricted maximum likelihood and marginal likelihood estimation of semiparametric generalized linear models. *J. R. Stat. Soc. Series B Stat. Methodol.* **73**, 3–36 (2011).

58. Chevan, A. & Sutherland, M. Hierarchical partitioning. *Am Stat.* **45**, 90–96 (1991).

59. Walsh, C. & Mac Nally, R. hier.part: Hierarchical Partitioning. R package version 1.0-4.Available at http://CRAN.R-project.org/package=hier.part (2013).

Acknowledgements

This work was part of an interdisciplinary effort by the NOAA Pacific Islands Fisheries Science Center's Coral Reef Ecosystem Program (CREP) to assess, understand and monitor coral reef ecosystems of the U.S. Pacific. The authors would like to thank Rusty Brainard, Principle Investigator of CREP, for his support of this research. We would also like to thank the NOAA ship Hi'ialakai for logistic support and field assistance. Funding for surveys (as part of the Pacific Reef Assessment and Monitoring Program, RAMP) was provided by NOAA's Coral Reef Conservation Program.

Author contributions

J.M.G. conceived the study with M.A.M., J.C.D., C.R.S., M.A.M. and A.M.F.; J.M.G., M.A.M., G.J.W. and A.B.N. developed and implemented the analysis; J.M.G., J.S.E. and C.W.Y. collected and contributed data; J.M.G. wrote the manuscript with assistance from M.A.M., A.B.N., J.J.P. and G.J.W. A.K.D. made substantive contributions to the manuscript.

Additional information

6

An extreme event of sea-level rise along the Northeast coast of North America in 2009–2010

Paul B. Goddard[1], Jianjun Yin[1], Stephen M. Griffies[2] & Shaoqing Zhang[2]

The coastal sea levels along the Northeast Coast of North America show significant year-to-year fluctuations in a general upward trend. The analysis of long-term tide gauge records identified an extreme sea-level rise (SLR) event during 2009–10. Within this 2-year period, the coastal sea level north of New York City jumped by 128 mm. This magnitude of interannual SLR is unprecedented (a 1-in-850 year event) during the entire history of the tide gauge records. Here we show that this extreme SLR event is a combined effect of two factors: an observed 30% downturn of the Atlantic meridional overturning circulation during 2009–10, and a significant negative North Atlantic Oscillation index. The extreme nature of the 2009–10 SLR event suggests that such a significant downturn of the Atlantic overturning circulation is very unusual. During the twenty-first century, climate models project an increase in magnitude and frequency of extreme interannual SLR events along this densely populated coast.

[1] Department of Geosciences, University of Arizona, Tucson, Arizona 85721, USA. [2] Geophysical Fluid Dynamics Laboratory, NOAA, Princeton, New Jersey 08540, USA. Correspondence and requests for materials should be addressed to J.Y. (email: yin@email.arizona.edu).

The Intergovernmental Panel on Climate Change Fifth Assessment Report[1,2] lists extreme sea levels among the top impacts of climate change. Hourly to daily extreme sea levels are typically associated with transient storms and eddies, tides and tsunamis. Once they occur and superimpose, these events pose a threat to coastal communities. On seasonal to interannual time scales, extreme sea-level events are usually linked to large-scale ocean dynamics and climate extremes, but have received little attention thus far.

In the North Atlantic, especially along the Northeast (NE) Coast of North America, sea levels are critically influenced by the Atlantic meridional overturning circulation (AMOC)[3–7]. Since 2004, the AMOC has been systematically monitored at 26.5°N in the North Atlantic[8,9]. The available data reveal a strength of $18.5 \pm 1.0 \, \mathrm{Sv}$ (mean $\pm 1\sigma$) and small interannual variability of the AMOC from 1 April 2004 to 31 March 2009. From April 2009 through March 2010, by contrast, the AMOC shows a significant 30% downturn to 12.8 Sv, followed by a second minimum during the winter of 2010–11 (refs 9–11). Additional observations at other latitudes in the Atlantic indicate that this downturn is a basin-wide phenomenon due to the spatial coherence of the AMOC[12,13].

Whether this downturn of the AMOC is a sign of its long-term trend[14] or a part of its natural variability[11,15], or both[16], will need further research with longer observational data. Regardless, this event provides a valuable opportunity to study the climate impact of the AMOC, quantify the AMOC–sea-level rise (SLR) relationship and test model simulation results. Here we analyse long-term tide gauge (TG) data and report an extreme and unprecedented SLR event in 2009–10 along the NE Coast of North America. It should be noted that we use the term SLR here to indicate interannual sea-level changes, while it is usually referred to the long-term and gradual trend of sea level in literature. With various observation and model data, we show that the 2009–10 SLR event was caused by the 30% downturn of

the AMOC and the wind stress anomalies associated with the significant negative North Atlantic Oscillation (NAO) index[17].

Results

TG records. Sea level along the East Coast of the United States and Canada exhibits interannual fluctuations superimposed on multi-decadal variations and a long-term upward trend[5,7,18–20]. Modelling and long-term TG data indicate that the behaviour of sea level is similar and highly correlated north or south of Cape Hatteras[4,5,7,18,20]. By taking the long-term rate and especially the 2009 SLR rate into account (Supplementary Fig. 1), we further divide the East Coast of North America into three SLR regimes: NE (North of New York City), Mid-Atlantic (New York City to Cape Hatteras) and Southeast (south of Cape Hatteras; Fig. 1a). To minimize the effect of local factors and reveal regionally coherent behaviour, we calculate the time series of the sea-level composite for the three SLR regimes (see the Methods section). From the noisy background, we separate an extreme SLR event having occurred between 2009 and 2010 along the NE Coast of North America.

The NE sea-level composite is calculated as the mean of maximum 18 TG stations from Montauk, New York, to Rimouski, Canada (Fig. 1b and Supplementary Table 1). Next, we calculate the yearly SLR rate by differentiating the annual mean time series of the sea-level composite (Fig. 2a) (see the Methods section). The results indicate that the s.d. (σ) of the yearly SLR rate is 14.5 mm per year since 1920, much larger than the secular trend of 2.5 mm per year. However, the composite SLR rate in 2009 is a remarkable outlier and reaches 46.9 mm per year ($>3\sigma$) (Fig. 2b). Between 2009 and 2010, the sea level in this region jumped by nearly 100 mm on average (Figs 1b and 3a), contributing significantly to the identified SLR acceleration during the recent decades[5,7,18]. To calculate the return period, we fit a Gaussian distribution to the yearly SLR rates from 1920 to 2011. The result shows that the 2009 SLR rate is a 1-in-850 year event (Fig. 2c).

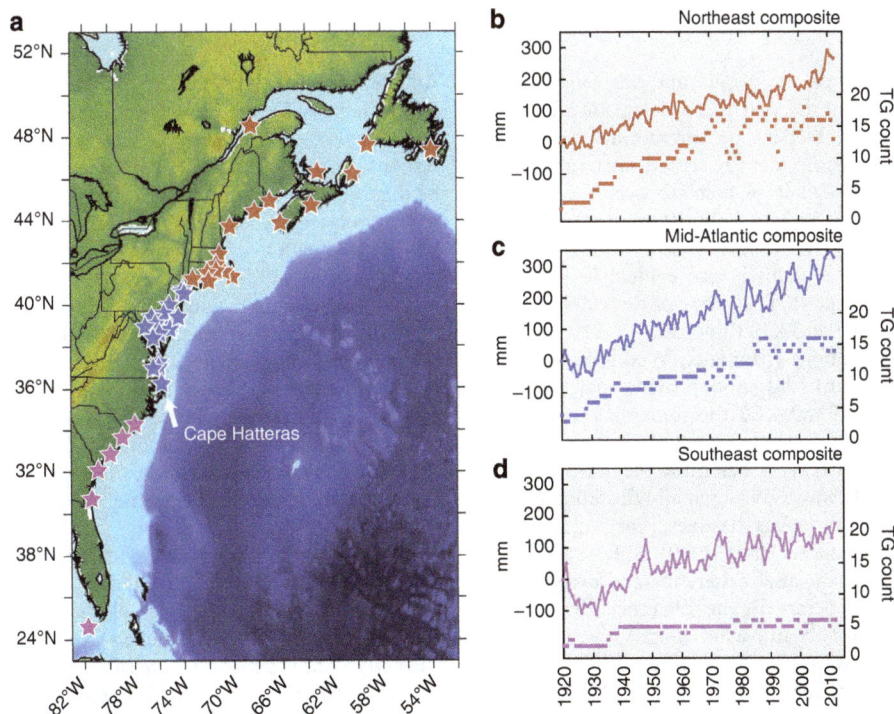

Figure 1 | Three SLR regimes along the East Coast of North America and the corresponding sea-level composite. (a) Three SLR regimes and locations of the 40 TG stations used by this study. Red, blue and purple colours indicate the Northeast, Mid-Atlantic and Southeast region, respectively. **(b–d)** Time series of sea-level composite (line) in the three regimes and counts of TG stations used in the composite calculation as a function of time (squares).

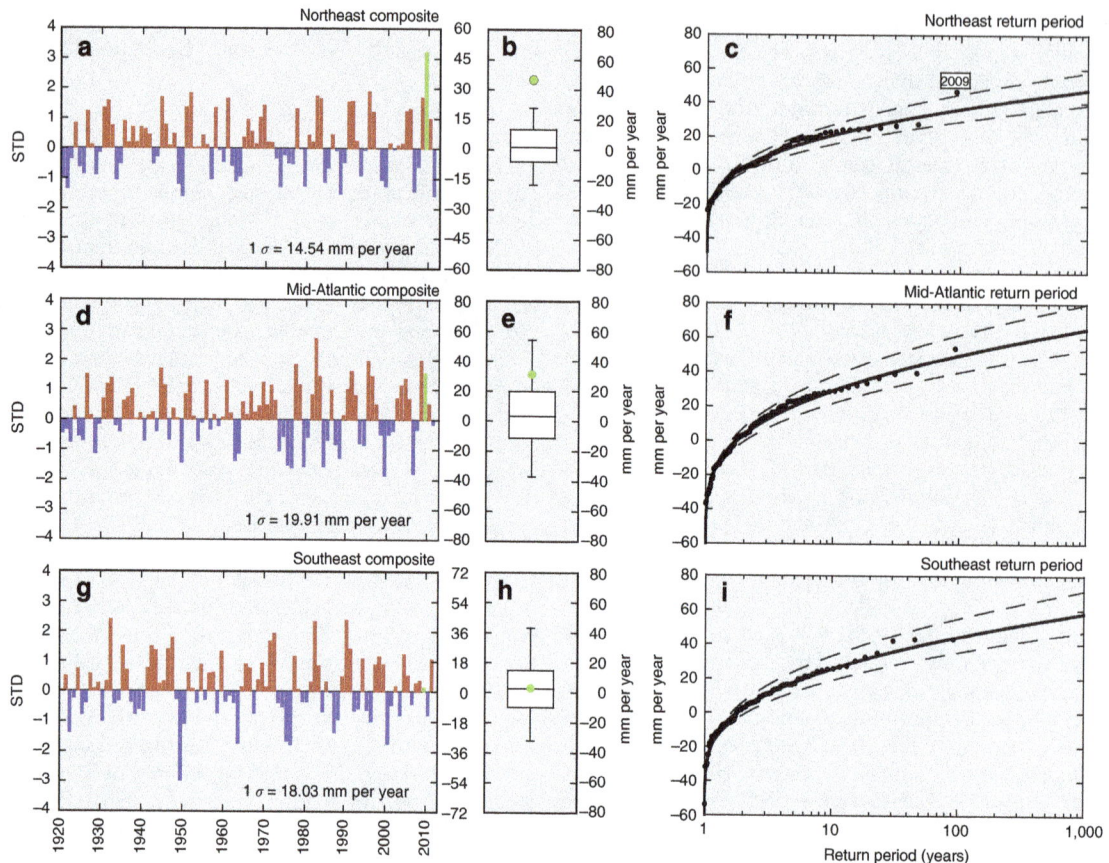

Figure 2 | Yearly SLR rates along the East Coast of North America. (a,d,g) Yearly SLR rates (mm per year) of the Northeast, Mid-Atlantic and Southeast composite, respectively. The absolute and relative (to 1σ) rates are shown by the right and left y axis, respectively. (**b,e,h**) Box and whisker plots indicate the position of the 2009 SLR rates for the three sea-level composites, respectively. The 2009 SLR rates are depicted by the green bars and dots. (**c,f,i**) Return period of the yearly SLR rates. Dots show the TG data during 1920-2011. Solid lines indicate the exceedance probability of the Gaussian distribution and the corresponding return period. Dotted lines represent the 5%-95% confidence interval of the Gaussian distribution fit. The 2009 box marks the NE 2009 SLR rate. See Methods section for the calculation of the return period.

To test the robustness of the result and its sensitivity to different methods, we repeat the yearly SLR calculation based on the linear fit to the monthly data on interannual time scales (Supplementary Fig. 2). We also identify record-breaking years of sea level and compare the sea-level increase over the previous records (Supplementary Fig. 3). The 2009–10 event stands out in all three methods (Table 1).

The 2009–10 extreme SLR event is also evident in individual TG data (Supplementary Fig. 4). The range of the 2009 SLR rate in the NE region varies from 32.5 (Rimouski) to 64 (Portland) mm per year. Four NE stations (Montauk, Woods Hole, North Sydney and Charlottetown) did not provide data for the calculation of the 2009 SLR rate. Of the remaining 14 stations in the NE region, 11 stations show that the 2009 SLR rate was the highest on record, with the other 3 being the second highest. In addition, the 2009 SLR rate was above 2σ at all NE stations except Rimouski. Five stations (Port-Aux-Basques, Eastport, Halifax, Portland and Boston) show the 2009 SLR rate $>3\sigma$ (Supplementary Fig. 4). The spatial coherency and extreme SLR rate suggest that large-scale ocean climate dynamics (for example, the AMOC[4,21], Gulf Stream[6,22] and wind effect[20,23]) rather than local mechanisms (for example, land subsidence[24]) is the main cause of the 2009 SLR event. Land subsidence is at least one order of magnitude smaller than the yearly SLR rates.

The signal of the 2009–10 extreme SLR event attenuates towards the south. Compared with the NE region, the amplitude of the yearly sea-level fluctuation is larger ($\sigma = 19.9$ mm per year) in the Mid-

Atlantic region, while the 2009 SLR rate is less extreme (Fig. 2d,e and Supplementary Fig. 4). The composite SLR rate is 31.2 mm per year ($>1\sigma$) in 2009, with a range of 22.0–41.5 mm per year at individual stations. In the Mid-Atlantic region, especially the Chesapeake Bay, land subsidence induced by glacial isostatic adjustment contributes to the long-term SLR[24]. The pronounced SLR rate during 1982–83 ($>2.5\sigma$ and with a return period of 150 years) is likely related to the strong El Niño in the Pacific and the resulting more coastal storms in the Mid-Atlantic region[23] (Fig. 2d,f).

South of Cape Hatteras, the 2009 SLR rate further reduces to 3.0 mm per year and falls within $\pm 1\sigma$ (Fig. 2g,h). The coastal sea level in this region is influenced by the North Atlantic subtropical gyre[25]. An extreme sea-level fall ($\sim -3\sigma$) occurred in 1949, following a rapid and continuous SLR during much of the 1940s. This decade was characterized by faster global SLR[26]. The detailed investigation about this extreme event is beyond the scope of this study.

The satellite altimetry data indicate that the most significant interannual variability of the dynamic sea level (DSL) occurs in the ocean interior, especially along the Gulf Stream and its extension (Fig. 3a). Along the East Coast of North America, the altimetry data is generally consistent with the TG data regarding the 2009–10 SLR event, but the exact magnitude differs.

Role of the AMOC. The 2009–10 sea-level spike in the NE region coincides with two significant ocean and climate events: a 30%

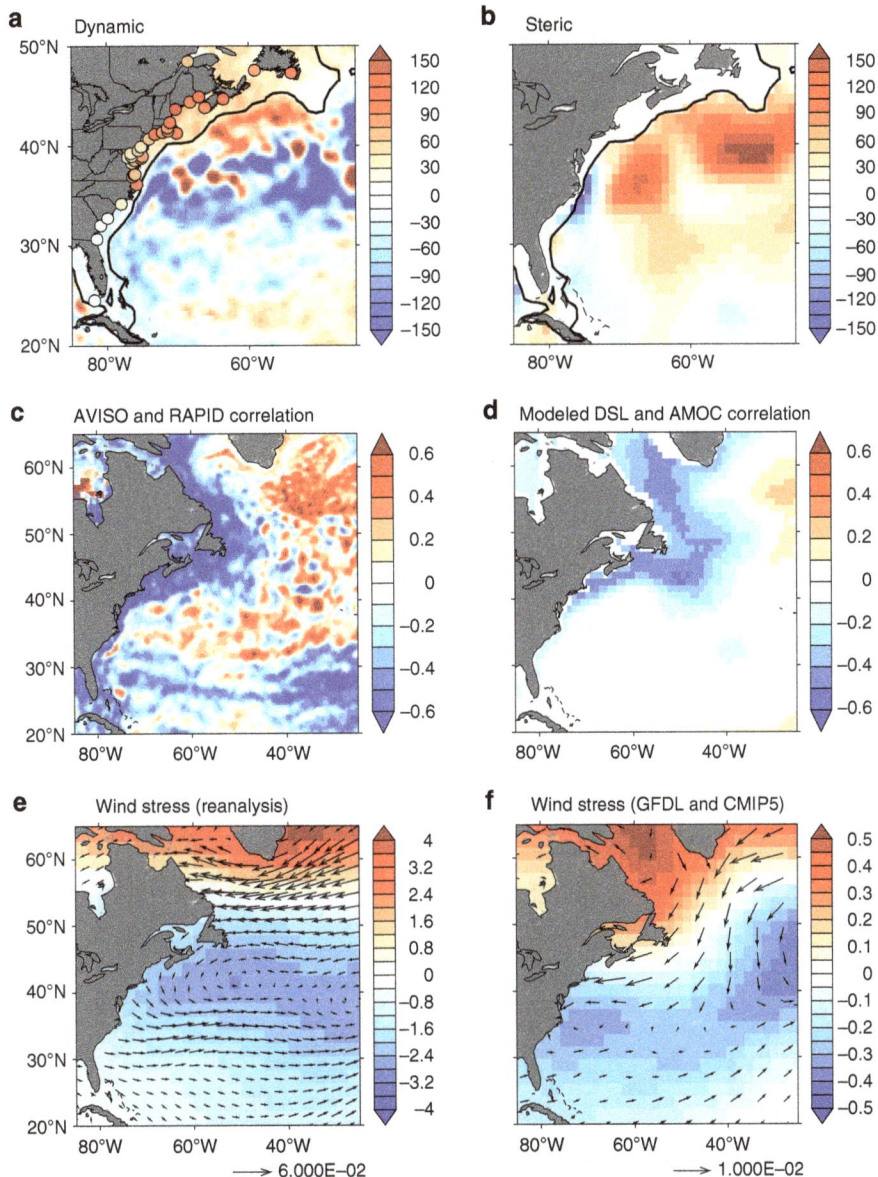

Figure 3 | Mechanisms of the 2009-2010 extreme SLR event. (**a**) Sea-level increase (mm) between 2008 and 2010 from the Archiving, Validation, and Interpretation of Satellite Oceanographic data (AVISO; shading) and TG stations (colour dots). The black line indicates the shelf break—500 m depth. (**b**) Steric sea-level anomalies (mm) in 2009 for the upper 2,000 m. (**c**) Correlation between the monthly AVISO and RAPID AMOC data for 2004-2012. (**d**) Correlation between the annual mean DSL and AMOC index (45°N) in the long-term control runs of the 10 GFDL and 14 CMIP5 models. The values show multi-model ensemble mean. See Supplementary Figs 7 and 8 for individual models. (**e**) Anomalies of sea-level pressure (shading; hPa) and wind stress (vector; N m^{-2}) in 2009-2010 from the GFDL reanalysis. (**f**) Difference in sea-level pressure (shading; hPa) and wind stress (vector; N m^{-2}) between the years of extreme positive and negative SLR at Boston. The results show the ensemble mean of the GFDL and CMIP5 models for 100-year control runs. See Supplementary Figs 10 and 11 for more details.

downturn of the AMOC[9] (Fig. 4a) and an extreme negative NAO index[27]. Both winds[20,23,25] and the AMOC[4,5,18,21] can cause sea-level variability and change along the NE coast of North America on various time scales. After removing the seasonal cycle, the monthly NE sea-level composite shows a good correlation ($R = 0.78$; 2-month lag) with the AMOC index (Fig. 4b). Especially, the two sea-level spikes during the winters of 2009–10 and 2010–11 coincide with the two AMOC minima, respectively. Regression based on all monthly data from 2004 to 2012 further reveals a 13.2-mm SLR along the NE coast in response to 1 Sv AMOC slowdown. After excluding the 30% downturn period of the AMOC, the AMOC-SLR ratio increases to 16.6 mm Sv^{-1}, reflecting the sea-level response to the gradual

decline of the AMOC during 2004–2012. In the altimetry data, the AMOC–DSL correlation extends far offshore, especially from the NE coast of North America (Fig. 3c). It changes sign across the Gulf Stream, implying that a decrease in the AMOC transport reduces the cross-current sea-level gradient and thus lifts sea level along the NE coast[5,6].

Interestingly, the 30% downturn of the AMOC in 2009–2010 followed a brief return to deep convection in the Labrador Sea during the winter of 2007–2008 (ref. 28). A data assimilation product by GFDL[29,30] (see the Methods section) shows that strong deep downwelling in 2007–2008 was mainly induced by a reduction in the freshwater input into the Labrador Sea (Supplementary Fig. 5). However, a record low of oceanic

Table 1 | Summary of the rise rates in the three SLR regimes.

	Northeast	Mid-Atlantic	Southeast
Sea level differentiation method for yearly SLR rate			
s.d. (mm per year)	14.5	19.9	18.0
2009 Rate (mm per year)	46.9	31.2	3.0
Range of the 2009 rate at different stations (mm per year)	32.5-64.0	22.0-41.5	− 7-16
Linear fit method for yearly SLR rate			
s.d. (mm per year)	14.1	20.3	17.7
2009 Rate (mm per year)	41.7	31.2	13.5
2009 Error bar (mm per year)	8.6	12.0	16.1
2010 Sea-level increase over previous record (mm)	59.4	17.1	—
Long-term trend (1920-2012) (mm per year)	2.5	3.5	2.3

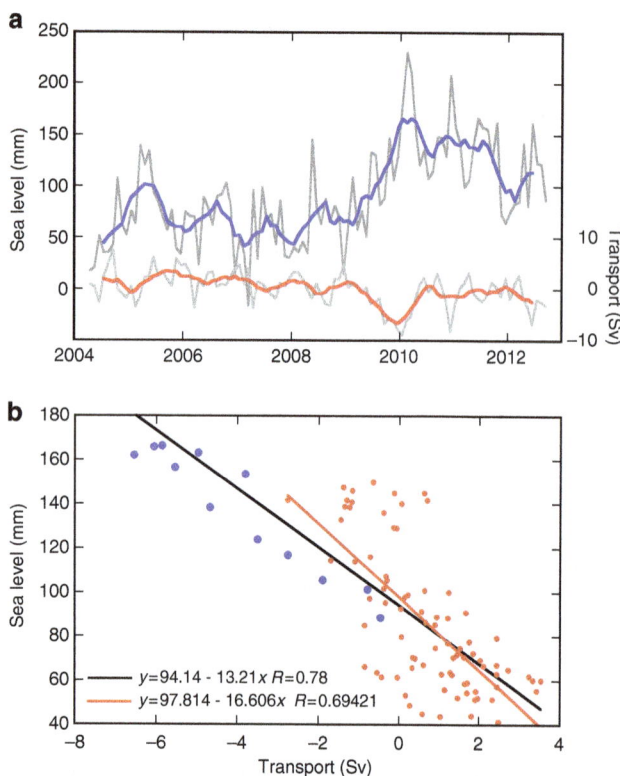

Figure 4 | Correlation between the AMOC and the sea-level composite along the NE coast. (a) The time series of the NE sea-level composite (monthly grey, filtered blue) and the AMOC strength at 26.5°N (Sv, monthly grey, filtered red). The seasonal cycle has been removed with a 6-month filter applied. **(b)** The monthly correlation and regression between the AMOC and NE sea-level composite (mm per Sv, 2-month lag with the AMOC leading the SLR). The blue dots highlight the data during the period of the 30% AMOC downturn. The linear fit in black is based on all monthly data during April 2004-September 2012. The linear fit in red is based on the red dots only and excluding the period of the 30% AMOC downtown. The data are from PSMSL and the RAPID-WATCH MOC project.

Table 2 | Ten global climate/Earth system models developed at GFDL and used in this study.

Model	Atmosphere	Ocean	Documentation
CM2.6	0.5°	0.1°, z* coordinate	Delworth et al.[44] Griffies et al.[32] Winton et al.[45]
CM2.5	0.5°	0.25°, z* coordinate	Delworth et al.[44] Griffies et al.[32] Winton et al.[45]
CM2.5FLORa6	0.5°	1°, z* coordinate	Vecchi et al.[46] Griffies et al.[32] Winton et al.[45]
CM2.5FLOR	0.5°	1°, z* coordinate	Vecchi et al.[46] Griffies et al.[32] Winton et al.[45]
CM2.0	2°	1°, z coordinate	Delworth et al.[47]
CM2.1	2°	1°, z coordinate	Delworth et al.[47]
CM3	2°	1°, z* coordinate	Griffies et al.[48]
ESM2M	2°	1°, z* coordinate	Dunne et al.[49]
ESM2preG	2°	1°, Isopycnal	Dunne et al.[49]
ESM2G	2°	1°, Isopycnal	Dunne et al.[49]

models developed at GFDL, including high-resolution models with eddying oceans[32] (Table 2), and the other set including 14 CMIP5 models with the AMOC and sea-level data available at the CMIP5 archive[33] (Table 3; also see the Methods section). The GFDL ensemble represents different model generations with progressive improvement and a systematic model development effort at one modelling centre. On the other hand, the CMIP5 ensemble represents similar generation models from different institutes, which bear significant difference in many aspects. Therefore, these models cover a wide range in the model formulation, parameterization and uncertainty space.

In the long-term control runs of these models without changing external forcing, the DSL along the NE coast of North America shows an instantaneous correlation with the AMOC on the interannual time scale (Fig. 3d and Supplementary Figs 7 and 8). This correlation suggests that a 30% weakening of the AMOC in 2009–2010 may have contributed to and increased the chance of extreme coastal SLR events. The two models with eddying oceans (GFDL CM2.6 and CM2.5) show weaker correlation, probably due to the relatively weak AMOC in these two models[32] (Supplementary Fig. 9).

Role of the NAO and associated winds. The significant negative NAO index during December 2009 through February 2010 (ref. 27) contributes to the extreme SLR event both remotely and locally. First, the negative NAO can cause an anomalous heat flux

heat and buoyancy loss, dominated by the sensible heat flux (Supplementary Fig. 6), occurred in the subsequent years 2008–2010. Associated with these changes is a significant positive steric sea-level anomaly in 2009 (> 100 mm), southeastward of the shelf break (Fig. 3b). The gradient across the shelf break can drive more water mass towards the shelf, thereby causing SLR along the NE coast of North America[4,31].

To better understand extreme SLR events in this region, we use two sets of state-of-the-art climate models: an ensemble of ten

Table 3 | 14 CMIP5 global climate/Earth system models used in this study.

Model	Institution
ACCESS1.0	Commonwealth Scientific and Industrial Research Organisation, Australia
ACCESS1.3	Commonwealth Scientific and Industrial Research Organisation, Australia
CCSM4	National Center for Atmospheric Research, USA
CESM1-BGC	National Center for Atmospheric Research, USA
CESM1-FASTCHEM	National Center for Atmospheric Research, USA
CESM1-WACCM	National Center for Atmospheric Research, USA
CMCC-CM	Centro Euro-Mediterraneo sui Cambiamenti Climatici, Italy
INM-CM4	Institute of Numerical Mathematics, Russia
MPI-ESM-LR	Max Planck Institute for Meteorology, Germany
MPI-ESM-MR	Max Planck Institute for Meteorology, Germany
MPI-ESM-P	Max Planck Institute for Meteorology, Germany
MRI-CGCM3	Meteorological Research Institute, Japan
NorESM1-M	Norwegian Climate Center, Norway
NorESM1-ME	Norwegian Climate Center, Norway

into the Labrador Sea, thereby influencing the AMOC and SLR along the NE coast of North America (Supplementary Fig. 5). Second, the NAO-induced wind stress anomalies can pile up waters directly against the NE coast or generate onshore Ekman transport.

The control runs with both the GFDL and other CMIP5 models indicate that extreme SLR events along the NE coast typically occur when the nearby wind stress shows an onshore (easterly) or alongshore (northeasterly) anomaly pattern (Fig. 3f and Supplementary Figs 10 and 11). Indeed, the lower sea-level pressure east of North America in 2009 results in northeasterly wind stress anomalies near the NE coastal regions (Fig. 3e). The anomalous Ekman transport contributes to the 2009–2010 SLR extreme. The negative NAO can also influence storminess, which contributes to higher sea levels due to more frequent storm surges[34].

The NE sea-level composite shows correlation ($R = 0.6$) with the NAO index during 2004–2012 (Supplementary Fig. 12). However, the correlation reduces significantly for the entire period of 1920–2012. In addition to 2009–2010, extreme negative NAO index also occurred in 1969 and other years (Supplementary Fig. 13). The lack of extreme SLR signal on the NE coast during these years indicates that the NAO is not the sole mechanism of the 2009–2010 SLR event. Finally, the lower sea-level pressure (by $\sim 1.6\,\mathrm{hPa}$) in 2009 could account for $\sim 15\%$ of the 2009–10 SLR event through the inverse barometer effect (Fig. 3e).

Future projections of extreme SLR events. Similar to extreme temperature and precipitation events, extreme SLR events on the interannual time scale may be also linked to human-induced climate change[35]. Increased greenhouse gas concentrations are likely to shift the probability density function towards more extremes. To study future changes of extreme SLR events, we consider the ten GFDL climate models and their long-term control runs, and idealized 1% per year CO_2 increase experiments for 100 years. Along the NE coast of North America, most of these models (CM2.6, CM2.5, CM2.5 FLORa6, CM2.5 FLOR, ESM2M and ESM2preG) suggest an increase in the magnitude and frequency of the extreme SLR events in response to the CO_2 increase (Fig. 5 and Supplementary Fig. 14). In five of these models (CM2.6, CM2.5 FLORa6, CM2.5 FLOR, ESM2M and ESM2preG), the increase in the yearly SLR extrema is

unproportionately larger than the increase in mean sea level (Supplementary Fig. 14).

There are several reasons for this increase in extremes: first, the global mean SLR from thermal expansion and land ice melt (note that the latter is not included in Fig. 5); second, the overall weakening of the AMOC in the CO_2 experiment leads to record lows of the circulation in some years (Supplementary Fig. 9), thereby facilitating SLR extremes along the NE coast of North America; and third, the NAO variability remains strong in the CO_2 experiments[36]. Recently, projections of the melting of glaciers[37], the Greenland[38] and Antarctic[39] ice sheets, have been made for the twenty-first century under greenhouse gas emission scenarios. Adding these contributions would further increase the probability of extreme SLR events along the East Coast of North America, especially its northeastern sector.

Discussion

In the present study, we focus on the extreme SLR event on interannual time scales. With long-term TG data, we calculate the sea-level composite and yearly SLR rates for three SLR regimes along the East Coast of North America. The resulting time series contain rich information about both climate variability and individual events. The extreme and unprecedented SLR event in 2009–2010 is particularly notable along the NE coast of North America. Our analysis suggests that this event was mainly caused by a 30% downturn of the AMOC and the wind stress anomalies associated with the negative NAO, although the two factors are inherently linked[17].

There is no direct observation of the AMOC before 2004. Some recent model hindcast suggests that similar downturns of the AMOC may have punctuated in the twentieth century[11]. Our analysis based on the long-term TG data and the AMOC–SLR relationship indicates that the 2009–2010 event is very unusual. In addition to internal variability, anthropogenic forcing could be another impact factor, as most climate models project a weakening of the AMOC during the twenty-first century in response to the increase in the atmospheric greenhouse gas concentrations[40]. Continuing observations of the AMOC is essential to confirm these modelling results.

In addition to the absolute SLR rate, the 2009–2010 event is very unusual also in the sense that it occurred during a short period of global sea-level fall[41]. Unlike storm surge, this event caused persistent and widespread coastal flooding[22] even without apparent weather processes. In terms of beach erosion, the impact of the 2009–2010 SLR event is almost as significant as some hurricane events[42]. For the twenty-first century, modelling results suggest that the increase in the greenhouse gas concentrations is likely to cause more extreme SLR events on the interannual time scale along this densely populated coast. Once coastal storms compound high sea levels, more damages will result.

Methods

TG data. The data are selected from the Permanent Service for Mean Sea Level global database for TG records[43] (http://www.psmsl.org/data/obtaining). We selected records with at least 25 years and 70% completeness that were up to date through 2012. This method provides 40 TG stations along the East Coast from Key West, Florida, to Rimouski, Canada (Supplementary Table 1).

Satellite altimetry data. For the absolute DSL, we use the Archiving, Validation, and Interpretation of Satellite Oceanographic data. This delayed time data set provides $1/4° \times 1/4°$ resolution in daily intervals from 1 January 1993 through 31 December 2012 (http://www.aviso.altimetry.fr/en/data/data-access.html).

AMOC time series. For *in-situ* observations of the AMOC, we use the data from the RAPID-WATCH MOC monitoring project (http://www.rapid.ac.uk/rapidmoc). This data set has a twice per day temporal resolution and ranges from April 2004 through October 2012 (at the time of download).

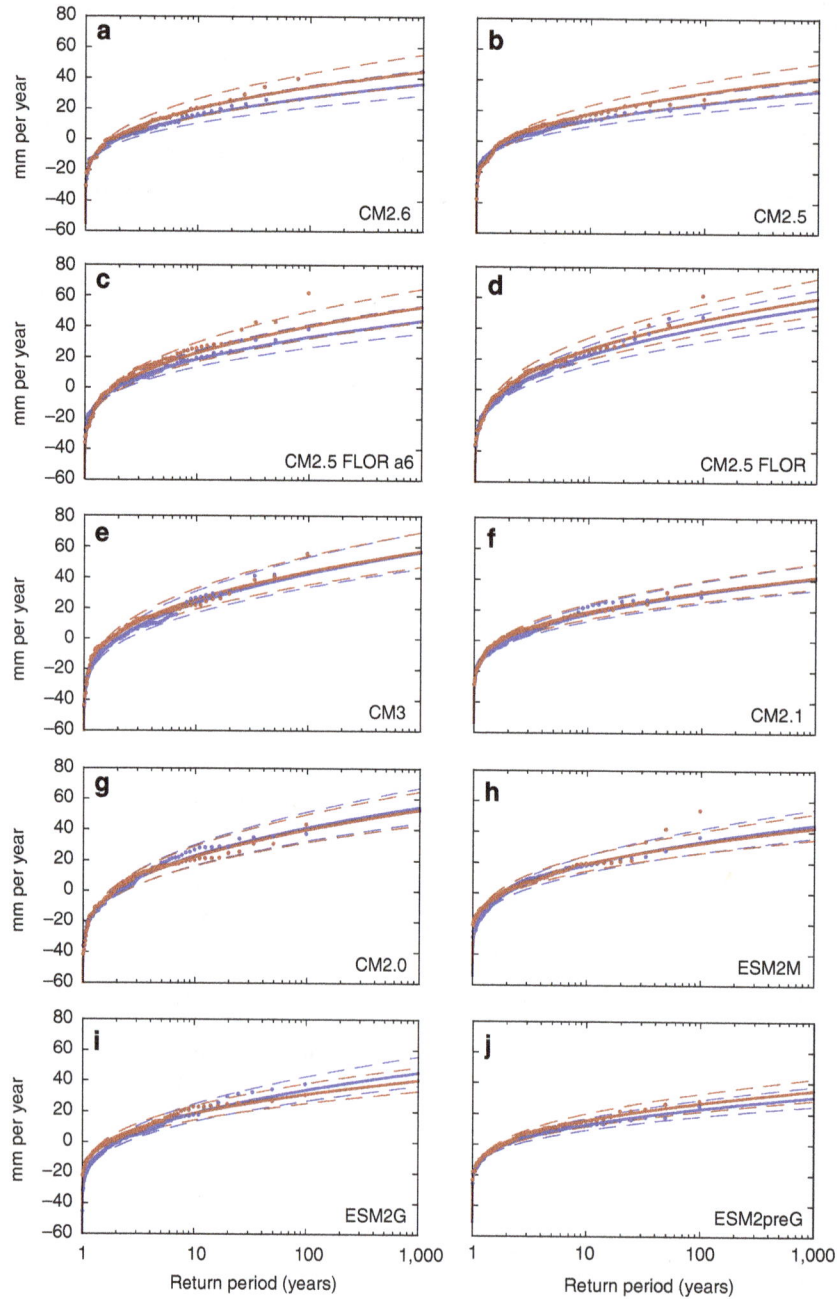

Figure 5 | Return periods of the yearly SLR rates in the control runs and 1% per year CO₂ increase experiments with the ten GFDL models. The values show the composite SLR rates along the NE Coast of North America and are calculated based on both the DSL changes and the global ocean thermal expansion. Both the control (blue colour) and CO₂ runs (red colour) are 100-year long. Dots show the model simulation results. Solid lines indicate the exceedance probability of the Gaussian distribution and the corresponding return period. Dotted lines represent the 5%–95% confidence interval of the Gaussian distribution fit. The red line above the blue line indicates an increase in the return level or a decrease in the return period.

Yearly SLR rate. We differentiate the sea-level time series to obtain the yearly SLR rates.

$$\text{SLR}(t) = \frac{\text{SL}(t+1) - \text{SL}(t-1)}{2}, \quad t = \text{second record,....2011} \quad (1)$$

$\text{SLR}(t)$ is the SLR rate (mm per year) for a particular year, and $\text{SL}(t+1)$ and $\text{SL}(t-1)$ denote the sea-level records for the previous and next year, respectively. It should be noted that six TG stations are not included in calculating the 2009 SLR rate due to missing data for 2008 or 2010. The results are compared with those based on the linear fit (Supplementary Fig. 2).

Return period. Based on the millennial time-scale control simulations of the ten GFDL models (not shown), we find the yearly SLR rates are well fitted with a Gaussian distribution. For a return level x_t, the probability of exceedance is

$Pr(x > x_t)$, where Pr is the cumulative density function associated with the Gaussian distribution. The return period (T) is calculated as $T = 1/Pr$.

GFDL ocean data assimilation data. The GFDL ocean data assimilation data are taken from the oceanic component of climate reanalysis by coupled data assimilation[30]. We use longwave (LW), shortwave (SW), sensible heat (SH), latent heat (LH), precipitation minus evaporation (P − E), and river runoff (R) fluxes from the GFDL climate reanalysis product[29], where Q_{HF} (W m⁻²) and Q_{WF} (m s⁻¹) are the net atmosphere–ocean heat (positive values indicate flux into ocean) and freshwater flux, respectively.

$$Q_{HF} = LW + SW + SH + LH \quad (2)$$

$$Q_{WF} = (P - E + R) \quad (3)$$

Ocean temperature, salinity and steric sea-level data. We use the data from the National Oceanographic Data Center (http://www.nodc.noaa.gov/OC5/3M_HEAT_CONTENT/). The steric sea-level anomaly data for the upper 2,000 m span from 2005–2013.

GFDL model suite. We consider ten coupled climate models built over the past decade at GFDL (Table 2). CM2.0 and CM2.1 were used as part of the CMIP3 projects, whereas CM3 and ESM2M/ESM2G were used as part of the CMIP5 project. ESM2preG is an early version of ESM2G. Each of these models uses a nominally 1° ocean, with various changes made to the atmosphere and ocean components, leading to the different configurations. Notably, the ESM2G and ESM2preG are based on an isopycnal ocean, whereas the other models use a level coordinate MOM ocean component. CM2.6, CM2.5, CM2.5 FLORa6 and CM2.5 FLOR all use the same 50-km finite volume atmospheric core, with CM2.6 using a 1/10° ocean, CM2.5 a 1/4° ocean and the FLOR simulations using a 1° ocean. The two FLOR simulations differ in their choice of ocean subgrid scale para-meterizations, with FLORa6 using a larger lateral viscosity than FLOR. CM2.5 FLOR is used for studies of tropical cyclones.

CMIP5 model suite. The data are downloaded from the CMIP5 archive[33] (Table 3). Detailed model description and experimental design can be found at http://cmip-pcmdi.llnl.gov/cmip5/.

References

1. Church, J. A. *et al.* in *Climate Change 2013: The Physical Science Basis* (eds Stocker, T. F. *et al.*) (Cambridge University Press, 2013).
2. Wong, P. P. *et al.* in *Climate Change 2014: Impacts, Adaptation, and Vulnerability. Part A: Global and Sectoral Aspects. Contribution of Working Group II to the Fifth Assessment Report of the Intergovernmental Panel on Climate Change* (eds Field, C. B. *et al.*) (Cambridge University Press, 2014).
3. Levermann, A., Griesel, A., Hofmann, M., Montoya, M. & Rahmstorf, S. Dynamic sea level changes following changes in the thermohaline circulation. *Clim. Dyn.* **24**, 347–354 (2005).
4. Yin, J., Schlesinger, M. E. & Stouffer, R. J. Model projections of rapid sea-level rise on the northeast coast of the United States. *Nat. Geosci.* **2**, 262–266 (2009).
5. Sallenger, A. H., Doran, K. S. & Howd, P. A. Hotspot of accelerated sea-level rise on the Atlantic coast of North America. *Nat. Clim. Change* **2**, 884–888 (2012).
6. Ezer, T., Atkinson, L. P., Corlett, W. B. & Blanco, J. L. Gulf Stream's induced sea level rise and variability along the U.S. mid-Atlantic coast. *J. Geophys. Res.* **118**, 685–697 (2013).
7. Boon, J. D. Evidence of sea level acceleration at U.S. and Canadian tide stations, Atlantic Coast, North America. *J. Coast. Res.* **28**, 1437–1445 (2012).
8. Cunningham, S. A. *et al.* Temporal variability of the Atlantic meridional overturning circulation at 26.5 degrees N. *Science* **317**, 935–938 (2007).
9. McCarthy, G. D. *et al.* Observed interannual variability of the atlantic meridional overturning circulation at 26.5°N. *Geophys. Res. Lett.* **39**, L19609 (2012).
10. Smeed, D. A. *et al.* Observed decline of the Atlantic meridional overturning circulation 2004–2012. *Ocean Sci.* **10**, 29–38 (2014).
11. Blaker, A. T. *et al.* Historical analogues of the recent extreme minima observed in the Atlantic meridional overturning circulation at 26° N. *Clim. Dyn.* **44**, 457–473 (2014).
12. Rhein, M. *et al.* in *Climate Change 2013: The Physical Science Basis. Contribution of Working Group I to the Fifth Assessment Report of the Intergovernmental Panel on Climate Change* (eds Stocker, T. F. *et al.*) (Cambridge University Press, 2013).
13. Newlin, I. L. & Gregg, M. C. Global Oceans [in 'State of the Climate in 2013']. *Bull. Am. Meteorol. Soc.* **95**, s51–s78 (2014).
14. Robson, J., Hodson, D., Hawkins, E. & Sutton, R. Atlantic overturning in decline? *Nat. Geosci.* **7**, 2–3 (2014).
15. Roberts, C. D., Jackson, L. & McNeall, D. Is the 2004–2012 reduction of the Atlantic meridional overturning circulation significant? *Geophys. Res. Lett.* **41**, 3204–3210 (2014).
16. Schiermeier, Q. OCEANOGRAPHY Atlantic current strength declines. *Nature* **509**, 270–271 (2014).
17. Bryden, H. L., King, B. A., McCarthy, G. D. & McDonagh, E. L. Impact of a 30% reduction in Atlantic meridional overturning during 2009–2010. *Ocean Sci.* **10**, 683–691 (2014).
18. Ezer, T. Sea level rise, spatially uneven and temporally unsteady: Why the US East Coast, the global tide gauge record, and the global altimeter data show different trends. *Geophys. Res. Lett.* **40**, 5439–5444 (2013).
19. Kopp, R. E. Does the mid-Atlantic United States sea level acceleration hot spot reflect ocean dynamic variability? *Geophys. Res. Lett.* **40**, 3981–3985 (2013).
20. Andres, M., Gawarkiewicz, G. G. & Toole, J. M. Interannual sea level variability in the western North Atlantic: regional forcing and remote response. *Geophys. Res. Lett.* **40**, 5915–5919 (2013).
21. Yin, J. & Goddard, P. B. Oceanic control of sea level rise patterns along the East Coast of the United States. *Geophys. Res. Lett.* **40**, 5514–5520 (2013).
22. Sweet, W., Zervas, C. & Gill, S. *Elevated East Coast Sea Level Anomaly: June-July 2009*. NOAA Technical Report NOS CO-OPS 051, 40pp (NOAA Natl. Ocean Service, Silver Spring, MD, USA, 2009).
23. Sweet, W. V. & Zervas, C. Cool-season sea level anomalies and storm surges along the US east coast: climatology and comparison with the 2009/10 El Nino. *Mon. Wea. Rev.* **139**, 2290–2299 (2011).
24. Boon, J. D., Brubaker, J. & Forrest, D. R. *Chesapeake Bay Land Subsidence and Sea Level Change: An Evaluation of Past and Present Trends and Future Outlook* (Virginia Institute of Marine Science, 2010).
25. Hong, B. G., Sturges, W. & Clarke, A. J. Sea level on the US East Coast: decadal variability caused by open ocean wind-curl forcing. *J. Phys. Oceanogr.* **30**, 2088–2098 (2000).
26. Church, J. A. & White, N. J. Sea-level rise from the late 19th to the early 21st century. *Surv. Geophys.* **32**, 585–602 (2011).
27. Hu, Z. Z. *et al.* Persistent atmospheric and oceanic anomalies in the North Atlantic from summer 2009 to summer 2010. *J. Clim.* **24**, 5812–5830 (2011).
28. Vage, K. *et al.* Surprising return of deep convection to the subpolar North Atlantic Ocean in winter 2007–2008. *Nat. Geosci.* **2**, 67–72 (2009).
29. Chang, Y.-S., Zhang, S., Rosati, A., Delworth, T. L. & Stern, W. F. An assessment of oceanic variability for 1960–2010 from the GFDL ensemble coupled data assimilation. *Clim. Dyn.* **40**, 775–803 (2013).
30. Zhang, S., Harrison, M. J., Rosati, A. & Wittenberg, A. System design and evaluation of coupled ensemble data assimilation for global oceanic climate studies. *Mon. Wea. Rev.* **135**, 3541–3564 (2007).
31. Griffies, S. M. *et al.* An assessment of global and regional sea level for years 1993–2007 in a suite of interannual CORE-II simulations. *Ocean Model.* **78**, 35–89 (2014).
32. Griffies, S. M. *et al.* Impacts on ocean heat from transient mesoscale eddies in a hierarchy of climate models. *J. Clim.* doi:10.1175/JCLI-D-14-00353.1 (2015).
33. Taylor, K. E., Stouffer, R. J. & Meehl, G. A. An overview of CMIP5 and the experiment design. *Bull. Am. Meteorol. Soc.* **93**, 485–498 (2012).
34. Ezer, T. & Atkinson, L. P. Accelerated flooding along the US East Coast: on the impact of sea-level rise, tides, storms, the Gulf Stream, and the North Atlantic oscillations. *Earth Future* **2**, 362–382 (2014).
35. Field, C. B. *Managing the Risks of Extreme Events and Disasters to Advance Climate Change Adaptation: Special Report of the Intergovernmental Panel on Climate Change* (Cambridge University Press, 2012).
36. Christensen, J. H. *et al.* in *Climate Change 2013: The Physical Science Basis. Contribution of Working Group I to the Fifth Assessment Report of the Intergovernmental Panel on Climate Change* (eds Stocker, T. F. *et al.*) (Cambridge University Press, 2013).
37. Marzeion, B., Jarosch, A. H. & Hofer, M. Past and future sea-level change from the surface mass balance of glaciers. *Cryosphere* **6**, 1295–1322 (2012).
38. Fettweis, X. *et al.* Estimating the Greenland ice sheet surface mass balance contribution to future sea level rise using the regional atmospheric climate model MAR. *Cryosphere* **7**, 469–489 (2013).
39. Levermann, A. *et al.* Projecting Antarctic ice discharge using response functions from SeaRISE ice-sheet models. *Earth Syst. Dyn.* **5**, 271–293 (2014).
40. Weaver, A. J. *et al.* Stability of the Atlantic meridional overturning circulation: a model intercomparison. *Geophys. Res. Lett.* **39**, L20709 (2012).
41. Boening, C., Willis, J. K., Landerer, F. W., Nerem, R. S. & Fasullo, J. The 2011 La Nina: so strong, the oceans fell. *Geophys. Res. Lett.* **39**, L19602 (2012).
42. Theuerkauf, E. J., Rodriguez, A. B., Fegley, S. R. & Luettich, R. A. Sea level anomalies exacerbate beach erosion. *Geophys. Res. Lett.* **41**, 5139–5147 (2014).
43. Holgate, S. J. *et al.* New data systems and products at the permanent service for mean sea level. *J. Coast. Res* **29**, 493–504 (2013).
44. Delworth, T. L. *et al.* Simulated climate and climate change in the GFDL CM2. 5 high-resolution coupled climate model. *J. Clim.* **25**, 2755–2781 (2012).
45. Winton, M. *et al.* Has coarse ocean resolution biased simulations of transient climate sensitivity? *Geophys. Res. Lett.* **41**, 8522–8529 (2014).
46. Vecchi, G. *et al.* On the seasonal forecasting of regional tropical cyclone activity. *J. Clim.* **27**, 7994–8016 (2014).
47. Delworth, T. L. *et al.* GFDL's CM2 global coupled climate models. Part I: formulation and simulation characteristics. *J. Clim.* **19**, 643–674 (2006).
48. Griffies, S. M. *et al.* The GFDL CM3 coupled climate model: characteristics of the ocean and sea ice simulations. *J. Clim.* **24**, 3520–3544 (2011).
49. Dunne, J. P. *et al.* GFDL's ESM2 global coupled climate-carbon Earth System Models Part I: physical formulation and baseline simulation characteristics. *J. Clim.* **25**, 6646–6665 (2012).

Acknowledgements

We thank Drs R. Stouffer and M. Winton for constructive comments. We thank Dr M. Winton for providing simulations of the ten GFDL models, and many research centres for providing the observation and modelling data. The work was supported by the NOAA Climate Program Office (grant number NA13OAR4310128).

Author contributions

P.G. and J.Y. designed the study, analysed the data and wrote the manuscript. S.M.G. provided the data and information of the ten GFDL models. S.Z. provided the GFDL reanalysis data. All authors made contributions to data interpretation and manuscript writing.

Additional information

7

Biological and physical controls in the Southern Ocean on past millennial-scale atmospheric CO_2 changes

Julia Gottschalk[1], Luke C. Skinner[1], Jörg Lippold[2], Hendrik Vogel[2], Norbert Frank[3], Samuel L. Jaccard[2] & Claire Waelbroeck[4]

Millennial-scale climate changes during the last glacial period and deglaciation were accompanied by rapid changes in atmospheric CO_2 that remain unexplained. While the role of the Southern Ocean as a 'control valve' on ocean–atmosphere CO_2 exchange has been emphasized, the exact nature of this role, in particular the relative contributions of physical (for example, ocean dynamics and air–sea gas exchange) versus biological processes (for example, export productivity), remains poorly constrained. Here we combine reconstructions of bottom-water $[O_2]$, export production and ^{14}C ventilation ages in the sub-Antarctic Atlantic, and show that atmospheric CO_2 pulses during the last glacial- and deglacial periods were consistently accompanied by decreases in the biological export of carbon and increases in deep-ocean ventilation via southern-sourced water masses. These findings demonstrate how the Southern Ocean's 'organic carbon pump' has exerted a tight control on atmospheric CO_2, and thus global climate, specifically via a synergy of both physical and biological processes.

[1] Godwin Laboratory for Palaeoclimate Research, Earth Sciences Department, University of Cambridge, Downing Street, Cambridge CB2 3EQ, UK. [2] Institute of Geological Sciences and Oeschger Center for Climate Change Research, University of Bern, Baltzerstr. 1-3, Bern 3012, Switzerland. [3] Institute of Environmental Physics, University of Heidelberg, Im Neuenheimer Feld 229, Heidelberg 69120, Germany. [4] Laboratoire des Sciences du Climat et de l'Environnement, LSCE/IPSL, CNRS-CEA-UVSQ, Université de Paris-Saclay, Domaine du CNRS, bât. 12, Gif-sur-Yvette 91198, France. Correspondence and requests for materials should be addressed to J.G. (email: jg619@cam.ac.uk).

The Southern Ocean is believed to play a key role in the global carbon cycle and millennial-scale variations in atmospheric CO_2 ($CO_{2,atm}$), which in turn may amplify the impacts of longer-term external climate forcing on global climate[1]. This role stems from the unique control the Southern Ocean is thought to exert on ocean–atmosphere CO_2 exchange[1-3] by both facilitating the upward transport of nutrient- and CO_2-rich water masses along outcropping density surfaces and their exposure to the atmosphere, and modulating the export of biologically fixed carbon into the ocean interior, where it is remineralized and may be effectively isolated from the atmosphere. It has been proposed that these two key aspects of the Southern Ocean's role in the marine carbon cycle may have exerted a dominant control on past $CO_{2,atm}$ change, for instance via variations of dust-driven biological carbon fixation in the sub-Antarctic[4], the extent of circum-Antarctic sea ice[5] impeding effective air–sea gas equilibration[6], and/or changes in the strength or position of southern hemisphere westerlies driving the residual overturning circulation in the Southern Ocean[7,8].

While all of these mechanisms for past $CO_{2,atm}$ change are compelling, observational evidence that might constrain the extent to which they have operated, in particular the balance of biological versus physical (that is, air–sea gas exchange or ocean dynamical) impacts, remains ambiguous. In the sub-Antarctic Atlantic north of the Polar Front (PF), decreased biological export production, along with a diminished aeolian supply of dust (and by inference iron) to the surface ocean, has been found to parallel millennial-scale increases in $CO_{2,atm}$. These observations suggest a significant impact of dust-driven variations of the strength of the 'organic carbon pump' on $CO_{2,atm}$ (refs 9–12). However, marked increases in $CO_{2,atm}$ are also accompanied by enhanced export productivity south of the PF (ref. 7). Polar- and sub-polar Southern Ocean export productivity changes thus appear to have opposed each other, raising questions concerning the overall magnitude and sign of the impact of Southern Ocean 'organic carbon pump' on $CO_{2,atm}$, when integrated across both regions[11,13]. On the other hand, while ^{14}C evidence has provided direct support for a link between Southern Ocean carbon sequestration (and millennial-scale $CO_{2,atm}$ variability) and physical/dynamical controls on air–sea CO_2 exchange[14], these data remain sparse and only extend across the last deglaciation.

Here we present sub-millennially resolved qualitative and quantitative proxy reconstructions of bottom-water [O_2] from sub-Antarctic Atlantic sediment core MD07-3076Q ($14°13.7'W$, $44°9.2'S$, 3,770 m water depth; Fig. 1) to estimate the apparent oxygen utilization (AOU) in deep waters, which is closely (stoichiometrically) related to the amount of remineralized dissolved inorganic carbon (DIC) because of the consumption of oxygen during the degradation of organic carbon. We use two independent proxy approaches: first, we determined the redox-sensitive enrichment of uranium and manganese in authigenic foraminifer coatings[15], and second, we measured the difference in carbon isotopic composition between pore waters at the zero-oxygen boundary and overlying bottom waters, which is assumed to be reflected in $\delta^{13}C$ of the benthic foraminifer *Globobulimina affinis* and *Cibicides kullenbergi*, respectively ($\Delta\delta^{13}C_{C.\ kullenbergi-G.\ affinis}$; refs 16,17). Our deep sub-Antarctic Atlantic [O_2] reconstructions show a close correlation to $CO_{2,atm}$ variations during the last deglacial- and glacial periods. The combination of our [O_2] reconstructions with analyses of [230]Th-normalized opal fluxes, an indicator of biological export production[7,18], and deep water ^{14}C ventilation ages, along with a robust age model for our study core[14,19,20] (Methods), highlights that carbon sequestration changes in the southern high latitudes cannot be attributed solely to changes in local biological export production. Instead, they involve significant changes in Southern

Ocean vertical mixing and air–sea gas exchange, having direct implications for millennial-scale $CO_{2,atm}$ variations, since 65,000 years before present (BP).

Results

Redox-sensitive U and Mn enrichment in foraminifer coatings. The uranium to calcium ratio in authigenic (that is, *in situ* generated) coatings (c), proposed to vary with changes in sedimentary redox-conditions, and therefore with bottom-water [O_2] (ref. 15), has been measured on weakly chemically cleaned ('host') calcium carbonate (cc) shells (hereafter referred to as U/Ca_{cc+c}) of the planktonic foraminifer *G. bulloides* and the benthic foraminifer *Uvigerina* spp. (Methods). The uranium concentration in the authigenic coatings of foraminifera strongly exceeds the concentration in the foraminiferal shell matrix[21,22]. Thus, the overall U/Ca_{cc+c} variability is marginally influenced by the uranium concentration in foraminifer shells, and has been proposed to primarily reflect coating-bound uranium variations instead that is inversely correlated with bottom-water oxygenation[15]. The co-variation of shell weights and U/Ca_{cc+c} levels of *G. bulloides*, however, indicates that shell size and/or wall thickness variations may bias U/Ca_{cc+c} ratios, via changes in the shell mass to surface-area ratio for example (Supplementary Fig. 1). The normalization of coating-bound uranium levels to manganese concentrations circumvents this bias for two reasons: manganese has generally an opposing redox-behaviour to that of uranium[23-25], and in particular manganese in weakly chemically cleaned foraminiferal tests mainly occurs in Fe-Mn-rich oxyhydroxides and/or Mn-rich carbonate overgrowths attached to the foraminiferal shell[22,26], which may be supported by the observed co-variation of Fe/Ca_{cc+c} and Mn/Ca_{cc+c} levels of *G. bulloides* (Supplementary Fig. 1). We propose that the U/Mn ratio of authigenic coatings in planktonic and benthic foraminifera, U/Mn_c, serves as reliable indicator of redox-conditions in marine sediments independent of shell matrix variations. The close agreement of planktonic and benthic foraminifer U/Mn_c suggests that it sensibly tracks early diagenetic redox-processes within the sediment consistent with previous findings[15] (Supplementary Fig. 2).

During the last glacial period, *G. bulloides* and *Uvigerina* spp. U/Mn_c are both found to vary with changes in $CO_{2,atm}$ (Fig. 2). During the last deglaciation, the large early deglacial decrease in U/Mn_c is clearly synchronous with the initial increase in $CO_{2,atm}$ before 15 kyr BP, while the second pulse in U/Mn_c in time with the $CO_{2,atm}$ increase during the following Antarctic warming period (that is, the northern-hemisphere Younger Dryas period) is more equivocal (Fig. 2). Our data are also in good agreement with changes in the authigenic enrichment of uranium in bulk sediments of Cape Basin core TN057-21 (ref. 27; location in Fig. 1), applying the most recent chronology of ref. 28 (Fig. 2). This suggests a basin-wide relevance of observed changes in sedimentary redox-conditions in the central sub-Antarctic Atlantic for variations in $CO_{2,atm}$.

Benthic foraminifer $\delta^{13}C$ gradients and bottom-water [O_2]. Redox-conditions in marine sediments generally reflect changes in organic carbon respiration within the sediment modulated by the downward diffusion of oxygen from bottom waters and/or the organic carbon supply to the sea floor[29]. Aerobic degradation of organic matter is the most efficient pathway of the respiration of organic carbon. Most of organic matter respiration therefore occurs above the sedimentary anoxic boundary. At the anoxic boundary, the diffusion of oxygen from the bottom water into the sediment is balanced by the rate of oxygen consumption during aerobic sedimentary organic carbon respiration in the sub-surface

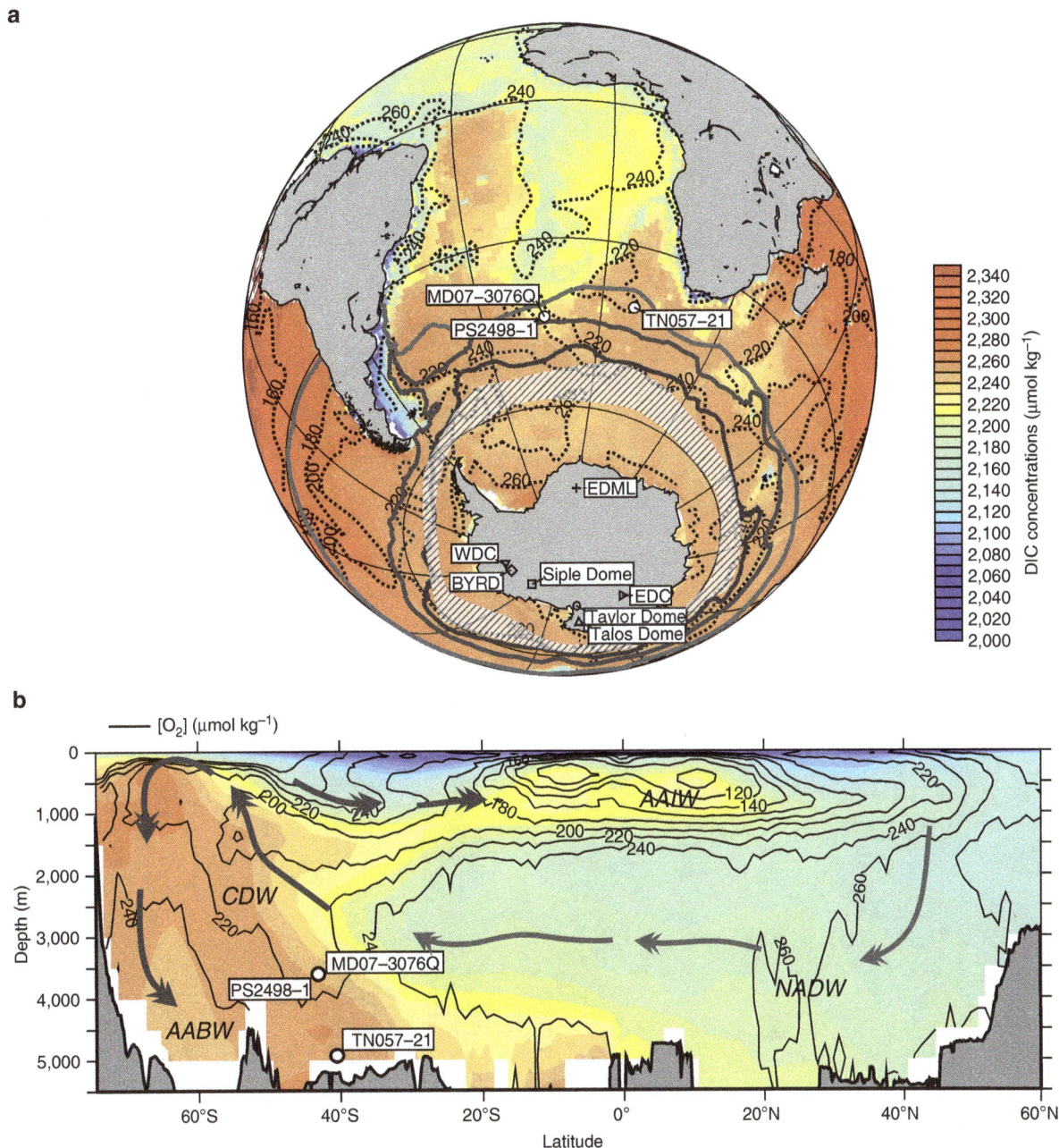

Figure 1 | Modern ocean DIC and oxygen concentrations. DIC levels (shaded) and [O$_2$] (contours, in µmol kg^{-1})[33,63] in (**a**) Southern Ocean- and Atlantic Ocean bottom waters and (**b**) in a meridional transect across the Atlantic (averaged between 70°W and 20°E). Hatched area broadly represents the region, where the deep DIC reservoir directly 'communicates' with the surface ocean and the atmosphere along steep density surfaces (equivalent to the area of strong positive CO$_2$ fluxes across the air–sea interface in austral winter in the Southern Ocean[64]), which is unique in the global ocean today. White circles show study cores and open symbols mark the location of ice cores that document past changes in atmospheric CO$_2$ (CO$_{2,atm}$; as in Figs 2 and 3). Thick lines show the modern positions of the PF, the sub-Antarctic Front (SAF) and the sub-Tropical Front (STF) (south to north)[65]. Arrows show general pathways of North Atlantic Deep Water (NADW), AABW (Antarctic Bottom Water), CDW (Circumpolar Deep Water) and Antarctic Intermediate Water (AAIW).

sediment column, such that [O$_2$] becomes zero. As organic carbon has typical δ^{13}C values of about −22 ‰, the release of ^{13}C-depleted carbon during the degradation of organic matter substantially drives the δ^{13}C gradient in marine sub-surface pore waters[30]. The total amount of aerobic sedimentary organic carbon respiration is thus a function of bottom-water [O$_2$] and is reflected in the δ^{13}C difference between bottom waters and pore waters at the zero-oxygen boundary[16,30].

The deep infaunal foraminifer *G. affinis* actively chooses the low-oxygen microhabitat near or at the anoxic boundary within

marine sub-surface sediments (in contrast to other benthic species)[31]. Assuming that *C. kullenbergi* δ^{13}C reflects bottom-water δ^{13}C (ref. 32), the offset of *G. affinis* δ^{13}C from bottom water (that is, *C. kullenbergi*) δ^{13}C thus sensitively records the relative depletion of pore-water δ^{13}C due to organic carbon respiraton[16,17,30] driven by the availability of oxygen in bottom waters. The occurrence of *G. affinis* in marine sediments may be in itself an indicator of an oxygen-limited sediment regime, where organic carbon is generally abundant and where the availability of oxygen is the main driver of organic matter respiration within the

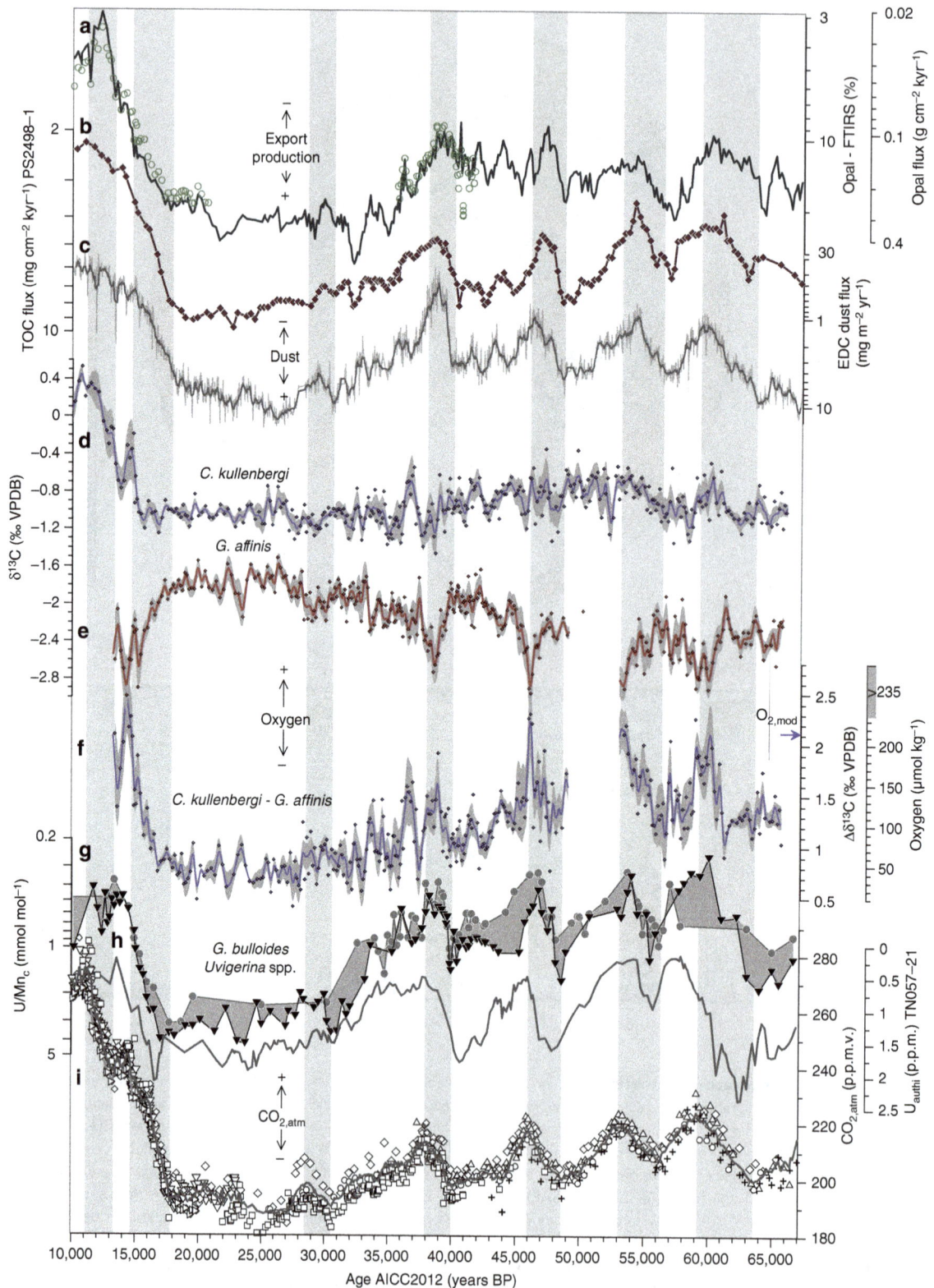

Figure 2 | Sub-Antarctic Atlantic bottom-water [O₂] and productivity changes during the last deglacial and glacial periods. (a) Sedimentary opal content (line) and ^{230}Thorium-normalized opal fluxes (circles), (b) flux of TOC in PS2498-1 (ref. 9; age scale adjusted as outlined in Methods), (c) Antarctic (EDC ice core) dust fluxes[66], (d) *C. kullenbergi* δ^{13}C (versus Vienna Pee Dee Belemnite (VPDB) standard), (e) *G. affinis* δ^{13}C (versus VPDB), (f) $\Delta\delta^{13}$C$_{C.\ kullenbergi-G.\ affinis}$ and corresponding bottom-water [O₂] (ref. 16), arrow shows modern [O₂] at the core site[33], (g) *G. bulloides* (circles) and *Uvigerina* spp. (triangles) U/Mn$_c$, (h) authigenic uranium concentrations in TN057-21 (ref. 27), (i) CO$_{2,atm}$ variations recorded in the Antarctic ice cores BYRD (diamonds)[67,68], EDML (crosses)[47,69], EDC (right pointed triangles)[70], Siple Dome (squares)[71], Talos Dome (triangles)[47], Taylor Dome (circles)[72] and WDC (inverted triangles)[73]. All data refer to the AICC2012 age scale[19,62]. Lines in d–f show 500 year-running averages with envelopes indicating the 500 year-window one-sigma standard deviation. Grey bars indicate periods of rising CO$_{2,atm}$.

sediment[29], because the characteristic zero-oxygen boundary in the shallow sub-surface of these sediments is the preferred habitat of *G. affinis*. The amount of pore water (that is, *G. affinis*) $\delta^{13}C$ depletion relative to bottom water (that is, *C. kullenbergi*) $\delta^{13}C$ is thus mostly insensitive to variations in organic carbon fluxes and scales instead with the amount of oxygen diffusing from the bottom water, allowing a quantification of bottom-water $[O_2]$ (refs 16,17).

In sediment core MD07-3076Q, *G. affinis* $\delta^{13}C$ becomes markedly depleted by up to 1‰ relative to bottom-water (*C. kullenbergi*) $\delta^{13}C$ during decreases in U/Mn_c (Fig. 2). The distinct negative offsets of *G. affinis* $\delta^{13}C$ from *C. kullenbergi* $\delta^{13}C$ mark millennial-scale increases in deep-water $[O_2]$ (ref. 16) in the deep sub-Antarctic Atlantic that closely track rises in $CO_{2,atm}$ during the last deglacial and glacial periods (Fig. 2).

According to the modern $\Delta\delta^{13}C$-$[O_2]$ calibration of ref. 16, bottom-water $[O_2]$ in the deep sub-Antarctic Atlantic would have reached a minimum of about $40 \pm 20\,\mu mol\,kg^{-1}$ during the peak glacial, which translates into a bottom-water $[O_2]$ reduction of $175 \pm 20\,\mu mol\,kg^{-1}$ from present-day levels of $\sim 215\,\mu mol\,kg^{-1}$ at the core site[33] (Fig. 2). During the last glacial period, that is, Marine Isotope Stage (MIS) 3, deep sub-Antarctic Atlantic $[O_2]$ would have varied between 90 ± 25 and $200 \pm 40\,\mu mol\,kg^{-1}$, in time with millennial-scale changes in $CO_{2,atm}$ (Fig. 2).

Our quantification of deep sub-Antarctic Atlantic $[O_2]$ relies on the assumption that bottom-water $\delta^{13}C$ is reliably reflected in *C. kullenbergi* $\delta^{13}C$. This species has mostly been employed to reconstruct bottom-water $\delta^{13}C$ in the southern high latitudes (because of the low abundance of other benthic epifaunal species); yet a difference of up to ~ 0.6 ‰ has been observed between sparse glacial *C. kullenbergi* $\delta^{13}C$- and glacial *C. wuellerstorfi* $\delta^{13}C$ measurements at ODP site 1090 in the Cape Basin[34]. This may imply that *C. kullenbergi* $\delta^{13}C$ is anomalously depleted, for example, due to a slight infaunal habitat during glacial times[34], and/or that $\delta^{13}C$ measured on episodically occurring *C. wuellerstorfi* is anomalously enriched, for example, due to an affinity to anomalously well-ventilated water masses[35] and/or low carbon fluxes. If *C. kullenbergi* $\delta^{13}C$ in MD07-3076Q does not adequately represent bottom-water $\delta^{13}C$ at our core site, then absolute bottom-water $[O_2]$ in the deep central sub-Antarctic Atlantic would be higher by up to $\sim 40\,\mu mol\,kg^{-1}$ per 0.3‰-deviation of glacial bottom-water $\delta^{13}C$ from glacial *C. kullenbergi* $\delta^{13}C$ observed in MD07-3076Q (Supplementary Fig. 3). However, our *C. kullenbergi* $\delta^{13}C$ data are consistent with glacial benthic foraminifer (*C. kullenbergi* and *Cibicidoides* spp.) $\delta^{13}C$ measurements from different locations throughout the South Atlantic[34,36,37], suggesting that they are representative of deep-water $\delta^{13}C$. Regardless of these quantitative uncertainties, the co-variation of the U/Mn_c- and $\Delta\delta^{13}C$-based $[O_2]$ reconstructions provides strong evidence for recurrent changes in deep sub-Antarctic oxygenation in parallel with $CO_{2,atm}$ over the last glacial and deglacial periods.

Changes in opal- and organic carbon fluxes. The flux of biogenic silica (opal) to marine sediments in the southern high latitudes is assumed to reflect changes in organic carbon flux to the sea floor and in the export of organic carbon from the euphotic zone (that is, export production)[9,38]. Variations in the weight percentages of opal observed in MD07-3076Q are tightly correlated with [230]Th-normalized opal fluxes ($R^2 = 0.94$, $P < 0.05$; Fig. 2; Supplementary Fig. 4), suggesting their accurate representation of past opal- (and therefore total organic carbon[9,38]; TOC) fluxes in the sub-Antarctic Atlantic. This is supported by synchronous variations in the TOC flux observed in the neighbouring core PS2498-1 (Fig. 2, location in Fig. 1), which

has been chronostratigraphically aligned to MD07-3076Q (Methods). As shown in Fig. 2, opal- and TOC fluxes in the sub-Antarctic Atlantic show a close link to dust flux variations in Antarctic ice cores and changes in dust supply to the sub-Antarctic region[9], which is consistent with earlier findings[9,10].

Estimates of radiocarbon ventilation ages. Two metrics for deep-water 'ventilation' (that is, deep ocean versus atmosphere gas/isotope equilibration) that provide a measure of the average time since carbon in the ocean interior last equilibrated with the atmosphere are considered here: [14]C age offsets between coexisting benthic (B) and planktonic (Pl) foraminifera (B-Pl ^{14}C ventilation ages), and benthic ^{14}C age offsets from contemporary atmospheric ^{14}C ages (B-Atm ^{14}C ventilation ages). While the first provide an estimate of deep-ocean ventilation relative to the local mixed layer, the latter provide a direct estimate of deep-ocean ventilation relative to the contemporary atmosphere. As shown in Fig. 3, B-Pl ^{14}C ventilation ages from sediment core MD07-3076Q broadly co-vary with changes in deep-ocean oxygenation (for example, with U/Mn_c: $R^2 = 0.31$, $P < 0.05$) and $CO_{2,atm}$ ($R^2 = 0.43$, $P < 0.05$), both statistically significant within the 95% significance interval (Supplementary Fig. 5). Parallel B-Atm ^{14}C ventilation age estimates agree with these observations, and confirm that B-Pl ^{14}C ventilation age fluctuations have not been significantly biased or masked by local surface-ocean radiocarbon disequilibrium effects (reservoir ages) (Fig. 3).

These findings are consistent with similar analyses in the central deep sub-Antarctic Atlantic for the last deglaciation[14]. Although B-Pl ^{14}C ventilation age variations are more strongly influenced by surface-ocean reservoir age variations during the last deglaciation, decreasing B-Atm ^{14}C age offsets are linked to deglacial increases in $CO_{2,atm}$, in particular during the early deglacial period[14].

Notably, absolute foraminifer ^{14}C ages appear to be slightly too young during the mid-glacial period, perhaps due to uncertainties associated with background corrections, which are especially important for old (> 30 kyr BP) sample material. In practice, these background corrections are based on one radiocarbon-dead spar calcite sample measured in each sample batch (that is, an accelerator mass spectrometry (AMS) sample carousel), whose apparent radiocarbon content is subtracted from the measured radiocarbon content of all the foraminifer samples measured in that sample carousel. If the true background deviates from the measured background in this single sample, then B-Atm ^{14}C and Pl-Atm ^{14}C age offsets may deviate significantly from their true absolute values. Godwin Radiocarbon Laboratory-internal backgrounds compiled for the 4 years from April 2011 to January 2015 amount to $^{14}C/^{12}C_0 = 5.3 \pm 1.5 \times 10^{-15}$ (Supplementary Fig. 6). Considering a one-off estimate of the background that is slightly smaller (that is, $^{14}C/^{12}C_0 = 4 \times 10^{-15}$; within 1 s.d. of the mean) or larger (that is, $^{14}C/^{12}C_0 = 6 \times 10^{-15}$; within 1 s.d. of the mean), this would result in B-Atm ^{14}C and Pl-Atm ^{14}C age offsets that are shifted towards slightly lower and higher absolute values respectively, without affecting the overall variability in each time-series (Supplementary Fig. 7). As benthic and planktonic ^{14}C ages have been obtained from the same AMS sample carousels in this study, B-Pl ^{14}C ventilation ages are not affected by these uncertainties and are essentially the same irrespective of the applied background correction (Supplementary Fig. 7). Therefore, while our absolute B-Atm ^{14}C and Pl-Atm ^{14}C age offsets are dependent on the accuracy of our background corrections (which are arguably difficult to assess), relative changes in B-Atm ^{14}C ventilation ages and absolute variations in B-Pl ^{14}C ventilation ages remain robust. As shown in Fig. 3, these clearly co-vary with our estimates of bottom-water

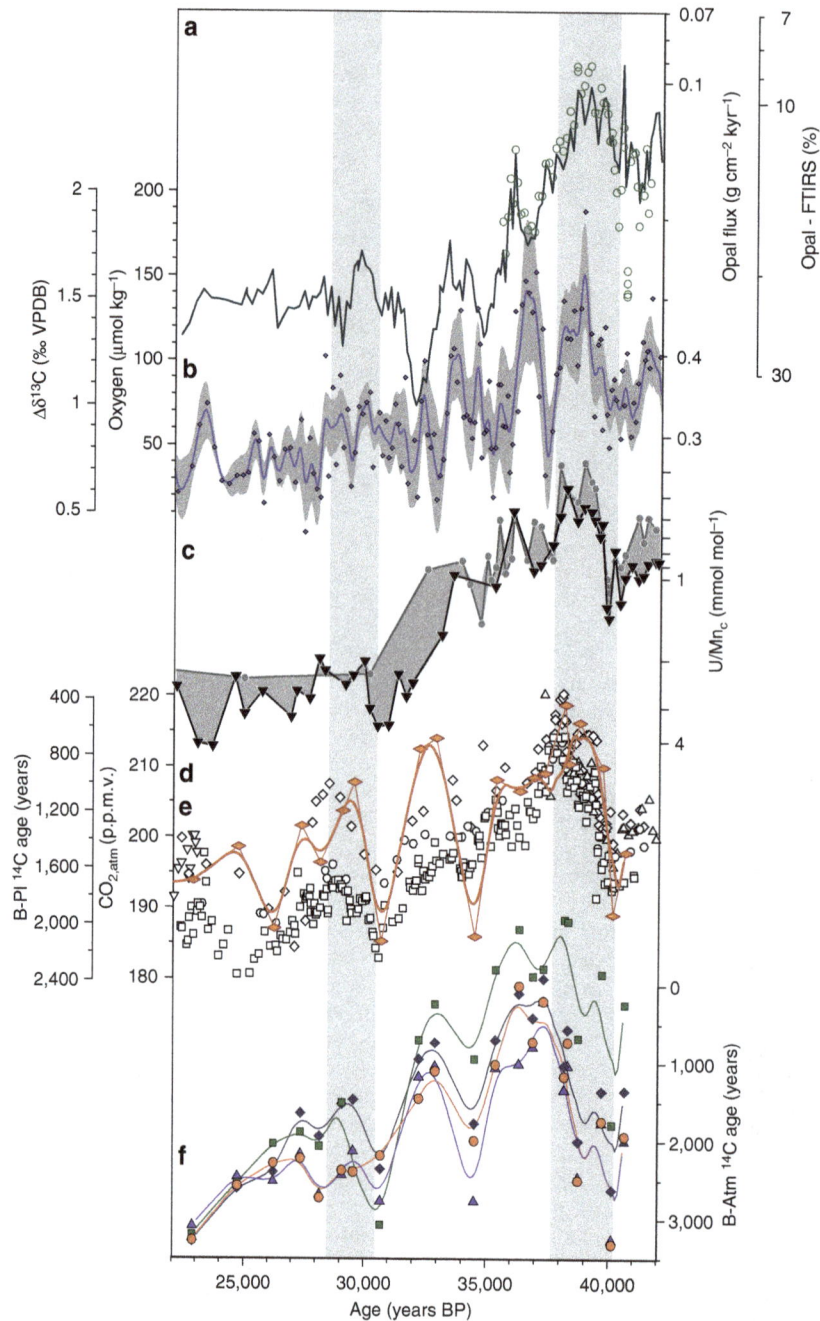

Figure 3 | Mid-glacial ventilation and carbon sequestration changes in the deep sub-Antarctic Atlantic. (a) Sedimentary opal content (line) and [230]Thorium-normalized opal fluxes (circles), (b) $\Delta\delta^{13}C_{C. kullenbergi-G. affinis}$ and corresponding bottom-water [O_2] (ref. 16), (c) *G. bulloides* (circles) and *Uvigerina* spp. (triangles) U/Mn$_c$, (d) Benthic-Planktonic (B-Pl) ^{14}C ventilation ages and the corresponding 1,000 years-running mean (thick line) plotted on top of (e) variations in $CO_{2,atm}$ recorded in the Antarctic ice cores (open symbols, refs as in Fig. 2), (f) benthic foraminifer ^{14}C age offset from atmospheric [14]C (Lake Suigetsu (green)[74], Cariaco Basin (orange)[75], Intcal09 (blue)[76] and Intcal13 (dark blue)[77]) shown as 1,000 years-running means (lines). Line and grey envelope in **b** show a 500 year-running average and the 500 year-window one-sigma standard deviation, respectively. Grey bars indicate periods of rising $CO_{2,atm}$.

oxygenation in the deep sub-Antarctic Atlantic (see also Supplementary Fig. 5).

Discussion

Changes in the elemental composition of foraminifer coatings and bottom-water versus pore-water $\delta^{13}C$ gradients, as described above, demonstrate that the amount of remineralized carbon sequestered in the deep sub-Antarctic Atlantic has varied

substantially and inversely with respect to millennial-scale $CO_{2,atm}$ changes (Fig. 2). These observations confirm a role for the Southern Ocean 'organic carbon pump' in regulating $CO_{2,atm}$ (refs 3,10,11). Below, we assess the quantitative impact of the inferred 'biological carbon pump' changes on $CO_{2,atm}$, as well as their governing biological and/or physical/dynamical controls.

Bottom-water [O_2] reconstructions at our core site via $\Delta\delta^{13}C_{C. kullenbergi-G. affinis}$ provide the basis for a quantification of the amount of respired carbon in the deep sub-Antarctic

Atlantic[16,30], provided the modern $\Delta\delta^{13}C$-$[O_2]$ relationship holds for the past. In principle, seawater $[O_2]$ consists of a saturated $[O_2]$ component ($[O_2]_{sat}$) arising from the solubility-controlled O_2 exchange between the atmosphere and the surface ocean, a biological $[O_2]$ component associated with the release and drawdown of $[O_2]$ during photosynthesis and respiration ($[O_2]_{bio}$), and a preformed disequilibrium $[O_2]$ component ($[O_2]_{diseq}$) due to inefficiencies in air–sea gas exchange ($[O_2]_{in\,situ} = [O_2]_{sat} + [O_2]_{bio} + [O_2]_{diseq}$)[39]. Assuming that ocean $[O_2]$ is in equilibrium with the atmosphere ($[O_2]_{diseq} \sim 0$) and that last glacial ocean $[O_2]_{sat}$ was slightly higher than today ($[O_2]_{sat,modern} = 345\,\mu mol\,kg^{-1}$) mostly due to a decrease in ocean temperature ($[O_2]_{sat,glacial} = 360\,\mu mol\,kg^{-1}$; Methods), the amount of $[O_2]$ depletion at ocean depth ($AOU = -[O_2]_{bio} = [O_2]_{sat} - [O_2]_{in\,situ}$) should scale with the formation of respired carbon according to a constant stoichiometric Redfield ratio of $C:[O_2] = 117 : -170$ (ref. 40). Changes in bottom-water $[O_2]$ (and AOU) in the sub-Antarctic Atlantic would therefore provide a direct quantitative measure of the amount of carbon sequestered in the southern high-latitude ocean, and thus the efficiency of the biological 'organic carbon pump'[16].

Converting our AOU estimates into respired carbon concentrations ($AOU_{Holocene} = (345 - 215)\,\mu mol\,kg^{-1}$, ($AOU_{MIS2} = (360 - 40) \pm 20\,\mu mol\,kg^{-1}$, $\Delta AOU_{Deglaciation} = 190 \pm 20\,\mu mol\,kg^{-1}$) based on the Redfield ratio of $C:[O_2] = 117 : -170$ (ref. 40) gives a respired DIC contribution of $220 \pm 14\,\mu mol\,kg^{-1}$ to the total DIC pool at the core site during the last glacial maximum (LGM). This is higher by $130 \pm 14\,\mu mol\,kg^{-1}$ compared with the Holocene[33], indicating greater respired carbon accumulation during the LGM. During millennial-scale variations in $CO_{2,atm}$ during the last glacial, respired carbon levels varied by $75 \pm 28\,\mu mol\,kg^{-1}$ between $110 \pm 28\,\mu mol\,kg^{-1}$ (during peak $CO_{2,atm}$ levels) and $185 \pm 14\,\mu mol\,kg^{-1}$ (during minimum $CO_{2,atm}$ levels) assuming that $[O_2]_{sat}$ was not significantly different from LGM levels (that is, $AOU_{MIS3\,'CO_2\,max'} = (360 - 200) \pm 40\,\mu mol\,kg^{-1}$, $AOU_{MIS3\,'CO_2\,min'} = (360 - 90) \pm 25\,\mu mol\,kg^{-1}$, $\Delta AOU_{MIS3} = 110 \pm 40\,\mu mol\,kg^{-1}$).

If we assume that the respired carbon lost from the deep sub-Antarctic Atlantic, where it was sequestered away from the atmosphere was transferred to a non-respired marine carbon pool that in turn equilibrated with the atmosphere via a surface-ocean DIC 'buffer factor' (that is, Revelle factor) of ~ 10 (ref. 41; Methods), our AOU and respired carbon estimates from the deep sub-Antarctic may only explain the full amplitude of observed $CO_{2,atm}$, if they are representative of a significant fraction of the global deep ocean, that is, at least $\sim 33\%$ during the mid-glacial period and $\sim 45\%$ during the early deglaciation (Methods). This would roughly correspond to the deep ocean below 2.9 and 2.3 km, respectively (Methods). These depths broadly agree with the depth of the putative glacial 'chemical divide' (~ 3 km water depth in the Atlantic)[42], and are supported by qualitative proxy data showing a decrease in oxygenation and radiocarbon ventilation in the global ocean below 2 km during the last peak glacial period[43,44]. The smaller the volume of the global deep ocean that experienced similar changes in AOU and respired carbon to our sub-Antarctic Atlantic site, the smaller the likely oceanic impact on $CO_{2,atm}$ concentrations.

Our calculations have two major caveats. First, we ignore possible open-system effects due to the interaction of deep waters with sediments, and second, we may have underestimated glacial deep sub-Antarctic Atlantic $[O_2]$, in the case that deep sub-Antarctic Atlantic $C.\ kullenbergi$ $\delta^{13}C$ values are strongly negatively biased versus bottom-water $\delta^{13}C$. Any open-system effects involving a degree of 'carbonate compensation' (on multi-millennial timescales) would tend to enhance the impact of

marine respired carbon inventory changes on $CO_{2,atm}$ (Methods). If true glacial bottom-water $[O_2]$ were higher than estimated in MD07-3076Q for instance via an anomalous depletion of $C.\ kullenbergi$ $\delta^{13}C$ from bottom-water $\delta^{13}C$ by 0.3‰, as mentioned above, deglacial changes in respired carbon and AOU (and thus the oceanic impact on $CO_{2,atm}$) would be reduced by $\sim 20\%$, as LGM AOU values would be lower (Supplementary Fig. 3). In contrast, our estimates of respired carbon changes during the last mid-glacial period remain to a large extent similar as they are based on relative $[O_2]$ changes (Supplementary Fig. 3). Our calculations are rough estimates that are intended only to provide a first indication of the potential impact of our observed marine carbon sequestration changes on $CO_{2,atm}$. To determine the full impact of changes in deep-ocean respired carbon levels on $CO_{2,atm}$ concentrations, our estimates would need to be corroborated by further reconstructions of past bottom-water oxygen- and DIC concentrations from throughout the global ocean, in particular in the volumetrically most significant Pacific Ocean.

The analysis above demonstrates the potential quantitative significance of the oxygenation changes that we observe, and more specifically of the role of the Southern Ocean 'organic carbon pump' in regulating $CO_{2,atm}$ (refs 3,10,11). However, it remains to be shown whether the observed decreases in 'organic carbon pump' efficiency resulted primarily from decreases in export productivity (allowing oxygen to increase due to reduced organic carbon remineralization in the ocean interior) or primarily from increases in ocean 'ventilation' (causing carbon loss to the atmosphere with direct oxygen gain of the ocean interior). Below, we address this question by reference to our export productivity- and ^{14}C ventilation age estimates.

The observed correlation between changes in the dust supply to the southern high-latitude regions and in export production in the central sub-Antarctic Atlantic (as recorded by variations in opal- and TOC fluxes[9]; Fig. 2) supports earlier findings of a dust-driven biological organic carbon pump in the sub-Antarctic Atlantic[9,10]. The close relationship between variations in sub-Antarctic export production and $CO_{2,atm}$ changes (Fig. 2) would be consistent with a significant impact of the efficiency of the sub-Antarctic biological organic carbon pump on surface-ocean DIC levels, and thus on $CO_{2,atm}$ (refs 9–11).

The correlation between opal- and TOC fluxes and bottom-water $[O_2]$ in the sub-Antarctic Atlantic (Fig. 2) may point to a role of organic carbon respiration at depth driving deep sub-Antarctic Atlantic bottom-water oxygenation. To test whether export production was the major driver of our observed deep-ocean $[O_2]$ changes (and therefore of the associated changes in deep-ocean respired carbon sequestration), we make use of the unique microhabitat of $G.\ affinis$ near the anoxic boundary in marine sediments and the associated mechanisms that drive its $\delta^{13}C$ signature. Notably, negative excursions of $G.\ affinis$ $\delta^{13}C$ are observed during each of the marked $CO_{2,atm}$ rises during MIS 3 and the last deglaciation. These excursions indicate that total organic carbon respiration within deep sub-Antarctic Atlantic sediments increased at times of reduced opal- and TOC fluxes, that is, reduced export production (Fig. 2). An increase in sedimentary organic carbon respiration (that is, pore water/ $G.\ affinis$ $\delta^{13}C$ depletions) would be driven by an increase in organic carbon flux, an increase in bottom-water $[O_2]$, or both of these together. As the first is evidently not the case (Fig. 2), we conclude that sedimentary carbon respiration must have instead been driven by enhanced deep-ocean 'ventilation' (that is, circulation/ convection rates and/or air–sea gas exchange) supplying oxygen to the deep sub-Antarctic Atlantic.

Alternatively, a decreased oxygen demand in bottom waters due to diminished organic carbon fluxes and less respiration of

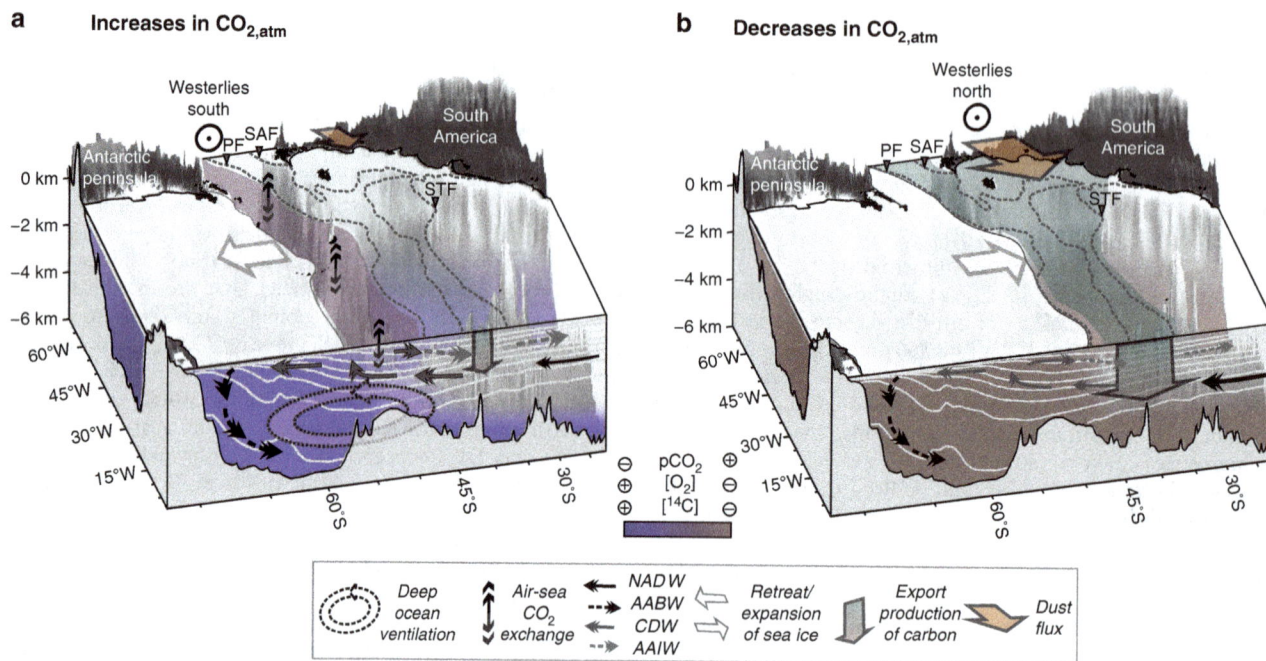

Figure 4 | Schematic view on the southern high-latitude Atlantic during millennial-scale $CO_{2,atm}$ variations based on new and existing proxy evidence.
(**a**) Dust-driven decreases of export production in the sub-Antarctic Atlantic[9,10] during the last glacial and deglacial periods were accompanied by decreases in deep carbon storage in the Southern Ocean (this study and ref. 48). The latter was further promoted by increases in the air–sea CO_2 exchange south of the PF and in the ventilation of the deep carbon pool (this study and ref. 48), causing millennial-scale increases in $CO_{2,atm}$, as postulated earlier[3,7]. (**b**) Enhanced dust-driven, biological export of carbon to the deep sub-Antarctic Atlantic[9,10] paralleled increases in deep Southern Ocean respired carbon levels during the last glacial period and the last deglaciation (this study and ref. 48). The enhanced Southern Ocean carbon pool was effectively isolated from the atmosphere by decreases in air–sea CO_2 equilibration in the Antarctic region and a poor 'ventilation' of the deep-ocean during these times (this study and ref. 48), leading to decreases in $CO_{2,atm}$ during the last 70,000 years, as proposed previously[3,7]. Accompanying changes in sea ice[5,6] and the westerly position/strength[7,8] are debated and remain speculative. The modern positions of ocean fronts (as in Fig. 1) and ocean density surfaces (white lines) are shown as reference.

organic matter in a benthic 'fluff' layer could facilitate the diffusion of oxygen into the sediment, and drive the *G. affinis* $\delta^{13}C$ signal more negative. However, a poor inverse correlation between epibenthic and deep infaunal benthic foraminifer $\delta^{13}C$ over past millennial timescales (Fig. 2; $R^2 = 0.0001$, $N = 258$, Supplementary Fig. 8) would appear to rule out this scenario. We therefore conclude that the observed changes in 'organic carbon pump' efficiency and deep sub-Antarctic carbon storage were not only controlled by changes in export productivity but must also have involved biology-independent processes that contributed to past $CO_{2,atm}$ changes specifically by enhancing ocean–atmosphere CO_2 exchange in the Antarctic region (Fig. 2).

Our interpretation is confirmed by parallel estimates of deep-water ^{14}C 'ventilation ages' (Fig. 3). We observe that the marked $CO_{2,atm}$ rise around 38 kyr BP is paralleled by a decrease in B-Atm ^{14}C ventilation ages of $\sim 2,000$ ^{14}C years. A consistent link between deep-ocean (B-Atm and B-Pl) ^{14}C ventilation and $CO_{2,atm}$ variability is further supported by a high and statistically significant correlation coefficient between them (up to $R^2 = 0.6$, $P < 0.05$; Supplementary Fig. 5). The good correlation between (B-Atm and B-Pl) ^{14}C ventilation ages, deep-water $[O_2]$ and $CO_{2,atm}$ provides strong independent support for changes in the air–sea equilibration of deep waters in the Southern Ocean and their link to changes in respired carbon storage.

It has previously been shown that the incursion of well-ventilated northern-sourced waters into the sub-Antarctic Atlantic was reduced during intervals of rising $CO_{2,atm}$ (refs 20,28). On this basis, the periods of increased ^{14}C ventilation that we observe would therefore specifically reflect periods of increased local dominance of southern-sourced deep

waters and an 'improvement' of their ventilation state. Numerous processes have been suggested to have caused changes in vertical mixing in the southern high latitudes, including for instance the intensity and/or the position of the southern hemisphere westerlies[7,8], a retreat of circum-Antarctic sea ice[6], a decline in the formation and advection of northern component waters[45] and/or changes in surface buoyancy fluxes[46]. It remains currently impossible to evaluate the relative importance of these specific processes and their controls on $CO_{2,atm}$. Nevertheless, the strong co-variations of our abyssal oxygenation and ventilation proxies with $CO_{2,atm}$ confirm that some combination of dynamical (that is, residual circulation and shallow mixing) and/or physical (gas exchange efficiency) processes in the southern high-latitude region indeed had a significant impact on deep-ocean carbon sequestration[3,7,19,45,47] (Fig. 4).

Furthermore, our findings are entirely consistent with recently published sedimentary redox-sensitive trace element data from the Antarctic Zone of the Atlantic Ocean[48]. These data show that the accumulation of authigenic uranium (and therefore oxygenation) in the Antarctic Atlantic is generally inversely correlated with opal fluxes (that is, organic carbon fluxes) over the past 80,000 years, ruling out a dominant control of local surface-ocean productivity on deep Antarctic Atlantic $[O_2]$ and deep-ocean respired carbon levels south of the PF (ref. 48). The combination of our sub-Antarctic study with the Antarctic study of ref. 48 provides strong evidence for millennial-scale changes in the respired carbon concentrations across the entire deep high-latitude South Atlantic, varying in parallel with $CO_{2,atm}$ during the last glacial period and deglaciation, and for a significant impact of physical 'ventilation' processes (that is, overturning circulation, mixing and/or air–sea gas exchange) on changes in deep-ocean respired carbon sequestration and millennial-scale $CO_{2,atm}$ in the past.

In conclusion, our results show that pulses of $CO_{2,atm}$ during the last glacial- and deglacial periods coincided with increases in the ventilation of the southern high-latitude deep ocean (specifically via regions of deep-water formation in the Southern Ocean[7,48]), in addition to reductions in sub-Antarctic export productivity. By ruling in a role for variations in both the strength and the efficiency of the biological carbon pump via changes in the biological carbon export as well as the air–sea CO_2 exchange and Southern Ocean vertical mixing, the findings reconcile two opposing theories for the Southern Ocean's role in past millennial-scale $CO_{2,atm}$ variability[3,7,10–12,47]. Further work, for example using numerical model simulations will be required to quantify more precisely the contributions of (sub-polar zone) biological export productivity changes and (polar zone) physical/dynamical changes to deep-ocean carbon sequestration, as well as their down-stream effects on low-latitude export production[49]. Nevertheless, our data emphasize that while biological carbon export to the deep ocean is ultimately what permits ocean dynamics and air–sea exchange to impact on $CO_{2,atm}$ by continually tending to 'recharge' the abyssal carbon pool, the rate of equilibration of the deep ocean with the atmosphere will ultimately determine whether or not the biological 'organic carbon pump' is efficient or not at sequestering CO_2 (Fig. 4). Thus, ocean physics and marine biology acted together, synergistically, to repeatedly nudge the Southern Ocean from carbon sink to carbon source, with a direct impact on global climate over the last $\sim 65,000$ years.

Methods

Regional setting and chronology. Sediment core MD07-3076Q (14°13.7'W, 44°9.2'S, 3,770 m water depth) is bathed in Lower Circumpolar Deep Water, which is formed by the entrainment of northward spreading DIC- and preformed nutrient-rich Circumpolar Deep Water into southward flowing DIC-low and regenerated nutrient-rich North Atlantic Deep Water[50]. Chronological control of sediment core MD07-3076Q is based on ^{14}C measurements of mono-specific planktonic foraminifer samples, which have been adjusted for variations in surface-ocean reservoir ages[14]. The ^{14}C-based age constraints are complemented by the stratigraphic alignment of abundance variations of the cold-water species *Neogloboquadrina pachyderma* (sinistral-coiling) with rate changes in Antarctic temperature over time[19]. Age model uncertainties, mainly a function of age marker density, amount to $1,600 \pm 500$ years during the last glacial period and to $1,200 \pm 400$ years after 27 kyr BP (ref. 19). Resulting sedimentation rates range between 5 cm kyr^{-1} during the last deglaciation and 15 cm kyr^{-1} during MIS 3.

Element composition of authigenic foraminifer coatings. Down-core measurements of U/Ca$_{cc+c}$ and U/Mn$_c$ have been made on 18–25 specimens of the planktonic foraminifer *G. bulloides* (250-300 µm size fraction) and the 5–13 specimens of the benthic infaunal foraminifer *Uvigerina* spp. (250-300 µm size fraction). Foraminifera have been weakly chemically cleaned (clay removal and silicate picking) to maintain foraminiferal coatings but to remove extraneous detritus[15]. Cleaned foraminifera have been dissolved in 0.1 M nitric acid for inductively coupled plasma (ICP)-atomic emission spectroscopy analyses. The samples were subsequently re-diluted to 10 p.p.m. Ca^{2+} concentration and elemental concentrations have been determined by ICP-mass spectrometry[15]. Mean s.d. of U/Mn$_c$ of six duplicate samples is 0.08 ± 0.06 mmol mol^{-1}. Given the high sedimentation rates of 15 cm kyr^{-1}, the impact of potential sedimentary re-oxidation processes ('burn-down' effects) of already precipitated uranium complexes is negligible for the interpretation of U/Ca$_{cc+c}$ and U/Mn$_c$ ratios.

Reconstruction of bottom- to pore-water δ^{13}C gradients. Stable isotopic analyses on *G. affinis* and *C. kullenbergi* have been performed on 1–4 specimens (>150 µm size fraction) on Finnigan $\Delta +$ and Elementar Isoprime mass spectrometers. The results are reported with reference to the international Vienna Pee Dee Belemnite (VPDB) standard. VPDB is defined with respect to the NBS-19 calcite standard. The mean external reproducibility of carbonate standards is $\sigma \pm 0.03$ ‰.

In MD07-3076Q, $\Delta\delta^{13}$C$_{C.\ kullenbergi-G.\ affinis}$ has been determined from δ^{13}C measurements of benthic foraminifera from the same sediment sample, and has been converted into bottom-water [O$_2$] after ref. 16. The calibration error

associated with bottom-water [O$_2$] reconstructions using this method is ± 17 µmol kg^{-1} (ref. 16). Analytical uncertainties of benthic δ^{13}C analyses (two-sigma) translate into a bottom-water [O$_2$] uncertainty of ± 8 µmol kg^{-1}. We have smoothed our high-resolution record by a running 500 year-window (solid line in Fig. 2) to reduce such biases and those from intra-specific δ^{13}C variations. Mean bottom-water [O$_2$] have been determined for the LGM (23-18 kyr BP) as well as CO$_2$ minima (40.2-39.9 kyr BP, 48.4-47.6 kyr BP, 56.7-55.7 kyr BP, 63.6-63.0 kyr BP) and -maxima (38.8-38.0 kyr BP, 46.3-45.8 kyr BP, 53.6-53.3 kyr BP and 59.3-58.8 kyr BP) during MIS 3. Errors reported in our study are one-sigma standard deviations of our bottom-water [O$_2$] estimates during these periods.

Calculation of deep-ocean and atmospheric carbon budgets. [O$_2$] saturation levels are calculated according to ref. 51 assuming a glacial increase in salinity from present-day (~ 35 p.s.u.) by ~ 2 p.s.u. and a decrease in deep-ocean temperatures from modern-day values (~ 1 °C) by 2 °C in the deep Southern Ocean[52]. [O$_2$] saturation in the glacial deep Southern Ocean increased by ~ 15 µmol kg^{-1} from modern-day levels (~ 345 µmol kg^{-1}) (ref. 33).

To estimate the amount of carbon that is transferred to the atmosphere from the ocean's remineralized carbon pool (sequestered in the deep ocean), via the ocean's non-remineralized carbon pool (in equilibrium with the atmosphere), we adopt the conceptual framework of ref. 41, whereby:

$$\frac{dpCO_2}{pCO_2} = -0.0053\,\Delta c_{soft} + 0.0034\,\Delta c_{carb} \tag{1}$$

Here, Δc_{soft} and Δc_{carb} are DIC changes for the ocean's total remineralized carbon pool (that is not in equilibrium with the atmosphere), due to changes in the soft-tissue pump and the carbonate pump (for instance via changes in the export of organic carbon and carbonate to the ocean interior), respectively. Our estimate of Δc_{DIC} during the last deglacial increase in $CO_{2,atm}$ ($\Delta c_{DIC} = 130 \pm 14$ µmol kg^{-1}) and during mid-glacial $CO_{2,atm}$ changes ($\Delta c_{DIC} = 75 \pm 28$ µmol kg^{-1}) determined above from oxygenation estimates provides an estimate of Δc_{soft}, during these time intervals, and we assume that the associated Δc_{soft} is approximately three times smaller (for example, as observed spatially in the modern ocean)[41], yielding:

$$\frac{dpCO_2}{pCO_2} = -0.004167\,\overline{\Delta c_{DIC}} \tag{2}$$

where $\overline{\Delta c_{DIC}}$ is the whole-ocean average change in remineralized carbon during the investigated time intervals. It is given by the product of the change observed at our core location and the fraction (f) of the total ocean volume that also experienced this magnitude of change:

$$\overline{\Delta c_{DIC}} = f\Delta c_{DIC} \tag{3}$$

Assuming that the rest of the ocean volume experienced no significant change in respired DIC, remaining well-equilibrated with the atmosphere, the fraction of the ocean f, and therefore the deep-ocean volume V_d and the upper 'boundary' of the deep-ocean z', may be calculated that would account for the last early deglacial and mid-glacial atmospheric pCO$_2$ changes of ~ 50 and ~ 20 p.p.m. (for glacial background pCO$_2$ levels of 190–200 p.p.m.), if affected by similar changes in AOU and respired DIC levels as our sub-Antarctic Atlantic core site.

We have calculated the deep-ocean volume V_d and z' based on the GEBCO bathymetric data set (excluding the Arctic Ocean) archived by the British Oceanographic Data Centre (http://www.gebco.net/), according to:

$$V_d = \sum\left(\left|\left(\frac{\pi\cos(\phi)r\Delta\phi}{180}\right)\right|*\left(\frac{\pi r\Delta\lambda}{180}\right)*(z-z')\right) \tag{4}$$

that is the sum of all volumes of grid boxes (distance in west-east direction (km) times distance north-south direction (km) times depth), where ϕ is latitude, λ is longitude, $\Delta\phi$ and $\Delta\lambda$ represent the grid spacing of the bathymetric data set, r is the Earth's radius, z the water depth and z' the upper limit of the deep ocean.

Opal measurements. Opal concentrations were measured on ~ 400 samples by means of Fourier transform infrared (FTIR) spectroscopy[53] using a Vertex 70 FTIR-spectrometer (Bruker Optics Inc.) at the Institute of Geological Sciences at the University of Bern (CH). The FTIR spectra have been independently calibrated based on FTIRS analyses of artificial sand/opal mixtures[54]. Opal concentrations determined by means of FTIR spectroscopy show excellent agreement with conventional photometric-based[55] opal concentration determinations ($R^2 = 0.91$; Supplementary Fig. 4) that have been performed on one quarter of the total number of samples ($N = 101$). However, an increasing offset between photometric and FTIR-based opal measurements towards increasing opal values (Supplementary Fig. 4) might point at incomplete alkaline opal dilution during photometric measurements[55], potentially caused by a significant fraction of radiolarian skeletons in MD07-3076Q sediments[56].

Opal fluxes have been determined by normalizing the opal data with measured ^{230}Th concentrations[57]. For these analyses, U- and Th- isotopes were analysed by means of ICP-quadrupole mass spectrometry (iCAP-Q ICP-MS, ThermoFisher) at the Institute for Environmental Physics in Heidelberg, Germany. The contribution of detrital ^{230}Th has been estimated by assuming a ^{238}U/^{232}Th ratio of 0.6 and a correction[58] for the detrital ^{234}U/^{238}U not in secular equilibrium of 0.96. The quality of the analyses and the sample digestion and purification process has been monitored by blanks, certified UREM-11 standard material and replicate measurements of samples. Full replicates ($N = 5$) yielded an average uncertainty of 2.8 % (two-sigma) of the excess ^{230}Th concentrations (Supplementary Table 1). The chosen parameter set for the measurements of marine sediments applied here for the first time using an iCAP-Q ICP-MS (Supplementary Tables 1 and 2) puts emphasis on time efficiency for high-matrix sample analyses.

Radiocarbon measurements. The previously published set of foraminiferal ^{14}C dates in sediment core MD07-3076Q (ref. 14) has been extended by additional paired ^{14}C measurements of mixed benthic and mono-specific planktonic foraminifera (*N. pachyderma* s.). The conventional ^{14}C ages are reported in Supplementary Tables 3 and 4. The mean ^{14}C age uncertainty of the new ^{14}C data set amounts to 650 ± 270 ^{14}C years (Supplementary Table 3).

Foraminifer samples had a mean weight of 5.1 ± 1.0 mg, and weighed always more than 3.4 mg. They have been gently cleaned in methanol, and were subsequently transferred to sealed septum vials after they were completely dry. After evacuation 0.5 ml dry phosphoric acid has been injected into the vials. The acid-carbonate reaction has been sustained for at least 0.5 h at 60 °C. The CO_2 samples were graphitized in the Godwin Radiocarbon Laboratory at the University of Cambridge (UK), along with standards and radiocarbon-dead spar calcite (backgrounds), following a standard hydrogen/iron catalyst protocol[59]. Pressed graphite targets were subsequently analysed by AMS at the ^{14}Chrono Centre, University of Belfast (UK). Measured ^{14}C ages have been corrected for mass-dependent fractionation (normalization to $\delta^{13}C = -25‰$) and the background radiocarbon content by analysing radiocarbon-dead spar calcite with each sample batch. Paired planktonic and benthic samples have been measured in the same AMS sample carousel.

Four paired measurements have resulted in younger benthic than planktonic foraminifera (Supplementary Fig. 5). We have omitted these data from the initial analyses, but including these samples does not alter the general trend of the data (Supplementary Fig. 5).

Correlation of marine proxy records with $CO_{2,atm}$ variations. To calculate correlation coefficients R^2 between $CO_{2,atm}$ variations and ^{14}C-based deep sub-Antarctic ventilation ages during the last glacial period, that is, 41-22 kyr BP (Supplementary Fig. 5e), we interpolated the mean $CO_{2,atm}$ record[19] at the sampling resolution of the ^{14}C proxy data. Similarly, the mean $CO_{2,atm}$ has been interpolated at the resolution of the mean U/Mn$_c$- and the $\Delta\delta^{13}$C-based [O_2] records in order to estimate the correlation (R^2) between changes in bottom-water oxygenation and ^{14}C ventilation in the deep sub-Antarctic Atlantic (Supplementary Fig. 5f,g). For these calculations, the mean U/Mn$_c$ has been obtained by averaging *G. bulloides* and *Uvigerina* spp. U/Mn$_c$ (stippled line in Supplementary Fig. 2a) and the $\Delta\delta^{13}$C-derived [O_2] record is based on a 500 year-running average (solid line in Fig. 2f).

Chronostratigraphy of other sub-Antarctic Atlantic cores. The most recent age model of sediment core PS2498-1 has been established based on an alignment of variations in lithogenic fluxes with the EPICA Dome C dust record[9]. Because sediment cores MD07-3076Q and PS2498-1 are in close proximity (Fig. 1), we have compared the magnetic susceptibility records and noticed stratigraphic offsets of ± 900 years. To allow a faithful inter-core comparison, we have adjusted the chronology of PS2498-1 by aligning the magnetic susceptibility record of PS2498-1 (ref. 60) to the magnetic susceptibility record of MD07-3076Q, which has been measured with the GEOTEK Multi-Sensor-Core-Logger aboard *R/V Marion Dufresne* using a low field susceptibility (Bartington) sensor. For TN057-21, we rely on the most recently established chronology of ref. 28, which is based on the GICC05 age scale[61] that is equivalent to the AICC2012 age scale used in this study within decades to few hundred years[62].

References

1. Sigman, D. M. & Boyle, E. A. Glacial/interglacial variations in atmospheric carbon dioxide. *Nature* **407**, 859–869 (2000).
2. Ito, T. & Follows, M. J. Preformed phosphate, soft tissue pump and atmospheric CO_2. *J. Mar. Res.* **63**, 813–839 (2005).
3. Schmittner, A. & Galbraith, E. D. Glacial greenhouse-gas fluctuations controlled by ocean circulation changes. *Nature* **456**, 373–376 (2008).
4. Martin, J. H. Glacial-interglacial CO_2 change: the iron hypothesis. *Paleoceanography* **5**, 1–13 (1990).
5. Ferrari, R. *et al.* Antarctic sea ice control on ocean circulation in present and glacial climates. *Proc. Natl. Acad. Sci. USA* **111**, 8753–8758 (2014).
6. Stephens, B. B. & Keeling, R. F. The influence of Antarctic sea ice on glacial-interglacial CO_2 variations. *Nature* **404**, 171–174 (2000).
7. Anderson, R. F. *et al.* Wind-driven upwelling in the Southern Ocean and the deglacial rise in atmospheric CO_2. *Science* **323**, 1443–1440 (2009).
8. Toggweiler, J. R., Russell, J. L. & Carson, S. R. Midlatitude westerlies, atmospheric CO_2, and climate change during the ice ages. *Paleoceanography* **21**, 2005 (2006).

9. Anderson, R. F. *et al.* Biological response to millennial variability of dust and nutrient supply in the Subantarctic South Atlantic Ocean. *Philos. Trans. R. A Math. Phys. Eng. Sci.* **372**, 20130054 (2014).
10. Martínez-García, A. *et al.* Iron fertilization of the Subantarctic Ocean during the last ice age. *Science* **343**, 1347–1350 (2014).
11. Jaccard, S. L. *et al.* Two modes of change in Southern Ocean productivity over the past million years. *Science* **339**, 1419–1423 (2013).
12. Ziegler, M., Diz, P., Hall, I. R. & Zahn, R. Millennial-scale changes in atmospheric CO_2 levels linked to the Southern Ocean carbon isotope gradient and dust flux. *Nat. Geosci.* **6**, 457–461 (2013).
13. Frank, M. *et al.* Similar glacial and interglacial export bioproductivity in the Atlantic sector of the Southern Ocean: multiproxy evidence and implications for glacial atmospheric CO_2. *Paleoceanography* **15**, 642–658 (2000).
14. Skinner, L. C., Fallon, S., Waelbroeck, C., Michel, E. & Barker, S. Ventilation of the deep Southern Ocean and deglacial CO_2 rise. *Science* **328**, 1147–1151 (2010).
15. Boiteau, R., Greaves, M. & Elderfield, H. Authigenic uranium in foraminiferal coatings: a proxy for ocean redox chemistry. *Paleoceanography* **27**, PA3227 (2012).
16. Hoogakker, B. A. A., Elderfield, H., Schmiedl, G., McCave, I. N. & Rickaby, R. E. M. Glacial – interglacial changes in bottom-water oxygen content on the Portuguese margin. *Nat. Geosci.* **8**, 40–43 (2015).
17. McCorkle, D. C., Keigwin, L. D., Corliss, B. H. & Emerson, S. R. The influence of microhabitats on the carbon isotopic composition of deep-sea benthic foraminifera. *Paleoceanography* **5**, 161–185 (1990).
18. Anderson, R. F. *et al.* Biological response to millennial variability of dust supply in the Subantarctic South Atlantic Ocean. *Philos. Trans. R. A Math. Phys. Eng. Sci.* **372**, 20130054 (2014).
19. Gottschalk, J., Skinner, L. C. & Waelbroeck, C. Contribution of seasonal sub-Antarctic surface water variability to millennial-scale changes in atmospheric CO_2 over the last deglaciation and Marine Isotope Stage 3. *Earth Planet. Sci. Lett.* **411**, 87–99 (2015).
20. Gottschalk, J. *et al.* Abrupt changes in the southern extent of North Atlantic Deep Water during Dansgaard-Oeschger events. *Nat. Geosci.* **8**, 950–955 (2015).
21. Russell, A. D., Hönisch, B., Spero, H. J. & Lea, D. W. Effects of seawater carbonate ion concentration and temperature on shell U, Mg, and Sr in cultured planktonic foraminifera. *Geochim. Cosmochim. Acta* **68**, 4347–4361 (2004).
22. Yu, J., Elderfield, H., Greaves, M. & Day, J. Preferential dissolution of benthic foraminiferal calcite during laboratory reductive cleaning. *Geochem. Geophys. Geosyst.* **8**, Q06016 (2007).
23. Klinkhammer, G. P. & Palmer, M. R. Uranium in the oceans: where it goes and why. *Geochim. Cosmochim. Acta* **55**, 1799–1806 (1991).
24. Froelich, P. N. *et al.* Early oxidation of organic matter in pelagic sediments of the eastern equatorial Atlantic: suboxic diagenesis. *Geochim. Cosmochim. Acta* **43**, 1075–1090 (1979).
25. Barnes, C. E. & Cochran, J. K. Uranium removal in oceanic sediments and the oceanic U balance. *Earth Planet. Sci. Lett.* **97**, 94–101 (1990).
26. Boyle, E. A. Manganese carbonate overgrowths on foraminifera tests. *Geochim. Cosmochim. Acta* **47**, 1815–1819 (1983).
27. Sachs, J. P. & Anderson, R. F. Fidelity of alkenone paleotemperatures in southern Cape Basin sediment drifts. *Paleoceanography* **18**, 1082 (2003).
28. Barker, S. & Diz, P. Timing of the descent into the last ice age determined by the bipolar seesaw. *Paleoceanography* **29**, 489–507 (2014).
29. Emerson, S., Fischer, K., Reimers, C. & Heggie, D. Organic carbon dynamics and preservation in deep-sea sediments. *Deep Sea Res.* **32**, 1–21 (1985).
30. McCorkle, D. C. & Emerson, S. R. The relationship between pore water carbon isotopic composition and bottom water oxygen concentration. *Geochim. Cosmochim. Acta* **52**, 1169–1178 (1988).
31. Geslin, E., Heinz, P., Jorissen, F. & Hemleben, C. Migratory responses of deep-sea benthic foraminifera to variable oxygen conditions: laboratory investigations. *Mar. Micropaleontol.* **53**, 227–243 (2004).
32. Duplessy, J.-C. *et al.* ^{13}C Record of benthic foraminifera in the last interglacial ocean: Implications for the carbon cycle and the global deep water circulation. *Quat. Res.* **21**, 225–243 (1984).
33. Garcia, H. E. *et al. World Ocean Atlas 2009* Vol. 3: Dissolved Oxygen, Apparent Oxygen Utilization, and Oxygen Saturation (Ed. Levitus, S.) 344 pp NOAA Atlas NESDIS 70, U.S. Government Printing Office, Washington, D.C. (2010).
34. Hodell, D. A., Venz, K. A., Charles, C. D. & Ninnemann, U. S. Pleistocene vertical carbon isotope and carbonate gradients in the South Atlantic sector of the Southern Ocean. *Geochem. Geophys. Geosyst.* **4**, 1–19 (2003).
35. Schmiedl, G. & Mackensen, A. Late quaternary paleoproductivity and deep water circulation in the eastern South Atlantic Ocean: evidence from benthic foraminifera. *Palaeogeogr. Palaeoclimatol. Palaeoecol.* **130**, 43–80 (1997).

36. Ninnemann, U. S. & Charles, C. D. Changes in the mode of Southern Ocean circulation over the last glacial cycle revealed by foraminiferal stable isotopic variability. *Earth Planet. Sci. Lett.* **201**, 383–396 (2002).

37. Mackensen, A., Rudolph, M. & Kuhn, G. Late Pleistocene deep-water circulation in the subantarctic eastern Atlantic. *Glob. Planet. Change* **30**, 197–229 (2001).

38. Ragueneau, O. *et al.* A review of the Si cycle in the modern ocean: recent progress and missing gaps in the application of biogenic opal as a paleoproductivity proxy. *Glob. Planet. Change* **26**, 317–365 (2000).

39. Jaccard, S. L., Galbraith, E. D., Frölicher, T. L. & Gruber, N. Ocean (de)oxygenation across the last deglaciation: insights for the future. *Oceanography* **27**, 26–35 (2014).

40. Anderson, L. A. & Sarmiento, J. L. Redfield ratios of remineralization determined by nutrient data analysis. *Global Biogeochem. Cycles* **8**, 65–80 (1994).

41. Kwon, E. Y., Sarmiento, J. L., Toggweiler, J. R. & DeVries, T. The control of atmospheric pCO$_2$ by ocean ventilation change: the effect of the oceanic storage of biogenic carbon. *Global Biogeochem. Cycles* **25**, GB3026 (2011).

42. Curry, W. B. & Oppo, D. W. Glacial water mass geometry and the distribution of δ^{13}C of ΣCO$_2$ in the western Atlantic Ocean. *Paleoceanography* **20**, PA1017 (2005).

43. Jaccard, S. L. & Galbraith, E. D. Large climate-driven changes of oceanic oxygen concentrations during the last deglaciation. *Nat. Geosci.* **5**, 151–156 (2012).

44. Sarnthein, M., Schneider, B. & Grootes, P. M. Peak glacial ^{14}C ventilation ages suggest major draw-down of carbon into the abyssal ocean. *Clim. Past* **9**, 2595–2614 (2013).

45. Schmittner, A., Brook, E. J. & Ahn, J. in *Ocean Circulation: Mechanisms and Impacts* (eds. Schmittner, A., Chiang, J. C. H. & Hemming, S. R.) **173**, 209–246 (American Geophysical Union, Geophysical Monograph Series, 2007).

46. Watson, A. J. & Naveira Garabato, A. C. The role of Southern Ocean mixing and upwelling in glacial-interglacial atmospheric CO$_2$ change. *Tellus B* **58**, 73–87 (2006).

47. Bereiter, B. *et al.* Mode change of millennial CO$_2$ variability during the last glacial cycle associated with a bipolar marine carbon seesaw. *Proc. Natl. Acad. Sci. USA* **109**, 9755–9760 (2012).

48. Jaccard, S. L., Galbraith, E. D., Martínez-Garcia, A. & Anderson, R. F. Covariation of abyssal Southern Ocean oxygenation and pCO$_2$ throughout the last ice age. *Nature* **530**, 207–210 (2016).

49. Sarmiento, J. L., Gruber, N., Brzezinski, M. A. & Dunne, J. P. High-latitude controls of thermocline nutrients and low latitude biological productivity. *Nature* **427**, 56–60 (2004).

50. Carter, L., McCave, I. N. & Williams, M. J. M. Circulation and water masses of the Southern Ocean: a review. *Dev. Earth Environ. Sci.* **8**, 85–114 (2009).

51. Weiss, R. F. The solubility of nitrogen, oxygen and argon in water and seawater. *Deep Sea Res.* **17**, 721–735 (1970).

52. Adkins, J. F., McIntyre, K. & Schrag, D. P. The salinity, temperature, and δ^{18}O of the glacial deep ocean. *Science* **298**, 1769–1773 (2002).

53. Vogel, H., Rosén, P., Wagner, B., Melles, M. & Persson, P. Fourier transform infrared spectroscopy, a new cost-effective tool for quantitative analysis of biogeochemical properties in long sediment records. *J. Paleolimnol.* **40**, 689–702 (2008).

54. Meyer-Jacob, C. *et al.* Independent measurement of biogenic silica in sediments by FTIR spectroscopy and PLS regression. *J. Paleolimnol.* **52**, 245–255 (2014).

55. DeMaster, D. J. The supply and accumulation of silica in the marine environment. *Geochim. Cosmochim. Acta* **45**, 1715–1732 (1981).

56. Mortlock, R. A. & Froelich, P. N. A simple method for the rapid determination of biogenic opal in pelagic marine sediments. *Deep Sea Res.* **36**, 1415–1426 (1989).

57. François, R., Frank, M., van der Loeff, M. M. R. & Bacon, M. P. ^{230}Th normalization: an essential tool for interpreting sedimentary fluxes during the late Quaternary. *Paleoceanography* **19**, 16 (2004).

58. Bourne, M. D., Thomas, A. L., Mac Niocaill, C. & Henderson, G. M. Improved determination of marine sedimentation rates using ^{230}Th$_{xs}$. *Geochemistry Geophys. Geosystems* **13**, Q09017 (2012).

59. Vogel, J. S., Southon, J. R., Nelson, D. E. & Brown, T. A. Performance of catalytically condensed carbon for use in accelerator mass spectrometry. *Nucl. Instrum. Methods Phys. Res.* **5**, 289–293 (1984).

60. Kuhn, G. Susceptibility raw data of sediment core PS2498-1. http://dx.doi.org/10.1594/PANGAEA.87282 (2002).

61. Svensson, A. *et al.* A 60000 year Greenland stratigraphic ice core chronology. *Clim. Past* **4**, 47–57 (2008).

62. Veres, D. *et al.* The Antarctic ice core chronology (AICC2012): an optimized multi-parameter and multi-site dating approach for the last 120 thousand years. *Clim. Past* **9**, 1733–1748 (2013).

63. Key, R. M. *et al.* A global ocean carbon climatology: Results from Global Data Analysis Project (GLODAP). *Global Biogeochem. Cycles* **18**, GB4031 (2004).

64. Takahashi, T. *et al.* Global sea-air CO$_2$ flux based on climatological surface ocean pCO$_2$, and seasonal biological and temperature effects. *Deep Sea Res.* **49**, 1601–1622 (2002).

65. Orsi, A. H., Whitworth, T. & Nowlin, W. D. On the meridional extent and fronts of the Antarctic Circumpolar Current. *Deep Sea Res.* **42**, 641–673 (1995).

66. Lambert, F. *et al.* Dust-climate couplings over the past 800,000 years from the EPICA Dome C ice core. *Nature* **452**, 616–619 (2008).

67. Ahn, J. & Brook, E. J. Atmospheric CO$_2$ and climate on millennial time scales during the last glacial period. *Science* **322**, 83–85 (2008).

68. Blunier, T. & Brook, E. J. Timing of millennial-scale climate change in Antarctica and Greenland during the last glacial period. *Science* **291**, 109 (2001).

69. Lüthi, D. *et al.* CO$_2$ and O$_2$/N$_2$ variations in and just below the bubble-clathrate transformation zone of Antarctic ice cores. *Earth Planet. Sci. Lett.* **297**, 226–233 (2010).

70. Monnin, E. *et al.* Atmospheric CO$_2$ concentrations over the last glacial termination. *Science* **291**, 112 (2001).

71. Ahn, J. & Brook, E. J. Siple Dome ice reveals two modes of millennial CO$_2$ change during the last ice age. *Nat. Commun.* **5**, 3723 (2014).

72. Indermühle, A., Monnin, E., Stauffer, B., Stocker, T. F. & Wahlen, M. Atmospheric CO$_2$ concentration from 60 to 20 kyr BP from the Taylor Dome ice core, Antarctica. *Geophys. Res. Lett.* **27**, 735–738 (2000).

73. Marcott, S. A. *et al.* Centennial-scale changes in the global carbon cycle during the last deglaciation. *Nature* **514**, 616–619 (2014).

74. Ramsey, C. B. *et al.* A complete terrestrial radiocarbon record for 11.2 to 52.8 kyr BP. *Science* **338**, 370–374 (2012).

75. Hughen, K., Southon, J., Lehman, S., Bertrand, C. & Turnbull, J. Marine-derived ^{14}C calibration and activity record for the past 50,000 years updated from the Cariaco Basin. *Quat. Sci. Rev.* **25**, 3216–3227 (2006).

76. Reimer, P. J. *et al.* IntCal09 and Marine09 radiocarbon age calibration curves, 0–50,000 years cal BP. *Radiocarbon* **51**, 1111–1150 (2009).

77. Reimer, P. J. *et al.* IntCal13 and Marine13 radiocarbon age calibration curves 0–50,000 years cal BP. *Radiocarbon* **55**, 1869–1887 (2013).

Acknowledgements

We are very indebted to Sambuddha Misra, Stephen Barker and Emma Freeman for fruitful discussions. Fabien Dewilde, Gülay Isguder, Margret Bayer, Verena Lanny, Lena Thöle, Emma Freeman, Ron Reimer, María de la Fuente and Benny Antz are thanked for the technical support. We also acknowledge Andreas Mackensen and Rainer Gersonde for sharing their expertise in South Atlantic coring locations. J.G. and L.C.S. acknowledge support from the Gates Cambridge Trust, the Royal Society, the Cambridge Newton Trust and NERC grant NE/J010545/1. J.L. was supported by Marie Curie Fellowship FP7-PEOPLE-2013-IEF (Marie Curie proposal 622483). S.L.J. was funded through the Swiss National Science Foundation (grant PP00P2-144811). C.W. acknowledges support from the European Research Council grant ACCLIMATE/no 339108. This is LSCE contribution no. 4488.

Author contributions

J.G. and L.C.S. designed the study. C.W. collected the core material. J.G., J.L. and H.V. performed the analyses with support from S.L.J. and N.F. J.G. and L.C.S. analysed the proxy data and wrote this manuscript with contributions from all authors.

Additional information

A role for diatom-like silicon transporters in calcifying coccolithophores

Grażyna M. Durak[1,*,†], Alison R. Taylor[2,*], Charlotte E. Walker[1], Ian Probert[3], Colomban de Vargas[3], Stephane Audic[3], Declan Schroeder[1], Colin Brownlee[1,4] & Glen L. Wheeler[1]

Biomineralization by marine phytoplankton, such as the silicifying diatoms and calcifying coccolithophores, plays an important role in carbon and nutrient cycling in the oceans. Silicification and calcification are distinct cellular processes with no known common mechanisms. It is thought that coccolithophores are able to outcompete diatoms in Si-depleted waters, which can contribute to the formation of coccolithophore blooms. Here we show that an expanded family of diatom-like silicon transporters (SITs) are present in both silicifying and calcifying haptophyte phytoplankton, including some globally important coccolithophores. Si is required for calcification in these coccolithophores, indicating that Si uptake contributes to the very different forms of biomineralization in diatoms and coccolithophores. Significantly, SITs and the requirement for Si are absent from highly abundant bloom-forming coccolithophores, such as *Emiliania huxleyi*. These very different requirements for Si in coccolithophores are likely to have major influence on their competitive interactions with diatoms and other siliceous phytoplankton.

[1] Marine Biological Association, The Laboratory, Citadel Hill, Plymouth, Devon PL1 2PB, UK. [2] Department of Biology and Marine Biology, University of North Carolina Wilmington, 601 South College Road, Wilmington, North Carolina, 28403-5915, USA. [3] Station Biologique de Roscoff, Place Georges Teissier, 29680 Roscoff, France. [4] School of Ocean and Earth Sciences, University of Southampton, National Oceanography Centre, Southampton SO14 3ZH, UK. * These authors contributed equally to this work. † Present address: University of Konstanz, Department of Chemistry, Physical Chemistry, Universitätsstr. 10, Box 714, D-78457 Konstanz, Germany. Correspondence and requests for materials should be addressed to C.B. (email: cbr@mba.ac.uk) or to G.W. (email: glw@mba.ac.uk).

The biomineralized phytoplankton are major contributors to marine primary productivity and play a major role in carbon export to the deep oceans by promoting the sinking of organic material from the photic zone[1,2]. The two primary forms of biomineralization found in marine plankton are the precipitation of silica (by diatoms, chrysophytes, synurophytes, dictyochophytes, choanoflagellates and radiolarians) and calcium carbonate (by coccolithophores, foraminifera, ciliates and dinoflagellates)[3]. These processes require very different chemistries and exhibit no known shared mechanisms. Both silification and calcification appear to have evolved independently on multiple occasions. However, since in many cases the underlying cellular mechanisms have not been elucidated, the evolutionary processes remain unclear. Improved knowledge of the cellular mechanisms of biomineralization will allow us to understand the impact of past climatic events on the major phytoplankton lineages and better predict their response to future environmental change.

The haptophyte algae are of particular interest in the evolution of biomineralization as they include closely related silicified and calcified representatives. The coccolithophores (Calcihaptophycidae)[4] produce an extracellular covering of ornate calcium carbonate plates (coccoliths) and are major contributors to biogenic calcification in the ocean[3]. The most abundant coccolithophore species in modern oceans are *Emiliania huxleyi* and *Gephyrocapsa oceanica*, which belong to the Noelarhabdaceae. These species have a small cell size and are able to form extensive blooms. Larger coccolithophores species such as *Coccolithus braarudii* and *Calcidiscus leptoporus* are less numerous, but as they are heavily calcified they are important contributors to global calcification[5]. Much of our understanding of coccolithophore biology comes from the study of *E. huxleyi*, but emerging evidence suggests that there is considerable physiological diversity among coccolithophores[6].

Though the biomineralized haptophytes are predominately calcified, a representative was recently described, *Prymnesium neolepis* (formerly *Hyalolithus neolepis*), which is covered with silica scales and resembles a 'silicified coccolithophore'[7-9]. The silica scales are produced intracellularly and then deposited outside the plasma membrane, in a manner analogous to coccolith secretion[8,10]. The Prymnesiales are estimated to have diverged from the coccolithophores around 280 Myr ago[11] and *P. neolepis* is the only known extensively silicified haptophyte. Understanding whether common cellular mechanisms contribute to silica scale production in *P. neolepis* and coccolith formation in the coccolithophores may help us to understand how these different forms of biomineralization have evolved in the haptophytes and also in other phytoplankton lineages.

Silicification by marine phytoplankton has both contributed to and been influenced by the marked changes in the biogeochemistry of Si in the surface ocean. The diatoms, representing the dominant silicifying phytoplankton in current oceans, appeared only relatively recently in the fossil record (120 Myr ago) and their expansion in the Cenozoic resulted in the extensive depletion of silicate from the surface ocean, leading to the decline of heavily silicified sponges and decreased silicification in radiolarians[12-15]. Si has therefore become a limiting nutrient for modern silicifying phytoplankton and is an important factor in competitive interactions with non-silicifying taxa. As the regeneration of available Si from silica dissolution is slow, diatom blooms can deplete Si in the surface ocean sufficiently to prevent further growth. If other nutrients such as nitrate or phosphate are still available, then Si limitation can contribute to seasonal succession, where an initial diatom spring bloom is followed by subsequent blooms of non-siliceous phytoplankton. There is evidence that the low availability of Si is an important contributory factor in the formation of some coccolithophore blooms. Major *E. huxleyi* blooms in areas such as the North Atlantic, the Black Sea and off the Patagonian shelf have been associated with low silicate availability[16-18]. These observations support the view that the ecological niche of coccolithophores is partly defined by conditions that reduce competition with the fast-growing resource-efficient diatoms, such as in areas of low silicate where other nutrients (for example, nitrate and phosphate) remain available[19].

To further understand the evolution of biomineralization in haptophytes, we characterized the cellular mechanisms underlying silica scale formation in *P. neolepis*. We examine commonalities with other silicified organisms and determine whether any common cellular mechanisms contribute to biomineralization in silicified and calcified haptophytes. Surprisingly, given that it is generally assumed that coccolithophores lack a requirement for Si, we identify that diatom-like Si transporters are present in haptophytes, not only in the silicified *P. neolepis* but also in some important calcifying coccolithophore species. We demonstrate that Si plays an important role in formation of calcite coccoliths in these coccolithophores, but that the requirement for Si is significantly absent from the most abundant species in present day oceans, *E. huxleyi*. The findings have important implications for the evolution of the biomineralized phytoplankton and their distribution in both past and modern oceans.

Results

Mechanisms of biomineralization in a silicifying haptophyte.

The known mechanisms of biosilicification in eukaryotes involve a number of common elements; a mechanism for Si uptake, an acidic silica deposition vesicle and an organic matrix for catalysing and organizing silica precipitation[20]. However, there is little evidence for shared mechanisms at the molecular level, suggesting that silicification has evolved independently in many lineages. We therefore examined the mechanisms of silicification in *P. neolepis*, using both molecular and physiological approaches. At low Si concentrations, Si uptake in diatoms is performed by a family of Na^+-coupled high-affinity Si transporters (SITs), although diatoms may also acquire Si by diffusive entry at higher Si concentrations[21,22]. Silicified sponges and land plants do not contain SITs, but use alternative mechanisms for Si transport[23,24]. A search for putative Si transporters in the transcriptome of *P. neolepis* strain PZ241 (Supplementary Fig. 1) identified a single gene bearing similarity to the SITs (*PnSIT1*). PnSIT1 exhibits 24.9–29.3% identity and 39.8–47.0% similarity to diatom SITs at the amino-acid level (sequences used for comparison were *Cylindrotheca. fusiformis* AAC49653.1, *Thalassiosira. pseudonana* ABB81826.1 and *Phaeodactylum tricornutum* ACJ65494.1). SITs have only previously been identified in siliceous stramenopiles (diatoms and chrysophytes) and choanoflagellates[25-27]. Many features of PnSIT1 are conserved with these SITs, including the 10 predicted transmembrane regions and the pair of motifs (EGxQ and GRQ) between TM2-3 and TM7-8 (ref. 27; Supplementary Fig. 2). We also identified a homologue of the Si efflux protein, Lsi2 in *P. neolepis* (Supplementary Table 1). Lsi2 is related to the bacterial arsenate transporter ArsB and mediates Si efflux in plant cells[28]. Lsi2 is also present in diatoms and its transcriptional regulation is highly similar to SIT2 in *Thalassiosira pseudonana*, although its cellular role has not yet been characterized[29]. The identification of Lsi2 in *P. neolepis* suggests that it may play a conserved role in siliceous phytoplankton.

We next determined the presence of an acidic silica deposition vesicle in *P. neolepis* using the fluorescent dye HCK-123, which partitions into acidic compartments and labels nascent silica

(Fig. 1a,b). We found that newly formed silica scales are secreted at the posterior pole of the cell, indicating that the principal cellular components involved in scale formation (silica precipitation in acidic non-Golgi-derived vesicles) are distinct from those involved in coccolith formation (calcite precipitation in alkaline Golgi-derived vesicles and secretion at the anterior pole of the cell)[10,30].

A search of the *P. neolepis* transcriptome for mechanisms involved in silica precipitation did not reveal homologues of any of the known silica-associated proteins from diatoms (silaffins, pleuralins and frustulins) or sponges (silicateins)[20,31]

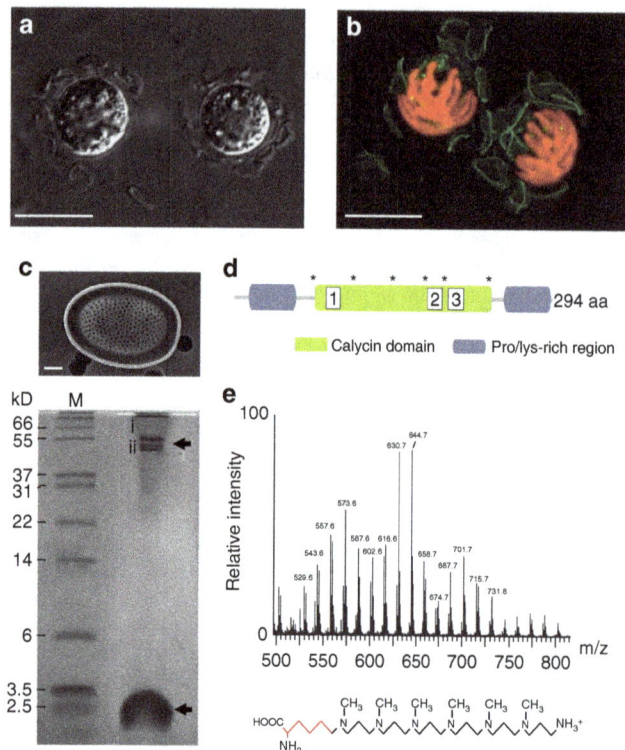

Figure 1 | Molecular mechanisms of silica scale production in P. neolepis.
(**a**) Differential interference contrast (DIC) microscopy image of *P. neolepis* cells displaying the loose covering of silica scales. Scale bar, 10 µm.
(**b**) Confocal microscopy of a *P. neolepis* cell showing incorporation of the fluorescent dye HCK-123 into newly formed silica scales (green). Chlorophyll autofluorescence is shown in red. The 3D-projection was generated from compiling a Z-stack of 15 images. Scale bar, 10 µm.
(**c**) Tricine/SDS–PAGE of organic components released after dissolution of silica scales with NH$_4$F. A SEM image of an isolated silica scale is also shown (Scale bar, 1 µm). The higher molecular weight component around 50 kDa is a single protein that runs as two bands (i, ii), whereas the low-molecular-weight components around 2.5 kDa are long-chain polyamines (LCPA). M, molecular-weight markers. (**d**) Domain organization of the lipocalin-like protein (LPCL1) identified from both protein bands in NH$_4$F extracted silica scales. The approximate positions of the proline/lysine-rich regions and the calycin domain (IPRO12674) are shown. Also shown are the positions of six highly conserved cysteines (asterisk) that may be involved in the formation of disulphide bridges. (**e**) Long-chain polyamines (LCPAs) from *P. neolepis* silica scales. Electrospray ionization mass spectrometry (ESI-MS) of the low-molecular-weight NH$_4$F-soluble fraction of silica scales revealed a series of mass peaks separated by 71 Da (highlighted in red), characteristic of *N*-methyl propyleneimine units. The additional mass peaks ±14 Da may indicate different methylation states, as is commonly observed in LCPAs. The proposed structure of the LCPAs in *P. neolepis* is shown with the putative lysine residue is highlighted in red.

(Supplementary Table 1). As some of these proteins have a low complexity amino-acid composition and may not be identified by sequence similarity searches, we directly analysed the organic components released by NH$_4$F dissolution of the silica scales. We identified two major organic components using Tricine/SDS–PAGE; a lipocalin-like protein and long-chain polyamines (LCPAs; Fig. 1c). The lipocalin-like protein contains two proline-/lysine-rich regions surrounding a lipocalin domain and represents a novel silica-associated protein (LPCL1, Fig. 1d, Supplementary Fig. 3). The LCPAs from *P. neolepis* are composed of *N*-methylated oligopropyleneimine repeats, similar to the silica-associated LCPAs previously characterized from diatoms and sponges[32,33], but differ from these LCPAs as the repeat units are linked to a lysine residue rather than putrescine, ornithine or spermidine (diatoms), or butaneamine (sponges) residues (Fig. 1e, Supplementary Fig. 4). Diatoms possess a series of unusual orthologues of the genes involved in polyamine synthesis that are proposed to play a specific role in the formation of LCPAs[34]. Homologues of these modified genes for polyamine synthesis were not found in the *P. neolepis* transcriptome, indicating that these modifications may be specific to diatoms.

An expanded family of SITs in haptophytes. Our analyses indicate that there are some similarities in the biosilicification mechanisms between *P. neolepis* and diatoms, including the silica deposition vesicle and the LCPAs. However, the silica-associated proteins bear no similarity and the only known silicification-related gene products common to both organisms are the Si transporters (SITs and Lsi2). To examine the origins of SITs and Lsi2 in *P. neolepis*, we performed sequence similarity searches of the *Emiliania huxleyi* genome and 24 other haptophyte transcriptomes (including six species of coccolithophore) from the Marine Microbial Eukaryote Transcriptome data set (http://marinemicroeukaryotes.org/)[35]. Homologues of the Si-associated protein LPCL1 from *P. neolepis* were not found in other haptophytes (Supplementary Tables 1 and 2). However, we identified a SIT homologue in the calcifying coccolithophore *Scyphosphaera apsteinii* that was highly similar to PnSIT1 (66% identity, 76.2% similarity at the amino-acid level). In addition, we found that three coccolithophores (*S. apsteinii*, *Coccolithus braarudii* and *Calcidiscus leptoporus*) possess a SIT-like protein that only contains five transmembrane regions (Fig. 2a,b). The Si efflux protein Lsi2 was not found in these coccolithophores, or in any other haptophyte, with the exception of the non-mineralized prymnesiophyte, *Haptolina ericina*.

Comparison of the two haptophyte SITs with 33 other SIT sequences originating from diatoms, chrysophytes and choanoflagellates indicated that all of the highly conserved amino-acid residues identified by Marron *et al.*[26] were also conserved in haptophytes (Supplementary Fig. 2). The single 5TM domain of the SIT-like (SITL) proteins displays a high sequence similarity to the N- and C-terminal 5TM domains of SITs. SITLs also possess the highly conserved EGxQ and GRQ motifs that are proposed to play a role in binding Si[26,27], as well as many of the other amino-acid residues that were identified as being highly conserved in SITs (Supplementary Fig. 2). The 5TM + 5TM inverted repeat topology of the SITs is characteristic of Na$^+$-coupled transporters with a LeuT fold and is also found in many other membrane transporters[36,37]. The inverted repeat topology in these transporters is thought to have evolved following gene duplication and fusion of a related transporter that initially existed as a homodimer with inverted symmetry[38]. Homodimerization of the SITLs may therefore result in a membrane transporter with similar properties to the SITs and it is likely that SITs evolved from a protein resembling the SITLs.

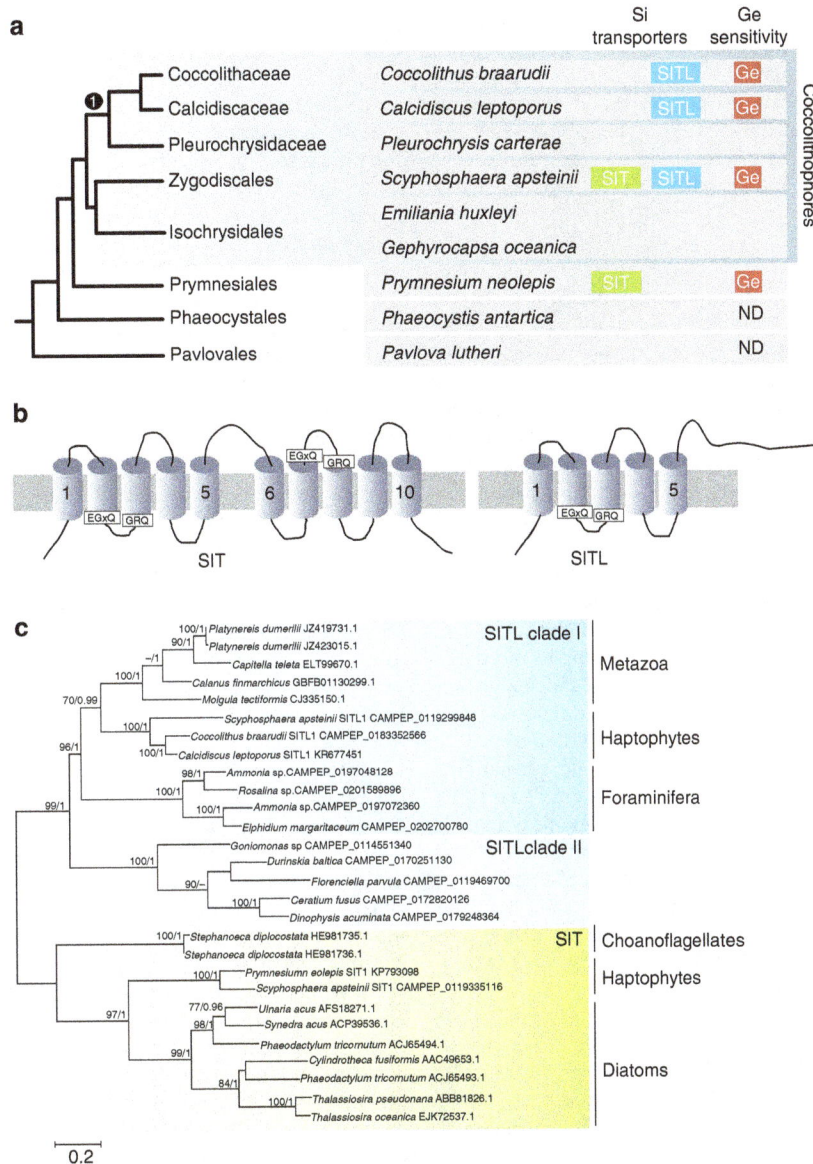

Figure 2 | An expanded family of diatom-like Si-transporters (SITs) in haptophytes. (**a**) Phylogenetic relationships between haptophytes. The schematic tree shows the currently accepted phylogenetic relationships of the major haptophyte lineages based on multigene phylogenies[11]. Representative species of each group are indicated, along with the presence of SITs or SITLs in these species. Sensitivity to Ge is shown in red, ND, not determined. ❶ Coccolithales. (**b**) A schematic image of the domain architecture of the SITs and the SITLs indicating the approximate position of the transmembrane domains and of the conserved motifs. (**c**) A maximum likelihood phylogenetic tree based on an alignment of selected SITL proteins with SITs (aligned to the N-terminal SIT domain). Final alignment size was 157 amino acids. The SITLs form a well-supported monophyletic clade. Within the SITLs two distinct clades can be observed. SITL clade I contains haptophytes, metazoa and foraminifera, whereas SITL clade II contains dinoflagellates, a cryptophyte and a dictyochophore. Bootstrap values >70% (100 bootstraps) and Bayesian posterior probabilities >0.95 (10,000,000 generations) are shown above nodes. Scale bar, substitutions per site.

SITLs were not found in any other haptophytes or in diatoms, but were present in a range of other eukaryotes, including foraminifera, dinoflagellates and metazoa (such as the polychaetes, *Capitella teleta* and *Platynereis dumerilii* and the copepod *Calanus finmarchicus*) (Fig. 2c). Many calanoid copepods have silicified teeth[39] and SITLs may provide a mechanism for Si transport in these ecologically important zooplankton. However, not all of the species that possess SITLs are silicified. The foraminifera and the coccolithophores are the predominant contributors to calcification in our oceans and so the identification of SITLs in these lineages is particularly intriguing.

The SITs from *P. neolepis* and *S. apsteinii* form a strongly supported monophyletic clade, suggesting a common evolutionary origin for the haptophyte SITs (Fig. 2c, Supplementary Fig. 5). To explain the limited distribution of SITs, Marron et al.[26] proposed that horizontal gene transfer (HGT) of SITs may have occurred between stramenopiles and choanoflagellates. However, there is no phylogenetic evidence to support recent HGT of SITs between stramenopiles, choanoflagellates or the haptophytes. The SITL proteins form a monophyletic clade distinct from true SITs, suggesting that they represent a novel but closely related group of transporter proteins (Fig. 2c). When aligned to the SITLs, the individual N- and C-terminal regions of SITs form strongly supported clades, suggesting that the SITs found in stramenopiles, choanoflagellates and haptophytes arose from a single gene duplication event, rather than from a series of more recent duplication events in each lineage (Supplementary Fig. 5). Phylogenetic analyses of Lsi2 provided no indication that haptophytes acquired this gene by recent HGT (Supplementary Fig. 6).

A novel role for Si in coccolithophore calcification. Calcified coccolithophores emerged in the early Mesozoic (c. 220 Myr ago)[11], when Si concentrations in the surface oceans were considerably higher than in present day. The distribution of the SITs and SITLs in haptophytes suggests that these transporters were present in ancestral haptophytes, including the last common ancestor of the coccolithophores. Although Si has not been generally identified as a component of calcite coccoliths, a recent study showed that Si is a minor component of the two forms of heterococcolith (muroliths and lopadoliths) found in *S. apsteinii*[40]. In many calcifying systems, calcite precipitation occurs by the crystallization of amorphous calcium carbonate (ACC). Recent evidence indicates that silica can modulate the crystallization of calcium carbonate *in vitro* by acting to modulate the metastability of ACC and facilitate ordered calcite crystal formation[41–44]. We therefore hypothesized that Si uptake via SITs or SITLs may contribute to calcification in coccolithophores.

To test this hypothesis, we used the Si analogue germanium (Ge), which competitively inhibits Si uptake in diatoms[22] and also prevents Si scale production in *P. neolepis* (Supplementary Fig. 7). In diatoms, Ge/Si ratios <0.01 do not have an inhibitory effect, but ratios >0.05 inhibit Si uptake and also disrupt Si metabolism within the cell[45–47]. Other silicifying algae, such as the chrysophytes *Synura petersenii* and *Paraphysomonas vestita,* are also sensitive to Ge, although growth in *Paraphysomonas* is only inhibited at much higher Ge/Si ratios than diatoms[48,49]. In contrast, non-silicified algae are reported to be largely unaffected by Ge[50]. Our initial experiments to screen for Ge sensitivity in coccolithophores were conducted in low-Si seawater ($<0.1\,\mu M$), to ensure high Ge:Si ratios (>1). Observations with light microscopy and scanning electron microscopy (SEM) identified that coccolith formation in *S. apsteinii* was severely disrupted by the addition of $1\,\mu M$ Ge, with 73% of cells displaying highly malformed coccoliths after 72 h (compared with 3.3% in Si-replete seawater; Fig. 3). The cup-shaped lopadoliths were severely misshapen, frequently exhibiting additional disorganized calcite precipitation at the apical rim, and the smaller disk-shaped muroliths also exhibited malformation. The addition of $5\,\mu M$ Ge to *C. braarudii* and *C. leptoporus* resulted in the production of severely malformed coccoliths that failed to integrate into the coccosphere and were shed into the surrounding seawater (Fig. 3). In all three species, addition of $100\,\mu M$ Si suppressed the disruptive effects of Ge on coccolith morphology, suggesting that Ge acts competitively with Si.

To examine the relationship between Ge and Si in greater detail, we grew *C. braarudii* cells at three different Si concentrations (2, 20 and $100\,\mu M$) and examined the effect of a range of Ge concentrations ($0.5–20\,\mu M$ Ge; Fig. 4). Because high Ge/Si ratios completely inhibit biosynthesis and growth in diatoms[46], we also assessed the physiological status of the Ge-treated coccolithophores. We found that the inhibitory effects of Ge on calcification (assessed by the accumulation of discarded coccoliths in the media) are dependent on the ratio of Ge/Si, rather than the absolute concentration of Ge. For example, $2\,\mu M$ Ge results in the production of many aberrant coccoliths at $2\,\mu M$ Si, but its impacts at 20 and $100\,\mu M$ Si are progressively reduced. The inhibitory effects of Ge on calcification in *C. leptoporus* and *S. apsteinii* were also dependent on the Ge/Si ratio (Supplementary Fig. 8). These data support the hypothesis that Ge is acting to competitively inhibit an aspect of Si uptake and/or metabolism that is required for production of coccoliths.

At high Ge/Si ratios (>1) both growth and calcification (accumulation of discarded coccoliths) were inhibited in *C. braarudii.* (Fig. 4). The maximum quantum yield of photosystem II (F_v/F_m) was only reduced at the very highest Ge/Si ratios. It is possible that the inhibition of growth results from the severe disruption of the calcification process. No effects on growth or photosynthetic efficiency were observed at low Ge/Si ratios, while coccolith defects were still observed, demonstrating that Ge had specifically disrupted calcification (Figs 4 and 5). The unique coccolith morphology of Ge-treated cells is distinct from defects in calcification caused by other stressors, such as nutrient limitation or high temperature[51].

Detailed examination of Si-limited coccolithophores provided direct evidence for a requirement for Si in the calcification process. In Si-replete cultures, defects in coccolith morphology were almost completely absent (Figs 3 and 5). However, highly aberrant coccoliths were consistently observed at a low frequency in both *C. braarudii* and *C. leptoporus* cultures after transfer to very low Si seawater (without Ge) for 72 h (Fig. 3a,b). In addition to the appearance of highly aberrant coccoliths, many cells exhibited more subtle but significant defects in coccolith morphology due to Si limitation, such as disorganization of the overlapping elements of the distal shield (termed 'blocky' morphology; Fig. 3a). Growth of *C. braarudii* was not inhibited after 8 days in very low Si ($<0.1\,\mu M$; Supplementary Fig. 9), indicating that the defects in calcification are not caused by a general disruption of cellular physiology. Defective coccolith morphology was also apparent in *C. braarudii* and *C. leptoporus* cultures grown at $2\,\mu M$ Si, compared to Si-replete cells grown at $100\,\mu M$ Si (Fig. 5, Supplementary Fig. 8). This is an important observation as it shows calcification defects may occur at ecologically relevant Si concentrations. The slower growing *S. apsteinii* did not exhibit obvious defects in calcification after transfer to low Si for 72 h, but after 8 days clear defects in coccolith formation were observed, such as missing muroliths or incomplete lopadoliths (Supplementary Fig. 10).

In combination, our results using Ge treatment and Si limitation strongly suggest that Si is required for calcification in certain coccolithophores. The dramatic effects of Ge on these species are surprising as most non-siliceous algae are considered to be insensitive to Ge[50,52]. However, many previous studies on coccolithophore physiology have focussed on *E. huxleyi*, a coccolithophore that lacks SITs or SITLs in its genome. When we examined the impact of Ge on *E. huxleyi* at very low Si ($<0.1\,\mu M$ Si), we found no effects on calcification, with normal coccospheres produced even in the presence of $20\,\mu M$ Ge (Fig. 6a). Concentrations of Ge up to $20\,\mu M$ also had no impact on the growth or photosynthetic efficiency of *E. huxleyi* at $2\,\mu M$ Si (Fig. 6b), in clear contrast to the marked effects of Ge on *C. braarudii*. Furthermore, no Ge sensitivity was observed in two further coccolithophore species in which SITs or SITLs appear absent (from their available transcriptome sequence data); *G. oceanica*, a coccolithophore that is closely related to *E. huxleyi*, and *Pleurochrysis carterae* (Fig. 6a, Supplementary Fig. 11). Our results suggest that Si plays an important role in calcification in coccolithophores that possess SITs and/or SITLs, but this requirement for Si is not universal and is notably absent from the abundant bloom-forming coccolithophore species in modern oceans (the Noelaerhabdaceae)[4].

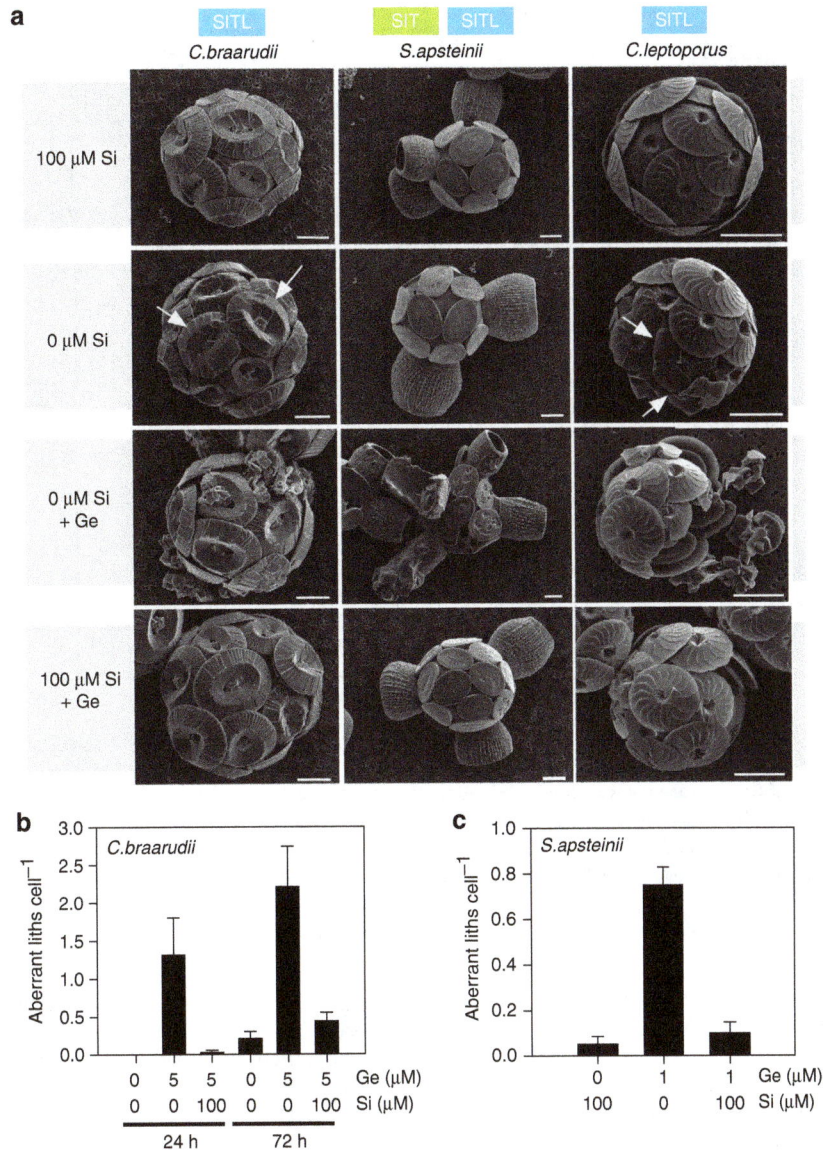

Figure 3 | A role for Si in coccolith formation. (a) Representative SEM micrographs demonstrating the effects of Si limitation and Ge addition on coccolith production. *C. braarudii*, *S. apsteinii* and *C. leptoporus* were incubated for 72 h in very low Si seawater (<0.1 μM), which was amended with Ge (1 μM for *S. apsteinii* or 5 μM for the other two species). *C. braarudii* and *C. leptoporus* cells grown in very low Si appeared superficially similar to cells grown in Si-replete seawater (100 μM Si), but closer inspection revealed that many 'blocky' coccoliths are apparent (arrowed), indicating a calcification defect related to the lack of Si. The addition of Ge resulted in the production of highly aberrant coccoliths in all three species. In *C. braarudii* these aberrant liths fail to integrate fully into the coccosphere and were often shed into the media. Both types of heterococcolith in *S. apsteinii* (the large cup-shaped lopadoliths and the small plate-like muroliths) exhibit extensive malformations. In *C. leptoporus* the aberrant coccoliths are all co-localized, suggesting that the newly formed liths in this species are secreted in a similar position in the coccosphere. The addition of 100 μM Si to Ge-treated cells markedly reduced the inhibitory effects on calcification. Scale bar, 5 μm. **(b)** Quantification of the production of aberrant coccoliths in *C. braarudii* grown in very low Si media for 24 and 72 h, amended with 5 μM Ge or 5 μM Ge + 100 μM Si. For this experiment, only highly aberrant coccoliths were scored and more subtle coccolith malformations such as the 'blocky' coccoliths observed under low Si were not scored. *n* = 40 cells. For discarded liths 4–7 fields of view were scored containing at least 40 cells. The experiment was repeated four times and representative results are shown. **(c)** Quantification of the production of aberrant lopadoliths in *S. apsteinii* grown in low Si media for 72 h, amended with 1 μM Ge, or 1 μM Ge + 100 μM Si. *n* = 40 cells. The experiment was repeated four times and representative results are shown. Error bars denote s.e.

Discussion

While *P. neolepis* is the only known haptophyte exhibiting extensive silicification, our results point towards a much broader role for Si in haptophyte physiology. *P. neolepis* exhibits key similarities with other silicifying eukaryotes, but there is no evidence that silicification in this lineage arose from recent HGT. The mechanisms for silicification in *P. neolepis* have most likely been assembled independently from existing cellular components.

Although *P. neolepis* contains a Si transporter belonging to the SIT family, the identification of a SIT in the coccolithophore *S. apsteinii* suggests that the presence of this family of Si transporters greatly predates the emergence of silicification in the haptophytes and may therefore have played an alternative role before being recruited for biomineralization. In diatoms, Thamatrakoln and Hildebrand[22] have proposed that SITs may have played an ancestral role in preventing

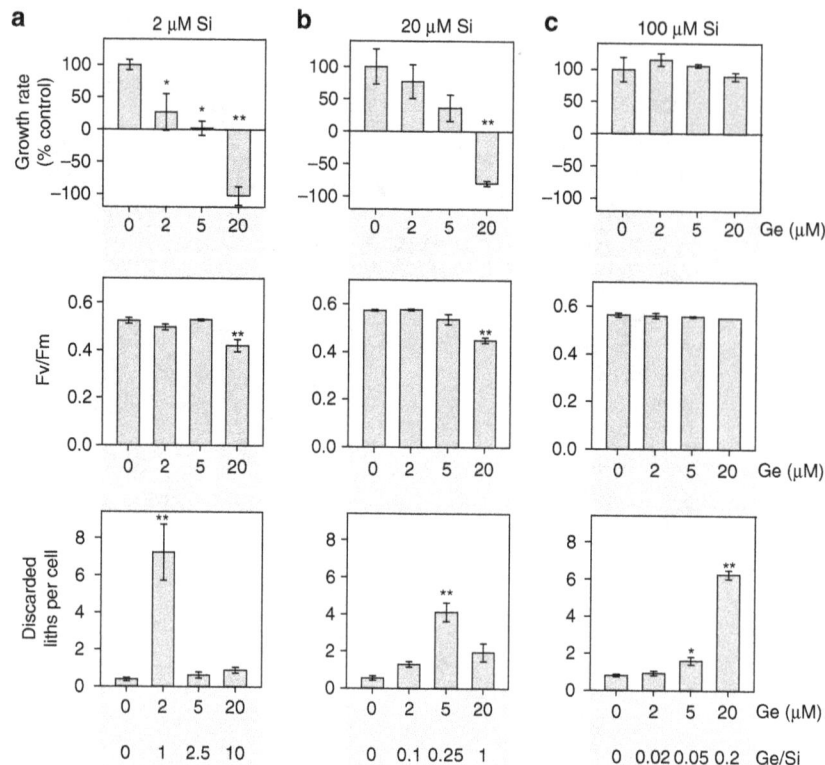

Figure 4 | The inhibitory effects of Ge are dependent on the Ge/Si ratio. (**a**) *C. braarudii* cells were treated with 0, 2, 5 or 20 μM Ge for 48 h in seawater containing 2 μM Si. Effects on coccolith morphology were determined by counting the mean number of discarded liths relative to the cell density. Specific growth rate (per day) and photosynthetic efficiency (the quantum yield of photosystem II, F_v/F_m) were also determined. *$P < 0.05$ and **$P < 0.01$ denote treatments that differ significantly from the 0 μM Ge control (one-way ANOVA with Holm-Sidak *post hoc* test, $n = 3$). Error bars denote standard errors. (**b**) *C. braarudii* cells treated as in **a** but in seawater containing 20 μM Si. (**c**) *C. braarudii* cells treated as in **a** but in seawater containing 100 μM Si. Ge had a much lower impact on coccolithophore physiology at higher Si concentrations, suggesting that Ge acts competitively with Si.

excessive accumulation of intracellular Si in the Si-rich waters of Mesozoic oceans, before they were recruited for frustule formation.

SITs exhibit a very limited distribution in eukaryotes (stramenopiles, haptophytes and choanoflagellates). In the absence of evidence for HGT, an alternative explanation is that this distribution results from multiple losses of a gene that was present in the last common ancestor of these lineages. However, as this ancestor was most likely close to the last common ancestor of all eukaryotes, this scenario requires gene loss of SITs on a massive scale. Two factors that may have contributed to the loss of SITs in eukaryotes are the potential functional redundancy between SITs and SITLs and the extensive depletion of Si from surface oceans in the Cenozoic. However, an alternative scenario that does not require such extensive gene loss is possible as the phylogenetic position of the haptophytes is not fully resolved. Recent phylogenomic evidence suggests a specific association between haptophytes and stramenopiles[53,54], with Stiller *et al.*[54] proposing that haptophytes acquired their plastids following endosymbiosis of a photosynthetic stramenopile belonging to the ochrophyta (which includes diatoms and chrysophytes). The associated endosymbiotic gene transfer therefore provides a mechanism through which the haptophytes may have acquired SITs from stramenopiles. The phylogeny of the SITs is not at odds with this scenario, as the proposed endosymbiosis would have occurred before the extensive radiation of the stramenopiles and the haptophytes, but it does infer that the SITs have been lost extensively in both of these taxonomic groups. This scenario does not explain the presence of SITs in choanoflagellates. Although HGT of SITs to or from choanoflagellates is not supported by the

phylogeny, it cannot be ruled out and there is evidence for extensive HGT from algae into choanoflagellates[55].

Clearly, there are broader evolutionary questions relating to the phylogeny of the haptophytes that must be addressed before we can fully determine the origins of the SITs. Further understanding of the function and roles of the SITLs may also provide important insight into these processes. Nevertheless, our results clearly suggest that an expanded family of SITs were present in ancestral haptophytes and that both SITs and SITLs were present in the ancestor of the calcifying coccolithophores. Therefore, it seems likely that this ancestor possessed the capacity for Si uptake. As Si exhibits the ability to modulate calcite precipitation *in vitro*[42,43], its presence in ancestral coccolithophores may even have facilitated the emergence of extensively calcified coccoliths. We have provided evidence of a role for Si in coccolith formation in *S. apsteinii*, *C. braarudii* and *C. leptoporus*. These results identify that Si uptake via SITs is an important common mechanism contributing to very different modes of biomineralization in two of the major phytoplankton lineages, the diatoms and the coccolithophores.

Lsi2 was not found in coccolithophores with SITs and SITLs, suggesting that its cellular role in *P. neolepis* and diatoms may relate to the process of silicification. In plants, Lsi2 is proposed to act as a H^+/silicic acid exchanger, using an inward H^+ gradient to drive the efflux of silicic acid across the plasma membrane[28]. In silicifying organisms, H^+/silicic acid exchangers could act to load the acidic silica deposition vesicle, using the H^+ gradient across the vesicle membrane to drive the accumulation of silicic acid. It will therefore be important to identify the cellular localization of Lsi2 in *P. neolepis* and diatoms.

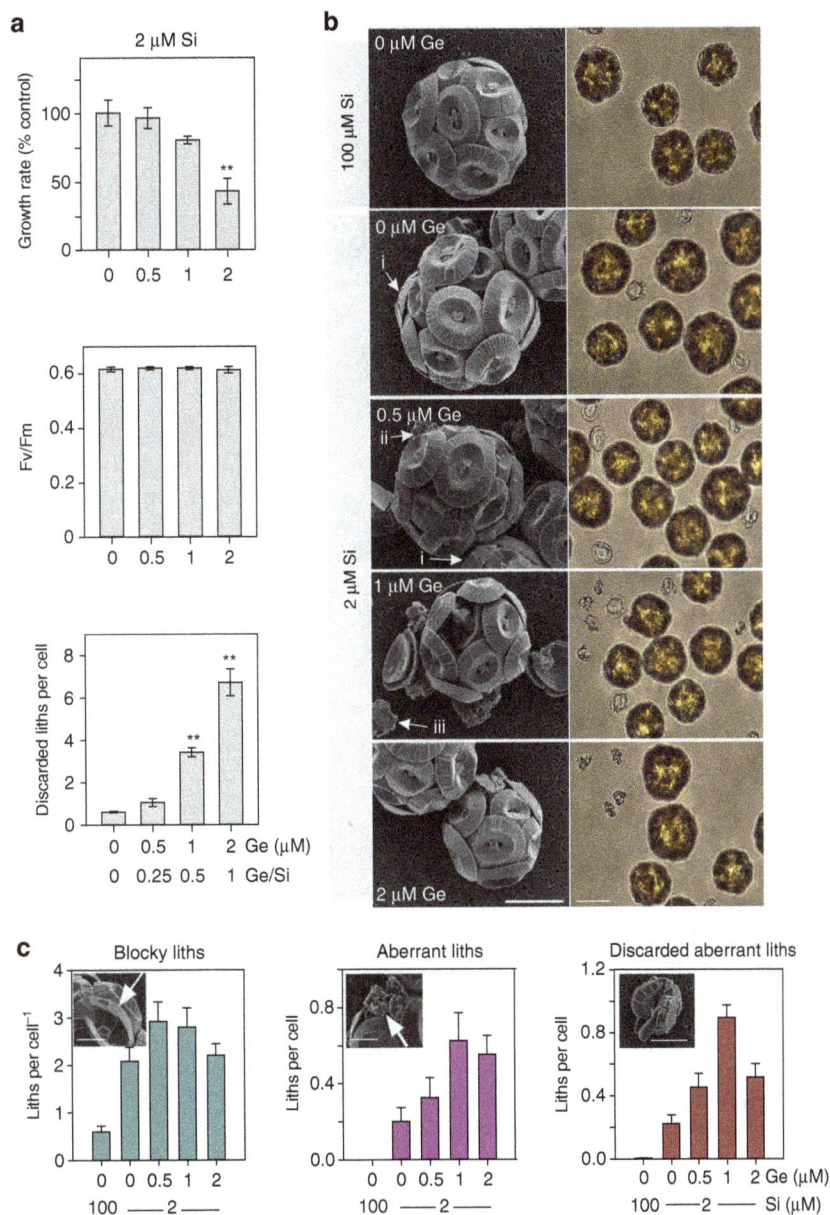

Figure 5 | Ge causes defects in calcification at low Ge/Si ratios. (a) *C. braarudii* cells were treated with 0, 0.5, 1 or 2 μM Ge for 48 h in seawater containing 2 μM Si. Growth, photosynthetic efficiency and the number of discarded liths were determined. *$P < 0.05$ and **$P < 0.01$ denote treatments that differ significantly from the 0 μM Ge control (one-way ANOVA with Holm–Sidak *post hoc* test, n = 3). (b) SEM (left panel) and bright-field microscopy (right panel) images of *C. braarudii* cells grown in Ge for 48 h (conditions described in a). Three classes of defective coccolith morphology were observed. (i) 'Blocky' coccoliths where the overlapping arrangement of the distal shield is disrupted, but the shape of the coccolith is preserved. (ii) Aberrant coccoliths with highly disrupted morphology (iii) Discarded aberrant coccoliths that are not successfully integrated into the coccosphere. Note that even without Ge treatment 'blocky' coccoliths can be observed at 2 μM Si, but these are not present at 100 μM Si. Scale bar, 10 μm. (c) Quantification of the defective coccolith morphology shown in b. At least 40 cells were scored for each treatment. For discarded liths four to seven fields of view were scored containing at least 40 cells. Error bars denote standard errors. Scale bar, 5 μm.

We do not yet know the cellular mechanisms through which Si contributes to the calcification process. Previous workers have identified a role for Si in bone formation in vertebrates[56,57]. However, the primary role of Si in bone formation appears to relate to the synthesis of collagen to form the underlying organic matrix, rather than a direct role in the mineralization process[58,59]. More recently, it has been demonstrated that silica plays an important role in formation of cystoliths, small calcium carbonate deposits that are found in the leaves of some land plants[41,60]. Although silica is only a minor component of cystoliths, it is essential for the formation of the amorphous calcium carbonate

phase that comprises the bulk of the structure[60]. These studies suggest that Si could act to modulate coccolith formation through a number of mechanisms. Further elucidation of its precise role will enable important insight into the cellular mechanisms of calcification in coccolithophores, which remain poorly understood.

Significantly, our results suggest that requirement for Si in coccolithophore calcification may have been lost by the Noelaerhabdaceae and Pleurochrysidaceae. There are other potential evolutionary scenarios that we cannot rule out at this stage, such as independent evolution of the Si requirement within the

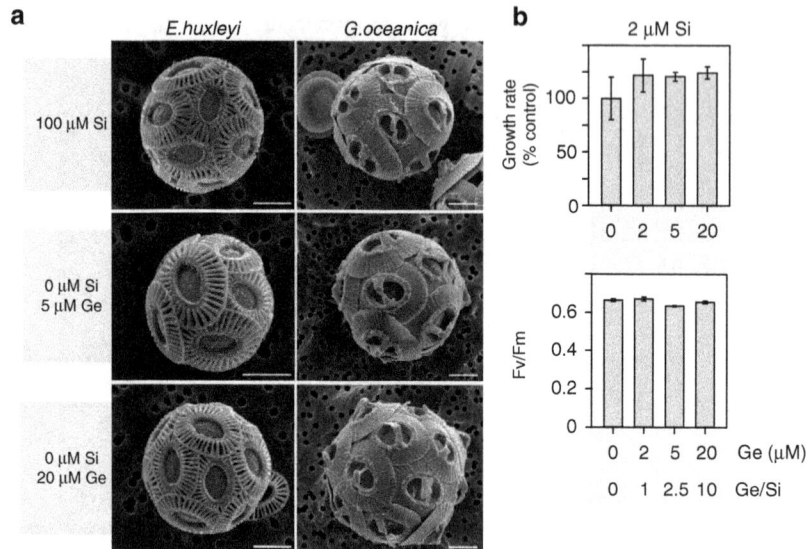

Figure 6 | *Emiliania huxleyi* and *Gephyrocapsa oceanica* are insensitive to Ge. (a) Representative SEM micrographs for *E. huxleyi* and *G. oceanica* following treatment for 72 h in low Si ($<0.1\,\mu M$) media with different additions of Ge (5 or $20\,\mu M$). No effect of Ge on coccolith morphology was observed in either species relative to the control grown in normal seawater media ($100\,\mu M$ Si). Si transporters (SIT/SITL) were not identified the genome of *E. huxleyi* or the transcriptome of *G. oceanica*. Scale bar, $2\,\mu m$. The results are representative of three independent experiments. **(b)** *E. huxleyi* cells were treated with 0, 0.5, 1 or $2\,\mu M$ Ge for 48 h in seawater containing $2\,\mu M$ Si. Mean specific growth rate and mean F_v/F_m as a measure of photosynthetic efficiency were determined. No significant differences were noted between treatments and the $0\,\mu M$ Ge control (one-way ANOVA, $n=3$). The results are representative of two independent experiments. Error bars denote s.e.

Zygodiscales and the Coccolithales, although these scenarios are less parsimonious. The marked decline of surface ocean silicate in the Cenozoic also suggests that loss of the requirement for Si would be more likely than gain. These evolutionary events have important implications for coccolithophore ecology and prompt a re-evaluation of the widely held view that the coccolithophores do not require Si. The Si-requiring coccolithophores identified in this study are important marine calcifiers, with *C. braarudii* and *C. leptoporus* contributing significantly to calcite flux to the deep ocean in large parts of the Atlantic Ocean[5,61]. Although the requirement of these coccolithophores for Si is likely to be considerably lower than that of extensively silicified organisms, Si limitation clearly impairs their ability to calcify. Whether these species encounter significant Si limitation in natural seawaters and can compete effectively for this resource with diatoms and other silicified plankton must be resolved. However, concentrations of silicate in the surface ocean can often reach very low levels, particularly after diatom blooms[62]. It is possible that small fast-growing coccolithophores, which are best suited to exploit the nutrient-depleted waters following a diatom bloom, may have encountered selective pressure to uncouple calcification from Si uptake to avoid Si limitation. The bloom-forming coccolithophores belonging to the Noelaerhabdaceae, such as *E. huxleyi* and *G. oceanica*, may therefore have developed alternative cellular mechanisms to replace the role of Si in coccolith formation. The Noelaerhabdaceae are the most abundant and broadly distributed coccolithophores in modern oceans and their ability to form extensive blooms (often in Si-depleted waters) has likely contributed to their considerable ecological success[16–18]. The differing requirements for Si may therefore have had a profound impact on the physiology of modern coccolithophores and contributed significantly to the evolution and global distribution of this important calcifying lineage.

Methods

Algal strains and culture growth. *Prymnesium neolepis* (NCBI Tax ID 284051) strains TMR5 (RCC3432—Sea of Japan) and PZ241 (RCC1453—Mediterranean

Sea) were obtained from the Roscoff Culture Collection. Strain TMR5 was used for all physiological analyses and for RT–PCR. Strain PZ241 was used to generate the transcriptome. Cultures of *P. neolepis* were maintained in filtered seawater (FSW) supplemented with f/2 nutrients (including $100\,\mu M$ $Na_2SiO_3.5H_2O$) under irradiance of $80–100\,\mu mol\,s^{-1}\,m^{-2}$ (18:6 h light:dark) at $18\,°C$. Stock cultures of the coccolithophores *Coccolithus braarudii* (formerly *Coccolithus pelagicus* ssp *braarudii*) (PLY182G), *Emiliania huxleyi* (PLY-B92/11) and *Pleurochrysis carterae* (PLY406) were maintained in FSW supplemented with f/2 nutrients (without added Si) and Guillard's vitamins as previously described[40]. *Calcidiscus leptoporus* (RCC1130), *Gephyrocapsa oceanica* (RCC1303) and *Scyphosphaera apsteinii* (RCC1456) were maintained in f/2 supplemented with 10% K medium. All coccolithophore cultures were grown at $15\,°C$ under $80–100\,\mu mol\,s^{-1}\,m^{-2}$ irradiance (14:10 h light:dark).

Manipulation of seawater Si and addition of Ge. To examine the effect of Si and Ge on coccolithophores, we used a batch of seawater from the Western English Channel in which Si was naturally low (measured at $2\,\mu M$ using the molybdate-ascorbate assay[63]). This batch of seawater was used for all subsequent analyses involving the effect of Ge on coccolithophores, except where very low Si concentrations were required (see below). Si concentration was amended by the addition of $Na_2SiO_3.5H_2O$. Ge was added in the form of GeO_2, to give concentrations ranging from $0.5–20\,\mu M$. f/2 nutrients (without Si) were added and all coccolithophore cultures were grown under identical conditions ($15\,°C$ under $80–100\,\mu mol\,s^{-1}\,m^{-2}$ irradiance, 14:10 h light:dark). For growth experiments, coccolithophore cultures were acclimated to $2\,\mu M$ Si for several generations (1–2 weeks) before the onset of the experimental period. For SEM analysis, all cultures were maintained in $100\,\mu M$ Si for 1–2 weeks before the onset of the experimental period to prevent accumulation of aberrant coccoliths in the control.

Very low Si seawater was prepared using diatoms to deplete Si as described previously[64]. One-litre batches of f/2 FSW (without added Si) were inoculated with the diatom *Thalassiosira weissflogii* and allowed to grow into stationary phase (6–10 days). Diatoms were removed by two passages through $0.2\text{-}\mu m$ filters and the Si concentration was verified on an autoanalyser (Bran + Luebb, Germany) using a molybdate-ascorbate assay[63]. The initial Si in Gulf Stream Seawater was $5.4\,\mu M$ and after diatom depletion the Si was below the level of detection ($<0.1\,\mu M$, hereto referred to as 0Si f/2 FSW). Before inoculation of treatment media, aliquots of cells were washed at least twice by allowing them to settle, drawing off the overlying media, and resuspending in 0Si f/2 FSW. An inoculum of 0Si f/2 FSW washed cells was then added to a tube of the treatment media and monitored over 72 h. Care was taken to ensure final cell numbers did not exceed 2×10^4 cells per ml for *E. huxleyi*, the most rapidly growing of the three species, thus avoiding any significant changes to the carbonate chemistry of the culture medium over the course of the experimental incubations.

Physiological measurements. Growth rates of coccolithophore cultures were determined by cell counts using a Sedgewick-Rafter counting chamber (*C. braarudii*, *P. carterae*) or a Neubauer improved haemocytometer (*E. huxleyi*). Specific growth rates (per day) were determined from the initial and final cell densities (N_{t0}, N_{t1}) using the formula $\mu = (\ln(N_{t1}) - \ln(N_{t0}))/t)$. For Ge-treated *C. braarudii* cultures, an initial cell density of 1.2×10^4 cells per ml was used to ensure sufficient biomass was available after 48 h for measurements of chlorophyll fluorimetry. For these short-term incubations, the control cultures exhibited a specific growth rate between 0.24–0.35 per day and growth of the Ge-treated cultures is shown as a percentage of the control. For Si-limited cultures, an initial cell density of 4.5×10^3 cells per ml was used and growth was monitored over 8 days. Discarded coccoliths of *C. braarudii* were also counted for selected experiments. As coccolith morphology can be difficult to determine accurately by light microscopy, we did not discriminate between intact and aberrant liths in these analyses. To assess the performance of the photosynthetic apparatus, the maximum quantum yield of photosystem II was determined using a Z985 AquaPen chlorophyll fluorimeter (Qubit Systems, Kingston, Canada). Statistical analyses of these data were performed in SigmaPlot v12.0 software (Systat Software Inc, London, UK).

Fluorescence microscopy of *P. neolepis* silica scales. One millilitre of *P. neolepis* cells was incubated with the fluorescent dye LysoTracker yellow HCK-123 or LysoSensor Yellow/Blue DND-160 (Invitrogen; 1 µM, 10 h). Fluorescently labelled scales were imaged by confocal laser scanning microscopy (Zeiss LSM 510 microscope). HCK-123 was viewed using excitation at 488 nm and emission at 500–550 nm. DND-160 was viewed using multiphoton excitation at 740 nm with emission at 435–485 nm and 500–550 nm. Chlorophyll autofluorescence was also detected (emission 650–710 nm).

Extraction of silica-associated organic components. Organic components were extracted from the silica scales of *P. neolepis* using a modified protocol for diatom frustules[65]. Cells in mid-exponential growth phase were harvested by low pressure filtration and pelleted by centrifugation ($500 \times g$, 5 min, Thermo Scientific, Waltham, MA). The cells were disrupted by the addition of 10 ml of lysis buffer (2% SDS, 100 mM EDTA, 0.1 M Tris pH 8.0), vortexed and centrifuged at 6,000g for 10 min. The pellet containing the silica scales was washed with lysis buffer a further five times to remove cellular organic material. The silica scales were further purified by centrifugation through a 50% glycerol cushion (3,200g, 2 min) to remove any traces of contaminating low-density organic material, such as the smaller organic scales. The purity of the silica scale preparation was assessed by light microscopy (Nikon Ti Eclipse, Tokyo, Japan) and electron microscopy (both SEM and transmission electron microscopy). No contamination with cell debris or organic scales was observed in the purified preparations of silica scales, although organic scales could clearly be viewed in crude cell extracts. To dissolve the silica component of scales, 2 ml of 10 M NH_4F was added to 30–100 mg biosilica sample and vortexed until the pellet was dissolved. 0.5 ml of 6 M HCl was then added to the mixture, vortexed, and the pH was adjusted to 4.5 with 6 M HCl. The sample was incubated at room temperature for 30 min before centrifugation (3,200g, 15 min) and the supernatant was transferred to a 3 kDa cut-off filtration column (Amicon) to concentrate and desalt protein. The concentrate was washed sequentially with 5 ml of 500 mM ammonium acetate, 5 ml of 200 mM ammonium acetate and three times with 5 ml of 50 mM ammonium acetate. The sample was then further concentrated to 150–400 µl and analysed using Tricine/SDS–PAGE with Coomassie Blue staining for both proteins and LCPAs[66]. Staining with silver stain or Stains-All (Sigma), which do not bind to LCPAs, was used to verify that the lower molecular weight component did not contain protein. A trypsin digest was also conducted, where 10 µl of the NH_4F soluble extract was incubated with 2 µg of TPCK (tosyl phenylalanyl chloromethyl ketone)-treated trypsin in 100 mM Tris-HCl at pH 8.8 at 37 °C (18 h). Analysis by Tricine/SDS–PAGE revealed that the higher molecular weight component had been removed by trypsin, but the low-molecular-weight component (LCPA) remained. To further confirm that silica scales were not contaminated with cellular debris or organic scales, the purified silica scales were extracted with 5 ml buffer (100 mM EDTA, 0.1 M Tris pH 8.0) in the absence of NH_4F dissolution. No organic components were observed following Tricine/SDS–PAGE analysis, indicating that the organic components observed following treatment with NH_4F are released by silica dissolution.

Protein identification from silica scale extract. Following Tricine/SDS–PAGE, protein bands were excised from the gel and analysed by peptide mass fingerprinting using a tryptic digest (Alta Bioscience, Abingdon, UK). The *P. neolepis* transcriptome was used to create a reference proteome. A single protein (LPCL1) was identified, with 8–16 unique peptides identified in each sample. The protein identification was repeated at an alternative facility (Mass Spectrometry Facility, Biosciences, University of Exeter, UK) using an independent protein extract. This gave an identical result identifying 8 unique peptides for LPCL1.

LCPA purification from silica scale extract. LCPAs were separated from the protein fraction by ultrafiltration of 500 µl of the NH_4F soluble extract through a 10 kD MW filtre. The LCPAs were then further purified by cation exchange through 2 ml of high S strong cation exchange resin (Bio-Rad, Hemel Hempstead, UK). The column was prepared by washing sequentially with 10 ml of deionised water, 10 ml of 2 M ammonium acetate and two further times with deionised water. The NH_4F extraction was diluted (4.5:100) with deionised water and passed through the column. The resin was then washed three times with 1 ml of 200 mM ammonium acetate and polyamines were eluted by 4 sequential additions of 1 ml of 2 M ammonium acetate. The eluant was neutralized with acetic acid and lyophilized. Long-chain polyamines were analysed by electrospray ionization mass spectrometry (ESI-MS) using an amaZon speed mass spectrometer (Bruker, Bremen). Samples were diluted in H_2O/CH_3CN (50/50), and injected by direct infusion at a flow rate of 500 nl min^{-1} using a Captive Spray ion source. MS and MSn spectra were acquired in positive ion mode.

Generation of the *P. neolepis* transcriptome. A 100-ml culture of *P. neolepis* strain PZ241 growing in standard conditions (mid-exponential phase, f/2 + Si, other growth conditions as described above) was used to generate the transcriptome. Cells were collected 4 h into the light cycle by centrifugation (500g, 5 min). RNA was extracted using the Trizol method (Invitrogen, Paisley, UK), with additional purification using an RNeasy kit (Qiagen, Venlo, Netherlands). Following reverse transcription using oligo-dT primers, *P. neolepis* complementary DNA was sequenced by Illumina technology, generating 64,548,084 paired end reads of 75 bp (Genoscope, Evry, France). The paired end reads were assembled by Trinity[67], producing 118,473 transcripts, including alternative forms of a total of 83,175 transcripts.

Reverse transcription PCR. Reverse transcription PCR (RT–PCR) was used to verify the expression of selected genes (*LPCL1*, *SIT* and *SITL*) identified in the haptophyte transcriptomes. Fifty-millilitre cultures of *P. neolepis* (TMR5), *C. braarudii*, *C. leptoporus* and *S. apsteinii* were grown in standard conditions (as described above). Cells were collected ~4 h into the light cycle by centrifugation (500g, 5 min). RNA was extracted using the Trizol method (Invitrogen), with additional purification using an RNeasy kit (Qiagen). Complementary DNA was synthesized using either oligo-dT primers (*PnSIT1*) or gene specific primers (all other products) using Superscript III reverse transcriptase (Invitrogen). Gene products were then amplified by PCR (95 °C for 30 s, 54 °C for 30 s, 72 °C for 60 s, 35 cycles) (Supplementary Table 3). PCR products were sequenced to confirm the amino-acid sequence of the predicted protein product (Source BioScience, Cambridge, UK). The nucleotide sequences of *P. neolepis* SIT1 and LPCL1 obtained from the TMR5 strain were 100% identical at the nucleotide level to those identified in the PZ241 transcriptome.

Bioinformatic analyses. Known proteins associated with silicification from diatoms, sponges and land plants were used to search the haptophyte transcriptomes (Supplementary Table S1). The additional haptophyte transcriptomes were obtained from the Marine Microbial Eukaryote Sequencing Project (MMETSP; http://camera.calit2.net/mmetsp/)[35]. The genomes of *Emiliania huxleyi* v1.0 and *Thalassiosira pseudonana* v3.0 were obtained from the Joint Genome Institute (JGI; http://genome.jgi.doe.gov/). Further searches were performed at NCBI (http://blast.ncbi.nlm.nih.gov/Blast.cgi), including Transcriptome Shotgun Assembly (TSA) and Expressed Sequence Tag (EST) databases. Databases were searched using BLASTP and TBLASTN. Position-specific iterative BLAST (PSI-BLAST) was used to identify highly conserved motifs in proteins that exhibit low levels of sequence identity (for example, lipocalins). Each potential hit was manually inspected using a multiple sequence alignment to identify conserved residues and then phylogenetic analyses were performed using both neighbour-joining and maximum likelihood methods within the MEGA5 software package to assess the relationship with known proteins[68]. For detailed phylogenetic analysis of SITs and SITLs, multiple sequence alignments were generated using MUSCLE and manually inspected for alignment quality. After manual refinement, GBLOCKS 0.91b was used to remove poorly aligned residues[69] and then ProtTest was used to determine the best substitution model (WAG with gamma and invariant). Maximum likelihood phylogenetic trees were generated using PhyML3.0 software with 100 bootstraps. Bayesian posterior probabilities were calculated using BEAST v1.8, running for 10,000,000 generations[70]. The identification of potential transmembrane domains in SITLs was performed using Phobius and TMHMM.

Electron microscopy. SEM images of *P. neolepis* scales were acquired with a JEOL 5000 and JEOL 7001 F microscopes (Jeol, Japan) at 15 keV accelerating voltage. Scales were collected using lysis buffer (2% SDS, 100 mM EDTA, 0.1 M Tris pH 8.00) as described above, but were additionally cleaned by heating at 95 °C for 10 min in the lysis buffer. Purified *P. neolepis* silica scale material was dried and sputter coated with gold or chromium before imaging. Samples of coccolithophores for SEM were collected by filtration onto a 13-mm 0.4-µm Isopore filter (Millpore EMD), followed by a rinse with 10 ml of 1 mM HEPES buffer (pH 8.2) to remove salts. Filters were air-dried, mounted onto an aluminium stub and sputter coated with 10 nm Pt/Pd (Cressington, USA). Samples were examined with a Phillips XL30S FEG SEM (FEI-Phillips, USA) and imaged in high-resolution secondary electron mode with beam acceleration of 5 kV. Three categories of coccolith morphology were scored. (i) 'Blocky' coccoliths where the overlapping arrangement of the distal shield is disrupted, but the overall shape of the coccolith is not disrupted (*C. braarudii* and *C. leptoporus* only). (ii) 'Aberrant' coccoliths were classified as coccoliths that clearly departed from the typical morphology for any given species. (iii) 'Discarded aberrant' coccoliths were classed as those aberrant coccoliths that failed to integrate into the coccosphere. To analyse coccolith morphology, at least 40 cells per treatment were scored for the number of malformed coccoliths present in the coccosphere. Coccoliths on the underside of cells could not be scored and the resultant underestimate of coccoliths per cell was assumed to be the same for any given species. Discarded aberrant coccoliths were counted in four to seven random fields of view in which both cells and loose aberrant coccoliths were scored (at least 40 cells in total were scored for each treatment). Ge-treated cultures for SEM analysis were grown in single replicates. Each experiment was repeated on multiple independent occasions and in each case the effects of Ge were highly reproducible. A representative example of each experiment is shown. Error bars denote standard error.

References

1. De La Rocha, C. L. & Passow, U. Factors influencing the sinking of POC and the efficiency of the biological carbon pump. *Deep Sea Res. Part II* **54**, 639–658 (2007).

2. Schmidt, K., De La Rocha, C. L., Gallinari, M. & Cortese, G. Not all calcite ballast is created equal: differing effects of foraminiferan and coccolith calcite on the formation and sinking of aggregates. *Biogeosciences* **11**, 135–145 (2014).

3. Raven, J. A. & Giordano, M. Biomineralization by photosynthetic organisms: evidence of coevolution of the organisms and their environment? *Geobiology* **7**, 140–154 (2009).

4. De Vargas, C., Aubry, M.-P., Probert, I. & Young, J. in *Evolution of Primary Producers in the Sea* (eds Falkowski, P. & Knoll, A.) (Elsevier, 2007).

5. Daniels, C. J., Sheward, R. M. & Poulton, A. J. Biogeochemical implications of comparative growth rates of *Emiliania huxleyi* and *Coccolithus* species. *Biogeosciences* **11**, 6915–6925 (2014).

6. Rickaby, R. E. M., Henderiks, J. & Young, J. N. Perturbing phytoplankton: response and isotopic fractionation with changing carbonate chemistry in two coccolithophore species. *Clim. Past* **6**, 771–785 (2010).

7. Edvardsen, B. *et al.* Ribosomal DNA phylogenies and a morphological revision provide the basis for a revised taxonomy of the Prymnesiales (Haptophyta). *Eur. J. Phycol.* **46**, 202–228 (2011).

8. Yoshida, M., Noel, M. H., Nakayama, T., Naganuma, T. & Inouye, I. A haptophyte bearing siliceous scales: Ultrastructure and phylogenetic position of *Hyalolithus neolepis* gen. et sp nov (Prymnesiophyceae, Haptophyta). *Protist* **157**, 213–234 (2006).

9. Patil, S., Mohan, R., Shetye, S., Gazi, S. & Jafar, S. A. *Prymnesium neolepis* (Prymnesiaceae), a siliceous Haptophyte from the Southern Indian Ocean. *Micropaleontology* **60**, 475–481 (2014).

10. Taylor, A. R., Russell, M. A., Harper, G. M., Collins, T. F. T. & Brownlee, C. Dynamics of the formation and secretion of heterococcoliths by *Coccolithus pelagicus* (ssp *braarudii*). *Eur. J. Phycol.* **42**, 125–1336 (2007).

11. Liu, H., Aris-Brosou, S., Probert, I. & de Vargas, C. A time line of the environmental genetics of the haptophytes. *Mol. Biol. Evol.* **27**, 161–176 (2010).

12. Harper, H. E. & Knoll, A. H. Silica, diatoms, and cenozoic radiolarian evolution. *Geology* **3**, 175–177 (1975).

13. Lazarus, D. B., Kotrc, B., Wulf, G. & Schmidt, D. N. Radiolarians decreased silicification as an evolutionary response to reduced Cenozoic ocean silica availability. *Proc. Natl Acad. Sci. USA* **106**, 9333–9338 (2009).

14. Maldonado, M., Carmona, M. G., Uriz, M. J. & Cruzado, A. Decline in Mesozoic reef-building sponges explained by silicon limitation. *Nature* **401**, 785–788 (1999).

15. Sims, P. A., Mann, D. G. & Medlin, L. K. Evolution of the diatoms: insights from fossil, biological and molecular data. *Phycologia* **45**, 361–402 (2006).

16. Balch, W. M. *et al.* Surface biological, chemical, and optical properties of the Patagonian Shelf coccolithophore bloom, the brightest waters of the Great Calcite Belt. *Limnol. Oceanogr.* **59**, 1715–1732 (2014).

17. Hopkins, J., Henson, S. A., Painter, S. C., Tyrrell, T. & Poulton, A. J. Phenological characteristics of global coccolithophore blooms. *Global Biogeochem. Cycles* **29**, 239–253 (2015).

18. Leblanc, K. *et al.* Distribution of calcifying and silicifying phytoplankton in relation to environmental and biogeochemical parameters during the late stages of the 2005 North East Atlantic Spring Bloom. *Biogeosciences* **6**, 2155–2179 (2009).

19. Tyrrell, T. & Merico, A. in *Coccolithophores: From Molecular Processes to Global Impact.* (eds Thierstein, H. R. & Young, J. R.) 75–97 (2004).

20. Kroger, N. & Poulson, N. Diatoms from cell wall biogenesis to nanotechnology. *Annu. Rev. Genet.* **42**, 83–107 (2008).

21. Hildebrand, M., Volcani, B. E., Gassmann, W. & Schroeder, J. I. A gene family of silicon transporters. *Nature* **385**, 688–689 (1997).

22. Thamatrakoln, K. & Hildebrand, M. Silicon uptake in diatoms revisited: a model for saturable and nonsaturable uptake kinetics and the role of silicon transporters. *Plant Physiol.* **146**, 1397–1407 (2008).

23. Ma, J. F. *et al.* A silicon transporter in rice. *Nature* **440**, 688–691 (2006).

24. Schroder, H. C. *et al.* Silica transport in the demosponge *Suberites domuncula*: fluorescence emission analysis using the PDMPO probe and cloning of a potential transporter. *Biochem. J.* **381**, 665–673 (2004).

25. Likhoshway, Y. V., Masyukova, Y. A., Sherbakova, T. A., Petrova, D. P. & Grachev, A. M. A. Detection of the gene responsible for silicic acid transport in chrysophycean algae. *Dokl. Biol. Sci.* **408**, 256–260 (2006).

26. Marron, A. O. *et al.* A family of diatom-like silicon transporters in the siliceous loricate choanoflagellates. *Proc. Biol. Sci.* **280**, 20122543 (2013).

27. Thamatrakoln, K., Alverson, A. J. & Hildebrand, M. Comparative sequence analysis of diatom silicon transporters: Toward a mechanistic model of silicon transport. *J. Phycol.* **42**, 822–834 (2006).

28. Ma, J. F. *et al.* An efflux transporter of silicon in rice. *Nature* **448**, 209–212 (2007).

29. Shrestha, R. P. *et al.* Whole transcriptome analysis of the silicon response of the diatom *Thalassiosira pseudonana*. *BMC Genomics* **13**, 499 (2012).

30. Manton, I. Further observations on the fine structure of *Chrysochromulina chiton* with special reference to the haptonema, 'peculiar' golgi structure and scale production. *J. Cell. Sci.* **2**, 265–272 (1967).

31. Shimizu, K., Cha, J., Stucky, G. D. & Morse, D. E. Silicatein alpha: cathepsin L-like protein in sponge biosilica. *Proc. Natl Acad. Sci. USA* **95**, 6234–6238 (1998).

32. Kroger, N., Deutzmann, R., Bergsdorf, C. & Sumper, M. Species-specific polyamines from diatoms control silica morphology. *Proc. Natl Acad. Sci. USA* **97**, 14133–14138 (2000).

33. Matsunaga, S., Sakai, R., Jimbo, M. & Kamiya, H. Long-chain polyamines (LCPAs) from marine sponge: possible implication in spicule formation. *Chembiochem* **8**, 1729–1735 (2007).

34. Michael, A. J. Molecular machines encoded by bacterially-derived multi-domain gene fusions that potentially synthesize, N-methylate and transfer long chain polyamines in diatoms. *FEBS Lett.* **585**, 2627–2634 (2011).

35. Keeling, P. J. *et al.* The Marine Microbial Eukaryote Transcriptome Sequencing Project (MMETSP): illuminating the functional diversity of eukaryotic life in the oceans through transcriptome sequencing. *PLoS Biol.* **12**, e1001889 (2014).

36. Abramson, J. & Wright, E. M. Structure and function of Na$^+$-symporters with inverted repeats. *Curr. Opin. Struct. Biol.* **19**, 425–432 (2009).

37. Khafizov, K., Staritzbichler, R., Stamm, M. & Forrest, L. R. A study of the evolution of inverted-topology repeats from LeuT-fold transporters using AlignMe. *Biochemistry* **49**, 10702–10713 (2010).

38. Duran, A. M. & Meiler, J. Inverted topologies in membrane proteins: a mini-review. *Comput. Struct. Biotechnol. J.* **8**, e201308004 (2013).

39. Michels, J., Vogt, J., Simon, P. & Gorb, S. N. New insights into the complex architecture of siliceous copepod teeth. *Zoology* **118**, 141–146 (2015).

40. Drescher, B., Dillaman, R. M. & Taylor, A. R. Coccolithogenesis in *Scyphosphaera apsteinii* (Prymnesiophyceae). *J. Phycol.* **48**, 1343–1361 (2012).

41. Gal, A., Weiner, S. & Addadi, L. The stabilizing effect of silicate on biogenic and synthetic amorphous calcium carbonate. *J. Am. Chem. Soc.* **132**, 13208–13211 (2010).

42. Ihli, J. *et al.* Dehydration and crystallization of amorphous calcium carbonate in solution and in air. *Nat. Commun.* **5**, 3169 (2014).

43. Kellermeier, M. *et al.* Stabilization of amorphous calcium carbonate in inorganic silica-rich environments. *J. Am. Chem. Soc.* **132**, 17859–17866 (2010).

44. Zhang, G., Delgado-Lopez, J. M., Choquesillo-Lazarte, D. & Garcia-Ruiz, J. M. Growth behavior of monohydrocalcite ($CaCO_3.H_2O$) in silica-rich alkaline solution. *Cryst. Growth Des.* **15**, 564–572 (2015).

45. Azam, F., Hemmings, B. b. & Volcani, B. E. Germanium incorporation into silica of diatom cell-walls. *Arch. Microbiol.* **92**, 11–20 (1973).

46. Darley, W. M. & Volcani, B. E. Role of silicon in diatom metabolism. A silicon requirement for deoxyribonucleic acid synthesis in the diatom *Cylindrotheca fusiformis* Reimann and Lewin. *Exp. Cell Res.* **58**, 334–342 (1969).

47. Azam, F. & Volcani, B. E. in *Silicon and siliceous structures in biological systems* (eds Simpson, T. L. & Volcani, B. E.) 43–68 (Springer-Velag, 1981).

48. Klaveness, D. & Guillard, R. R. L. Requirement for silicon in *Synura petersenii* (Chrysophyceae). *J. Phycol.* **11**, 349–355 (1975).

49. Lee, R. E. Formation of scales in *Paraphysomonas vestita* and inhibition of growth by germanium dioxide. *J. Protozool.* **25**, 163–166 (1978).

50. Lewin, J. C. Silicon metabolism in diatoms. V. germanium dioxide, a specific inhibitor of diatom growth. *Phycologia* **6**, 1–12 (1966).

51. Gerecht, A. C., Supraha, L., Edvardsen, B., Probert, I. & Henderiks, J. High temperature decreases the PIC/POC ratio and increases phosphorus requirements in *Coccolithus pelagicus* (Haptophyta). *Biogeosciences* **11**, 3531–3545 (2014).

52. Probert, I. & Houdan, A. in *Coccolithophores: From Molecular Processes to Global Impact.* (eds Thierstein, H. R. & Young, J. R.) 217–249 (2004).

53. Miller, J. J. & Delwiche, C. F. Phylogenomic analysis of *Emiliania huxleyi* provides evidence for haptophyte-stramenopile association and a chimeric haptophyte nuclear genome. *Marine Genomics* **21**, 31–42 (2015).

54. Stiller, J. W. *et al.* The evolution of photosynthesis in chromist algae through serial endosymbioses. *Nat. Commun.* **5**, 5764 (2014).

55. Yue, J., Sun, G., Hu, X. & Huang, J. The scale and evolutionary significance of horizontal gene transfer in the choanoflagellate *Monosiga brevicollis*. *BMC Genomics* **14**, 729 (2013).

56. Schwarz, K. & Milne, D. B. Growth-promoting effects of silicon in rats. *Nature* **239**, 333 (1972).

57. Carlisle, E. M. Silicon. A possible factor in bone calcification. *Science* **167**, 279 (1970).

58. Reffitt, D. M. *et al.* Orthosilicic acid stimulates collagen type 1 synthesis and osteoblastic differentiation in human osteoblast-like cells *in vitro*. *Bone* **32**, 127–135 (2003).

59. Carlisle, E. M., Berger, J. W. & Alpenfels, W. F. A silicon requirement for prolyl hydroxylase-activity. *Fed. Proc.* **40**, 886–886 (1981).

60. Gal, A. *et al.* Plant cystoliths: a complex functional biocomposite of four distinct silica and amorphous calcium carbonate phases. *Chemistry* **18**, 10262–10270 (2012).

61. Ziveri, P., Broerse, A. T. C., van Hinte, J. E., Westbroek, P. & Honjo, S. The fate of coccoliths at 48 degrees N 21 degrees W, northeastern Atlantic. *Deep Sea Res. Part II* **47**, 1853–1875 (2000).

62. Yool, A. & Tyrrell, T. Role of diatoms in regulating the ocean's silicon cycle. *Global Biogeochem. Cycles* **17**, 1103 (2003).

63. Zhang, J. Z. & Berberian, G. A. in *Methods for the Determination Of Chemical Substances In Marine And Estuarine Environmental Matrices* (ed Arar, E. J.) 366.360-361–366.360-313 (U.S. Environmental Protection Agency, 1997).

64. Timmermans, K. R., Veldhuis, M. J. W. & Brussaard, C. P. D. Cell death in three marine diatom species in response to different irradiance levels, silicate, or iron concentrations. *Aquat. Microb. Ecol.* **46**, 253–261 (2007).

65. Kroger, N., Deutzmann, R. & Sumper, M. Polycationic peptides from diatom biosilica that direct silica nanosphere formation. *Science* **286**, 1129–1132 (1999).

66. Schagger, H. Tricine-SDS-PAGE. *Nat. Protoc.* **1**, 16–22 d (2006).

67. Haas, B. J. *et al.* De novo transcript sequence reconstruction from RNA-seq using the Trinity platform for reference generation and analysis. *Nat. Protoc.* **8**, 1494–1512 (2013).

68. Tamura, K. *et al.* MEGA5: molecular evolutionary genetics analysis using maximum likelihood, evolutionary distance, and maximum parsimony methods. *Mol. Biol. Evol.* **28**, 2731–2739 (2011).

69. Talavera, G. & Castresana, J. Improvement of phylogenies after removing divergent and ambiguously aligned blocks from protein sequence alignments. *Syst. Biol.* **56**, 564–577 (2007).

70. Drummond, A. J., Suchard, M. A., Xie, D. & Rambaut, A. Bayesian phylogenetics with BEAUti and the BEAST 1.7. *Mol. Biol. Evol.* **29**, 1969–1973 (2012).

Acknowledgements

We thank Roy Moate, Glenn Harper and Peter Bond (University of Plymouth, UK) for help with electron microscopy imaging, Malcolm Woodward (Plymouth Marine Laboratory, UK) for help with silicate analyses, Nicholas Smirnoff and Hannah Florance (University of Exeter, UK) for help with protein sequencing and Nils Kroger and Nicole Poulsen (TU Dresden, Germany) for their guidance on analysing the organic component of the silica scales. This study was funded by EU Interreg IV *Marinexus* grant. GW and CB acknowledge support from NERC grant (NE/J021954/1). ART acknowledges NSF grant IOS 0949744 and the University of North Carolina Wilmington Microscopy Facility.

Author contributions

G.M.D. characterized the mechanisms of silicification in *P. neolepis* and identified the requirement for Si in coccolithophores. A.R.T. performed detailed analysis of coccolithophore calcification. C.E.W. performed additional experiments. S.A., I.P., C.d.V. sequenced the transcriptome of *P. neolepis*. G.M.D., A.R.T., C.B. and G.L.W. designed the study and analysed the data. G.M.D., A.R.T., D.C.S., C.B. and G.L.W. wrote the paper.

Additional information

9

Deep-reaching thermocline mixing in the equatorial pacific cold tongue

Chuanyu Liu[1,2,3], Armin Köhl[1], Zhiyu Liu[4], Fan Wang[2,3] & Detlef Stammer[1]

Vertical mixing is an important factor in determining the temperature, sharpness and depth of the equatorial Pacific thermocline, which are critical to the development of El Ninõ and Southern Oscillation (ENSO). Yet, properties, dynamical causes and large-scale impacts of vertical mixing in the thermocline are much less understood than that nearer the surface. Here, based on Argo float and the Tropical Ocean and Atmosphere (TAO) mooring measurements, we identify a large number of thermocline mixing events occurring down to the lower half of the thermocline and the lower flank of the Equatorial Undercurrent (EUC), in particular in summer to winter. The deep-reaching mixing events occur more often and much deeper during periods with tropical instability waves (TIWs) than those without and under La Niña than under El Niño conditions. We demonstrate that the mixing events are caused by lower Richardson numbers resulting from shear of both TIWs and the EUC.

[1] Institute of Oceanography, Center for Earth System Research and Sustainability (CEN), University of Hamburg (UHH), Hamburg 20146, Germany. [2] Key Lab of Ocean Circulation and Waves (KLOCAW), Institute of Oceanology, Chinese Academy of Sciences (IOCAS), Nanhai Road 7, Qingdao 266071, China. [3] Function Laboratory for Ocean and Climate Dynamics, Qingdao National Laboratory for Marine Science and Technology (QNLM), Qingdao 266237, China. [4] State Key Laboratory of Marine Environmental Science (MEL) and Department of Physical Oceanography, College of Ocean and Earth Sciences, Xiamen University, Xiamen 361102, China. Correspondence and requests for materials should be addressed to C.L. (email: chuanyu.liu@qdio.ac.cn).

The maintenance of the equatorial Pacific thermocline relies either on high-latitude buoyancy forcing or on extra-tropical wind and buoyancy forcing[1,2] at annual to inter-annual time scales, but is modulated by local Kelvin waves[3] and wind stress curl[4] at intra-seasonal to seasonal time scales. Numerical experiments suggest that the sharpness and depth of the thermocline is also determined by vertical mixing within it[5,6]. Measurements and model studies suggests that turbulence and mixing below the mixed layer base of the equatorial Pacific are attributed to the vertical velocity gradient (shear) between the eastward flowing Equatorial Undercurrent (EUC) and the westward flowing South Equatorial Current[7-15], which is likely to be further modulated by the wind stress[16-18]. Mixing or instabilities in layers further below, ranging from the upper[13] to the lower[14,19] parts of the thermocline, are also observed from limited measurements. The instabilities in the lower part of the thermocline may be caused by absorption and saturation of wave energy at critical levels[19], whereas the mixing in the upper part of the thermocline is found to be related to baroclinic inertial-gravity waves[20], Kelvin waves[14] and, in particular, the tropical instability waves (TIWs)[13].

TIWs refer to energetic meanders frequently emerging in the middle and eastern equatorial ocean. They have long been proposed to be a combination of a Yanai(-like) wave on the Equator and a first-meridional-mode Rossby wave just north of the Equator, with periods of 12–40 days and wavelengths of 700–1,600 km (refs 21–26). Alternatively, TIWs are also suggested to be manifestations of tropical vortices or highly nonlinear waves[27,28].

A prominent feature of TIWs is the large meridional velocity ranging from the surface to the core of the EUC, providing the potential for vigorous interactions with the already energetic equatorial current system. Turbulence measurements taken by a Lagrangian float encountering a TIW[29] and modelling studies of the impact of TIWs[30] in the eastern/middle equatorial Pacific both found strong vertical mixing at the base of the surface mixed layer, which induces intensive cooling of the sea surface[30,31]. The measurements[29] suggest that the strong mixing can be explained by the enhancement of shear modulated by the TIW[29]. Direct turbulence measurements at 0°, 140° W encountering a TIW further confirmed the enhancement of mixing by TIW both in and below the surface mixed layer; in particular, the measurements also revealed a tenfold increase in turbulent heat flux in the upper half of the thermocline[13]. The resulting mixing was accompanied with a significant temperature change in the upper 150 m within a cycle of the TIW[32]. The vigorous deep-reaching mixing are also attributed to additional shear provided by the meridional velocity of the TIW above the EUC core[13].

If this identified relationship between TIW and enhanced deep thermocline mixing is largely representative, it implies that after a long duration of TIWs the associated thermocline mixing may have the potential to alter the structure of the thermocline and the subsurface temperature of the Pacific cold tongue, which may further have an impact on the large-scale oceanic-atmospheric dynamics, such as El Niño and Southern Oscillation (ENSO)[6] and the global climate at large[33].

However, observational evidence for the link between the TIWs and enhanced thermocline mixing is far from adequate. To date, direct turbulence measurements were confined to a few specific locations and covered only short time spans. Whether the thermocline mixing is organized in seasonal or longer-period cycles that are mechanistically related to variations of TIWs at the same periods and to what depths the TIWs may have an impact on the vertical mixing need to be explored.

Two databases could be employed to investigate the both issues. One is the Argo float database[34]. More than 3,000 freely drifting Argo floats continuously provide millions of profiles of temperature and salinity in the upper ~2,000 m ocean. The measurements offer a great opportunity to shed light on vertical mixing, because many profiles possess fine resolution, that is, small enough sample spacing ($O(1\,\text{m})$), to resolve turbulent mixing processes in the ocean interior. Mapping the global distribution of vertical mixing based on the Argo observations[35] with a fine-scale parameterization method[36] has demonstrated the usefulness of Argo measurements in ocean mixing studies. However, such estimation so far has been restricted to extra-equatorial regions below the thermocline due to limitations of the employed method[36,37]. Alternatively, the Thorpe method (see Methods and ref. 38) is suitable in the thermocline and could be applied to fine resolution Argo profiles, to detect mixing events in the equatorial thermocline.

The second database is the Tropical Atmosphere and Ocean (TAO) mooring observations[39]. The TAO array has provided continuous and high-quality oceanographic data including velocity, temperature and salinity in the upper 500 m over the last two decades. The method of linear stability analysis (LSA; see Methods and refs 18,40–42) is applied to the hourly profiles of density and velocity at a location in the middle equatorial Pacific. This method enables to detect potential instabilities occurring in the thermocline.

From both databases, we obtained large amount of possible mixing events (featured as density overturns and potential instabilities). We show that the mixing events occurred not only in the upper part of the thermocline but also deep down to the centre and lower part of it. We also show that the mixing events occurred more often and much deeper during periods of TIWs and of La Niña conditions because of stronger shear instabilities.

Results

Deep-reaching density overturns in Argo float measurements. The equatorial Pacific is a region with accumulated Argo float observations. Among all observed profiles, there exist ~20,000 fine resolution profiles that are with a sample spacing of no more than 2 m (see Methods) in the upper 200 m and covering 10° S to 10° N and 180 to 80° W over the period of January 2000 to June 2014 (Supplementary Fig. 1). The Thorpe method (see Methods) is applied to the fine resolution profiles and eventually ~800 density overturns are identified. Among them, a large portion of the overturns occurred after 2008, a period when most of the fine-resolution profiles exist. The horizontal distributions of the detected overturns are shown in Fig. 1a. In particular, in the region between 160° and 100° W, most overturns are confined to the equatorial band, ranging from 3° S to 6° N (about 400 overturns are found between 160°–110° W and 3° S–6° N); east of 100° W, overturns extend meridionally to ± 10°. Away from these regions, fewer overturns are detected.

Overlapped by the overturns in Fig. 1a is the occurrence probability of overturns calculated in 2° × 2° bins. The occurrence probability refers to the ratio of the number of Argo profiles that contain overturn(s) to the number of total qualified Argo profiles (see Methods) in a given area. Here, the time span is January 2000–June 2014. The occurrence probability ranges from 1 to 20% and peaks at 3%. It is noteworthy that the small values may not reflect the real occurrence probability of turbulent overturns, because the real overturns may have sizes of 10 cm to several metres, whereas here only the overturns larger than the sample spacing of the Argo profiles (2 m) were detected. Despite the scale selection, the results are still indicative for inferring the relationship between mixing events and TIWs. For example, a prominent feature of the horizontal distribution of the overturns is that they are concentrated in a band across the Equator but display a meridional asymmetry: 3° S–6° N.

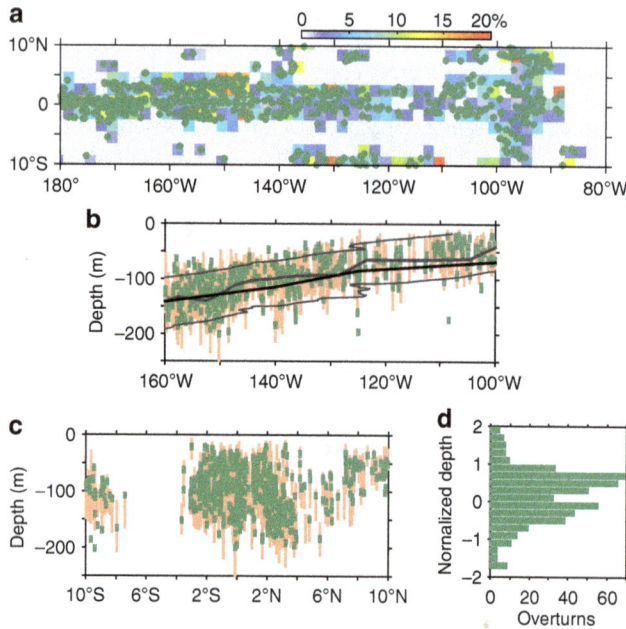

Figure 1 | Spatial distribution of detected density overturns in the equatorial Pacific cold tongue. (**a**) Occurrence probability (colour) and horizontal locations of overturns (dark green dots). (**b**) Depth and sizes (in metres) of the overturns (dark green bars) occurred between 3° S and 6° N, and the corresponding PLs (light orange bars) from a latitudinal view. The blue curve denotes the mean depth of the EUC core (averaged over ±1°; data are obtained over the 1990s (ref. 70)). The thick black curve and thin black curves denote the centre (depth of maximum N^2; N^2_{max}) and bounds (depths of half N^2_{max}) of the mean pycnocline. N^2 is calculated from sample mean density that is meridionally averaged from all fine-resolution Argo profiles over ±1°. (**c**) The same as in **b** but assembled from data between 160 and 100° W (curves of average variables are not added due to large zonal variation). (**d**) Histogram of overturns at the referenced and normalized vertical coordinate.

This band is within the region of the South Equatorial Current and EUC, and matches well the region of the TIWs, which are concentrated at the Equator and have two centres of high temperature variability at 5° N and 2° S (ref. 21).

We further graph the overturns with their corresponding instantaneous pycnocline layers (PLs) against both longitudes (Fig. 1b) and latitudes (Fig. 1c). Here, the centre of a PL is defined as the depth of the maximum N^2 (hereafter N^2_{max}, $N^2 = -g\rho_z/\rho_0$ is the buoyancy frequency squared, $\rho = \rho(z)$ is monotonically sorted potential density and is smoothed by a 40-m running mean, to remove influences of noises or intermittent internal waves, $\rho_0 = 1,000$ kg m^{-3} is the reference potential density and g is the gravitational acceleration). The upper and lower bounds of a PL are defined as the depths where $N^2 = 0.5 \times N^2_{max}$ above and below the PL centre (but is additionally bounded by the depth of $N^2 = 0.625 \times 10^{-4}$ s^{-2}). The size of the most detected overturns is 6 m (Supplementary Fig. 2), whereas the thicknesses of instantaneous PL vary from 50 to 100 m and are generally larger in the western than in the eastern equatorial Pacific (Fig. 1b). The synoptic overturns are confined to the temporally averaged PL (Fig. 1b; thin black curves) of the Equator. It is noticeable that a large fraction occurred below the centre of the average pycnocline (Fig. 1b; the thick black curve); they reached as deep as ~ -200 and ~ -100 m in the western and eastern equatorial Pacific, respectively.

We emphasize that the deep-reaching overturns in the pycnocline in the meantime also reached to the lower flank of the EUC. This can be inferred from Fig. 1b, where the average

core of the EUC (the depth of maximum eastward velocity; Fig. 1b; the blue curve) is more or less coincident with the centre of the pycnocline. This result is confirmed in the following by LSA examinations.

Figure 1c shows the meridional distribution of the detected overturns. It confirms the feature that the equatorial overturns are confined between 3° S and 6° N, the regime of the TIWs. Overturns outside this region are mainly found in the region east of 100° W.

The overturns and PLs shown on the physical depths (Fig. 1b,c) are subject to spatial and temporal variations. To provide an overview of the vertical distribution of the overturns relative to their corresponding pycnocline, we redistribute the overturns with respect to a transformed and normalized vertical coordinate (Fig. 1d). This coordinate is referred to the depth of N^2_{max}, and normalized in the upper (lower) half of the pycnocline by the thicknesses of the upper (lower) half of each PL. As such, in this coordinate, 0 represents the PL centre, while 1 and -1 represent the upper and lower bounds of the PL, respectively.

It shows that overturns occur not only in the upper part of the PL but also below the centre of the PL. Most overturns occur in the upper three quarters of the PL (between -0.5 and 1). Although in Fig. 1d the overturns peak at ~ 0.7, that is, near the upper bound of the pycnocline, it may not mean that the overturns in the ocean really peak here. This is because the prescribed cutoff buoyancy frequency (minimum of $N^2 = 0.5 \times N^2_{max}$ and $N^2 = 0.625 \times 10^{-4}$ s^{-2}) in the Thorpe method may have omitted overturns in weak-stratification layers, including the mixed layer and the layer just below. Nevertheless, the overturns peaking at ~ 0.7 needs an interpretation. Taking the location 0°, 140° W for reference, the centre and upper bound of the temporally averaged pycnocline are at -100 and -60 m, respectively (Fig. 1b); in consequence, depth 0.7 of the normalized coordinate corresponds to 28 m above the pycnocline centre, that is, at the physical depth of -72 m. According to direct turbulence measurements[13,32], this depth mostly belongs to the upper core layer, which refers to a layer that is located above the EUC core and accompanied with strong TIW-induced turbulence. In the depths above 0.7, the overturns may come from the deep cycle layer[7,8,10,11,14], which refers to a layer several tens of metres below the base of the surface mixed layer that undergoes a nighttime enhancement of turbulence; this layer is dynamically related to the diurnal varying surface buoyancy and wind forcing. It is noteworthy that the TIW-related upper core layer is seemingly separated from the surface-driven deep cycle layer[32]. Between the depths -0.5 and 0.7, more than a half overturns as those at depth 0.7 are found, indicating that intensive turbulence extends into the deep pycnocline.

Relationship between the deep-reaching overturns and TIWs. In general, TIWs are active from boreal summer to winter, while inactive in boreal spring[26]. To investigate whether the occurrence of overturns is also organized in such a seasonal cycle, the monthly occurrence probability of overturns in the region of active TIWs, 3° S and 6° N, and 160° W and 100° W, is calculated over the period between January 2005 and December 2013 (Fig. 2b). It is shown that the occurrence probability is indeed subject to similar seasonal variation: they peak in August and December, and have minimum values in boreal spring (April to June) and October. This seasonality is statistically significant. The maximum in August is twice the minima in October and April; the secondary maximum in December is >50% larger than the minima. From direct turbulence measurements over a 6-year span at 0°, 140° W, the vertical heat flux in the subsurface layers (-60 to -20 m) is found to be largest in boreal August, second largest in December, and least in spring and second least in

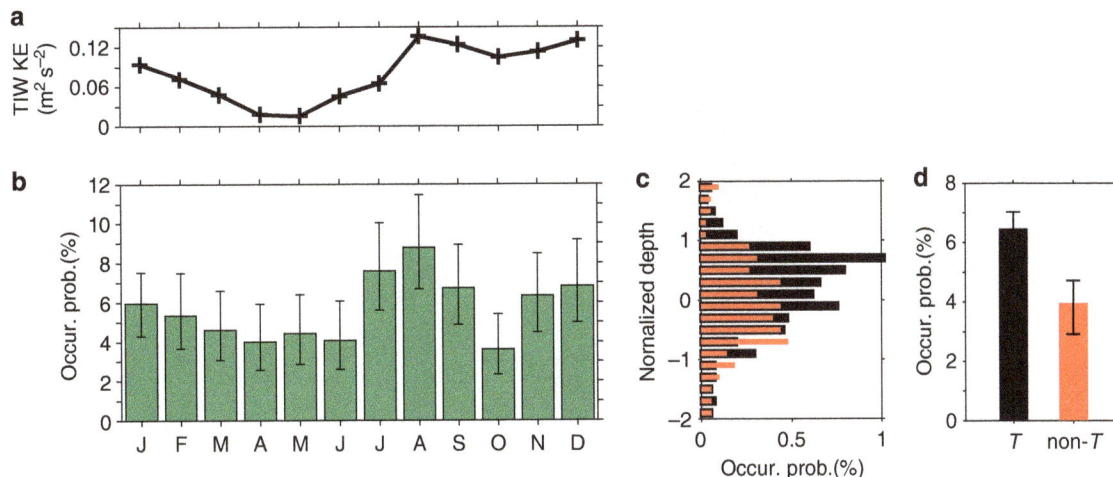

Figure 2 | Occurrence probability of detected overturns and its relation to tropical instability waves. (a) Monthly climatology of TIWs KE (averaged over years 2005-2011). **(b)** Monthly occurrence probability of overturns between 3°S and 6°N over 160 and 270°W; error bars are 95% bootstrap CIs. Peak at August is significantly different from surrounding troughs in June and October at the 95% bootstrap confidence level; the peak in December is significantly different from the trough at October at the 95% bootstrap confidence level and from troughs in April and June at the 90% bootstrap confidence level. The correlation coefficient between monthly TIWs KE in **a** and occurrence probability in **b** is $r = 0.66$, with P-value $= 0.020$ and 95% CI $= (0.14, 0.89)$, that is, statistically significant. **(c)** Histogram of the occurrence probability in the normalized coordinate for periods of TIW (blue) and non-TIW (red) (see text). **(d)** Total occurrence probability of overturns during TIW (6.34%, blue) and non-TIW (3.76%, red) periods; error bars are 95% bootstrap CIs (blue: (5.67%, 7.07%) and red: (2.92%, 4.71%)). In **b,c** and **d**, the occurrence probability is calculated over the years 2005-2013.

October and November[43], consistent with the occurrence probability of overturns shown here. Specifically, the monthly occurrence probability is significantly correlated with the multi-year (2005–2010) averaged monthly TIW kinetic energy (KE) at 0°, 140°W (Fig. 2a). (The TIW KE is calculated as $\overline{(\langle \widehat{u} \rangle^2_{30-70} + \langle \widehat{v} \rangle^2_{30-70})}$, where u and v are the eastward and northward components, respectively, of velocity observed by TAO moorings, $\widehat{u}$ and $\widehat{v}$ are the 12–40 days band-pass filtered, $\langle \ \rangle_{30-70}$ denotes the vertical mean over -70 to ~ -30 m and $\overline{(\ )}$ denotes a 40-day low-pass filtering.)

The results strongly indicate the modulating effect of TIWs on the occurrence of deep-reaching overturning in the pycnocline. Given that overturning and mixing usually accompany with each other, the result not only confirms the notion that TIWs lead to enhanced turbulence and mixing in the upper part of the thermocline[13] but also implies that the modulation effects of TIWs on mixing can reach deeper depths of the pycnocline. (As density here is dominated by temperature[32], in the remainder of the study we focus on the thermocline instead of the pycnocline.)

In the following, we will further verify the impact of TIWs in enhancing the occurrence of overturns by comparing overturn properties of TIW periods with those of non-TIW periods. To this end, we followed ref. 44 and defined TIW periods and non-TIW periods based on meridional sea surface temperature (SST) gradient. The reason why we adopted this strategy is because a large portion of the fine resolution Argo profiles and overturns are found between the years 2011 and 2014, whereas the TAO velocity measurement and hence the TIW KE index are not available since 2011. SST[45] (see Methods for data source) is first averaged in longitude spanning 12° centred at two latitudes of 140°W, 4.5°N and 0.5°N, and then the averaged SSTs are 140-day low-pass filtered; finally, the meridional gradient of the filtered SSTs (SST_y) is calculated. TIW periods are defined as the periods when the SST_y is $> 0.25 \times 10^{-2}$ °C km^{-1}; other periods are defined as non-TIW periods (Fig. 3). This proxy matches the TIW KE index well (Fig. 3).

Overall, about two-thirds of the total time periods belong to TIW periods and one-third of them belong to non-TIW periods (Fig. 3). The numbers of overturns within 160°–110°W and 3°S–6°N over the years 2008–2013 are 314 and 67, whereas the numbers of fine-resolution Argo profiles are 4,952 and 1,783, in TIW and non-TIW periods, respectively. This leads to the occurrence probability of 6.34% for TIW periods (the 95% bootstrap confidence interval (95% CI) $= (5.67\%, 7.07\%)$) and of 3.76% for non-TIW periods (the 95% bootstrap CI $= (2.92\%, 4.71\%)$). The former is 69% larger than the latter (Fig. 2d); in addition, the occurrence probability for TIW periods is larger at almost every depth than non-TIW periods within the upper and centre of the thermocline (depths -0.5 to ~ 1; Fig. 2c). The results demonstrate again that TIWs are associated with a higher occurrence of overturns.

Link the overturns with TIWs via shear instability. The observed higher occurrence of deep-reaching overturns during TIWs, so far established in the seasonal and period-to-period cycles, calls for a physical interpretation. The overturns are indicative of breaking of internal waves and/or turbulence generated by shear instability. Two ways may be employed to demonstrate this physical interpretation. One is the LSA, which can determine the potential instabilities of an observed flow by providing locations and other detailed properties of the exponentially growing unstable modes (see Methods and refs 18,40–42). The LSA is applied to $\sim 8 \times 10^4$ hourly TAO profiles of years 2000–2010 at 0°, 140°W (this site locates meridionally at the centre of Pacific TIWs and thus is representative for TIW studies[21]).

The monthly counts of the potential instabilities (in terms of the critical levels of the detected unstable modes; see Methods) is shown on physical depth in Fig. 4a and on referenced depths in Fig. 4b,c. In Fig. 4b, the depth is referenced to the thermocline centre of each profile, which is defined as the depth of maximum vertical temperature gradient (before calculation, temperature is 40-m running smoothed, to remove effects of noises and intermittent waves). In Fig. 4c, the depth is referenced to the

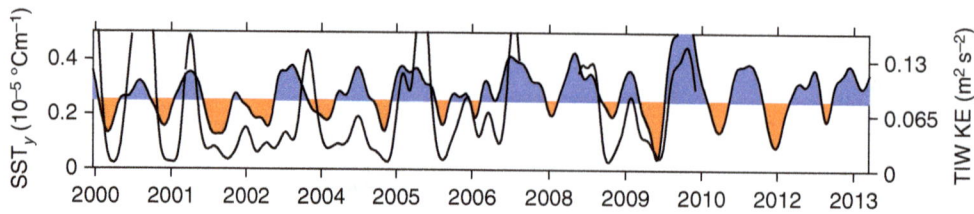

Figure 3 | Separating TIW and non-TIW periods with meridional SST gradient. The blue curve denotes the meridional SST gradient (SST_y). The TIW (non-TIW) periods are depicted with blue (red) shading. The black curve denotes the 140-day low-pass-filtered TIW KE. The correlation coefficient between the filtered TIW KE and the SST_y over 2000–2010 is $r = 0.47$, with $P < 0.001$ and 95% CI = (0.44, 0.49).

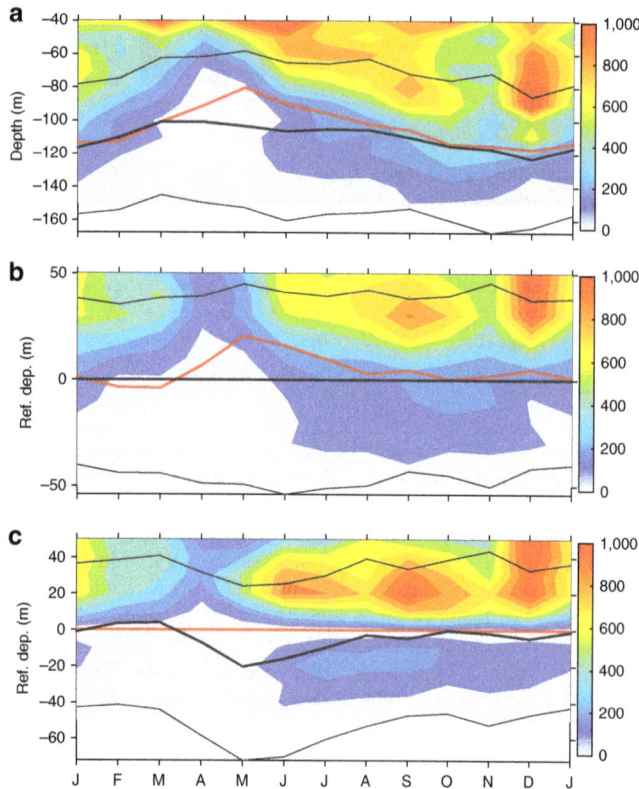

Figure 4 | Monthly climatology of count of critical levels at 140° W of the Equator. (**a**) On physical depths. (**b**) On depths that is referenced to hourly centres of the thermocline (defined as the depth of maximum vertical gradient of 40-m running-averaged temperature). (**c**) On depth that is referenced to hourly centres of the EUC (defined as the depth of maximum eastward velocity). Shown are counts in 10-m bins. In each panel, the red curve denotes the average depth of EUC core, the black thick curve denotes the average depth of the thermocline centre and the two black thin curves denote the average upper and lower bounds of the thermocline.

EUC core of each profile, which is defined as the depth of maximum eastward velocity. In addition, the long-term averaged monthly depths of the EUC core, the thermocline centre and the upper and lower thermocline bounds (defined as the depths of half the maximum vertical temperature gradient), are overlaid on the three panels of Fig. 4.

A prominent feature is the seasonal variation of the potential instabilities (Fig. 4a). Within the thermocline, more potential instabilities were found in boreal summer to winter, while relatively few potential instabilities were found in boreal spring. There were also fewer potential instabilities occurring during October and November. This feature is consistent with the seasonal variations of both TIW KE and occurrence probability of Argo-determined overturns (Fig. 2a,b).

Another distinguished characteristic is the deep-reaching nature of the potential instabilities. From boreal summer to winter, the potential instabilities may reach down to -120 m, with deepest depths of -150 m in fall. Potential instabilities occurring below the upper thermocline bound are as many as those above it. As the depth of the upper thermocline bound varies from -80 to -60 m and roughly coincides with the top of the observed upper core layer[13], it demonstrates that the upper core layer mixing is a remarkably persistent phenomenon in the study site. Moreover, $\sim 15\%$ of all the determined instabilities locate well below the average centres of both the EUC and the thermocline, which coincide with each other in boreal June to March (Fig. 4a). By contrast, in spring, instabilities can only occur within the upper 75 m, which is ~ 25 m above the thermocline centre.

As the numbers of hourly profiles in each month and at each depth are nearly the same, the occurrence probability of the instabilities (Fig. 5a) displays a similar pattern as the counts of potential instabilities shown in Fig. 4a.

When all the potential instabilities are redistributed on the depth that is referenced to the thermocline centre of each profile (Fig. 4b), a centre of potential instabilities emerges ~ 26 m above the thermocline centre (the averaged thickness of the upper flank of the thermocline is ~ 40 m), in particular in summer to fall. The centre coincides well with the peak at 0.7 on the normalized coordinate of the Argo-detected overturns (Fig. 1d). In addition, potential instabilities occur also in and below the centres of the thermocline. Alternatively, when the potential instabilities are redistributed on the vertical coordinate that is referenced to the EUC core of each profile (Fig. 4c), a striking feature is clearly observed: in addition to those in the upper core layer, an isolated region of potential instabilities stand out in the lower flank of the EUC. These potential instabilities occur only in summer to winter and accounts for $\sim 20\%$ of those in the upper core layer. Instabilities peaked ~ 20 m above and below the EUC core, but no instabilities were found in the EUC core.

The relation of the deep-reaching potential instabilities to TIWs is further illustrated from a period-to-period point of view in Fig. 5b,c. In these two panels, we show the occurrence probability of potential instabilities in the TIWs and non-TIWs periods, respectively, on the depth referenced to the EUC core. Here, the TIW periods are defined as periods when the 140-day low-passed TIW KE (Fig. 3) is larger than 4×10^{-2} m^2 s^{-1} (corresponding to a characteristic horizontal velocity of 20 cm s^{-1}); the other periods are defined as non-TIW periods. These newly defined periods of TIWs and non-TIWs are consistent with those defined based on the SST gradient (Fig. 3).

The occurrence probability is ~ 50 to $\sim 100\%$ larger, almost at every depth during TIW periods than non-TIW periods in the upper core layer (except in February). In particular, the instabilities of the lower flank of the EUC can only occur with the existence of TIWs. Consequently, the results clearly demonstrate the enhancement effect of TIWs on the occurrence of potential instabilities in both the upper and lower flanks of the EUC.

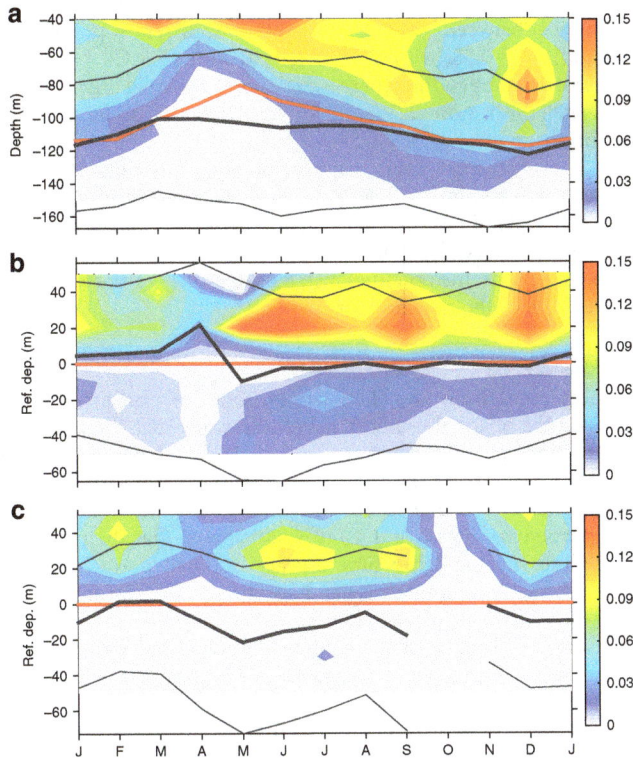

Figure 5 | Monthly climatology of occurrence probability of critical levels at 140° W of the Equator. (**a**) On physical depths. (**b**) For periods of TIWs but on the depth that is referenced to instantons EUC cores (see caption of Fig. 4). (**c**) The same as in **b**, but for periods of non-TIWs. TIW (non-TIW) periods are defined when the TIW KE is larger (less) than $0.04\,m^2\,s^{-2}$. The occurrence probability is defined as the ratio of the number of unstable modes over the number of profiles in 10-m bins. The red curve denotes the average depth of EUC core, the black thick curve denotes the average depth of the thermocline centre and the black thin curves denote the average depths of the thermocline bounds.

Link the instabilities to low Richardson numbers. The other way to link the shear instability to the deep-reaching feature of the detected overturns in the Argo profiles (as well as the TAO-determined potential instabilities) is to examine the Richardson number, Ri $(= N^2/S^2$, where $S^2 = |\partial u/\partial z|^2 + |\partial v/\partial z|^2$ is the shear squared). The shear instability (in particular of the Kelvin–Helmholtz type) is dynamically related to the local Richardson number, a critical value of which Ri$_c$ is ~ 0.25. Ri $=$ Ri$_c$ is an equilibrium state for turbulence in a stratified shear flow. When Ri $<$ Ri$_c$, turbulence may be initiated or continue to grow due to shear instabilities. When Ri $>$ Ri$_c$, the flow is dynamically stable and any turbulence will decay[46,47]; however, when Ri is close to Ri$_c$, the flow may lie in the regime subject to marginal instability[48]. For example, turbulence can persist up to a Ri value typically near 1/3 (ref. 49). The marginal instability is well identified in the upper layer (upper $\sim 75\,m$) of TIW periods at 0°, 140° W[50]. Accordingly, based on the hourly TAO measurements over the years 2000–2010, the occurrence frequency of Ri ≤ 0.35 was computed to represent the possibility of instabilities (Fig. 6).

The high occurrence frequency of Ri ≤ 0.35 is roughly associated with the high-occurrence probability of potential instabilities as shown in Fig. 5a, although they match well only in their main structure, rather than in details. The pattern of higher occurrence frequency (say ≥ 0.25) includes a deep extension to $\sim -100\,m$ in winter and summer months, and a subsurface centre (at $-50\,m$) from February to September. The lower bound

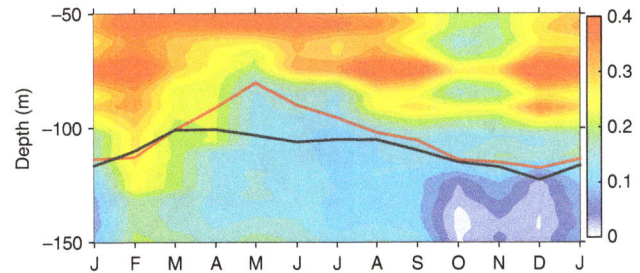

Figure 6 | Monthly occurrence frequency of low Richardson number. The occurrence frequency is calculated as the ratio of numbers of Ri ≤ 0.35 over numbers of all Ri in 10-m bins. The red curve denotes the average depth of EUC core and the black curve denotes the average depth of the thermocline centre.

of the higher-occurrence frequency is generally confined to the centres of the thermocline and the EUC. This feature is consistent with the occurrence of the potential instabilities.

The inconsistence in detailed structures between the occurrence frequency of Ri ≤ 0.35 and the occurrence probability of potential instabilities is explainable. In particular, from February to June, relatively high occurrence frequency of Ri ≤ 0.35 is found below $-100\,m$, where fewer potential instabilities were determined here (Fig. 5a). This may be because the shear in the depths is weak (Fig. 7). Hence, although a large portion of Ri is small (resulted from weak stratification), there was not enough KE available in the mean flow to drive unstable modes that have high-enough growth rate[18] that could pass the growth rate criterion used in the LSA (see Methods).

As mentioned, the annual cycle of the occurrence frequency of low Ri should have resulted from not only the shear of EUC and the TIWs, but also the thermal structure of the upper ocean. All the processes and properties are ultimately also related to wind stresses and exhibit seasonal variations. For example, in boreal spring the wind reduces, the shear weakens and the water warms with increased stratification in the subsurface layers; since late summer, the wind stress increases, the shear strengthens and the surface water cools down with decreased stratification.

Nevertheless, the contribution of TIWs to the low Ri and therefore the generation of potential instabilities could be roughly isolated from the EUC. This was done by separating the individual shear they induce. The shear squared induced by the background EUC is calculated as $S_0^2 = |\partial \bar{u}/\partial z|^2 + |\partial \bar{v}/\partial z|^2$, where $\bar{u}, \bar{v}$ are the 40-day low-pass-filtered velocities, representing the background flows, whereas the shear squared associated with the TIWs is estimated as the difference between the original and the background shear squared: $S_{tiw}^2 = S^2 - S_0^2 = (|\partial u/\partial z|^2 + |\partial v/\partial z|^2) - (|\partial \bar{u}/\partial z|^2 + |\partial \bar{v}/\partial z|^2)$ (Fig. 7a,b). In general, the EUC is associated with stronger shear squared, which centres ~ 20–$50\,m$ above the seasonally varying EUC core (Fig. 7a), whereas the TIWs are associated with weaker shear (Fig. 7b).

However, the magnitude of TIW-induced shear squared could reach half of that induced by the EUC in a thick layer. In addition, as a prominent feature, the TIW-induced shear is centred just above the EUC core and covers both the upper core layer and the layers immediately below the EUC core, in particular during TIW seasons (boreal summer to winter). The TIW shear covering the EUC core adds to the EUC-induced shear and provides the conditions favourable for instability; besides, the strong velocity of TIWs provides necessary KE for the instability to grow fast. This explains the occurrence of potential instabilities occurring below the centres of both the EUC and thermocline (Fig. 4).

The portion of the TIW-induced shear is calculated as S_{tiw}^2/S^2 (Fig. 7c). The TIW-induced shear accounts for $30 \sim 50\%$ for most

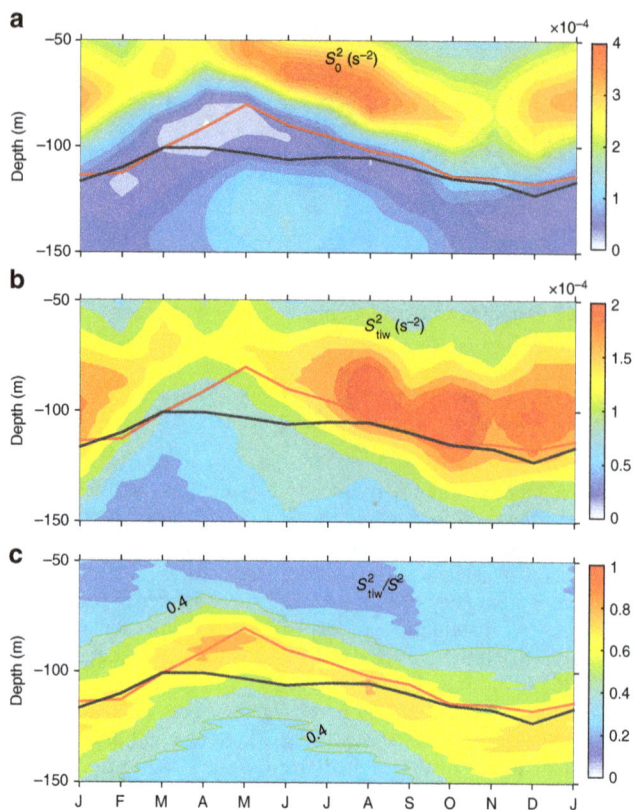

Figure 7 | Shear of background flows and TIWs. (**a**) Shear squared induced by the background flow, S_0^2. (**b**) Shear squared associated with TIWs, S_{tiw}^2. (**c**) The proportion of the shear squared associated with TIWs, S_{tiw}^2/S^2. In **a**,**b** and **c**, the red curve denotes the average depth of the EUC core and the black curve denotes the average depth of the thermocline centre. In **c**, contour of 0.4 is highlighted for reference. The different colour scales in **a** and **b** are noteworthy.

of the upper layer. This percentage is consistent with the direct measurements that shows ~30% larger of shear induced by TIW[13]. In particular, it accounts for 60~80% just above and below the EUC core. These results indicate that the TIWs provide a modulating effect on the generation of unstable disturbances.

TIWs and instabilities at ENSO timescales. In Fig. 8a,d we show the monthly TIW KE and the occurrence probability of unstable modes within the thermocline (− 50 to ~ − 150 m) for the years 2000–2010. In years of stronger TIWs, larger occurrence probability of unstable modes are observed. The high correlation between them (correlation coefficient $r = 0.71$, P-value < 0.001, 95% CI = (0.62, 0.79)) further demonstrates that TIWs are associated with higher thermocline instability occurrence also at the inter-annual timescale.

Previous studies, based on modelling results and a TIW proxy in terms of the SST variance, found that the activity of TIWs is larger under La Niña conditions and smaller under El Niño conditions, because the former are associated with stronger latitudinal gradient of SST immediately north of the Equator and thus more occurrence of baroclinic instability[51]. Using the monthly TIW KE and the Oceanic Niño Index (ONI) calculated from the monthly Optimum Interpolation Sea Surface Temperature (ref. 52) (Fig. 8a,b), we confirmed such a significantly negative correlation (correlation coefficient $r = -0.69$, with P-value < 0.001 and 95% CI = (− 0.77, − 0.59)).

The implication of the relation is that the inter-annual variation of occurrence probability of instabilities could also be

related to El Niño and La Niña conditions. The correlation coefficient between the ONI and the occurrence probability (Fig. 8d) is − 0.58, with P-value < 0.001 and 95% CI = (− 0.68, − 0.45). It implies that there were more potential instabilities, associated with more TIWs, under La Niña than under El Niño conditions.

Moreover, the extension range of potential instabilities differs between two conditions (Fig. 8c). Under El Niño conditions, the potential instabilities are mainly confined to the upper flank of the thermocline, except for stronger TIWs. By contrast, under La Niña conditions, the potential instabilities mostly can reach to the lower flank of the thermocline.

Discussion

In the present study, we show the existence of overturns in the deep depths of the thermocline. We also show that the potential instabilities are organized in a physically quite reasonable structure. Given the good coincidence of the determined potential instabilities and the measured mixing during November 2008 (Supplementary Figs 3 and 4), it is anticipated that the potential instabilities during other time are also associated with mixing, although the mixing intensity and accompanying heat fluxes can not be correctly estimated yet.

In the cold tongue of the equatorial Pacific, maintaining cool SSTs in the presence of intense solar heating requires a combination of subsurface mixing and vertical advection to transport surface heat downward[43,53–56]. Analyses of direct turbulence measurements have demonstrated that the subsurface mixing (over − 60 to ~ − 20 m) reduces SST during a particular season—boreal summer[43]. If mixing is indeed associated with the detected overturns and potential instabilities in the deep depths, it could also blend water between the upper part and middle/lower part of the thermocline, resulting in cooling of the upper thermocline, and further cooling of the surface.

The TIW-related mixing during La Niña conditions may have rich implications for ENSO dynamics. It has been found that incorporating TIWs in the ocean–atmosphere coupled models results in a significant asymmetric negative feedback to ENSO[57–59] (anomalously heating the Equator under La Niña conditions and cooling it under El Niño conditions via horizontal advection). Accordingly, the asymmetric negative feedback is argued to explain the observed asymmetric feature of a stronger-amplitude El Niño and weaker-amplitude La Niña relative to the models. However, the cooling effect via vertical mixing associated with TIWs, in particular during La Niña conditions, was missed or under-represented by the numerical models due to underestimation by the parameterizations[60]. Therefore, the effects of TIWs on ENSO development requires to be re-examined.

To best simulate the oceans, numerical models need to reproduce or properly represent the TIWs and the associated turbulence. Although the main structure of TIWs can be reproduced in some coarse resolution ocean general circulation models (OGCMs)[30], the small-scale structures of the frontal areas of TIWs, which are key regions of turbulence generation[28,29,61,62], remain unresolved by coarse resolution OGCMs and ocean-atmospheric coupled models. This shortage may lead to underestimates of thermocline mixing by the oversimplified vertical mixing parameterizations incorporated in coarse resolution OGCMs[60] and hence to model-data deviations not only in the equatorial ocean but also in mid-latitudes[63,64].

Methods

Data processing. The Argo data (see below) covers the period from 2000 till June 2014 and only profiles with vertical sample resolution of at least 2 m and with maximal sampling depth deeper than − 200 m are used to detect overturns. Both

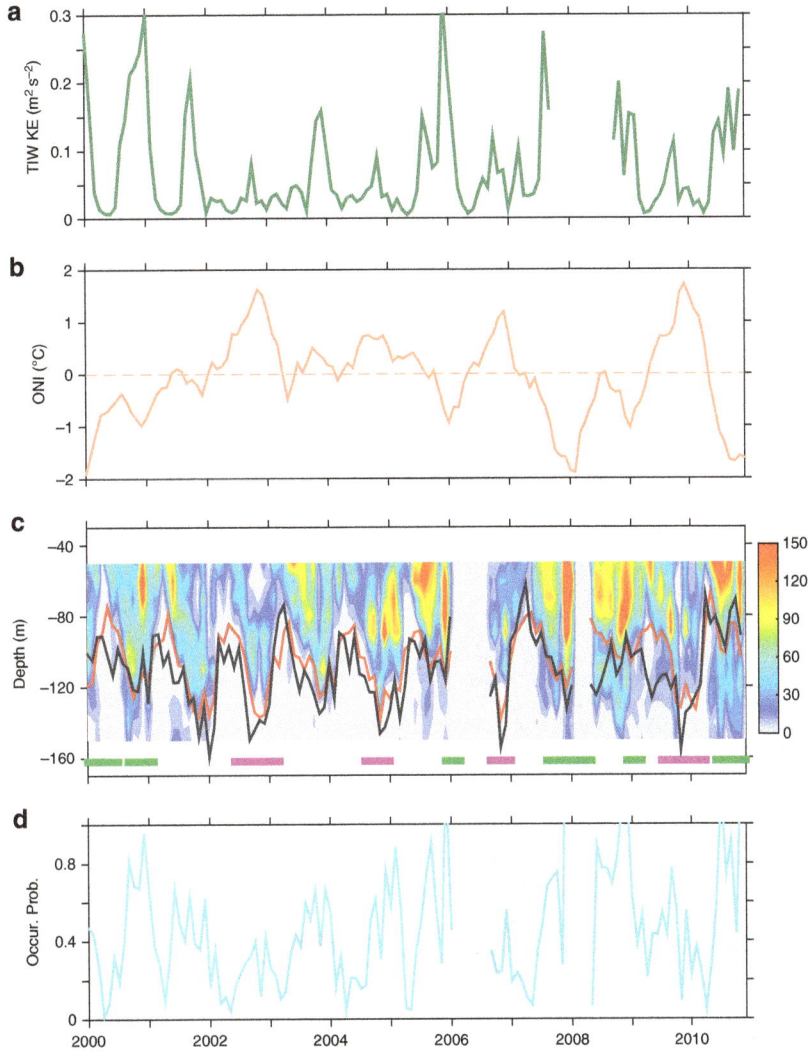

Figure 8 | Relation among instabilities, TIWs and large-scale processes on the inter-annual time scale. (a) Monthly TIW KE. (b) ONI, showing the Niño 3.4 (5° S–5° N, 170–120° W) SST anomaly (1981–2010 mean removed) calculated from v2 of the Optimum Interpolation Sea Surface Temperature (OISST). The dashed line denotes zero. (c) Monthly count of unstable modes in 10-m bins. The green and magenta bars on the bottom denote periods of the La Niña and El Niño conditions, defined when the ONI is $\leq -0.5\,°C$ and $\geq 0.5\,°C$, respectively. The red curve denotes the average depth of EUC core and the black curve denotes the average depth of the thermocline centre. (d) Monthly occurrence probability of unstable modes, defined as the ratio of counts of critical levels occurring between -50 and $-150\,m$ of a month over the number of profiles of the given month.

hourly temperature and velocity from TAO mooring at 0°, 140° W is interpolated (extrapolated if needed) at 1 m spacing grids for both Ri calculation and the LSA (see below). As salinity observations of TAO are not sufficient and their contribution to density in this region is minor, salinity needed for real-time density calculation was often replaced by its temporal average[50]; here we use the salinity climatology averaged from 251 Argo and CTD (CTD is obtained from the same source of Argo) profiles falling into the range of $0° \pm 0.5°$, $140° W \pm 0.5°$.

The Thorpe method. The Thorpe method[38] is commonly used for estimating dissipation rate and vertical turbulence diffusivities, which is based on the size of detected density overturn patch and the stratification intensity over the patch, of a measured potential density profile. In this work the Thorpe method is applied only for overturn detection, rather than diffusivity estimation (because the resolution of the Argo profiles is still too low for such estimation). The Thorpe method is suitable to be applied in the thermocline where the stratification is strong; in contrast, care should be taken in layers of low stratification due to its sensitivity to noise[65]; therefore, both the upper and lower layers of low stratifications defined by $N^2 < 0.625 \times 10^{-4}\,s^2$ are omitted from our analysis. The size of any overturn should not be smaller than three times the profile resolution, a minimum criteria for the overturn size[66]. Profiles with large spikes from any of the properties (temperature, salinity and pressure) were removed from the data set. With the above criteria, few unreasonably large overturns are still found (see Supplementary Fig. 2). Therefore, each detected overturn and the corresponding density profile were carefully

examined afterwards, to guarantee that it is physically sensible. In particular, the detected overturns that have sizes $> 30\,m$ were removed.

Linear stability analysis. The LSA is designed to detect instabilities potentially occurring in an observed flow. The stability of an inviscid, incompressible, stratified, unidirectional shear flow to small disturbances is determined by the solutions of the Taylor–Goldstein equation:

$$d^2\varphi/dz^2 + \{N^2/(U-c)^2 - k^2 - d^2U/dz^2/(U-c)\}\varphi = 0, 0 \leq z \leq h \quad (1)$$

where $U(z)$ and $N(z)$ are the profiles of z-dependent mean velocities and buoyancy frequency, respectively, and $\varphi(z) = \varphi_0(z) \exp[ik(l-ct)]$ is the z-dependent stream function of a disturbance with real horizontal wavenumber k and complex phase speed $c = c_r + ic_i$. Here, $\varphi_0(z)$ is the amplitude of the stream function, t is time and l is along the direction of the perturbation wave vector. For non-parallel flows, the stability can be examined by taking U as the velocity component in the direction (α) of the disturbance wave vector: $U = u\cos(\alpha) + v\sin(\alpha)$, where (u, v) is the measured eastward and northward components of the velocity vector.

The Taylor–Goldstein equation (equation (1)) is a linear eigenvalue problem that can be solved numerically using matrix method[40–42,67] subject to prescribed boundary conditions. In this study, zero condition is applied at both surface ($z = 0$) and the lower boundary $z = -200\,m$. Unstable modes should at least have the property $kc_i > 0$, to ensure that the disturbance grows exponentially. Useful quantities derived through solving the problem include the following: the wave vector direction of the instability perturbation wave, α; the perturbation wave length, $2\pi/k$; the horizontal phase speed of the perturbation wave, c_r; the growth rate of the perturbation wave, kc_i and the critical level of it, z_c, which is defined as the depth where $U(z_c) - c_r = 0$; the critical level is also considered as the position of the unstable mode.

Before applying the LSA, the data are carefully processed. The hourly TAO velocities are interpolated to a 1-m grid using cubic splines from surface to -200 m. Specifically, in the upper layers where velocity data are not available (the upper -25 m or -37.5 m, depending on data quality), the profiles of u and v are extrapolated using a polynomial fit, which should meet the requirement that the first derivative is continuous between -200 m and the surface, and approaches zero at the top two grids (-1 m and 0).

The hourly temperature of TAO observations is inter/extrapolated into the same 1-m spacing grids. As the temperature sample spacing is sparse (see Supplementary Fig. 3), a special strategy is applied. First, the raw data are interpolated into 5-m (above -50 m), 10-m (between -50 and -150 m) and 50-m (between -150 and -500) grids using a linear method. Second, the inter/extrapolated data are further inter/extrapolated into the 1-m resolution grids using a cubic spline method (other grid spacing and inter/extrapolated methods are tested; Supplementary Table 1). If data at the lowest sample grids (-300 m and -500 m; sometimes also at -180 m) are not available, their first derivatives are required to smoothly approach climatologic values, a similar manner as used in dealing with (u, v) on the top grids. We found that if the unavailable temperature data in the lowest sample grids are not constrained by prescribed values (such as by nudging their first derivatives to climatology), extrapolation may produce unphysically extreme values out of the data ranges and lead to obviously artificial unstable modes. Apart from this, the LSA results seem only weakly sensitive to grid spacing and inter/extrapolation methods (see Supplementary Fig. 3 and Supplementary Table 1). Nevertheless, all detected unstable modes that are below -150 m are rejected for accuracy.

Salinity measurements are sparse, so a temporal and spatial mean that is averaged into the 1-m grid over all fine resolution Argo measurements over the region of $0° \pm 0.5°$, $140° W \pm 0.5°$ is employed instead. This substitution of salinity does not induce problems in N^2 calculation and in the LSA, because the effect of salinity on density is small[32]. A similar treatment is adopted in ref. 50 at the same location.

As no prior information of the perturbation waves exist, the LSA scans both the wave vector directions and the wave numbers for each observed flow. For computational efficiency, the disturbance wave vector direction α is scanned from $0°$ to $180°$ with interval $15°$ (direction 0 represents east) in this study (direction 0 to $\sim -180°$ is symmetry to 0 to $\sim 180°$ and need not be scanned); the hourly mean component of TAO velocities (u, v) at α direction, that is, U, is calculated subsequently. The wave number k of the perturbation wave is scanned over 85 values ranging from $2\pi/60$ to $2\pi/1{,}000$ m^{-1}. For a given wave vector (k, α), the number of unstable modes, which are determined based on criteria of ref. 18 (see below), ranges from 0 to ~ 10. Furthermore, all potential unstable modes of the flow (in terms of critical levels) constitute several mode families.

The idea of mode family is based on the nature that the critical level z_c of the flow is relatively consistent, even though most mode properties profile vary with k and α. A histogram of z_c obtained from all (k, α) vectors is constructed and peaks identified. All modes close to a given peak (that is, having critical levels between the adjacent minima of the histogram) are considered part of the same mode family, which usually focus in different depth ranges[16,18] (Supplementary Fig. 3). For each mode family, the fastest growing mode (should satisfy additional criteria; see below) is defined as the unstable modes of the flow (Supplementary Table 1 and Supplementary Fig. 4).

The criteria (see ref. 18) adopted to determine all possible unstable modes (which are in terms of critical levels) and reasonable mode families include cutoff depths, cutoff growth rate of instability, cutoff wavelengths, Ri criteria, critical layer criteria and others, which are described in details below.

The cutoff depths are -40 m and -150 m in the present study. This criterion rejects potentially unphysical modes that are induced by extrapolation near the boundaries as mentioned above. Previous sensitivity studies also demonstrated that the location of detected unstable mode near boundary may depart to some extent from its real position[42].

The cutoff growth rate of instability is 1 h^{-1}. This criterion guarantees that the instability grows faster than hourly variation of the mean flow.

The cutoff wavelengths depend on the sampling spacing of the temperature measurements. Instabilities of inviscid, non-diffusive, stratified shear layers typically have wavelength around 2π times the thickness of the shear layer[18,41,68]. Based on this, modes with wavelength 65 m are likely to grow from layers of thickness 10 m, the Nyquist wavelength of the ~ 5-m vertical bins of TAO temperature data in the above -60 m depth; similarly, modes with wavelength 250 m (500 m) are likely to grow from the ~ 20-m (~ 40cm) vertical bins of TAO temperature data over -60 to ~ -140-m (-140 to ~ -200) depths. We remove modes of wavelengths <65 m above -60 m depth and of wavelengths <250 m between -60 and -140 m, and of wavelengths smaller than 500 m between -140 and -200 m, to avoid possibly unphysical modes that are resulted from interpolation. The bins of TAO velocity data are constantly 5 m and thus do not require extra limitations of wavelengths.

The lowest Ri of the profile should be <0.25. Moreover, in the vicinity ($\pm 1/14$ wavelength) of the critical level of the potential unstable mode, lowest Ri is required to be <0.25, to assure typical Kelvin–Helmholtz instability of this unstable mode[17].

Any mode family must include at least one resolved mode with a larger wavelength than the fastest growing mode and at least one with smaller wavelength. This effectively rejects modes whose true maxima lie outside the range of wavelengths tested.

In addition, the critical level(s) is determined as the depth(s) where $|U(z) - c_r| \leq 0.01$ (m s^{-1}) in the present study. Why we added such a criterion is because the 1-m spacing, inter/extrapolated velocity profile is still discrete so that it may not guarantee a depth that meets the restrict definition of critical level: $U(z) - c_r = 0$. Under such a criterion, there may exist more than one critical level of a (k, α, c_r) vector that satisfy the above criteria. All are retained for further analysis.

(It is noteworthy that the LSA performed here differs in physics from ref. 18 in that we did not include effect of eddy viscosity in equation (1), while the referred work did. In ref. 18, the authors demonstrated that the addition of eddy viscosity to equation (1) damped the generation of instabilities mainly at night when turbulence is strongest. However, the effect of eddy viscosity could be subtle under different conditions, that is, it may also destabilize a stratified shear flow[69].)

Supplementary Figs 3 and 4 show detailed results of LSA that is applied to an example profile and to consecutive profiles over a period of ~ 8 days, respectively. Based on the flow shown on Supplementary Fig. 3, we also discuss the sensitivity of the LSA to the inter/extrapolation method (Supplementary Table 1). In Supplementary Note 1 we describe the details of both analyses. In summary, the sensitivity study suggests that the unstable modes occur in vicinity of low Ri, and as long as this region of low Ri is accurately solved and not close to the boundary, reasonable unstable mode can be detected. By Supplementary Fig. 4, we demonstrate the usefulness of LSA via showing the coincidence of the detected unstable modes with the direct turbulence measurements.

References

1. Shin, S. I. & Liu, Z. Y. Response of the equatorial thermocline to extratropical buoyancy forcing. *J. Phys. Oceanogr.* **30**, 2883–2905 (2000).
2. Huang, R. X. & Pedlosky, J. Climate variablility of the equatorial thermocline inferred from a two-moving-layer model of the ventilated thermocline. *J. Phys. Oceanogr.* **30**, 2610–2626 (2000).
3. Kessler, W. S., Mcphaden, M. J. & Weickmann, K. M. Forcing of intraseasonal Kelvin waves in the equatorial Pacific. *J. Geophys. Res. Oceans* **100**, 10613–10631 (1995).
4. Wang, B., Wu, R. G. & Lukas, R. Annual adjustment of the thermocline in the tropical Pacific Ocean. *J. Climate* **13**, 596–616 (2000).
5. Li, X. J., Chao, Y., McWilliams, J. C. & Fu, L. L. A comparison of two vertical-mixing schemes in a Pacific Ocean general circulation model. *J. Climate* **14**, 1377–1398 (2001).
6. Meehl, G. A. *et al.* Factors that affect the amplitude of El Nino in global coupled climate models. *Clim. Dynam.* **17**, 515–526 (2001).
7. Gregg, M. C., Peters, H., Wesson, J. C., Oakey, N. S. & Shay, T. J. Intensive measurements of turbulence and shear in the equatorial undercurrent. *Nature* **318**, 140–144 (1985).
8. Moum, J. N. & Caldwell, D. R. Local influences on shear-flow turbulence in the equatorial ocean. *Science* **230**, 315–316 (1985).
9. Peters, H., Gregg, M. C. & Toole, J. M. On the parameterization of equatorial turbulence. *J. Geophys. Res. Oceans* **93**, 1199–1218 (1988).
10. Peters, H., Gregg, M. C. & Sanford, T. B. The diurnal cycle of the upper equatorial ocean - turbulence, fine-scale shear, and mean shear. *J. Geophys. Res. Oceans* **99**, 7707–7723 (1994).
11. Moum, J. N., Caldwell, D. R. & Paulson, C. A. Mixing in the equatorial surface-layer and thermocline. *J. Geophys. Res. Oceans* **94**, 2005–2021 (1989).
12. Moum, J. N. & Nash, J. D. Mixing measurements on an equatorial ocean mooring. *J. Atmos. Ocean Technol.* **26**, 317–336 (2009).
13. Moum, J. N. *et al.* Sea surface cooling at the Equator by subsurface mixing in tropical instability waves. *Nat. Geosci.* **2**, 761–765 (2009).
14. Lien, R. C., Caldwell, D. R., Gregg, M. C. & Moum, J. N. Turbulence variability at the equator in the Central Pacific at the beginning of the 1991-1993 El-Nino. *J. Geophys. Res. Oceans* **100**, 6881–6898 (1995).
15. Wang, D. L. & Muller, P. Effects of equatorial undercurrent shear on upper-ocean mixing and internal waves. *J. Phys. Oceanogr.* **32**, 1041–1057 (2002).

16. Sun, C. J., Smyth, W. D. & Moum, J. N. Dynamic instability of stratified shear flow in the upper equatorial Pacific. *J. Geophys. Res. Oceans* **103**, 10323–10337 (1998).

17. Moum, J. N., Nash, J. D. & Smyth, W. D. Narrowband oscillations in the upper equatorial ocean. Part I: interpretation as shear instabilities. *J. Phys. Oceanogr.* **41**, 397–411 (2011).

18. Smyth, W. D., Moum, J. N., Li, L. & Thorpe, S. A. Diurnal shear instability, the descent of the surface shear layer, and the deep cycle of equatorial turbulence. *J. Phys. Oceanogr.* **43**, 2432–2455 (2013).

19. Smyth, W. D. & Moum, J. N. Shear instability and gravity wave saturation in an asymmetrically stratified jet. *Dynam. Atmos. Oceans* **35**, 265–294 (2002).

20. Peters, H., Gregg, M. C. & Sanford, T. B. Equatorial and off-equatorial fine-scale and large-scale shear variability at 140-degrees-W. *J. Geophys. Res. Oceans* **96**, 16913–16928 (1991).

21. Lyman, J. M., Johnson, G. C. & Kessler, W. S. Distinct 17-and 33-day tropical instability waves in subsurface observations. *J. Phys. Oceanogr.* **37**, 855–872 (2007).

22. Legeckis, R. Long waves in eastern equatorial Pacific ocean - view from a geostationary satellite. *Science* **197**, 1179–1181 (1977).

23. Miller, L., Watts, D. R. & Wimbush, M. Oscillations of dynamic topography in the eastern equatorial Pacific. *J. Phys. Oceanogr.* **15**, 1759–1770 (1985).

24. Strutton, P. G., Ryan, J. P. & Chavez, F. P. Enhanced chlorophyll associated with tropical instability waves in the equatorial Pacific. *Geophys. Res. Lett.* **28**, 2005–2008 (2001).

25. McPhaden, M. J. Monthly period oscillations in the Pacific North equatorial countercurrent. *J. Geophys. Res. Oceans* **101**, 6337–6359 (1996).

26. Halpern, D., Knox, R. A. & Luther, D. S. Observations of 20-day period meridional current oscillations in the upper ocean along the Pacific Equator. *J. Phys. Oceanogr.* **18**, 1514–1534 (1988).

27. Lyman, J. M., Chelton, D. B., deSzoeke, R. A. & Samelson, R. M. Tropical instability waves as a resonance between equatorial Rossby waves. *J. Phys. Oceanogr.* **35**, 232–254 (2005).

28. Kennan, S. C. & Flament, P. J. Observations of a tropical instability vortex. *J. Phys. Oceanogr.* **30**, 2277–2301 (2000).

29. Lien, R. C., D'Asaro, E. A. & Menkes, C. E. Modulation of equatorial turbulence by tropical instability waves. *Geophys. Res. Lett.* **35**, L24607 (2008).

30. Menkes, C. E. R., Vialard, J. G., Kennan, S. C., Boulanger, J. P. & Madec, G. V. A modeling study of the impact of tropical instability waves on the heat budget of the eastern equatorial Pacific. *J. Phys. Oceanogr.* **36**, 847–865 (2006).

31. Jochum, M. & Murtugudde, R. Temperature advection by tropical instability waves. *J. Phys. Oceanogr.* **36**, 592–605 (2006).

32. Inoue, R., Lien, R. C. & Moum, J. N. Modulation of equatorial turbulence by a tropical instability wave. *J. Geophys. Res. Oceans* **117**, C10009 (2012).

33. Xie, S. P. Climate science unequal equinoxes. *Nature* **500**, 33–34 (2013).

34. Argo. Argo float data and metadata from Global Data Assembly Centre (Argo GDAC). *Ifremer* (2000).

35. Whalen, C. B., Talley, L. D. & MacKinnon, J. A. Spatial and temporal variability of global ocean mixing inferred from Argo profiles. *Geophys. Res. Lett.* **39**, L18612 (2012).

36. Kunze, E., Firing, E., Hummon, J. M., Chereskin, T. K. & Thurnherr, A. M. Global abyssal mixing inferred from lowered ADCP shear and CTD strain profiles. *J. Phys. Oceanogr.* **36**, 1553–1576 (2006).

37. Gregg, M. C., Sanford, T. B. & Winkel, D. P. Reduced mixing from the breaking of internal waves in equatorial waters. *Nature* **422**, 513–515 (2003).

38. Thorpe, S. A. Turbulence and mixing in a Scottish Loch. *Philos. Trans. R. Soc. Lond. A Math. Phys. Sci.* **286**, 125–181 (1977).

39. McPhaden, M. J. The tropical atmosphere ocean array is completed. *Bull. Am. Meteorol. Soc.* **76**, 739–741 (1995).

40. Moum, J. N., Farmer, D. M., Smyth, W. D., Armi, L. & Vagle, S. Structure and generation of turbulence at interfaces strained by internal solitary waves propagating shoreward over the continental shelf. *J. Phys. Oceanogr.* **33**, 2093–2112 (2003).

41. Smyth, W. D., Moum, J. N. & Nash, J. D. Narrowband oscillations in the upper equatorial ocean. Part II: properties of shear instabilities. *J. Phys. Oceanogr.* **41**, 412–428 (2011).

42. Liu, Z. Y. Instability of Baroclinic tidal flow in a stratified Fjord. *J. Phys. Oceanogr.* **40**, 139–154 (2010).

43. Moum, J. N., Perlin, A., Nash, J. D. & McPhaden, M. J. Seasonal sea surface cooling in the equatorial Pacific cold tongue controlled by ocean mixing. *Nature* **500**, 64–67 (2013).

44. Contreras, R. F. Long-term observations of tropical instability waves. *J. Phys. Oceanogr.* **32**, 2715–2722 (2002).

45. Reynolds, R. W. et al. Daily high-resolution-blended analyses for sea surface temperature. *J. Climate* **20**, 5473–5496 (2007).

46. Miles, J. W. On the stability of heterogeneous shear flows. *J. Fluid Mech.* **10**, 496–508 (1961).

47. Rohr, J. J., Itsweire, E. C., Helland, K. N. & Vanatta, C. W. Growth and decay of turbulence in a stably stratified shear-flow. *J. Fluid. Mech.* **195**, 77–111 (1988).

48. Thorpe, S. A. & Liu, Z. Y. Marginal instability? *J. Phys. Oceanogr.* **39**, 2373–2381 (2009).

49. Smyth, W. D. & Moum, J. N. Length scales of turbulence in stably stratified mixing layers. *Phys. Fluids* **12**, 1327–1342 (2000).

50. Smyth, W. D. & Moum, J. N. Marginal instability and deep cycle turbulence in the eastern equatorial Pacific Ocean. *Geophys. Res. Lett.* **40**, 6181–6185 (2013).

51. Yu, J. Y. & Lui, W. T. A linear relationship between ENSO intensity and tropical instability wave activity in the eastern Pacific Ocean. *Geophys. Res. Lett.* **30**, 1735 (2003).

52. Reynolds, R. W., Rayner, N. A., Smith, T. M., Stokes, D. C. & Wang, W. Q. An improved in situ and satellite SST analysis for climate. *J. Climate* **15**, 1609–1625 (2002).

53. Wang, W. M. & McPhaden, M. J. The surface-layer heat balance in the equatorial Pacific Ocean. Part I: mean seasonal cycle. *J. Phys. Oceanogr.* **29**, 1812–1831 (1999).

54. Wang, B. & Fu, X. H. Processes determining the rapid reestablishment of the equatorial Pacific cold tongue/ITCZ complex. *J. Climate* **14**, 2250–2265 (2001).

55. Jouanno, J., Marin, F., du Penhoat, Y., Sheinbaum, J. & Molines, J. M. Seasonal heat balance in the upper 100 m of the equatorial Atlantic Ocean. *J. Geophys. Res. Oceans* **116**, C09003 (2011).

56. Mcphaden, M. J., Cronin, M. F. & Mcclurg, D. C. Meridional structure of the seasonally varying mixed layer temperature balance in the eastern tropical Pacific. *J. Climate* **21**, 3240–3260 (2008).

57. Ham, Y. G. & Kang, I. S. Improvement of seasonal forecasts with inclusion of tropical instability waves on initial conditions. *Clim. Dynam.* **36**, 1277–1290 (2011).

58. Imada, Y. & Kimoto, M. Parameterization of tropical instability waves and examination of their impact on ENSO characteristics. *J. Climate* **25**, 4568–4581 (2012).

59. An, S. I. Interannual variations of the Tropical Ocean instability wave and ENSO. *J. Climate* **21**, 3680–3686 (2008).

60. Zaron, E. D. & Moum, J. N. A new look at Richardson number mixing schemes for equatorial ocean modeling. *J. Phys. Oceanogr.* **39**, 2652–2664 (2009).

61. Johnson, E. S. A convergent instability wave front in the central tropical Pacific. *Deep Sea Res. Pt II* **43**, 753–778 (1996).

62. Flament, P. J., Kennan, S. C., Knox, R. A., Niiler, P. P. & Bernstein, R. L. The three-dimensional structure of an upper ocean vortex in the tropical Pacific Ocean. *Nature* **383**, 610–613 (1996).

63. Furue, R. et al. Impacts of regional mixing on the temperature structure of the equatorial Pacific Ocean. Part 1: vertically uniform vertical diffusion. *Ocean Model* **91**, 91–111 (2015).

64. Jia, Y. L., Furue, R. & McCreary, J. P. Impacts of regional mixing on the temperature structure of the equatorial Pacific Ocean. Part 2: depth-dependent vertical diffusion. *Ocean Model* **91**, 112–127 (2015).

65. Gargett, A. & Garner, T. Determining Thorpe scales from ship-lowered CTD density profiles. *J. Atmos. Ocean Technol.* **25**, 1657–1670 (2008).

66. Galbraith, P. S. & Kelley, D. E. Identifying overturns in CTD profiles. *J. Atmos. Ocean Technol.* **13**, 688–702 (1996).

67. Liu, Z., Thorpe, S. A. & Smyth, W. D. Instability and hydraulics of turbulent stratified shear flows. *J. Fluid Mech.* **695**, 235–256 (2012).

68. Hazel, P. Numerical studies of stability of inviscid stratified shear flows. *J. Fluid Mech.* **51**, 39–61 (1972).

69. Li, L., Smyth, W. D. & Thorpe, S. A. Destabilization of a stratified shear layer by ambient turbulence. *J. Fluid Mech.* **771**, 1–15 (2015).

70. Johnson, G. C., Sloyan, B. M., Kessler, W. S. & McTaggart, K. E. Direct measurements of upper ocean currents and water properties across the tropical Pacific during the 1990s. *Prog. Oceanogr.* **52**, 31–61 (2002).

Acknowledgements

C.L., A.K. and D.S. acknowledge funding by the German Federal Ministry for Education and Research via the project RACE (FZ 03F0651A). Contribution to the DFG funded CliSAP Excellence initiative of the University of Hamburg. C.L. was also supported by the Knowledge Innovation Program of the Chinese Academy of Sciences (Y62114101Q). Z.L. was funded by the National Basic Research Program of China (2012CB417402), the National Natural Science Foundation of China (NSFC) (41476006) and the Natural Science Foundation of Fujian Province of China (2015J06010). F.W. was funded by the Strategic Priority Research Program of the Chinese Academy of Sciences (XDA11010201), the NSFC Innovative Group Grant (41421005) and the NSFC-Shandong Joint Fund for Marine Science Research Centers (U1406401). We are grateful to three anonymous reviewers who provided instructive suggestions that greatly improved the manuscript. We thank all the data providers. We thank M. Carson for proofreading.

Author contributions

C.L. and A.K. designed the research and conducted data analysis. Z.L. proposed and C.L. conducted the LSA. C.L. and A.K. wrote the first draft of the paper with all the authors' contribution to the revisions. The idea, analysis and manuscript were motivated, performed and written in UHH. The first and later revisions were made in UHH and IOCAS, respectively.

Additional information

Competing financial interests: The authors declare no competing financial interests.

10

A better-ventilated ocean triggered by Late Cretaceous changes in continental configuration

Yannick Donnadieu[1], Emmanuelle Pucéat[2], Mathieu Moiroud[2], François Guillocheau[3] & Jean- François Deconinck[2]

Oceanic anoxic events (OAEs) are large-scale events of oxygen depletion in the deep ocean that happened during pre-Cenozoic periods of extreme warmth. Here, to assess the role of major continental configuration changes occurring during the Late Cretaceous on oceanic circulation modes, which in turn influence the oxygenation level of the deep ocean, we use a coupled ocean atmosphere climate model. We simulate ocean dynamics during two different time slices and compare these with existing neodymium isotope data (ε_{Nd}). Although deep-water production in the North Pacific is continuous, the simulations at 94 and 71 Ma show a shift in southern deep-water production sites from South Pacific to South Atlantic and Indian Ocean locations. Our modelling results support the hypothesis that an intensification of southern Atlantic deep-water production and a reversal of deep-water fluxes through the Caribbean Seaway were the main causes of the decrease in ε_{Nd} values recorded in the Atlantic and Indian deep waters during the Late Cretaceous.

[1] Laboratoire des Sciences du Climat et de l'Environnement, LSCE-IPSL, CEA-CNRS-UVSQ, Université Paris-Saclay, 91191 Gif sur Yvette, France. [2] Biogéosciences Dijon, Université Bourgogne-Franche-Comté, UMR CNRS 6282, Dijon 21000, France. [3] Géosciences Rennes, Université de Rennes, UMR CNRS 6118, Rennes 35042, France. Correspondence and requests for materials should be addressed to Y.D. (email: yannick.donnadieu@lsce.ipsl.fr).

In the context of recent warming, modern ocean de-oxygenation that occurs not only on continental margins but also in the tropical oceans worldwide[1] resembles the model initially invoked for the onset of oceanic anoxic events (OAEs)[2]. Recent studies point to a major role of increased nutrient inputs as a trigger for OAEs[3–5]. Nevertheless, ocean circulation, through its impact on oxygen concentration in deep waters, may have affected the thresholds required to trigger an OAE. It has been suggested that Late Cretaceous changes in climate and continental configuration[6–8], namely the widening of the Atlantic Ocean and the deepening of the Central Atlantic (CA) gateway[9], could have induced major changes in oceanic circulation that may have had an impact on the general oxygenation state of the oceanic basins and contributed to the conclusion of these large-scale anoxic events in the deep ocean[10,11].

Nevertheless, no consensus exists on oceanic circulation modes and their possible evolution during the Cretaceous, despite recent improvements of the spatial and temporal coverage of neodymium isotopic data (ε_{Nd}), a proxy for oceanic circulation[11–15]. For instance, a decrease in bottom water ε_{Nd} values during the Late Cretaceous in the Atlantic and Southern Oceans has been interpreted either as reflecting the initiation or intensification of deep-water production in the northern Atlantic[15] or in the South Atlantic (SA) and in the Indian Ocean[10,11,13]. Additional sources for deep waters have been suggested in the North or South Pacific, or even at low latitudes, based on ε_{Nd} or oxygen isotope data[16–18].

General circulation models have also been used to study oceanic circulation during the Cretaceous[7,19,20]. To our knowledge, no modelling studies have reconstructed the evolution of ocean dynamics resulting from the widening of the South and CA Ocean occurring between the Cenomanian and the Maastrichtian. Published simulations either focus on a specific time period[20,21] or are devoted to the impact of the CA gateway opening between the Albian and the Cenomanian[22].

Here we use the fast ocean atmosphere model (FOAM), to explore the evolution of oceanic circulation occurring during the Late Cretaceous (see Methods). Our simulations highlight an evolution from a sluggish circulation in the South and CA using a Cenomanian/Turonian land–sea mask towards a much more active circulation in these basins with an early Maastrichtian land-sea mask.

Results

Changes in oceanic circulation at 1,120 p.p.m. The 94 Ma simulation at a CO_2 concentration of 1,120 p.p.m. (four times the pre-industrial atmospheric level) displays a bipolar oceanic circulation characterized by large areas of deep-water formation located both in the north and in the south of the Pacific Ocean (Fig. 1). The production of deep waters along the northwest boundary of the Pacific Ocean is common to many numerical Cretaceous simulations[4,20,22] but not all[23]. The sinking of water masses occurs over the cyclonic subpolar gyres and results from the winter cooling of warm and salty surface waters located within this large region[20]. The use of the early Maastrichtian land–sea mask for the same atmospheric CO_2 level induces substantial changes in the location of deep-water production in the model. The north Pacific area of deep-water formation is reduced to the northwestern Pacific area only and sinking in the south Pacific area completely disappears. Conversely, the Atlantic and the Indian sectors of the Austral Ocean become larger contributors of deep waters and a small area of intermediate-water formation appears along the northeast American margin in the North Atlantic (Fig. 1).

Sensitivity to atmospheric CO_2 levels. Changes in atmospheric CO_2 level from 8 to 2 pre-industrial atmospheric level have a small effect on the distribution of deep-water formation areas in our study (Supplementary Figs 1 and 2). The main features of oceanic circulation modes remain quite similar, with a disappearance of deep-water convection in the southern Pacific and a larger area of deep-water formation appearing in the southern Atlantic when using early Maastrichtian palaeogeography.

Nevertheless, the sensitivity of overturning to CO_2 level remains an open debate, as some models[24,25] found that ocean dynamics vary with CO_2 while others did not[4,26]. We emphasize that all simulations presented here represent an equilibrated climate. The ocean can go through a transient stratified state resulting from a faster warming of the oceanic upper layers when confronted with a CO_2 increase imposed on a short time scale (a few hundred to a few thousand years). Winter convection at high latitudes would then stop, owing to the decrease in vertical density gradient, and the classical thermohaline circulation may stop as well, but only transiently. We thus recognize that there is room for improvement in this area of research and await a modelling inter-comparison project for warm climates such as the one that has been initiated on the warm Eocene climate[27]; however, such a comparison is beyond the scope of this study.

The results of our model suggest at this point that the changes in ocean dynamics simulated between 94 and 71 Ma are driven by changes in palaeogeography rather than by the cooling recorded between the Cenomanian and the Maastrichtian. As only minor changes occur with varying CO_2 levels, only the simulations at 1,120 p.p.m. will be further described and compared for each palaeogeography in the remainder of the study. Our main aim in this study is to describe and understand the changes in ocean dynamics that occur. It is however interesting to note that the atmospheric CO_2 concentration range used in our simulations produces sea-surface and deep-water temperatures that are comparable to oceanic latitudinal thermal gradients reconstructed

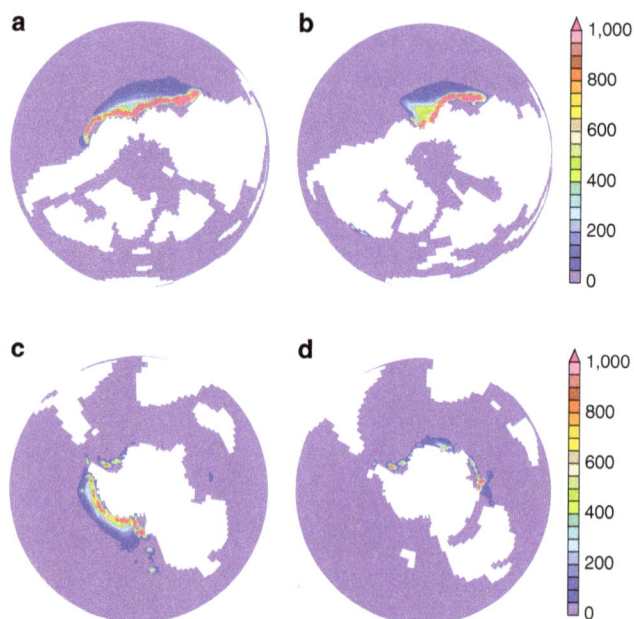

Figure 1 | Annual distribution of convective adjustments across the water column. (**a**) For 94 Ma, 4 × CO_2 from the North Pole. (**b**) For 71 Ma, 4 × CO_2 from the North Pole. (**c**) For 94 Ma, 4 × CO_2 from the South Pole. (**d**) For 71 Ma, 4 × CO_2 from the South Pole. The shading signifies the number of times the water column has undergone convective mixing summed over a year. Regions experiencing a large occurrence of convective adjustments are interpreted to represent site of intermediate and deep-water formation.

from palaeoceanographic data for the Cenomanian/Turonian boundary and for the early Maastrichtian (Supplementary Fig. 7).

Pathways of water masses between the oceanic basins. To go a step beyond this first-order description of ocean dynamics, we have computed zonal and/or meridional water transport across several key sections of the ocean basins (Figs 2 and 3). In most previous modelling studies, ocean dynamics have been described through a global overturning function. However, a global overturning function may be dominated by the largest oceanic basin, that is, the Pacific Ocean, and does not allow for the identification of ocean dynamics occurring in the still relatively narrow Atlantic Ocean, in the Tethys or in the Indian Ocean. Our aim here is to provide a better three-dimensional picture of water flow within and between the different oceanic basins.

In the CTRL94Ma run, 7 Sv are transported from the South Pacific to the SA across the D section (Fig. 3c and Table 1). The circulation simulated within the SA Ocean appears quite sluggish at depth, with a very limited northward flow of 4.2 Sv throughout the SA section between 1,200 and 2,750 m (Fig. 3c and Table 1). Conversely, across the East India section, water flow is directed northward down to a depth of 3,000 m and transports 25 Sv to the

North (Fig. 3c and Table 1). In the CTRL71Ma run, deep-water production intensifies in the southern Atlantic, extends eastward to the Indian sector and does not occur in the southern Pacific sector any more (Figs 1 and 3). Figure 3d shows a drastic intensification of the northern flow across the SA section for depth below 800 m. These northward-flowing deep waters reach the North Atlantic, as testified by the large flux of water across the CA section (Fig. 3b). Surface-to-intermediate waters across the Central to the Southern Atlantic are also flowing more intensively from the North to the South.

In the CTRL94Ma run and in the North Atlantic, the vertical flux profiles computed for the Caribbean Seaway section, CA section and Mediterranean (Med) section indicate that the surface and intermediate waters down to about 1,000 m flow from the Tethys to the North Atlantic, and then southward across the CA Gateway and westward into the equatorial Pacific via the Caribbean Seaway. Across the Caribbean Seaway below 1,000 m, waters flow from the Pacific to the Atlantic as shown by the eastward water flux from 1,000 to ~2,500 m (Fig. 3). The existence of an estuarine circulation pattern, with the surface layers flowing from the Tethys to the Pacific and the deeper layers flowing in the opposite direction, has been noticed several times in previous modelling studies[4,19,21]. In the CTRL71Ma runs, the model simulates an increase in circum-equatorial transport across

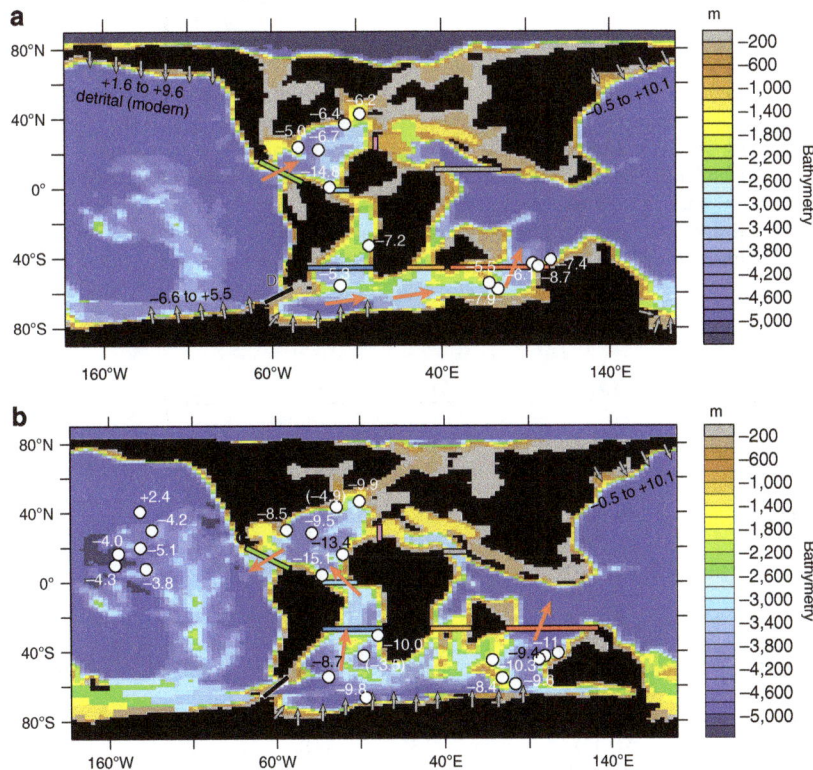

Figure 2 | Land–sea distribution and bathymetry specified in the FOAM simulations. (**a**) Cenomanian/Turonian boundary (94 Ma). (**b**) Early Maastrichtian (71 Ma). Latitudinal and meridional sections across which water transport has been calculated for Fig. 3 are shown. C, Caribbean Seaway; CA, Central Atlantic; D, Drake Passage; EI, East India; Med, Mediterranean; SA, South Atlantic; Tet, Tethys; WI, West India. Supplementary Figs 3 and 4 show the latitude/longitude—depth profile for each section except for the c and d sections for which the bathymetry of the area of interest is shown. The red arrows schematically show the deep-water pathways deduced from our modelling experiments. $\varepsilon_{Nd}(t)$ values of intermediate and deep waters (in white) inferred from fish remains and oxide coatings and averaged for the Turonian (**a**) and Maastrichtian (**b**) are reported at available ODP sites (see Supplementary Fig. 5 for the names of each ODP site). Details on the calculation of average $\varepsilon_{Nd}(t)$ values can be found in Supplementary Data 2. For ODP sites at Demerara Rise (Central Atlantic), $\varepsilon_{Nd}(t)$ values within OAE2 have not been included in the calculation. Values in parentheses indicate a contamination of the samples or a contamination of deep waters by volcanogenic particles[11,12]. Values in orange on the Maastrichtian map (**b**) are from the late Campanian. ε_{Nd} values of the modern detrital fraction (core-top detrital fraction and continental lithogenic sources) in the regions of modelled deep-water production are reported in grey for discussion (details on the specified range of values are available in Supplementary Data 1 and 2, and Supplementary Figs 5 and 6 with associated references).

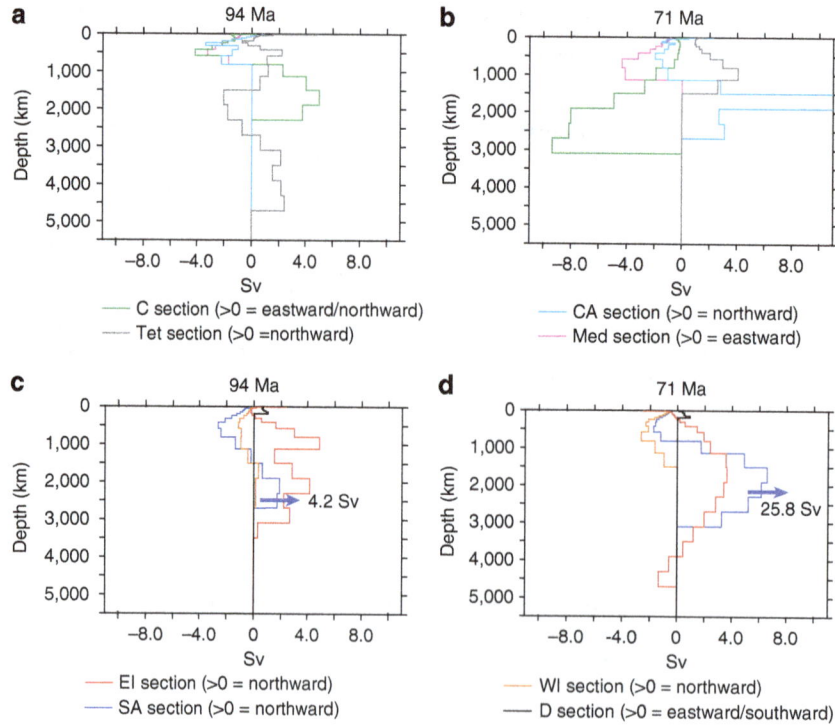

Figure 3 | Water transports. As computed for the segments defined in Fig. 2 for the North Atlantic and the Tethys basins (**a,b**) and for the SA and Indian basins (**c,d**). The units are in Sv, that is, $10^6\,m^3\,s^{-1}$. Sv is used to quantify the volumetric rate of transport of ocean currents. Values corresponding to the water flow for each vertical level of the ocean model are plotted. The variable corresponds to the integration of the velocity V (U) on the x axis (or y axis depending on the orientation of the section) and on the z axis. Calculated water fluxes account for both the velocity and the area of the section at a given level of the model. For zonal water transports such as the one calculated for the Med segment, positive values indicate eastward transport. For meridional water transports such as those calculated for the Tethyan (Tet), Central Atlantic (CA), South Atlantic (SA), East Indian (EI) and West Indian (WI) segments, positive values indicate northward transport; negative values indicate southward transport. For the specific case of the D section, positive directions are Southward and Eastward, and for the **c** section, positive directions are Northward and Eastward. Blue arrows in **c,d** represent water fluxes computed across the SA section for the deeper vertical levels where the transport is northward.

Table 1 | Water fluxes integrated over the water column for each section defined in Figure 1.

	Drake	SA	WI	EI	C	CA	Med	Tet
94 Ma								
DRAKE CLOSED	0	− 15 (5.8)	− 5	20	− 2.3	− 15.1	− 17.4	17.3
CTRL	**7**	**− 11.6 (4.2)**	**− 6.4**	**25**	**− 5.6**	**− 11.7**	**− 17.4**	**17.3**
DRAKE408m	16	− 9 (3.6)	− 5.2	30.2	− 7.2	− 9.2	− 16.5	16.3
SEWALL	17.6	− 7 (3.7)	− 8.	32.6	− 9	− 7.	− 16.	16
CAS560m	17.2	− 4.6 (3.9)	− 8.4	30.2	− 16.2	− 4.7	− 21.1	21
CAS-Islands	16	− 8.8 (3.8)	− 4.5	29.3	− 8.1	− 8.9	− 17	16.9
71 Ma								
DRAKE CLOSED	0	12.3 (35.4)	− 9.5	21.8	− 34.9	12.3	− 22.5	21.7
CTRL	**6.25**	**14.9 (25.8)**	**− 21.9**	**13.3**	**− 39.5**	**14.9**	**− 24.8**	**23.8**
DRAKE408m	12.4	18.7 (23.7)	− 19.8	13.5	− 41.1	18.7	− 22.4	22.1
SEWALL	14.4	15.6 (16.2)	− 15.3	14.1	− 33.6	15.6	− 18.1	18.6
CAS560m	10.1	− 14 (1.1)	− 9.8	33.8	2	− 14	− 12.1	11.6
CAS-Islands	14.6	17.7 (22.1)	− 18.5	15.4	− 41.1	17.7	− 23.4	23.5

C, Caribbean Seaway; CA, Central Atlantic; CAS560m, Caribbean gateway at 560 m; D, Drake Passage; DRAKE408m, Drake Passage at 408 m; EI, East India; Med, Mediterranean; SA, South Atlantic; Tet, Tethys; WI, West India.
Additional numbers in SA column correspond to the deep flow going northward in the South Atlantic across the SA section. Numbers in bold put forward the main simulations discussed in the paper, that is, with a shallow Drake Passage at 145 m.

the Caribbean Seaway and the Med gateway. Deep waters are now flowing from the Atlantic to the Pacific at all depths across the Caribbean Seaway, which marks the end of estuarine circulation in the Atlantic.

Comparison with available ε_{Nd}. Nd in seawater ultimately derives from continents and the rocks eroded around an area of

deep-water production imprint the surface waters with a distinct isotopic composition that depends on the age and lithology of the rock[28]. This signature is then exported to depth as the water sinks. The residence time of Nd in the ocean is shorter than the oceanic mixing time, which prevents complete homogenization but is long enough for Nd to be transported by deep-water masses along their pathways[28]. The Nd isotope composition of deep waters has thus been used to track oceanic circulation patterns in both modern and ancient oceans[10,28–30].

Available data for the Late Cretaceous highlight a major decrease in deep-water ε_{Nd} recorded both in the North and South Atlantic and in the Indian Ocean[10,14,31], with more unradiogenic (lower) ε_{Nd} values recorded during the Maastrichtian (on average in the -8.5 to -11 ε-units range) compared with the Turonian (on average in the -5 to -8.5 ε-units; Fig. 2). This decrease has been interpreted to reflect initiation of deep-water production in an area receiving unradiogenic Nd inputs from the nearby continents, with the northern Atlantic[15], or the Atlantic or Indian sector of the Southern Ocean proposed as possible sources[10,11]. Indeed at present, ε_{Nd} values as low as -14 to -26 ε-units recorded in Baffin Bay and Labrador Sea waters reflect the unradiogenic Nd continental supply from nearby Archean terranes[32–34] that were already present during the Cretaceous. In the southern Ocean, unradiogenic detrital inputs occur at present (Fig. 2, Supplementary Fig. 5 and Supplementary Data 1). The presence of similar unradiogenic inputs during the Late Cretaceous is supported by Campanian–Maastrichtian ε_{Nd} values

of detrital material of around -10 ε-units at ocean drilling program (ODP) site 690 (ref. 12). Because of these unradiogenic detrital inputs from the Antarctic continent, the more radiogenic ε_{Nd} values (higher) recorded in the SA and Indian deep waters during the mid-Cretaceous have been interpreted[10] to reflect at that time additional inputs of radiogenic Nd from abundant volcanic dust in the context of a sluggish circulation within a restricted basin.

It has also been suggested that the subsidence of Rio Grande Rise and other large igneous provinces (for example, Kerguelen or the Madagascar Plateau) during the Late Cretaceous could have diminished such inputs of radiogenic Nd to the southern Atlantic and Indian surface waters, contributing to decrease the ε_{Nd} values of the deep waters sourced in this region[31].

The oceanic circulation modelled here is in agreement with existing ε_{Nd} and part of the proposed scenarios. In the CTRL94Ma runs, the model simulates deep-water formation in the northern and southern Pacific. In the modern ocean, the southern Pacific receives inputs of quite radiogenic Nd eroded from the nearby Antarctic continent (Fig. 2, Supplementary Fig. 5 and Supplementary Data 1). $^{40}Ar/^{39}Ar$ ages of detrital hornblende grains from West Antarctica[35] support the presence of intrusive and volcanic rocks during the Cretaceous (and as early as the Jurassic) that probably provided radiogenic material to the nearby surface waters[36], imprinting the deep water produced there with a radiogenic signature. Deep water produced in the North Pacific region should also have a radiogenic composition, because weathering of radiogenic young circum-Pacific volcanic arcs linked to the subduction of the Kula and Farallon plates was already active during the mid-Cretaceous[37]. This is supported by the high ε_{Nd} values of neritic seawater inferred from Late Cretaceous fish remains recovered from the northwestern Pacific, (typically in the -3 to $+1$ ε-units range)[38]. In the North Atlantic, deep waters below 1,000 m are conveyed in our simulation from the equatorial eastern Pacific through the Caribbean Seaway and result from the mixing of radiogenic deep waters formed both in the North and in the South Pacific. Therefore, the ocean dynamics simulated by the model are expected to result in quite radiogenic ε_{Nd} values of deep waters in the North Atlantic, in agreement with existing ε_{Nd} (Fig. 2 and Supplementary Data 2).

In the CTRL94Ma run, the deep waters bathing the southern Atlantic and Indian Oceans originate in the area located near the modern Weddell Sea, a region that most probably received unradiogenic continental supply from Antarctica as discussed above (Fig. 2). It has been suggested[10,31] that the southern Atlantic may have received during the mid-Cretaceous additional inputs of radiogenic Nd from subaerially exposed continental volcanic provinces, exposure of submarine Large Igneous Provinces and hot spots, and from associated volcanic activity. The sluggish circulation simulated in this work in the SA in the CTRL94Ma run (Fig. 3 and Table 1) would have favoured seawater-particle exchange processes with such radiogenic volcanic particles[10], resulting in a quite radiogenic composition of SA and Indian deep waters as depicted in the available Nd data set (Fig. 2).

In the CTRL71Ma run, convection in the southern Pacific Ocean completely disappears and is reduced in the North Pacific to the northwest area, whereas convection in the SA intensifies and extends eastward to Australia (Figs 2 and 3). The development of sites of deep-water formation near areas of probable unradiogenic continental supply in the southern Atlantic is expected to result in a relatively unradiogenic composition of SA and Indian Ocean waters. This is in agreement with available ε_{Nd} that display quite negative ε_{Nd} values of deep waters during the latest Cretaceous in the southern Atlantic and Indian Oceans, in the range of about -9.5 to -10.5 ε-units at ODP site 690 for the Late Campanian–Maastrichtian interval[12]

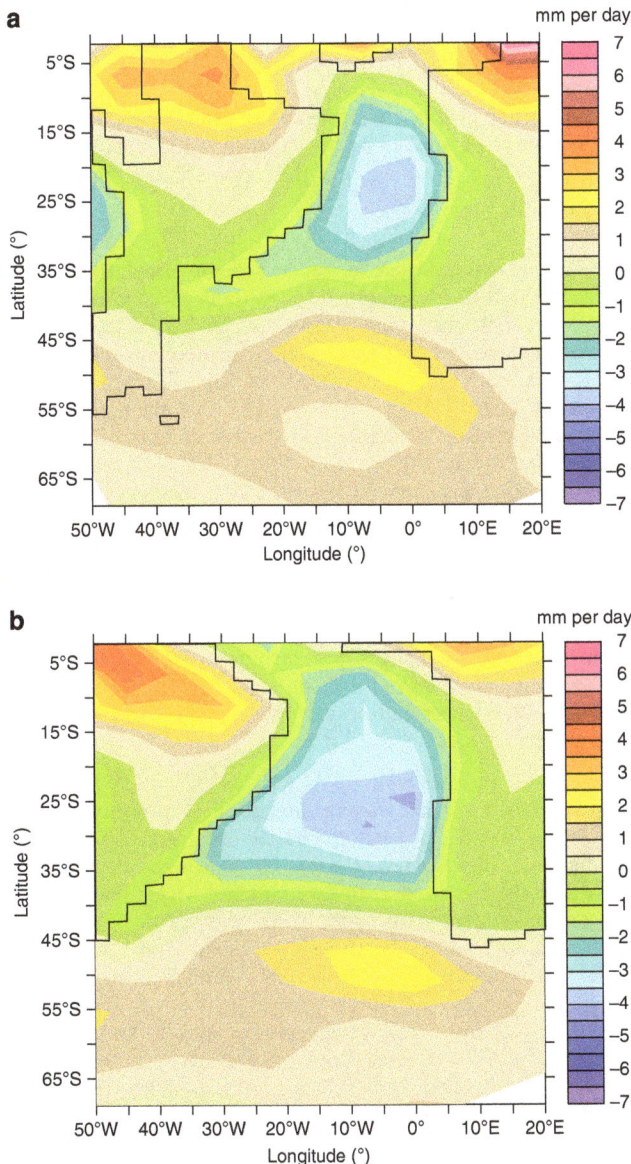

Figure 4 | Hydrological budget over the Central and South Atlantic.
Precipitation minus evaporation for 94 (**a**) and 71 Ma (**b**) runs.

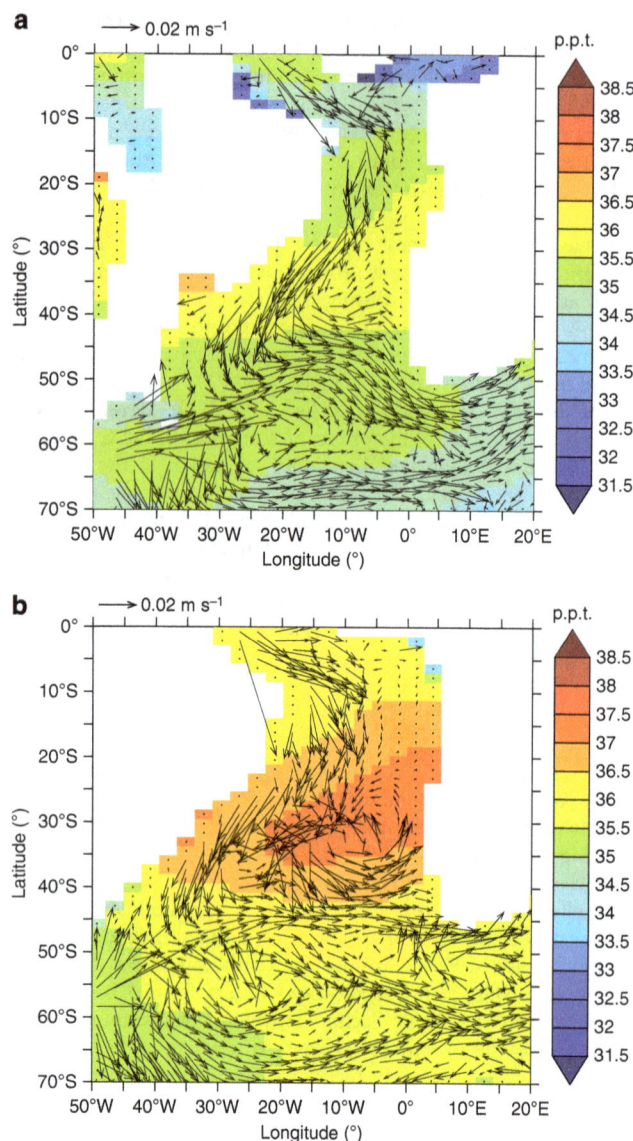

Figure 5 | Absolute salinity fields. Salinity and current velocity (m s^{-1}) averaged over the first 300 m of the ocean for 94 (**a**) and 71 Ma (**b**) runs.

and of -8.4 to -11 ε-units on average for the Maastrichtian at several Indian Ocean ODP sites[10,12,13] (Fig. 2).

Following the simulated circulation patterns, this unradiogenic deep-water composition would then be exported to the North Atlantic through the deeper CA gateway. In addition, with the inversion of the deep-water flux through the Caribbean Seaway in the model, the radiogenic Pacific deep waters would no longer enter the North Atlantic. A westward flux of deep waters throughout the Caribbean Seaway during the Campanian and Maastrichtian is supported by ε_{Nd} values of the detrital fraction higher than that of the bottom waters at site 152 (Nicaragua Rise)[39]. The oceanic circulation simulated for the Maastrichtian should thus generate a much less radiogenic Nd isotope signature of North and SA deep waters, as well as in the Indian Ocean down to 3,000 m of depth, in agreement with ε_{Nd} data (Fig. 2).

Our work thus supports an intensification of deep-water production in the southern Atlantic and Indian Oceans along with a reversal of the deep-water flux through the Caribbean Seaway as a driver for the decrease in deep-water ε_{Nd} depicted during the Late Cretaceous. The subsidence of large igneous provinces such as the Kerguelen plateau could also have contributed to reducing the

inputs of radiogenic Nd to the surface waters of the southern Atlantic and Indian Oceans[31]. As Nd is not incorporated as a tracer into our model, the impact of reduced inputs of radiogenic Nd linked to this mechanism on the decrease recorded by deep-water ε_{Nd} values cannot be further discussed here.

The simulated circulation during the Maastrichtian remains difficult to reconcile with the highly non-radiogenic values recorded at Cape Verde and Demerara Rise[14,16] in the low-latitude Atlantic Ocean. It has been suggested that intermediate to deep waters could have been generated in the area of Demerara Rise ('Demerara Rise Bottom Waters')[16]. Nevertheless, the Late Cretaceous ε_{Nd} from Demerara Rise stands in marked contrast to the data from other North Atlantic sites[11]. Demerara Rise bottom waters may have been restricted to intermediate depths, similar to Med outflow water, and thus would not have greatly influenced deep-water masses in the North Atlantic[11]. Our results show no evidence for water sinking in this area (Fig. 1). One way to reconcile the ε_{Nd} of Demerara Rise and Cape Verde with the modelled circulation could be to invoke the impact of boundary exchange processes that are known to occur along continental margins and can modify the Nd isotope composition of the local bottom waters[32,40]. Considering the proximity of the two sites to very old, unradiogenic terranes (Guyana and African Shields), such a process may have locally lowered the Nd isotopic composition of the water masses. The increase in Nd isotope values recorded at this site at the end of the Maastrichtian when they reach values similar to the other Atlantic sites may then reflect the progressive opening of the Demerara Rise region to the remaining of the Atlantic.

Mechanisms driving the changes in oceanic circulation. In the model, the shift in the location of deep-water formation from the Pacific to the Atlantic/Indian sector is clearly due to modifications of the land–sea mask. In detail, it is the hydrological cycle over the SA basin that seems to be responsible for the shift in the deep-water formation area between the Cenomanian and the early Maastrichtian. During the early Maastrichtian, the fall in sea level and the retreat of the sea from the tropical continents induce more tropical monsoons and less vertical advection of moist air, which result in a more negative P − E (precipitation minus evaporation) budget in the CA Ocean (Fig. 4). When integrated over the latitudes 40°S–10°N, corresponding to the Central and SA, the P − E budget is more negative in the 71 Ma run with a value around −1.5 mm per day (−0.14 mm per day for the 94 Ma run). These changes will favour convection in the SA during the early Maastrichtian, because the saltier the surface waters are, the denser and heavier they will be when they reach the southern high latitudes (Fig. 5).

Sensitivity to the Drake Passage and to the Caribbean Seaway. The results presented here are based on the two palaeogeographies published in Sewall et al.[9], in which the depth of the Drake Passage has been reduced to 145 m. Major uncertainties exist on the configuration of the Drake Passage during the Late Cretaceous, with reported depths ranging from over 1,000 m[9] to a shallow[41] or completely closed passage[42]. Similarly, the depth and configuration of the Caribbean gateway is not well constrained. Island arc volcanism was already present across the Caribbean Seaway during the Late Cretaceous, possibly partly impeding deep or intermediate water communications across the seaway[43,44]. Nevertheless, the configuration of these two seaways have been shown to have important implications for oceanic circulation in more recent periods[45,46].

To test the impact of the configuration of the Drake Passage and of the Caribbean Seaway on oceanic circulation,

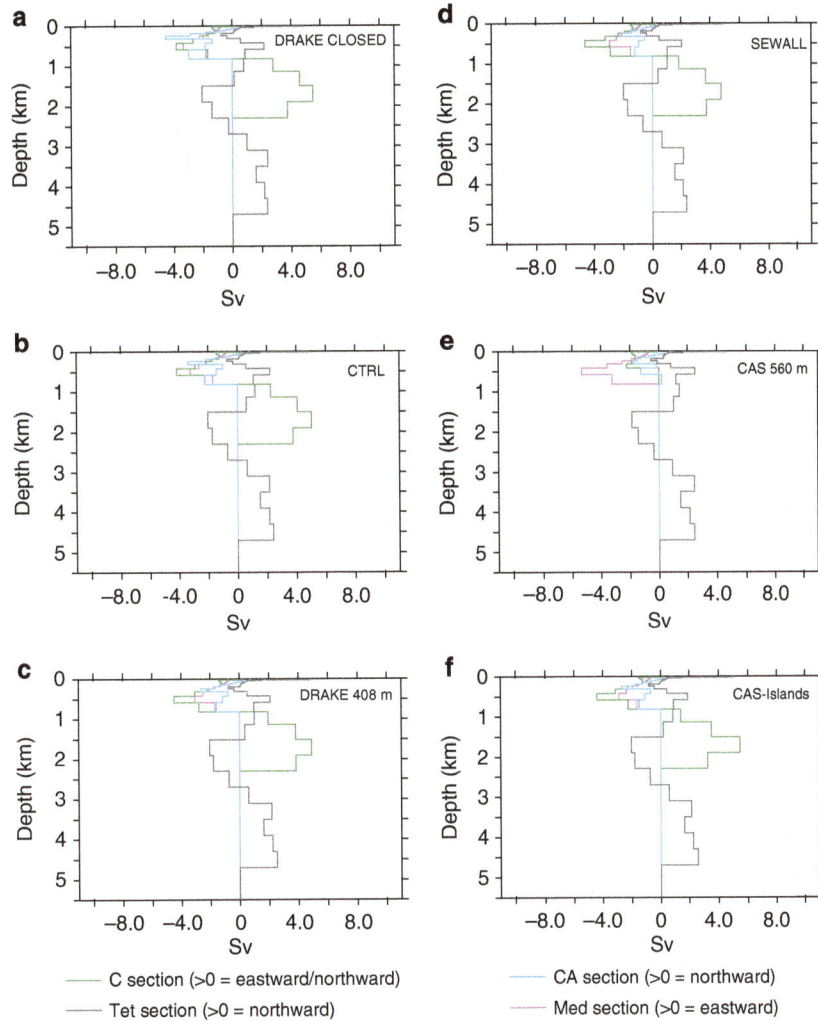

Figure 6 | Water transports for the North Atlantic and the Tethys basins at 94 Ma. (**a**) Drake Passage closed (DRAKE CLOSED), (**b**) Drake Passage at 145 m (CTRL), (**c**) Drake Passage at 408 m (DRAKE408m), (**d**) Drake Passage as in Sewall et al.[9] (SEWALL), (**e**) Caribbean gateway at 560 m (CAS560m) and (**f**) continental plate across the Caribbean gateway (CAS-Islands).

we performed a series of sensitivity experiments. In a first set of simulations, we modified the Cenomanian/Turonian and the early Maastrichtian palaeogeographies of Sewall et al.[9] using the same horizontal configuration but varying the depth of the Drake Passage, using the depth originally specified by Sewall et al.[9], which is around 1,000 m (SEWALL simulations), a depth of 408 m (DRAKE408m simulations), a depth of 145 m (CTRL simulations) and a closed Drake Passage (DRAKE CLOSED simulation). In a second set of simulations, two runs for each palaeogeography were performed with a very shallow Caribbean Gateway, fixed at 560 m depth (CAS560m simulations) and a small continental plate across the Caribbean Seaway (Supplementary Fig. 8) to simulate the presence of islands (CAS-Islands simulation). The results of these experiments are presented in Figs 6–9 and in Table 1.

Most features described above for the two land–sea masks in the CTRL runs remain the same independently of the depth of the Drake Passage (Figs 6–9). These features are the inversion of the direction of deep-water flow across the Caribbean Seaway for depths > 1,000 m and the increase in northward water fluxes at depth across the SA basin. Summed over the whole water column, water flows globally from the North Atlantic to the SA during the Cenomanian–Turonian and in the opposite direction during the early Maastrichtian. However, the depth of the Drake Passage

substantially affects the intensity of these flows. The main changes occur when going from a closed passage to a depth of 408 m. Most notably, the deeper branch of the thermohaline circulation coming from the Weddell Sea decreases its volume transport from 5.8 to 3.7 Sv for the 94 Ma experiments and from 35.4 to 16.2 Sv for the 71 Ma experiments (Table 1). Nevertheless, the trend from a sluggish circulation to a more active circulation at depth in the South and CA with the early Maastrichtian palaeogeography holds independently of the configuration of the Drake Passage.

Sensitivity experiments assessing the effect of Caribbean Seaway depth on circulation yield contrasting results, especially for the early Maastrichtian simulations. A shallow gateway across the Caribbean Seaway (CAS560m simulations) completely modifies deep circulation and results in a shut down of thermohaline circulation in the SA (Figs 8 and 9). By contrast, the same configuration but for the 94 Ma experiments has a minor impact on the general pattern of oceanic circulation (Figs 6 and 7). Finally, experiments with a small continental plate across the Caribbean Seaway result in only minor changes in the modelled oceanic circulation for both palaeogeographies.

The circulation modelled with a shallow Caribbean Seaway (CAS560 m) for the Cenomanian/Turonian is less consistent with

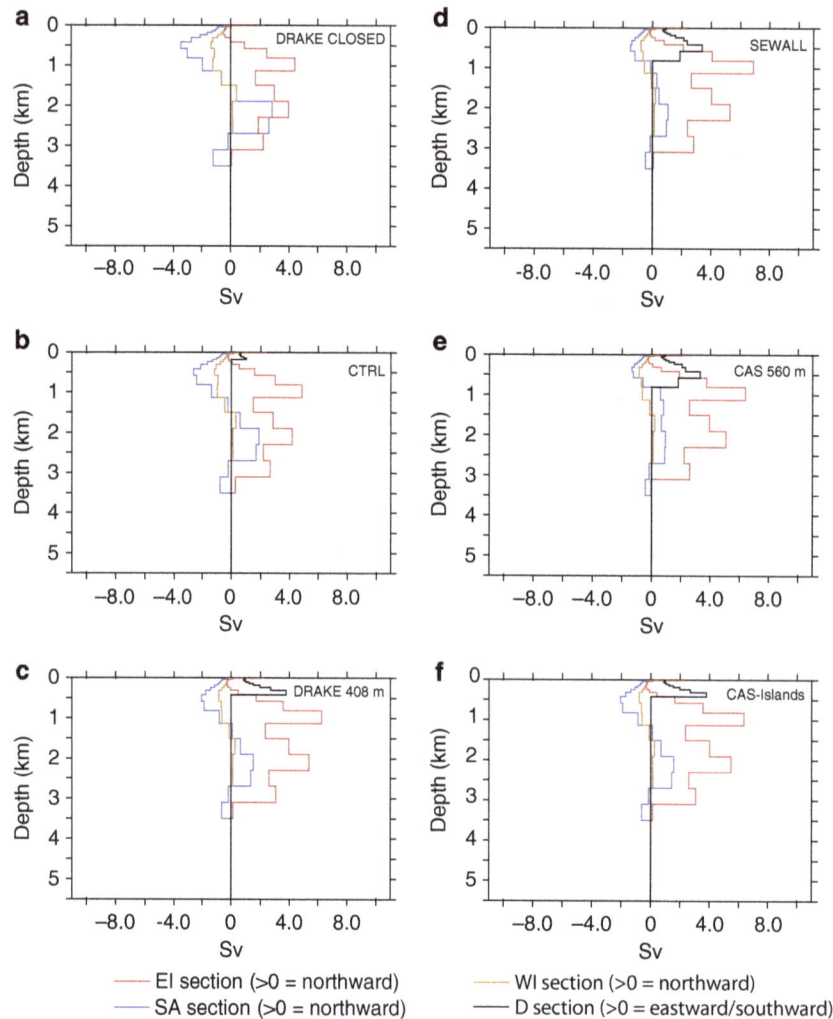

Figure 7 | Water transport for the SA and Indian basins at 94 Ma. (a) Drake Passage closed (DRAKE CLOSED), (b) Drake Passage at 145 m (CTRL), (c) Drake Passage at 408 m (DRAKE408m), (d) Drake Passage as in Sewall et al.[9] (SEWALL), (e) Caribbean gateway at 560 m (CAS560m) and (f) Continental plate across the Caribbean gateway (CAS-Islands).

the ε_{Nd}, because a deep-water influx of radiogenic Pacific waters through the Caribbean Seaway as observed in the other simulations is in better agreement with the radiogenic composition of North Atlantic deep waters. In the early Maastrichtian, the weaker fluxes of surface and intermediate waters across the Tethys, Med and Caribbean Seaway modelled in the CAS560m experiment are also less consistent with an intense equatorial circumglobal current that has been suggested based on the large-scale deposition of phosphorites along the southern Tethyan margin[47].

These additional sensitivity experiments reveal a major role of the circulation pattern across the Caribbean Seaway on the overall modelled thermohaline circulation and shed new light on the mechanisms driving the change in oceanic circulation observed between the simulations using the Cenomanian/Turonian versus early Maastrichtian palaeogeographies. The hydrological cycle changes described above contribute to the intensification of deep-water production in the southern Atlantic and Indian Oceans. Changes in land–sea configuration and the associated atmospheric feedbacks induce a saltier subtropical SA Ocean during the early Maastrichtian, which in turn results in a larger area of deep-water formation around the eastern Antarctic (Fig. 1). Indeed, the saltier the surface waters are, the denser and heavier they will be when they reach the southern high latitudes.

An additional important factor appears to be the strong flow of Atlantic water into the Pacific at all depths during the Maastrichtian, observed in all experiments, except the CAS 560 m simulation. This flow creates a large divergence area in the CA, drawing in waters from the Med gateway and from the SA (Table 1). The transition from an estuarine circulation during the Cenomanian–Turonian to the one characterizing the early Maastrichtian as simulated here is probably the main mechanism explaining the shift from a sluggish to an active deep circulation. This transition occurs in response to continental drift and the subsequent modification of the geometry of the Caribbean Seaways. The sensitivity of oceanic circulation across the Caribbean Seaways has already been investigated many times and results from a complex interplay between local processes, such as wind-driven circulation and salinity contrast[45], and remote processes such as the geometry of other gateways[46,48].

Discussion

Our simulations highlight an evolution from a sluggish circulation in the SA and CA using the Cenomanian/Turonian land–sea mask towards a much more active circulation in these basins with the early Maastrichtian land–sea mask. Alongside changes in the location of deep-water production areas at high latitudes, these

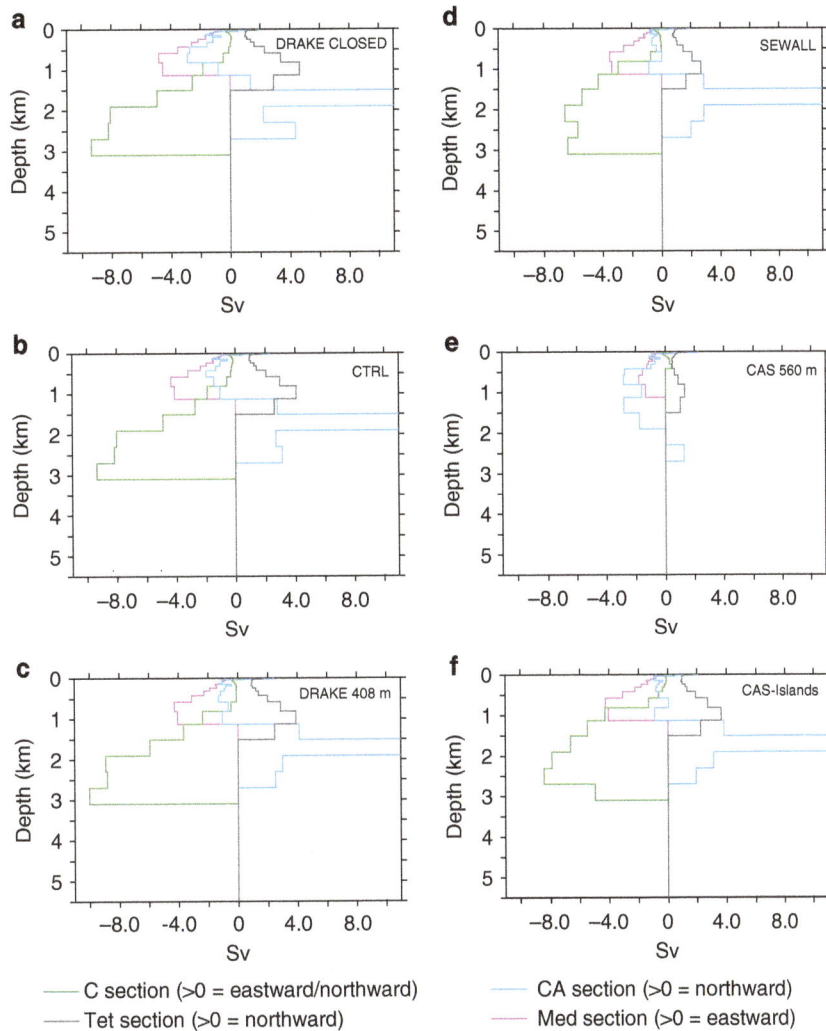

Figure 8 | Water transport for the North Atlantic and the Tethys basins at 71 Ma. (**a**) Drake Passage closed (DRAKE CLOSED), (**b**) Drake Passage at 145 m (CTRL), (**c**) Drake Passage at 408 m (DRAKE408m), (**d**) Drake Passage as in Sewall et al.[9] (SEWALL), (**e**) Caribbean gateway at 560 m (CAS560m) and (**f**) continental plate across the Caribbean gateway (CAS-Islands).

features may have paved the way for a well-oxygenated deep ocean. The absence of large-scale anoxia in deep waters during abrupt warming events after the Late Cretaceous, such as the Paleocene–Eocene transition[49] (~55 Ma), supports this hypothesis. The experiments conducted here emphasize the potentially important role of land–sea configuration as a pre-conditioning factor that affects the ease with which OAEs can develop. In that sense, palaeogeography represents one of the required conditions for the occurrence of OAEs but would not be sufficient on its own to trigger these events, implying additional causal factors[50,51]. As a result of the palaeogeographic changes occurring during the Late Cretaceous, our results suggest that worldwide anoxic conditions in deep waters were more likely to develop as a result of the same triggering factor (for example, increased nutrient input into the oceans[2–4]) during the Cenomanian/Turonian compared with the early Maastrichtian. Two recent geochemical modelling studies have emphasized the need for a sluggish circulation in the Atlantic basin, to explain the large spatial extent of OAE-2 (ref. 52) and a substantial sensitivity of the onset of anoxia/euxinia to oceanic overturning[53]. Based on the latter study, we predict that in a sluggish state such as the one simulated at the Cenomanian/Turonian boundary with <5 Sv of deep water flowing northward in the SA, a 20–40% increase in nutrient input to the ocean could generate deep-water euxinia. In

contrast, a 225% increase in nutrient input is required to provoke euxinia with the early Maastrichtian palaeogeography in which 25.8 Sv feed the deep-water flowing northward (Fig. 3c,d). Our results imply that thresholds for the ocean-climate system to shift towards a state of global anoxia in deep waters are likely to be much higher at present than during the Cenomanian/Turonian because of the strength of the modern thermohaline circulation.

In summary, this study provides a clear description of oceanic circulation changes occurring during the Late Cretaceous and is able to broadly explain available published ε_{Nd} data. Our work points to changes in continental configuration as the major driver for the depicted ocean circulation change, through a modification of the hydrological cycle over the SA basin and the development of a strong westward flow of water at all depths throughout the Caribbean Seaway, that both favour an intensification of deep-water production in the southern Atlantic and Indian region. We note that our knowledge of the geometry of Cretaceous oceanic gateways is contentious and a definitive history of ocean circulation during this time must await further constraints on the state of the various oceanic connections. We also speculate on the consequences of long-term oceanic circulation changes on the likelihood that the Earth System will be affected by an OAE. More explicit modelling including

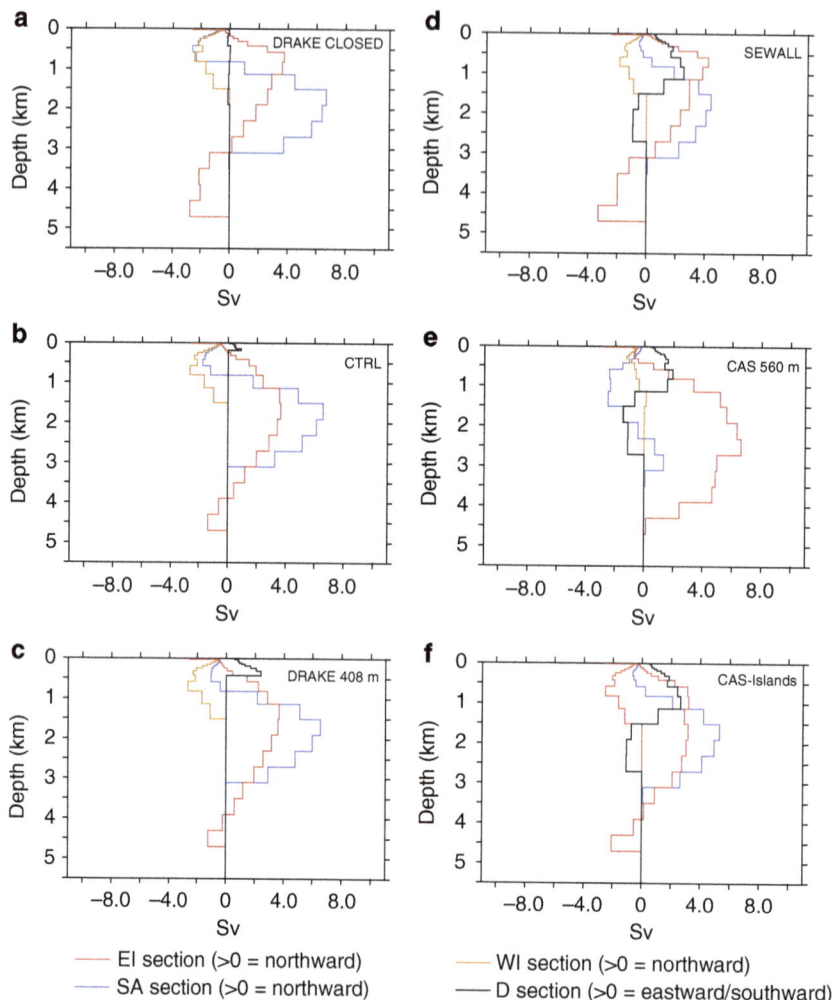

Figure 9 | Water transport for the SA and Indian basins at 71 Ma. (a) Drake Passage closed (DRAKE CLOSED), (b) Drake Passage at 145 m (CTRL), (c) Drake Passage at 408 m (DRAKE408m), (d) Drake Passage as in Sewall et al.[9] (SEWALL), (e) Caribbean gateway at 560 m (CAS560m) and (f) continental plate across the Caribbean gateway (CAS-Islands).

complex climate models and biogeochemistry will help reveal the extent to which the tectonic and climatic background state can influence the development of oceanic anoxic conditions.

Methods

Code availability. The code of the model FOAM is available on request by e-mail to the first author.

Description of the model. The model experiments were completed using the FOAM, developed by Jacob[54]. This model combines a low spectral resolution R15 (48 × 40 grid) atmosphere model counting 18 altimetric levels with a highly efficient medium resolution (128 × 128 grid) ocean module composed by 24 bathymetric levels. FOAM successfully simulates many aspects of the present-day climate and compares well with other contemporary medium-resolution climate models. It was previously used to investigate numerous past climate changes, ranging from the Neoproterozoic glaciations to the onset of the ACC at the Eocene–Oligocene boundary[55,56].

Description of the experiments. The two sets of simulations (called hereafter CTRL94Ma and CTRL71Ma) developed for this study are based on existing palaeogeographic reconstructions for the Cenomanian/Turonian boundary and for the early Maastrichtian[9] (Fig. 2). For the sake of simplicity, we choose to use 94 Ma though the accurate age of the boundary is 93.9 Ma[57]. Based on existing palaeogeographies[41,58,59], the depth of the Drake Passage in the CTRL simulations has been restricted to a shallower depth (that is, 145 m) than in the original bathymetry in which the depth reached more than a thousand metres. Indeed, first evidences for the deepening of the Drake Passage correspond to the creation of oceanic crust and are generally dated between the Eocene and the Oligocene[60,61].

During the Late Cretaceous, pull-apart basins were formed between the Antarctica Peninsula and southern South America[62], which represent the onset of the opening of the Scotia Sea. Therefore, the Drake Passage most probably formed a narrow and shallow marine domain at that time[63]. To account for existing uncertainties on the configuration of the Drake Passage and of the Caribbean Seaway, sensitivity experiments assessing the effect of the depth specified for these two marine gateways are also presented in the discussion. Data and model-based atmospheric CO_2 level estimates for the Late Cretaceous are between two and eight times the pre-industrial atmospheric CO_2 concentration (280 p.p.m.)[64–69]. Accordingly, three atmospheric CO_2 concentrations were tested for the two palaeogeographies: 560, 1,120 and 2,240 p.p.m. The solar luminosity was specified as 99% of modern. Orbital parameters and other greenhouse gases were set to present-day values. River routing is specified using the Sewall et al. reconstruction.

The experiments were integrated for 2,000 years without flux corrections or deep ocean acceleration. During the last 100 years of model integration, there is no apparent drift in the upper ocean (between the surface and 300 m depth) and <0.1 °C per century change in globally averaged ocean temperature. All model results have been averaged over the last 50 model years.

References

1. Keeling, R. E., Körtzinger, A. & Gruber, N. Ocean deoxygenation in a warming world. *Annu. Rev. Marine Sci.* **2**, 199–229 (2010).
2. Jenkyns, H. C. Geochemistry of oceanic anoxic events. *Geochem. Geophys. Geosyst.* **11** doi:10.1029/2009GC002788 (2010).
3. Meyer, K. M. & Kump, L. R. Oceanic euxinia in earth history: causes and consequences. *Annu. Rev. Earth Planet Sci.* **36**, 251–288 (2008).
4. Monteiro, F. M., Pancost, R. D., Ridgwell, A. & Donnadieu, Y. Nutrients as the dominant control on the spread of anoxia and euxinia across the

Cenomanian-Turonian oceanic anoxic event (OAE2): model-data comparison. *Paleoceanography* **27** doi:10.1029/2012PA002351 (2012).

5. Föllmi, K. B. Early Cretaceous life, climate and anoxia. *Cretaceous Res.* **35**, 230–257 (2012).

6. Friedrich, O., Norris, R. D. & Erbacher, J. Evolution of middle to Late Cretaceous oceans--A 55 m.y. record of Earth's temperature and carbon cycle. *Geology* **40**, 107–110 (2011).

7. Poulsen, C. J., Gendaszek, A. S. & Jacob, R. L. Did the rifting of the Atlantic Ocean cause the Cretaceous thermal maximum? *Geology* **31**, 115–118 (2003).

8. Pucéat, E. *et al*. Thermal evolution of Cretaceous Tethyan marine waters inferred from oxygen isotope composition of fish tooth enamels. *Paleoceanography* **18** doi:10.1029/2002PA000823 (2003).

9. Sewall, J. O. *et al*. Climate model boundary conditions for four Cretaceous time slices. *Climate of the Past* **3**, 647–657 (2007).

10. Robinson, S. A., Murphy, D. P., Vance, D. & Thomas, D. J. Formation of 'Southern Component Water' in the Late Cretaceous: evidence from Nd-isotopes. *Geology* **38**, 871–874 (2010).

11. Robinson, S. A. & Vance, D. Widespread and synchronous change in deep-ocean circulation in the North and South Atlantic during the Late Cretaceous. *Paleoceanography* **27** doi:10.1029/2011PA002240 (2012).

12. Voigt, S. *et al*. Tectonically restricted deep-ocean circulation at the end of the Cretaceous greenhouse. *Earth Planet Sci. Lett.* **369-370**, 169–177 (2013).

13. Murphy, D. P. & Thomas, D. J. Cretaceous deep-water formation in the Indian sector of the Southern Ocean. *Paleoceanography* **27**, PA1211 (2012).

14. Martin, E. E., MacLeod, K. G., Berrocoso, A. J. & Bourbon, E. Water mass circulation on Demerara Rise during the Late Cretaceous based on Nd isotopes. *Earth Planet Sci. Lett.* **327**, 111–120 (2012).

15. MacLeod, K. G., Londono, C. I., Martin, E. E., Berrocoso, A. J. & Basak, C. Changes in North Atlantic circulation at the end of the Cretaceous greenhouse interval. *Nat. Geosci.* **4**, 779–782 (2011).

16. MacLeod, K. G., Martin, E. E. & Blair, S. W. Nd isotopic excursion across Cretaceous ocean anoxic event 2 (Cenomanian-Turonian) in the tropical North Atlantic. *Geology* **36**, 811–814 (2008).

17. Friedrich, O., Erbacher, J., Moriya, K., Wilson, P. A. & Kuhnert, H. Warm saline intermediate waters in the Cretaceous tropical Atlantic Ocean. *Nat. Geosci.* **1**, 453–457 (2008).

18. Hague, A. M. *et al*. Convection of North Pacific deep water during the early Cenozoic. *Geology* **40**, 527–530 (2012).

19. Poulsen, C. J., Seidov, D., Barron, E. J. & Peterson, W. H. The impact of paleoceanographic evolution on the surface oceanic circulation and the marine environment within the mid-Cretaceous Tethys. *Paleoceanography* **13**, 546–559 (1998).

20. Otto-Bliesner, B. L., Brady, E. C. & Shields, C. Late Cretaceous ocean: coupled simulations with the national center for atmospheric research climate system model. *J. Geophys. Res.* **107**, Art. No. 4019 (2002).

21. Trabucho Alexandre, J. *et al*. The mide-Cretaceous North Atlantic nutrient trap: Black shales and OAEs. *Paleoceanography* **25** doi:10.1029/2010PA001925 (2010).

22. Poulsen, C. J., Barron, E. J., Arthur, M. A. & Peterson, W. H. Response of the mid-Cretaceous global oceanic circulation to tectonic and CO_2 forcings. *Paleoceanography* **16**, 576–592 (2001).

23. Brady, E. C., DeConto, R. & Thompson, S. L. Deep water formation and poleward ocean heat transport in the warm climate extreme of the Cretaceous (80 Ma). *Geophys. Res. Lett.* **25**, 4205–4208 (1998).

24. Lunt, D. J. *et al*. CO2-driven ocean circulation changes as an amplifier of Paleocene-Eocene thermal maximum hydrate destabilization. *Geology* **38**, 875–878 (2010).

25. Poulsen, C. J. & Zhou, J. Sensitivity of Arctic climate variability to mean state: insights from the Cretaceous. *J. Clim.* **26**, 7003–7022 (2013).

26. Flögel, S. *et al*. Simulating the biogeochemical effects of volcanic CO2 degassing on the oxygen-state of the deep ocean during the Cenomanian/Turonian Anoxic Event (OAE2). *Earth Planet Sci. Lett.* **305**, 371–384 (2011).

27. Lunt, D. J. *et al*. A model-data comparison for a multi-model ensemble of early Eocene atmosphere-ocean simulations: EoMIP. *Clim. Past Discuss.* **8**, 1229–1273 (2012).

28. Frank, M. Radiogenic isotopes: tracers of past ocean circulation and erosional input. *Rev. Geophys.* **40** doi:10.1029/2000RG000094 (2002).

29. Goldstein, S. L. & Jacobsen, S. B. The Nd and Sr isotopic systematics of river-water dissolved material: Implications for the sources of Nd and Sr in seawater. *Chem. Geol.* **66**, 245–272 (1987).

30. Thomas, D. J. Evidence for deep-water production in the North Pacific Ocean during the early Cenozoic warm interval. *Nature* **430**, 65–68 (2004).

31. Murphy, D. P. & Thomas, D. J. The evolution of Late Cretaceous deep-ocean circulation in the Atlantic basins: Neodymium isotope evidence from South Atlantic drill sites for tectonic controls. *Geochem. Geophys. Geosyst.* **14** doi:1002/2013GC004889 (2013).

32. Jeandel, C., Arsouze, T., Lacan, F., Téchiné, P. & Dutay, J. -C. Isotopic Nd compositions and concentrations of the lithogenic inputs into the ocean: A compilation, with an emphasis on the margins. *Chem. Geol.* **239**, 156–164 (2007).

33. Lacan, F. & Jeandel, C. Acquisition of the neodymium isotopic composition of the North Atlantic Deep Water. *Geochem. Geophys. Geosyst.* **6** doi:10.1029/2005GC000956 (2005).

34. Vance, D. & Burton, K. Neodymium isotopes in planktonic foraminifera: a record of the response of continental weathering and ocean circulation rates to climate change. *Earth Planet Sci. Lett.* **173**, 365–379 (1999).

35. Roy, M., van de Flierdt, T., Hemming, S. R. & Goldstein, S. L. 40Ar/39Ar ages of hornblende grains and bulk Sm/Nd isotopes of circum-Antarctic glacio-marine sediments: implications for sediment provenance in the southern ocean. *Chem. Geol.* **244**, 507–519 (2007).

36. Dalziel, I. W. D. Antarctica; a tale of two supercontinents? *Annu. Rev. Earth Planet Sci.* **20**, 501–526 (1992).

37. Larson, R. L. & Pitman, W. C. World-wide correlation of Mesozoic magnetic anomalies, and its implications. *Geol. Soc. Am. Bull.* **83**, 3645–3662 (1972).

38. Moiroud, M. *et al*. Evolution of neodymium isotopic signature of neritic seawater on a northwestern Pacific margin: new constraints on possible end-members for the composition of deep-water masses in the Late Cretaceous ocean. *Chem. Geol.* **356**, 160–170 (2013).

39. Moiroud, M. *et al*. Evolution of neodymium isotopic signature of seawater during the Late Cretaceous: implications for intermediate and deep circulation. *Gond. Res.* http://dx.doi.org/10.1016/j.gr.2015.08.005 (2015).

40. Carter, P., Vance, D., Hillenbrand, C. D., Smith, J. A. & Shoosmith, D. R. The neodymium isotopic composition of water masses in the eastern Pacific sector of the Southern Ocean. *Geochem. Cosmochim. Acta* **79**, 41–59 (2012).

41. Vrielynck, B. & Bouysse, P. *The Changing Face of the Earth: The Breakup of Pangea and Continental Drift Over the Past 250 Million Years in Ten Steps* (Commission de la Carte Géologique du Monde and Unesco, 2003).

42. Hay, W. W. *et al*.in *Evolution of the Cretaceous Ocean-Climate System* Vol. 332 (eds Barrera, E. & Johnson, C. C.) 1–47 (Geological Society of America Special Paper, 1999).

43. Giunta, G., Marroni, E., Padoa, E. & Pandolfi, L. in *The Circum-Gulf of Mexico and the Carribean: Hydrocarbon habitats, basin formation, and plate tectonics* Vol. 79 (eds Bartolini, C., Buffer, R. & Blickwede, J.) 104–125 (2003).

44. Itturalde-Vinent, M. in *From Greenhouse to Icehouse: The Marine Eocene-Oligocene Transition.* (eds Prothero, D., Ivany, L. C. & Nesbitt, E.) **Ch. 22**, 386–396 (Colombia Univ. Press, 2003).

45. Sepulchre, P. *et al*. Consequences of shoaling of the Central American Seaway determined from modeling Nd isotopes. *Paleoceanography* **29** doi:10.1002/2013PA002501 (2014).

46. von der Heydt, A. S. & Dijkstra, H. A. Effect of ocean gateways on the global ocean circulation in the late Oligocene and early Miocene. *Paleoceanography* **21** doi:10.1029/2005PA001149 (2006).

47. Soudry, D., Glenn, C. R., Nathan, Y., Segal, I. & VonderHaar, D. L. Evolution of Tethyan phosphogenesis along the northern edges of the Arabian-African shield during the Cretaceous-Eocene as deduced from temporal variations of Ca and Nd isotopes and rates of P accumulation. *Earth Sci. Rev.* **78**, 27–57 (2006).

48. Omta, A. W. & Dijkstra, H. A. A physical mechanism for the Atlantic-Pacific flow reversal in the early Miocene. *Glob. Planet Change* **36**, 265–276 (2003).

49. Zachos, J. C., Dickens, G. R. & Zeebe, R. E. An early Cenozoic perspective on greenhouse warming and carbon-cycle dynamics. *Nature* **451**, 279–283 (2008).

50. Tejada, M. L. G. *et al*. Ontong Java Plateau eruption as a trigger for the early Aptian oceanic anoxic event. *Geology* **37**, 855–858 (2009).

51. Turgeon, S. C. & Creaser, R. A. Cretaceous oceanic anoxic event 2 triggered by a massive magmatic episode. *Nature* **454**, 323–327 (2008).

52. Ruvalcaba Baroni, I., Topper, R. P. M., van Helmond, N. A. G. M., Brinkhuis, H. & Slomp, C. P. Biogeochemistry of the North Atlantic during oceanic anoxic event 2: role of changes in ocean circulation and phosphorus input. *Biogeosciences* **11**, 977–993 (2014).

53. Ozaki, K., Tajima, S. & Tajika, E. Conditions required for oceanic anoxia/euxinia: constraints from a one-dimensional ocean biogeochemical cycle model. *Earth Planet Sci. Lett.* **304**, 270–279 (2011).

54. Jacob, R. L. *Low Frequency Variability in a Simulated Atmosphere Ocean System* (Wisconsin, 1997).

55. Poulsen, C. J. & Jacob, R. L. Factors that inhibit snowball Earth simulation. *Paleoceanography* **19** doi:10.1029/2004PA001056 (2004).

56. Lefebvre, V., Donnadieu, Y., Sepulchre, P., Swingedouw, D. & Zhang, Z. S. Deciphering the role of southern gateways and carbon dioxide on the onset of the Antarctic Circumpolar Current. *Paleoceanography* **27** doi:10.1029/2012PA002345 (2012).

57. Gradstein, F. M., Ogg, J. G., Schmitz, M. D. & Ogg, G. M. *The Geologic Time Scale 2012* Vol. 2 (Elsevier, 2012).

58. Hay, W. W. Cretaceous paleoceanography. *Geol. Carp.* **46**, 257–266 (1995).

59. Vérard, C., Flores, K. & Stampfli, G. Geodynamic reconstructions of the South America-Antarctica plate system. *J. Geodyn.* **53**, 43–60 (2012).

60. Lagabrielle, Y., Godderis, Y., Donnadieu, Y., Malavieille, J. & Suarez, M. The tectonic history of Drake Passage and its possible impacts on global climate. *Earth Planet Sci. Lett.* **279**, 197–211 (2009).

61. Maldonado, A. *et al.* A model of oceanic development by ridge jumping: opening of the Scotia Sea. *Glob. Planet Change* **123**, 152–173 (2014).

62. Martos, Y. M., Catalan, M., Galindo-Zaldivar, J., Maldonado, A. & Bohoyo, F. Insights about the structure and evolution of the Scotia Arc from a new magnetic data compilation. *Glob. Planet Change* **123**, 239–248 (2014).

63. Eagles, G. The age and origin of the central Scotia Sea. *Geophys. J. Int.* **183**, 587–600 (2010).

64. Royer, D. L., Pagani, M. & Beerling, D. J. Geobiological constraints on Earth system sensitivity to CO2 during the Cretaceous and Cenozoic. *Geobiology* **10**, 298–310 (2012).

65. Donnadieu, Y., Godderis, Y. & Bouttes, N. Exploring the climatic impact of the continental vegetation on the Mezosoic atmospheric CO2 and climate history. *Clim. Past* **5**, 85–96 (2009).

66. Fletcher, B. J., Beerling, D. J., Brentnall, S. J. & Royer, D. L. Fossil bryophytes as recorders of ancient CO2 levels: experimental evidence and a Cretaceous case study. *Global Biogeochem. Cycles* **19** doi:10.1029/2005GB002495 (2005).

67. Franks, P. J., Royer, D. L., Johnson, K. R., Miller, I. & Enquist, B. J. New constraints on atmopsheric CO2 concentration for the Phanerozoic. *Geophys. Res. Lett.* **41**, 4685–4694 (2014).

68. Haworth, M., Hesselbo, S. P., McElwain, J. C., Robinson, S. A. & Brunt, J. W. Mid-Cretaceous pCO(2) based on stomata of the extinct conifer *Pseudofrenelopsis* (Cheirolepidiaceae). *Geology* **33**, 749–752 (2005).

69. Passalia, M. G. Cretaceous pCO2 estimation from stomatal frequency analysis of gymnosperm leaves of Patagonia, Argentina. *Palaeogeogr. Palaeoclimatol. Palaeoecol.* **273**, 17–24 (2009).

Acknowledgements

This work was supported by a funding from the ANR project Anox-Sea. We thank three anonymous reviewers for their constructive comments that greatly contributed to improve this manuscript. We thank the CEA/CCRT for providing access to the HPC resources of TGCC under the allocation 2014–012212 made by GENCI.

Author contributions

Y.D. and E.P. conceived the project. E.P., M.M. and J.-F.D. built the neodymium database and performed the analysis. Y.D. conducted all numerical climate modelling. Y.D. and E.P. wrote the manuscript with contributions from all authors.

Additional information

Tropical Atlantic temperature seasonality at the end of the last interglacial

Thomas Felis[1], Cyril Giry[1], Denis Scholz[2], Gerrit Lohmann[1,3], Madlene Pfeiffer[3], Jürgen Pätzold[1], Martin Kölling[1] & Sander R. Scheffers[4]

The end of the last interglacial period, $\sim$118 kyr ago, was characterized by substantial ocean circulation and climate perturbations resulting from instabilities of polar ice sheets. These perturbations are crucial for a better understanding of future climate change. The seasonal temperature changes of the tropical ocean, however, which play an important role in seasonal climate extremes such as hurricanes, floods and droughts at the present day, are not well known for this period that led into the last glacial. Here we present a monthly resolved snapshot of reconstructed sea surface temperature in the tropical North Atlantic Ocean for 117.7 ± 0.8 kyr ago, using coral Sr/Ca and $\delta^{18}O$ records. We find that temperature seasonality was similar to today, which is consistent with the orbital insolation forcing. Our coral and climate model results suggest that temperature seasonality of the tropical surface ocean is controlled mainly by orbital insolation changes during interglacials.

[1] MARUM—Center for Marine Environmental Sciences, University of Bremen, 28359 Bremen, Germany. [2] Institute for Geosciences, Johannes Gutenberg University Mainz, 55099 Mainz, Germany. [3] Alfred Wegener Institute, Helmholtz Centre for Polar and Marine Research (AWI), 27570 Bremerhaven, Germany. [4] Marine Ecology Research Centre, Southern Cross University, Lismore, New South Wales 2480, Australia. Correspondence and requests for materials should be addressed to T.F. (email: tfelis@marum.de).

The last interglacial, although not a direct analogue for future climate, has received much attention in the climate-modelling community[1-3] and has been suggested as a test bed for models developed for future climate prediction[2,4]. This period (~127–117 kyr ago) was characterized by strong orbital insolation forcing[5], relative warmth[6] and high sea level[7]. In the Northern Hemisphere, changes in the Earth's orbit around the sun led to a stronger seasonality of insolation compared to today[5], which resulted in increased temperature seasonality at the Earth's surface as inferred from proxy records[8-10] that commonly represent the time interval of maximum seasonal insolation forcing[5] between ~127 and ~124 kyr ago. In contrast, the temperature seasonality at the end of the last interglacial (~118 kyr ago), when Northern Hemisphere insolation seasonality was close to today's value[5], is not well known. This period that led into the last glacial is particularly interesting as it was characterized by catastrophic collapse of polar ice sheets and substantial sea-level rise[11,12], abrupt changes in ocean circulation[13,14] and large-scale climate perturbations[15]. It has been suggested that the end of the last interglacial may provide clues to a better understanding of the potential for rapid ice-sheet collapse and sea-level rise and, consequently, for abrupt perturbations of the ocean–atmosphere system, under future climate change[11,12,14]. At the present day, the seasonal temperature changes of the tropical ocean play an important role in seasonal climate extremes such as hurricanes, floods and droughts[16-19]. A better understanding of the temperature seasonality ~118 kyr ago is, thus, essential to establish a baseline to evaluate the seasonal response in climate model simulations, for both the end of the last interglacial and for projections of future climate change.

Here we investigate the monthly resolved Sr/Ca and $\delta^{18}O$ environmental proxy signals in a precisely dated shallow-water fossil coral recovered from the southern Caribbean and reconstruct the temperature seasonality in the surface waters of the tropical North Atlantic Ocean at the end of the last interglacial. Sr/Ca variations in aragonitic coral skeletons are a proxy for temperature variability[20], which has previously been successfully applied to last interglacial fossil corals[9,10,21]. Coral $\delta^{18}O$, a proxy that reflects both temperature and seawater $\delta^{18}O$ variations, is used to support our reconstruction. The $^{230}Th/U$ method allows precise dating of corals that grew during the last interglacial period[22]. Our findings indicate that temperature seasonality in the southern Caribbean Sea at 118 kyr ago was similar to today. Our coral records and simulations with a coupled atmosphere–ocean general circulation model indicate an orbital control on temperature seasonality in the tropical North Atlantic at the end of the last interglacial, despite the large-scale perturbations of ocean circulation and climate during this period, and suggest that temperature seasonality of the tropical surface ocean is controlled mainly by orbital insolation changes during interglacials.

Results

Coral preservation and age. The fossil shallow-water coral (*Diploria strigosa*) was recovered at Bonaire, an open-ocean island in the southern Caribbean Sea, located ~100 km north of South America and ~300 km northwest of the Cariaco Basin (Fig. 1). Bonaire is situated off the South American continental shelf in the northwestward-flowing Caribbean Current, an extension of the Guyana Current that transports equatorial Atlantic surface waters along northeastern South America towards the Caribbean Sea. Thus, sea surface temperature (SST) at Bonaire is representative for a large area of the tropical North Atlantic Ocean[23]. Bonaire is influenced by the easterly trade

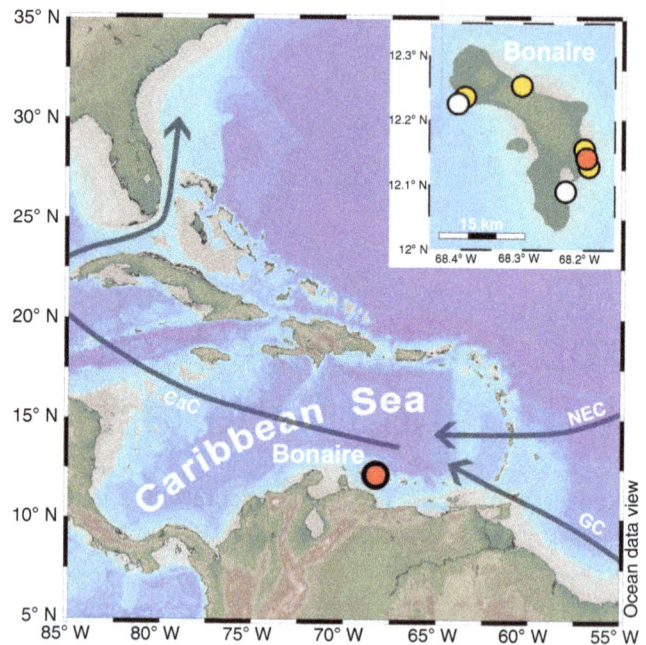

Figure 1 | Map of the western tropical North Atlantic Ocean. The location of our coral site at Bonaire in the southern Caribbean Sea and surface ocean circulation patterns in the study area (Guyana Current, GC; Caribbean Current, CaC; North Equatorial Current, NEC) are indicated. Bonaire is situated off the continental shelf of South America in open-ocean waters. The inset shows the locations of our last interglacial (red circle, this study), Holocene[23,27] (orange circles) and modern[23,27] (white circles) coral sites at Bonaire.

winds, and its present-day climate is semi-arid with an annual precipitation of ~550 mm and the main rainy season during boreal winter[24]. Bonaire is not influenced by the seasonally migrating Intertropical Convergence Zone (ITCZ) because the northernmost ITCZ position that is reached during boreal summer is located south of Bonaire, over northern South America and the Cariaco Basin[25]. The fossil coral colony (BON-5-D) was drilled in growth position on top of an elevated reef terrace at the eastern coast of Bonaire (Washikemba). The coral site (68° 11.765′ W, 12° 8.246′ N) is at ~1.5 to ~2.0 m above present sea level, in a distance of ~50 m from the present-day sea cliff. Nearby *D. strigosa* colonies in growth position (<6 m distance) suggest that this fossil coral community is preserved *in situ*. X-radiography, powder X-ray diffraction, thin-section petrography and scanning electron microscope analysis indicate that the fossil coral is very well preserved (Methods and Supplementary Figs 1–3). $^{230}Th/U$ dating yielded a coral age of 117.7 ± 0.8 kyr, showing that the colony grew at the end of the last interglacial period. The initial ($^{234}U/^{238}U$) activity ratio is in agreement with the ($^{234}U/^{238}U$) of modern seawater, providing strong confidence for the reliability of the coral age (Methods and Supplementary Table 1).

Coral-based SST seasonality reconstruction. The 118-kyr-old Bonaire coral provides a monthly resolved snapshot of tropical Atlantic SST variability for a time window of 20 years at the end of the last interglacial. This is substantially longer than the only other seasonally resolved snapshot of tropical Atlantic SST for the last interglacial, an ~5-year record of a 127-kyr-old coral from Isla de Mona (67.9° W, 18.1° N) in the northern Caribbean Sea[9]. Our Bonaire monthly resolved coral Sr/Ca- and $\delta^{18}O$-SST reconstructions show clear annual cycles in both proxies

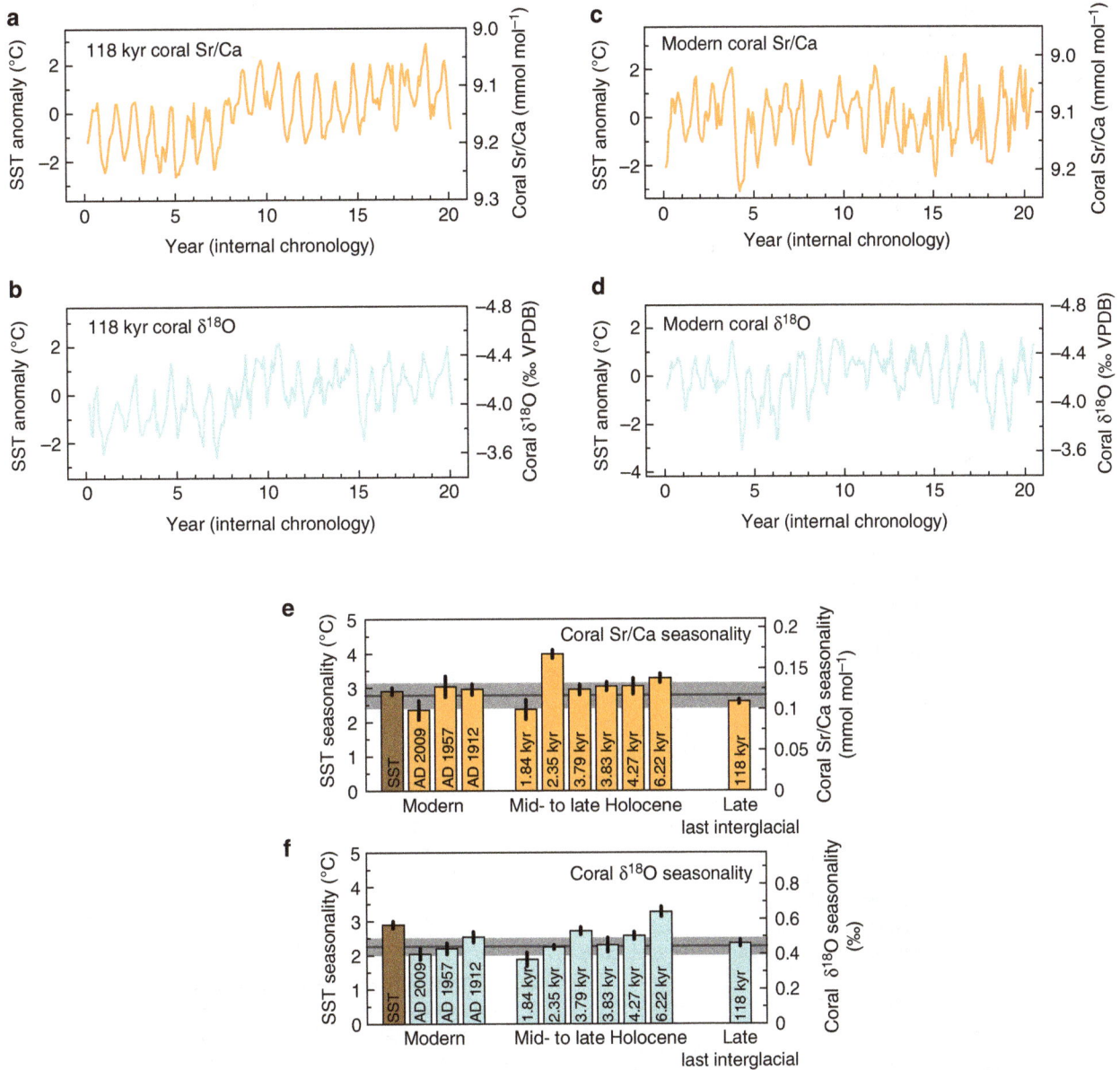

Figure 2 | Tropical North Atlantic coral-based temperature seasonality. (**a**) Monthly Sr/Ca record of a fossil Bonaire *Diploria strigosa* coral that grew at 117.7 ± 0.8 kyr ago for 20 years in southern Caribbean Sea surface waters. (**b**) The monthly coral $\delta^{18}O$ record. (**c**) Monthly Sr/Ca record of a modern Bonaire *D. strigosa* coral that grew around AD 1912. (**d**) The monthly coral $\delta^{18}O$ record. (**e**) Sr/Ca-based sea surface temperature (SST) seasonality from Bonaire *D. strigosa* corals for snapshots since 118 kyr ago, based on monthly records comprising a total of 315 years, and Bonaire instrumental SST seasonality (1910–2000, 2° × 2° gridbox centred at 12° N, 68° W, ERSST.v3b)[26]. The dark grey line represents the reconstructed modern mean SST seasonality based on three modern corals and the light grey bar the ±1 s.d. around this mean. (**f**) The coral $\delta^{18}O$-based SST seasonality. Deviations from Sr/Ca- and instrument-based estimates are due to seasonal seawater $\delta^{18}O$ effects. Coral-based SST anomalies (corresponding mean value was subtracted) (**a**–**d**) and SST seasonalities (**e**,**f**) are derived from seasonal Sr/Ca-SST (-0.042 mmol mol^{-1} per °C) and $\delta^{18}O$-SST relationships (-0.196‰ per °C) for *D. strigosa*[37]. The uncertainty assigned to each SST seasonality estimate is the ±1 s.e. Holocene and modern coral data are from refs 23,27.

(Fig. 2a,b), giving additional confidence that the analysed coral skeleton was not subject to diagenetic alteration. The Sr/Ca-SST reconstruction indicates a seasonality of 2.6 ± 0.1 °C (±1 s.e.) at 118 kyr ago (Fig. 2a,e). Monthly resolved records of three modern Bonaire *D. strigosa* corals satisfactorily document the instrumental SST[26] seasonality of 2.9 ± 0.1 °C (±1 s.e.; 1910–2000), indicating a reconstructed modern Sr/Ca-SST seasonality that ranges from 2.4 ± 0.3 °C (±1 s.e.) to 3.0 ± 0.3 °C (±1 s.e.) for time intervals of the last century, resulting in a reconstructed modern mean seasonality of 2.8 ± 0.4 °C (±1 s.d.; ref. 23; Fig. 2c,e). Taking into account these differences in the reconstructed SST seasonality among the

three modern corals indicates that the reconstructed SST seasonality of 2.6 ± 0.1 °C (±1 s.e.) at 118 kyr ago, at the end of the last interglacial, is not significantly different from today (Methods and Supplementary Note 1).

The coral $\delta^{18}O$-SST reconstruction for 118 kyr ago indicates a seasonality of 2.4 ± 0.1 °C (±1 s.e.), which is very similar to the Sr/Ca-based seasonality estimate of 2.6 ± 0.1 °C (±1 s.e.; Fig. 2a,b,e,f). Thus, the coral $\delta^{18}O$ seasonality at 118 kyr ago may be attributed mainly to the seasonality of SST. This is broadly in line with the modern situation[27], where the mean SST seasonality reconstructed by coral $\delta^{18}O$ of 2.3 ± 0.3 °C (±1 s.d.) is slightly reduced (by ~0.5 °C, not correcting for seasonal seawater $\delta^{18}O$ changes) relative to the Sr/Ca- and instrument-based estimates (Fig. 2e,f), most likely owing to hydrologic cycle effects such as the Bonaire winter rainfall regime[24]. The coral $\delta^{18}O$-SST reconstruction supports our major finding based on coral Sr/Ca, and both proxies indicate SST seasonality in the southern Caribbean Sea at the end of the last interglacial similar to today. Consequently, both proxies may also suggest a Bonaire hydrologic cycle similar to today at 118 kyr ago. Crucially, our results are robust towards the choice of the coral Sr/Ca-SST and $\delta^{18}O$-SST relationships, which affect mainly the absolute magnitude of reconstructed SST seasonality but have only minor effect on the relative seasonality estimates among corals, and we would have reached identical conclusions using other relationships (Supplementary Fig. 4).

Discussion

The annual SST cycle in the Caribbean Sea, with a minimum in boreal winter/spring and a maximum in boreal summer/fall, follows primarily the annual cycle of insolation[28,29]. Bonaire monthly coral Sr/Ca records for snapshots since the mid-Holocene, comprising a total length of 295 years, suggest that the SST annual cycle in the southern Caribbean Sea has not substantially changed, with the exception of a time interval at 2.35 kyr ago[23] (Fig. 2e). Disregarding the 2.35 kyr coral, a trend towards lower SST seasonality during the time interval 6.22–1.84 kyr ago may be inferred from the coral Sr/Ca records, as well as a slightly but significantly higher SST seasonality than the present day at 6.22 kyr ago. Such an evolution through time would be consistent with an orbital insolation control on Holocene SST seasonality in the southern Caribbean Sea (Fig. 3a), which is supported by simulations with a coupled atmosphere–ocean general circulation model (Community Earth System Models, COSMOS; Methods and Fig. 3b). However, we note that the magnitude of the trend in the fossil coral data is minor, close to the ± 1 s.d. range of the modern mean Sr/Ca-SST seasonality reconstructed from three modern corals (Fig. 2e), which may also reflect the relatively small magnitude of insolation-controlled SST seasonality changes at lower latitudes throughout the Holocene (Fig. 3a,b).

Similarly, the coral $\delta^{18}O$-SST reconstruction[27] reveals a trend towards lower seasonality during the time interval 6.22–1.84 kyr ago, which is more pronounced compared with the trend that may be inferred from coral Sr/Ca, as well as a substantially and significantly higher seasonality than the present day at 6.22 kyr ago (Fig. 2f). This evolution of coral $\delta^{18}O$ seasonality through time is consistent with an insolation control on Holocene SST seasonality in the southern Caribbean Sea (Fig. 3a,b). Differences between the coral $\delta^{18}O$-SST and Sr/Ca-SST seasonality estimates (Fig. 2e,f) reflect primarily seasonal changes in seawater $\delta^{18}O$; however, we note that reconstructions of seawater $\delta^{18}O$ seasonality can be sensitive towards the choice of the coral $\delta^{18}O$-SST and Sr/Ca-SST relationships (Supplementary Fig. 4). However, for 6.22 kyr ago, an anomalous seawater $\delta^{18}O$ seasonality may be inferred from the coral records that could be explained by hydrologic cycle effects such as, among others, Bonaire summer rainfall[27], which would be contrary to the present-day winter rainfall regime[24] (Supplementary Fig. 4). This interpretation would be in line with reconstructions of increased summer rainfall over northernmost South America during the early to mid-Holocene, owing to a more northerly position of the boreal-summer ITCZ[30] and possibly paired with a thermodynamic increase in rainfall because of strengthening local summer insolation[31]. We note that the subsequent southward migration of the boreal-summer ITCZ over the course of the Holocene that was controlled by orbital insolation

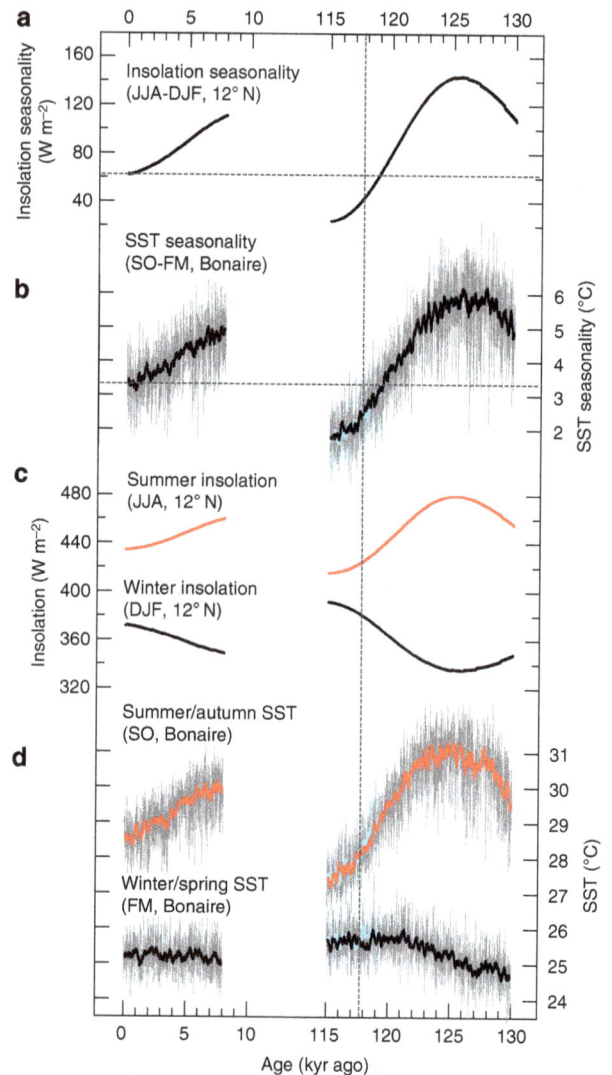

Figure 3 | Tropical North Atlantic insolation and temperature changes. (a) Insolation seasonality[5] at the latitude of Bonaire, calculated as difference of boreal summer (June–July–August, JJA) minus winter insolation (December–January–February, DJF). (b) Sea surface temperature (SST) seasonality at Bonaire simulated by the coupled atmosphere-ocean general circulation model COSMOS (1° × 1° gridbox centred at 12.5° N, 68° W), derived from the difference of simulated summer/autumn (September–October, SO) minus winter/spring (February–March, FM) SST. The SST seasonality evolution is very similar to that derived from the difference of warmest minus coolest SST (Supplementary Fig. 7). (c) Summer (JJA) and winter (DJF) insolation[5] at the latitude of Bonaire. (d) Summer/autumn (SO) and winter/spring (FM) SST at Bonaire simulated by COSMOS. Bold line (b,d) represents a 21-point running average, representing an average of 210 calendar years. Results of the freshwater hosing experiment are also shown (light blue). Dashed horizontal lines (a,b) indicate the modern value for insolation and simulated SST seasonality. Dashed vertical line indicates the Bonaire coral age (117.7 ± 0.8 kyr).

changes[30,31] is also in line with the trend towards lower coral $\delta^{18}O$ seasonality over this time interval (Fig. 2f).

The significantly increased SST seasonality at 2.35 kyr ago, indicated by coral Sr/Ca (Fig. 2e), may be related to internal climate variability and is interpreted to reflect a time interval of strengthened El Niño-Southern Oscillation (ENSO) teleconnections to the Caribbean region[23], probably modulated by the North Atlantic Oscillation (NAO). This interpretation is broadly in line with the present-day modulation of southern Caribbean SST seasonality by ENSO teleconnections[23,32], which vary in strength on interdecadal timescales and are modulated by the NAO[33]. Indeed, pronounced interannual variability at a period of 5.7 years in the Sr/Ca record of the 2.35 kyr coral[23], the most prominent period in the cospectrum of the instrumental indices of ENSO and NAO[34,35], may be indicative of pronounced ENSO–NAO interactions at that time[23]. Importantly, the strength of the ENSO phenomenon in the tropical Pacific did not change markedly around 2.3 kyr ago[36]. We note that the increased coral Sr/Ca-SST seasonality at 2.35 kyr ago is not accompanied by an increased coral δ^{18}O-SST seasonality (Fig. 2e,f), which suggests an anomalous seawater δ^{18}O seasonality that could be explained by hydrologic cycle effects such as, among others, increased Bonaire winter rainfall (Supplementary Fig. 4). This interpretation would be broadly in line with the present-day modulation of Bonaire climate by ENSO teleconnections, where La Niña events result in increased SST seasonality through anomalous winter cooling[23] as well as in increased winter rainfall[24].

Our coral-based finding of SST seasonality similar to today in the southern Caribbean Sea at 118 kyr ago (Fig. 2e) is consistent with an insolation seasonality at the latitude of Bonaire that was close to today's value (Fig. 3a). This result could be interpreted in a way that southern Caribbean SST seasonality at that time was controlled mainly by orbital insolation changes. Simulations performed with a coupled atmosphere–ocean general circulation model (COSMOS) support this interpretation (Methods). The modelled changes in southern Caribbean SST seasonality at Bonaire throughout the last interglacial follow largely the variations in insolation forcing over the time interval 130–115 kyr ago (Fig. 3). Moreover, the modelled global surface air temperature anomaly indicates that temperature seasonality in the southern Caribbean at 118 kyr ago is part of a hemisphere-scale pattern that can be attributed largely to insolation forcing (Supplementary Fig. 5). Additional model simulations with freshwater forcing to mimic an abrupt ice-sheet collapse and weakening of the North Atlantic thermohaline circulation at 118 kyr ago or with reduced Greenland ice sheet and dynamic vegetation reveal very similar results (Methods), indicating no significant impact on southern Caribbean SST seasonality (Fig. 3 and Supplementary Figs 5 and 6). Thus, our model-based results strongly suggest that SST seasonality in the tropical North Atlantic Ocean at the end of the last interglacial was controlled mainly by orbital insolation changes. Although the slightly lower modelled SST seasonality at 118 kyr ago relative to today (Fig. 3b) appears to be consistent with the coral Sr/Ca-SST seasonality estimate for the end of the last interglacial (Fig. 2e), we consider the latter as similar to today as a result of our uncertainty assignments that take into account the differences in the seasonality estimates among the three modern corals (Methods).

The relatively stable SST seasonality in the tropical North Atlantic Ocean at the end of the last interglacial and its inferred orbital control is remarkable as this period was characterized by large-scale perturbations of ocean circulation and climate resulting from instabilities of polar ice sheets[11–15]. Results from Western Australia suggest that, after a prolonged period of stable sea level at ~3–4 m above present sea level between 127 and 119 kyr ago, eustatic sea level rose rapidly to ~8 m above present at the end of the last interglacial, peaking at 118.1 ± 1.4 kyr ago[12], which is contemporaneous with the age of our southern Caribbean coral (117.7 ± 0.8 kyr; Fig. 4b). It has been suggested that this substantial jump in sea level at the end of the last interglacial resulted from collapse of the Greenland and particularly Antarctic ice sheets, after a critical ice-sheet stability threshold was crossed[12]. Such an event may have had

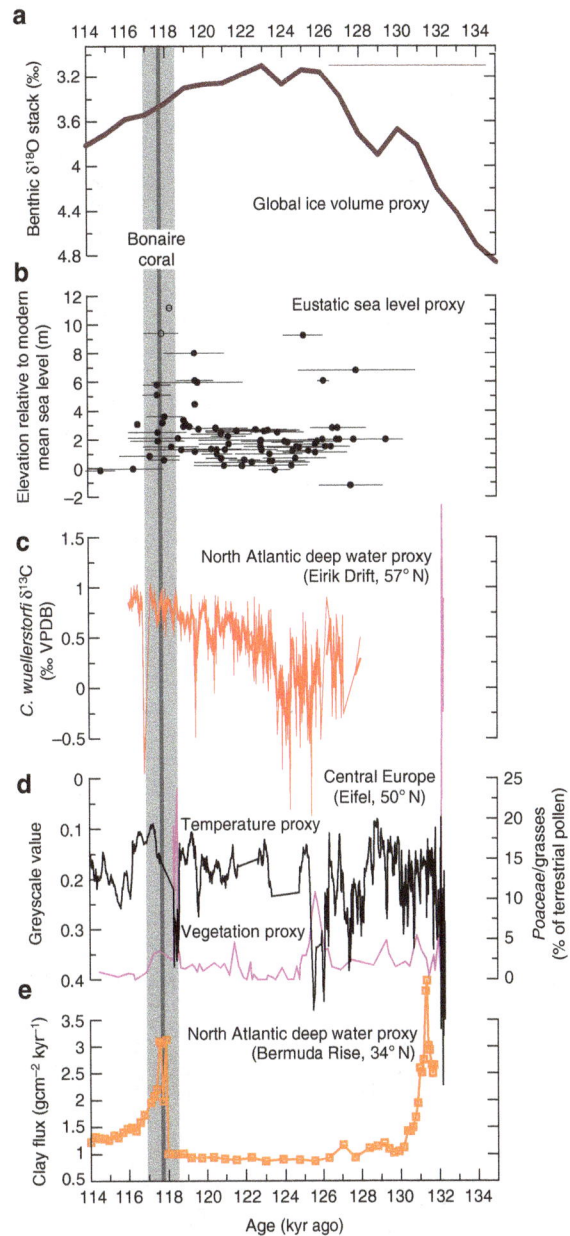

Figure 4 | Bonaire coral age and last interglacial sea level and climate change. (a) LR04 stack of globally distributed benthic δ^{18}O records, reflecting global ice volume changes[66]. **(b)** Relative sea level from Western Australian corals, indicating eustatic sea level rose to ~8 m above present at 118.1 ± 1.4 kyr ago[12]. Open symbols indicate corals collected not *in situ* or affected by tectonic uplift[12]. **(c)** North Atlantic epibenthic foraminiferal δ^{13}C record, indicating pronounced reductions in North Atlantic Deep Water production (bottom water δ^{13}C reductions) at ~119.5 and ~116.8 kyr ago[14]. Bold line indicates a 3-point running average. **(d)** Eifel Laminated Sediment Archive greyscale stack from maar lakes in Germany, indicating a prominent cold and arid event at 118 kyr ago that was accompanied by high grass pollen abundance[15]. **(e)** Clay flux record from excess ^{230}Th-measurements in North Atlantic sediments indicating a rapid increase in recirculation-derived clay supply (and the proportion of southern source water) at ~118 kyr ago, associated with a cessation in North Atlantic deep water flow[13]. The dark grey line indicates the Bonaire coral age (117.7 kyr) and the light grey shading the corresponding 2σ uncertainty (± 0.8 kyr). Both Bonaire coral and sea-level jump[12] were dated by the ^{230}Th/U-method, whereas the sediment records[13–15,66] were not absolutely dated. Age uncertainty is shown as reported in original publication, if available.

substantial impacts on global ocean circulation and climate. Interestingly, varved lake sediments in central Europe indicate an extreme 468-year arid and cold event at 118 kyr ago (Fig. 4d), which has been interpreted to result from a sudden southward shift of the warm North Atlantic drift[15]. Furthermore, western North Atlantic sediments indicate an abrupt ~400-year deep-water reorganization event at ~118 kyr ago associated with changes in the thermohaline circulation[13] (Fig. 4e), which has been interpreted to mark the beginning of climate deterioration at the end of the last interglacial[13]. Recent evidence suggests even two events of substantial North Atlantic deep-water reduction at the end of the last interglacial, at ~119.5 and ~116.8 kyr ago[14] (Fig. 4c).

Our findings based on combining coral proxy records with climate model simulations indicate that northern tropical Atlantic SST seasonality at 118 kyr ago was similar to today and controlled mainly by orbital insolation changes, despite dramatic ocean circulation and climate perturbations resulting from instabilities of polar ice sheets that characterized the end of the last interglacial[11-15]. Today, tropical Atlantic SST plays a major role in seasonal climate extremes, such as hurricanes, flashfloods and droughts[16-19], which cause severe socioeconomic damage on the adjacent continents. Our results indicate that SST seasonality in the tropical Atlantic did not substantially change during a period of abrupt high-latitude ice sheet, ocean and climate perturbations at the end of the last interglacial, and, thus, suggest that tropical SST seasonality is controlled mainly by orbital insolation changes during interglacials. However, more seasonally resolved proxy records of SST are needed to better constrain both the climate sensitivity of the tropical ocean in the past and the seasonal response in model-based scenarios of past and future climate change.

Methods

Screening for diagenesis. The fossil *D. strigosa* coral (BON-5-D) was screened for potential diagenetic alteration of its skeleton using X-radiography, powder X-ray diffraction, thin-section petrography and scanning electron microscope (SEM) analysis. X-radiography reveals a well-preserved skeleton, a clear pattern of alternating bands of high and low skeletal density and continuous upward growth at a rate of 0.68 ± 0.15 cm per year (± 1 s.d.) (Supplementary Fig. 1), similar to the annual density-band pairs and growth rates reported for Holocene *D. strigosa* corals from Bonaire[23]. Powder X-ray diffraction analysis indicates that the aragonitic skeleton has a calcite content of <1%. Petrographic thin sections indicate excellent preservation of primary porosity, with no evidence for significant amounts of secondary aragonite or calcite cements (Supplementary Fig. 2). SEM analysis indicates slight dissolution of more fragile skeletal elements such as septa and columella; however, the dense theca walls that are the target for our geochemical analysis[37,38] are unaffected by these subtle diagenetic alterations (Supplementary Fig. 3). Overall, the fossil coral is very well preserved.

^{230}Th/U dating. The age of the fossil *D. strigosa* coral (BON-5-D) was determined by thermal ionization mass spectrometry ^{230}Th/U dating carried out at the Heidelberg Academy of Sciences, Heidelberg, Germany[39,40]. The ^{230}Th/U age of 117.7 ± 0.8 is reported with its 2σ error in kyr before the year of measurement, which is AD 2009 (Supplementary Table 1), a common procedure in studies of ^{230}Th/U-dated last interglacial corals[7,11,12,21,22]. The age was calculated using the half-lives of ref. 41 and corrected for the effect of detrital contamination assuming a bulk earth ^{232}Th/^{238}U weight ratio of 3.8 and secular equilibrium between ^{238}U, ^{234}U and ^{230}Th. However, this correction is insignificant for the Bonaire coral. The reliability of the determined ^{230}Th/U age was checked using established criteria[42], such as initial (^{234}U/^{238}U) in agreement with the value of modern seawater (that is, 1.1466 ± 0.0025 (ref. 43)), ^{238}U concentrations comparable to modern corals of the same species, a ^{232}Th content lower than 2 p.p.b. and negligible calcite content[42]. All these criteria are fulfilled for coral BON-5-D. The coral's ^{230}Th/U-age is, thus, considered as strictly reliable. The 2σ uncertainties of the ^{230}Th/U ages (Fig. 2e,f) are 0.8 kyr for the late last interglacial coral, ~0.03 kyr for the Holocene corals[23] (ages are given relative to AD 1950) and <0.01 kyr for the modern corals[23].

Microsampling. *D. strigosa* coral BON-5-D was microsampled along its major growth axis by carefully drilling continuously along the centre of the dense theca walls using a 0.6-mm-diameter drill bit following established methods[38] (Supplementary Fig. 1). The methodology is identical to our microsampling of

modern and fossil Holocene *D. strigosa* corals from Bonaire[23,27]. An average of 11.4 samples per year was obtained, which is in the range of sampling resolutions reported for our modern and fossil Holocene *D. strigosa* corals from Bonaire (10.8–15.3 samples per year; ref. 23).

Geochemical and isotopic analyses. Coral Sr/Ca and δ^{18}O were analysed on splits of the same powder samples at MARUM (University of Bremen) as previously described[23,38]. Twenty-five splits of the coral reference material JCp-1 (ref. 44) were treated like samples, and the average Sr/Ca value obtained in this study was 8.919 ± 0.008 mmol mol^{-1}, which is the same JCp-1 reference composition as reported in our Holocene Bonaire coral study[23].

Coral record. The internal chronology of the BON-5-D coral record is based on counting the clear annual cycles in Sr/Ca and δ^{18}O that reflect the SST seasonality. This age model is corroborated by the skeletal pattern of annual density-band pairs as revealed by X-radiographs. For the construction of the chronology, annual Sr/Ca maxima were set to February/March (on average the coolest months) and annual Sr/Ca minima to September/October (on average the warmest months) using the present-day SST climatology[26] as a benchmark. The coral δ^{18}O chronology uses the tie points of the coral Sr/Ca chronology. The resulting records were interpolated to monthly resolution. The methodology is similar as described for our Holocene Bonaire coral records[23,27]. We note that the shift in the mean coral Sr/Ca and δ^{18}O that occurs between years 7 and 8 of the internal chronology (Fig. 2a,b) is not related to a change of the microsampling transect nor to any shifts in extension rate or coral δ^{13}C and, consequently, likely reflects a climatic shift. Importantly, this shift does not affect the amplitude of our SST seasonality calculations described below.

Coral seasonality. Coral Sr/Ca (δ^{18}O) seasonality was calculated following established methods[10] and similar to our Holocene Bonaire coral study[23,27]. Seasonality is calculated as the difference between the maximum and the minimum monthly coral Sr/Ca (δ^{18}O) value of a given year. The mean seasonality is calculated by averaging the seasonality of all years of a given coral record. The uncertainty assigned to each coral-SST seasonality estimate for a given snapshot is ± 1 s.e. (Fig. 2e,f). The fossil coral-SST seasonality estimates are then compared with the ± 1 s.d. around the reconstructed modern mean SST seasonality based on the three modern corals. In addition, the combined error[45] (± 1 CE) is considered for each fossil coral-SST seasonality estimate (Supplementary Fig. 4), which is derived from the combination (root of the sum of the squares) of (1) the s.d. (2 s.d.) around the reconstructed modern mean SST seasonality based on the three modern corals and (2) the s.e. (2 s.e.) of the mean of multiple SST seasonality estimates for each fossil coral[23,27,45].

Climate model simulations. The state-of-the-art coupled atmosphere–ocean general circulation model COSMOS is applied[46-48], which is also used in the 5th Assessment Report (AR5) of the Intergovernmental Panel on Climate Change. COSMOS consists of the atmosphere model ECHAM5 (ref. 49), the land–surface model JSBACH[50], the general ocean circulation model MPIOM[51] and the OASIS3 coupler[52]. The land-surface and vegetation model JSBACH comprise a dynamic vegetation module[53], which enables the plant cover to adjust to a change in the climate state. The model has been tested and applied for early and mid-Holocene climates[47,48], glacial climates[54-56], as well as for Cenozoic climates[57-59]. The resolution used in our simulations is T31 (3.75°) in the atmosphere with 19 vertical levels and a horizontal resolution of $3° \times 1.8°$ in the ocean with 40 vertical levels. The ocean grid has an effective higher resolution in the polar regions[51,58]. Fixed modern distributions of continental ice sheets (except for the experiment with reduced Greenland ice sheet), sea level and distribution of land were used throughout the simulations. For the transient simulations, orbital acceleration[60] with a factor of 10 has been applied to simulate the time interval from 130 to 115 kyr ago and the last 8 kyr.

The last interglacial transient simulation starts from a quasi-equilibrated time slice run for 130 kyr ago that was spun up for 1,000 years, which has previously been analysed in a multimodel assessment of last interglacial temperatures[2], using the greenhouse gas boundary conditions specified by the third phase of the Paleoclimate Model Intercomparison Project (PMIP3): 257 ppmv for CO_2, 512 ppbv for CH_4 and 239 ppbv for N_2O. Throughout the last interglacial transient simulation, the greenhouse gas concentrations varied according to the values specified by PMIP3 (refs 61–63), which were interpolated to 0.01 kyr resolution. In accordance with PMIP3, both the spin-up and the transient simulation were performed with fixed pre-industrial vegetation.

In addition, a freshwater perturbation experiment based on the last interglacial transient simulation was performed with an identical set-up, by distributing a freshwater flux anomaly (0.02 Sv) over the North Atlantic Ocean (45° W–20° W, 40° N–55° N) that starts at 119.25 kyr ago and ends 117.76 kyr ago, representing 1,500 calendar years (150 model years). This time interval was chosen in order to have the last 500 calendar years (50 model years) of the freshwater perturbation centred at 118.00 kyr ago. The transient simulation then continues from 117.75 to 115.0 kyr ago without freshwater forcing.

Moreover, a last interglacial transient simulation with reduced Greenland ice sheet was performed. The simulation starts from a quasi-equilibrated time slice run for 130 kyr ago that was spun up for 1,500 years, which has previously been analysed in a sensitivity study on the influence of the Greenland ice sheet on last interglacial climate[64]. Throughout the spin-up and last interglacial transient simulations with reduced Greenland ice sheet, the greenhouse gas boundary conditions were fixed to pre-industrial values: 278 ppmv for CO_2, 650 ppbv for CH_4 and 270 ppbv for N_2O. In the simulations, the Greenland ice-sheet elevation was reduced by subtracting 1,300 m from each grid point over Greenland. Areas where the present elevation is lower than 1,300 m were defined as ice-free and the albedo was adjusted accordingly. The spin-up and the transient simulation were performed with dynamic vegetation.

The Holocene transient simulation has previously been analysed in terms of Southern Hemisphere westerly winds evolution[65]. The Holocene transient simulation starts from a quasi-equilibrated time slice run for 8.1 kyr ago, using greenhouse gas concentrations at pre-industrial levels: 278 ppmv for CO_2, 650 ppbv for CH_4 and 270 ppbv for N_2O. Throughout the Holocene transient simulation the greenhouse gas concentrations were fixed. The spin-up and the transient simulation were performed with dynamic vegetation.

References

1. Bakker, P. et al. Last interglacial temperature evolution - a model inter-comparison. Clim. Past 9, 605–619 (2013).
2. Lunt, D. J. et al. A multi-model assessment of last interglacial temperatures. Clim. Past 9, 699–717 (2013).
3. Nikolova, I., Yin, Q., Berger, A., Singh, U. K. & Karami, M. P. The last interglacial (Eemian) climate simulated by LOVECLIM and CCSM3. Clim. Past 9, 1789–1806 (2013).
4. Braconnot, P. et al. Evaluation of climate models using palaeoclimatic data. Nat. Clim. Change 2, 417–424 (2012).
5. Berger, A. L. Long-term variations of daily insolation and Quaternary climatic changes. J. Atmos. Sci. 35, 2362–2367 (1978).
6. Kukla, G. J. et al. Last interglacial climates. Quatern. Res. 58, 2–13 (2002).
7. Dutton, A. & Lambeck, K. Ice volume and sea level during the last interglacial. Science 337, 216–219 (2012).
8. Suzuki, A. et al. Last interglacial coral record of enhanced insolation seasonality and seawater ^{18}O enrichment in the Ryukyu Islands, northwest Pacific. Geophys. Res. Lett. 28, 3685–3688 (2001).
9. Winter, A. et al. Orbital control of low-latitude seasonality during the Eemian. Geophys. Res. Lett. 30, 1163 (2003).
10. Felis, T. et al. Increased seasonality in Middle East temperatures during the last interglacial period. Nature 429, 164–168 (2004).
11. Blanchon, P., Eisenhauer, A., Fietzke, J. & Liebetrau, V. Rapid sea-level rise and reef back-stepping at the close of the last interglacial highstand. Nature 458, 881–884 (2009).
12. O'Leary, M. J. et al. Ice sheet collapse following a prolonged period of stable sea level during the last interglacial. Nat. Geosci. 6, 796–800 (2013).
13. Adkins, J. F., Boyle, E. A., Keigwin, L. & Cortijo, E. Variability of the North Atlantic thermohaline circulation during the last interglacial period. Nature 390, 154–156 (1997).
14. Galaasen, E. V. et al. Rapid reductions in North Atlantic deep water during the peak of the last interglacial period. Science 343, 1129–1132 (2014).
15. Sirocko, F. et al. A late Eemian aridity pulse in central Europe during the last glacial inception. Nature 436, 833–836 (2005).
16. Wang, C., Enfield, D. B., Lee, S.-k. & Landsea, C. W. Influences of the Atlantic Warm Pool on Western Hemisphere summer rainfall and Atlantic hurricanes. J. Clim. 19, 3011–3028 (2006).
17. Smith, D. M. et al. Skilful multi-year predictions of Atlantic hurricane frequency. Nat. Geosci. 3, 846–849 (2010).
18. Misra, V. & Li, H. The seasonal climate predictability of the Atlantic Warm Pool and its teleconnections. Geophys. Res. Lett. 41, 661–666 (2014).
19. Rodrigues, R. R. & McPhaden, M. J. Why did the 2011–2012 La Niña cause a severe drought in the Brazilian Northeast? Geophys. Res. Lett. 41, 1012–1018 (2014).
20. Beck, J. W. et al. Sea-surface temperature from coral skeletal strontium/calcium ratios. Science 257, 644–647 (1992).
21. McCulloch, M. T. & Esat, T. The coral record of last interglacial sea levels and sea surface temperatures. Chem. Geol. 169, 107–129 (2000).
22. Edwards, R. L., Chen, J. H., Ku, T.-L. & Wasserburg, G. J. Precise timing of the last interglacial period from mass spectrometric determination of Thorium-230 in corals. Science 236, 1547–1553 (1987).
23. Giry, C. et al. Mid- to late Holocene changes in tropical Atlantic temperature seasonality and interannual to multidecadal variability documented in southern Caribbean corals. Earth Planet Sci. Lett. 331-332, 187–200 (2012).
24. Martis, A., van Oldenborgh, G. J. & Burgers, G. Predicting rainfall in the Dutch Caribbean—more than El Niño? Int. J. Climatol. 22, 1219–1234 (2002).
25. Peterson, L. C. & Haug, G. H. Variability in the mean latitude of the Atlantic Intertropical Convergence Zone as recorded by riverine input of sediments to

26. the Cariaco Basin (Venezuela). Palaeogeogr. Palaeoclimatol. Palaeoecol. 234, 97–113 (2006).
26. Smith, T. M., Reynolds, R. W., Peterson, T. C. & Lawrimore, J. Improvements to NOAA's historical merged land-ocean surface temperature analysis (1880–2006). J. Clim. 21, 2283–2296 (2008).
27. Giry, C. et al. Controls of Caribbean surface hydrology during the mid- to late Holocene: insights from monthly resolved coral records. Clim. Past 9, 841–858 (2013).
28. Carton, J. A. & Zhou, Z. Annual cycle of sea surface temperature in the tropical Atlantic Ocean. J. Geophys. Res. Oceans 102, 27813–27824 (1997).
29. Chollett, I., Müller-Karger, F. E., Heron, S. F., Skirving, W. & Mumby, P. J. Seasonal and spatial heterogeneity of recent sea surface temperature trends in the Caribbean Sea and southeast Gulf of Mexico. Mar. Pollut. Bull. 64, 956–965 (2012).
30. Haug, G. H., Hughen, K. A., Sigman, D. M., Peterson, L. C. & Röhl, U. Southward migration of the intertropical convergence zone through the Holocene. Science 293, 1304–1308 (2001).
31. Schneider, T., Bischoff, T. & Haug, G. H. Migrations and dynamics of the intertropical convergence zone. Nature 513, 45–53 (2014).
32. Giannini, A., Chiang, J. C. H., Cane, M. A., Kushnir, Y. & Seager, R. The ENSO teleconnection to the tropical Atlantic Ocean: contributions of the remote and local SSTs to rainfall variability in the tropical Americas. J. Clim. 14, 4530–4544 (2001).
33. Giannini, A., Cane, M. A. & Kushnir, Y. Interdecadal changes in the ENSO teleconnection to the Caribbean region and the North Atlantic oscillation. J. Clim 14, 2867–2879 (2001).
34. Rogers, J. C. The association between the North Atlantic Oscillation and the Southern Oscillation in the Northern Hemisphere. Mon. Wea. Rev. 112, 1999–2015 (1984).
35. Huang, J., Higuchi, K. & Shabbar, A. The relationship between the North Atlantic Oscillation and El Niño-Southern Oscillation. Geophys. Res. Lett. 25, 2707–2710 (1998).
36. Cobb, K. M. et al. Highly variable El Niño–Southern Oscillation throughout the Holocene. Science 339, 67–70 (2013).
37. Hetzinger, S., Pfeiffer, M., Dullo, W.-C., Ruprecht, E. & Garbe-Schönberg, D. Sr/Ca and $\delta^{18}O$ in a fast-growing Diploria strigosa coral: Evaluation of a new climate archive for the tropical Atlantic. Geochem. Geophys. Geosyst. 7, Q10002 (2006).
38. Giry, C., Felis, T., Kölling, M. & Scheffers, S. Geochemistry and skeletal structure of Diploria strigosa, implications for coral-based climate reconstruction. Palaeogeogr. Palaeoclimatol. Palaeoecol. 298, 378–387 (2010).
39. Scholz, D., Mangini, A. & Felis, T. U-series dating of diagenetically altered fossil reef corals. Earth Planet Sci. Lett. 218, 163–178 (2004).
40. Hoffmann, D. L. et al. Procedures for accurate U and Th isotope measurements by high precision MC-ICPMS. Int. J. Mass Spectrom. 264, 97–109 (2007).
41. Cheng, H. et al. The half-lives of uranium-234 and thorium-230. Chem. Geol. 169, 17–33 (2000).
42. Scholz, D. & Mangini, A. How precise are U-series coral ages? Geochim. Cosmochim. Acta 71, 1935–1948 (2007).
43. Robinson, L. F., Belshaw, N. S. & Henderson, G. M. U and Th concentrations and isotope ratios in modern carbonates and waters from the Bahamas. Geochim. Cosmochim. Acta 68, 1777–1789 (2004).
44. Hathorne, E. C. et al. Interlaboratory study for coral Sr/Ca and other element/Ca measurements. Geochem. Geophys. Geosyst. 14, 3730–3750 (2013).
45. Abram, N. J., McGregor, H. V., Gagan, M. K., Hantoro, W. S. & Suwargadi, B. W. Oscillations in the southern extent of the Indo-Pacific Warm Pool during the mid-Holocene. Quat. Sci. Rev. 28, 2794–2803 (2009).
46. Jungclaus, J. H. et al. Ocean circulation and tropical variability in the coupled model ECHAM5/MPI-OM. J. Clim. 19, 3952–3972 (2006).
47. Wei, W. & Lohmann, G. Simulated Atlantic Multidecadal Oscillation during the Holocene. J. Clim. 25, 6989–7002 (2012).
48. Wei, W., Lohmann, G. & Dima, M. Distinct modes of internal variability in the global meridional overturning circulation associated with the Southern Hemisphere westerly winds. J. Phys. Oceanogr. 42, 785–801 (2012).
49. Roeckner, E. et al. The Atmospheric General Circulation Model ECHAM5. Part I: Model description. Report No. 349, 127 (Max Planck Institute for Meteorology, 2003).
50. Raddatz, T. J. et al. Will the tropical land biosphere dominate the climate–carbon cycle feedback during the twenty-first century? Clim. Dyn. 29, 565–574 (2007).
51. Marsland, S. J., Haak, H., Jungclaus, J. H., Latif, M. & Röske, F. The Max-Planck-Institute global ocean/sea ice model with orthogonal curvilinear coordinates. Ocean Model. 5, 91–127 (2003).
52. Valcke, S., Caubel, A., Declat, D. & Terray, L. OASIS3 Ocean Atmosphere Sea Ice Soil User's Guide. Report No. TR/CMGC/03/69, 57 (CERFACS, 2003).
53. Brovkin, V., Raddatz, T., Reick, C. H., Claussen, M. & Gayler, V. Global biogeophysical interactions between forest and climate. Geophys. Res. Lett. 36, L07405 (2009).

54. Zhang, X., Lohmann, G., Knorr, G. & Xu, X. Different ocean states and transient characteristics in Last Glacial Maximum simulations and implications for deglaciation. *Clim. Past* **9**, 2319–2333 (2013).

55. Gong, X., Knorr, G., Lohmann, G. & Zhang, X. Dependence of abrupt Atlantic meridional ocean circulation changes on climate background states. *Geophys. Res. Lett.* **40**, 3698–3704 (2013).

56. Zhang, X., Lohmann, G., Knorr, G. & Purcell, C. Abrupt glacial climate shifts controlled by ice sheet changes. *Nature* **512**, 290–294 (2014).

57. Knorr, G., Butzin, M., Micheels, A. & Lohmann, G. A warm Miocene climate at low atmospheric CO_2 levels. *Geophys. Res. Lett.* **38**, L20701 (2011).

58. Stepanek, C. & Lohmann, G. Modelling mid-Pliocene climate with COSMOS. *Geosci. Model Dev.* **5**, 1221–1243 (2012).

59. Knorr, G. & Lohmann, G. Climate warming during Antarctic ice sheet expansion at the Middle Miocene transition. *Nat. Geosci.* **7**, 376–381 (2014).

60. Lorenz, S. J. & Lohmann, G. Acceleration technique for Milankovitch type forcing in a coupled atmosphere-ocean circulation model: method and application for the Holocene. *Clim. Dyn.* **23**, 727–743 (2004).

61. Lüthi, D. *et al.* High-resolution carbon dioxide concentration record 650,000-800,000 years before present. *Nature* **453**, 379–382 (2008).

62. Loulergue, L. *et al.* Orbital and millennial-scale features of atmospheric CH_4 over the past 800,000 years. *Nature* **453**, 383–386 (2008).

63. Spahni, R. *et al.* Atmospheric methane and nitrous oxide of the late Pleistocene from Antarctic ice cores. *Science* **310**, 1317–1321 (2005).

64. Pfeiffer, M. & Lohmann, G. in *Earth System Science: Bridging the Gaps between Disciplines SpringerBriefs in Earth System Sciences* (eds Lohmann, G. *et al.*) 57–64 (Springer, 2013).

65. Varma, V. *et al.* Holocene evolution of the Southern Hemisphere westerly winds in transient simulations with global climate models. *Clim. Past* **8**, 391–402 (2012).

66. Lisiecki, L. E. & Raymo, M. E. A Pliocene-Pleistocene stack of 57 globally distributed benthic $\delta^{18}O$ records. *Paleoceanography* **20**, PA1003 (2005).

Acknowledgements

We thank the Government of the Island Territory of Bonaire of the former Netherlands Antilles (now Caribbean Netherlands) for research and fieldwork permissions and E. Beukenboom (STINAPA Bonaire National Parks Foundation) for support. This study was funded by the Deutsche Forschungsgemeinschaft through grants FE 615/3-2, 3-4 to T.F., SCHO 1274/4-4 to D.S. and LO 895/9-4 to G.L. (DFG Priority Programme INTERDYNAMIK - SPP 1266), and FE 615/5-1 to T.F. We thank S. Pape, M. Segl, M. Zuther, K.-H. Baumann and O. Mund for analytical and technical support, and C. Fensterer and R. Eichstädter for assistance with ^{230}Th/U-dating in the laboratory of A. Mangini (Heidelberg). T.F. is supported through the DFG-Research Center/Cluster of Excellence 'The Ocean in the Earth System' at the University of Bremen.

Author contributions

T.F. designed the study and wrote the manuscript; C.G. performed coral microsampling, generated coral geochemical time series and was responsible for quantification of seasonality and screening for diagenesis; D.S. was responsible for coral ^{230}Th/U dating; G.L. and M.P. performed model simulations; J.P., T.F. and S.R.S. discovered the coral and drilled the core; M.K. was responsible for coral Sr/Ca analysis; T.F. was responsible for the Bonaire 2006 Expedition; S.R.S. was responsible for local logistics, local field expertise and permissions. All authors contributed to data interpretation and manuscript preparation.

Additional information

Initialized decadal prediction for transition to positive phase of the Interdecadal Pacific Oscillation

Gerald A. Meehl[1], Aixue Hu[1] & Haiyan Teng[1]

The negative phase of the Interdecadal Pacific Oscillation (IPO), a dominant mode of multidecadal variability of sea surface temperatures (SSTs) in the Pacific, contributed to the reduced rate of global surface temperature warming in the early 2000s. A proposed mechanism for IPO multidecadal variability indicates that the presence of decadal timescale upper ocean heat content in the off-equatorial western tropical Pacific can provide conditions for an interannual El Niño/Southern Oscillation event to trigger a transition of tropical Pacific SSTs to the opposite IPO phase. Here we show that a decadal prediction initialized in 2013 simulates predicted Niño3.4 SSTs that have qualitatively tracked the observations through 2015. The year three to seven average prediction (2015-2019) from the 2013 initial state shows a transition to the positive phase of the IPO from the previous negative phase and a resumption of larger rates of global warming over the 2013-2022 period consistent with a positive IPO phase.

[1] National Center for Atmospheric Research, 3090 Center Green Drive, Boulder, Colorado 80301, USA. Correspondence and requests for materials should be addressed to G.A.M. (email: meehl@ucar.edu).

Retrospective hindcasts have shown skill in simulating the two phase changes of the Interdecadal Pacific Oscillation (IPO) since 1960, the first from negative to positive in the 1970s and the second from positive to negative in the late 1990s[1-5]. A proposed mechanism for IPO multidecadal variability indicates that for such decadal transitions to occur, there should be a build-up of upper ocean heat content in the off-equatorial western tropical Pacific such that an El Niño/Southern Oscillation (ENSO) event can then trigger a transition of tropical Pacific SSTs to the opposite IPO phase[5-8].

This mechanism is a variant of the delayed-action oscillator[9] that depends on off-equatorial wind forcing[10,11], which could be produced by tropical-midlatitude interaction. Convective heating anomalies in the tropical Pacific, associated with precipitation and SST anomalies there, produce anomalous Rossby waves in the atmosphere and sea level pressure signals in the mid-latitude North and South Pacific. These sea level pressure anomalies are associated with surface wind stress curl anomalies that force anomalous ocean Rossby waves near 25°N and 25°S. These slow-moving ocean Rossby waves subsequently produce, on decadal timescales, off-equatorial ocean heat content anomalies in the western tropical Pacific. The build-up of off-equatorial heat content that is necessary for an IPO transition contributes to a 15- to 20-year timescale for the IPO[6]. The interannual variability associated with ENSO, combined with the off-equatorial ocean heat content anomalies at the western boundary, then produce equatorial ocean Kelvin waves that change the equatorial thermocline depth in such a way as to reverse the sign of equatorial Pacific SST anomalies to the opposite phase of the IPO. The mechanism, as originally formulated[6], has been independently confirmed[7] and generalized in the context of the subtropical gyre and the Pacific subtropical cells (STCs)[12]. This is a variant of a previous mechanism for Pacific decadal variability that postulated IPO-like variability could be associated with amplitude modulation of ENSO events, thus also connecting interannual and decadal timescale variability in the Pacific[13].

Transitions in the sign of the IPO have been shown to be simulated in single and multi-model initialized decadal hindcasts and verified both through epoch differences and empirical orthogonal function metrics[1-5]. For the negative IPO from roughly 2000–2013, there was a slowdown or reduced rate of global surface warming (also sometimes referred to as the early-2000s hiatus[14-17]). The previous positive phase of the IPO from the 1970s to late 1990s was associated with an accelerated rate of global warming and has been documented in climate models[17-19].

Here we connect these transitions to the mechanism described above[6,7] and use those results to provide the context for a decadal prediction initialized in 2013, which shows a transition of the IPO from its recent negative state back to positive. Although there have been earlier attempts to predict a recent IPO transition from negative to positive[20], here we provide a process-based mechanism on which to base the credibility of the prediction of the transition to the positive phase of the IPO. We choose the CCSM4, because it is the only current climate model with an extensive documentation with regards to processes and initialized predictions involved with IPO transitions[1,2,17], as well as initialized predictions of decadal climate variability in the North Atlantic[21] (see Methods). There are ten ensemble members run for each initial year.

Results

Off-equatorial heat content. As noted above, there are two crucial elements of the IPO mechanism outlined above[6]: a build-up of off-equatorial ocean heat content anomalies in the western tropical Pacific over at least a 15-year period before a transition and an interannual ENSO event that triggers a transition (El Niño

for transition from negative to positive and La Niña for transition from positive to negative). For the IPO transitions during the period of the CMIP5 decadal hindcasts from 1960 to the early 2000s[22], the ocean states used to initialize the hindcasts for CCSM4 are shown in Fig. 1. For the tropical northwestern Pacific region (5°N–30°N, 125°E–180°E), the vertically integrated heat content to the depth of the 20 °C isotherm in the ocean initial state is $\sim 2.7 \times 10^{24}$ J just before the 1976–77 El Niño event, with roughly 2.2×10^{24} J in the tropical southwestern Pacific region (5°S–30°S, 150°E–160°W) (Fig. 1a). The 1972–73 El Niño, evidenced by an increase in vertically integrated heat content to the depth of the 18 °C isotherm in the eastern equatorial Pacific (5°S–5°N, 160°W–80°W) to values near 1.5×10^{24} J (Fig. 1b) is followed by a subsequent decrease over the next few years in off-equatorial southwestern tropical Pacific heat content by nearly 0.1×10^{24} J (Fig. 1a). However, the subsequent 1976–77 El Niño, with an increase in vertically integrated heat content in the eastern equatorial Pacific of $\sim 1.4 \times 10^{24}$ J (Fig. 1b), is associated with a subsequent drop in off-equatorial northwestern and

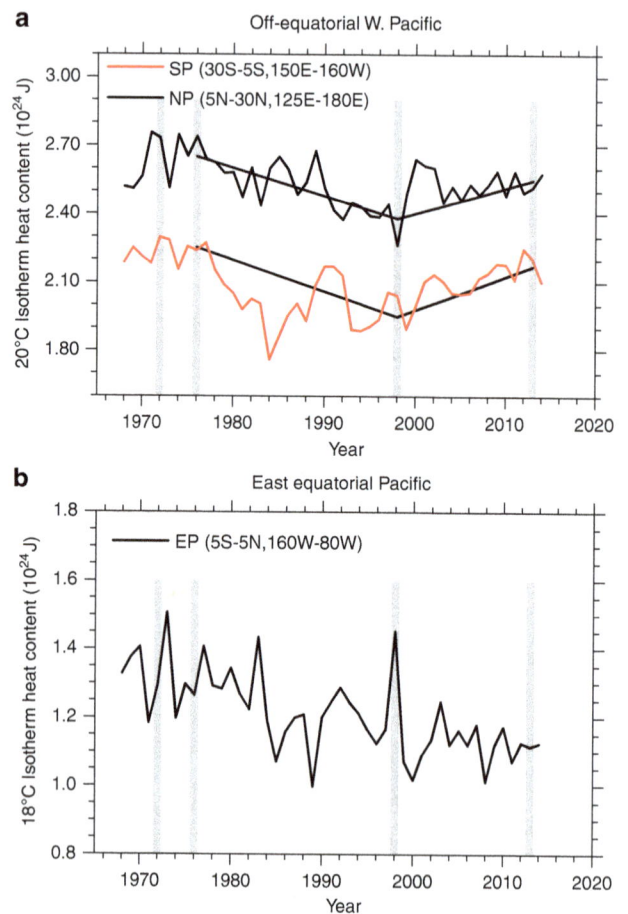

Figure 1 | Off-equatorial western Pacific and east equatorial Pacific heat content. (**a**) Annual mean time series of vertically integrated ocean heat content ($\times 10^{24}$ J) to the depth of the 20 °C isotherm for regions in the off-equatorial western Pacific from the ocean initial states used for the initialized model hindcasts; blue line for an area in the northwest tropical Pacific (5°N–30°N, 125°E–180°E); red line for an area in the southwest tropical Pacific (5°S–30°S, 150°E–160°W); shaded grey vertical lines indicate start points for the 1972-73 El Niño, 1976-77 El Niño, 1998-2000 La Niña and 2014-2015 El Niño, respectively; solid lines are linear trends from 1976 to 1998 and from 1998 to 2013; (**b**) same as in **a**, except for vertically integrated ocean heat content down to the depth of the 18 °C isotherm for an area in the equatorial eastern Pacific (5°N–5°S, 160°W–80°W).

southwestern tropical Pacific heat content for the next decade and longer (Fig. 1a). With the caveat that the initial ocean state values are less certain before the satellite era that began in the late 1970s (see Methods), this sequence suggests that the 1972–73 El Niño started the discharge of off-equatorial western Pacific heat content that was then given a boost by the 1976–77 El Niño, which closely followed, and thus contributed to the onset of the positive phase of the IPO in the mid-1970s[19].

Values for the southwestern Pacific region bottom out after the 1982–83 El Niño, which occurred while the IPO was still in a positive phase. The weak La Niña in 1984–85 contributed to a subsequent increase in off-equatorial heat content in the southwestern area of $\sim 0.3 \times 10^{24}$ J, but the northwestern area heat content continued to decline. Thus, the changes in the southwestern area were necessary but not sufficient to produce an IPO transition and the positive IPO phase continued until the late 1990s.

After the partial recovery around 1990, the southwestern Pacific region heat content declined again until a minimum value of $\sim 2.0 \times 10^{24}$ J was reached in 1997, whereas there was a minimum value of roughly 2.3×10^{24} J in the northwestern Pacific region around 1998, coincident with the large 1997–98 El Niño (Fig. 1a). These low heat content values in the off-equatorial western Pacific in the ocean initial states were then associated with the 2-year La Niña event (1998–2000) with low eastern equatorial Pacific heat content values down to near 1.0×10^{24} J (Fig. 1b) and a transition to the negative phase of the IPO. Next, off-equatorial western tropical Pacific vertically integrated heat content values started to climb again until the initial-state heat content in the northwestern tropical Pacific region in 2013 was nearly 2.6×10^{24} J, close to its value in the 1976 initial state when the IPO previously transitioned to positive. The southwestern Pacific region heat content in 2012 and 2013 was around 2.1–2.2 $\times 10^{24}$ J, also its highest value since the mid-1970s. Indications from these ocean heat content values from the initial states for the initialized hindcasts suggest that if even a moderate El Niño were to occur shortly after 2013, there could be an IPO transition from negative to positive.

It would be tempting to speculate, based on this small sample, that threshold values amounting to around 2.6×10^{24} J in the northwestern area and 2.2×10^{24} J in the southwestern area would be necessary for an El Niño event to trigger a transition to positive IPO. Conversely, heat content values of roughly 2.4×10^{24} and 2.0×10^{24} J in the northwestern and southwestern areas, respectively, could be necessary for a La Niña event to trigger a transition to the negative phase of the IPO. However, based on the small sample size and the limitations of the ocean initialization scheme (see Methods), these values can only be considered as instructive at this time, pending further research with more models.

Hindcast skill. Credibility for an initialized prediction in the future must be evaluated in the context of past predictions or hindcasts. Thus, it is necessary to quantify IPO hindcast skill in CCSM4 to see whether there are false IPO transitions that appear in initialized hindcasts with previous El Niño or La Niña events. Figure 2 shows hindcast verification in terms of pattern correlation for Pacific SSTs (Pacific basin, ocean grid points 40°S–70°N, 100°E–80°W) for CCSM4 for the ensemble average of ten ensemble members for each initial year starting in 1960 (that is, the correlation is calculated using the ensemble average[4]). This is a version of a similar figure shown for initialized hindcasts for all the CMIP5 models[4]. Hindcast skill for the year three to seven predictions (5 year average over years 3–7, hereafter referred to as '3- to 7-year' hindcasts or predictions) is defined as a pattern correlation for Pacific SSTs above the 95% significance level

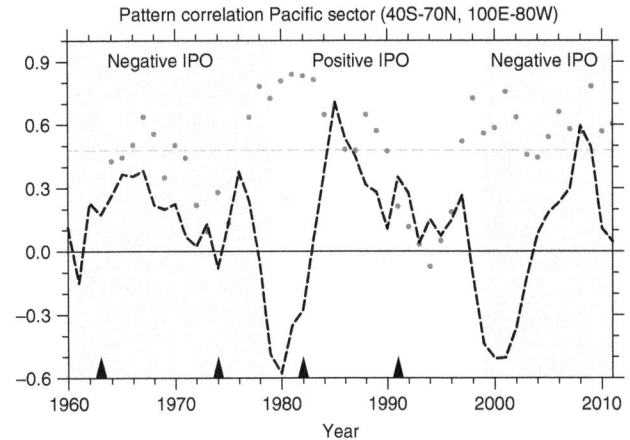

Pattern correlation Pacific sector (40S-70N, 100E-80W)

Figure 2 | Hindcast skill for the IPO. Anomaly pattern correlation for the Pacific SSTs in hindcasts initialized every year from 1960 for CCSM4 (ocean points only, 40°S–70°N, 100°E–80°W; same as in Fig. 2e in ref. 4, but just for CCSM4 see Methods). Horizontal dashed line is 95% significance. Blue dashed line is persistence prediction defined as the average of years three to seven in the future having the same SST anomalies as the 5-year average before the start of the prediction. Each red dot represents the central year of the 3- to 7-year average hindcast; thus, the value for the initial year 1960 for the prediction averaged for 1962–66 is plotted for 1964 and so on. Phase changes of the IPO from negative to positive in the late 1970s and from positive to negative in the late 1990s are denoted by shading. Hindcast skill notably drops below 95% when influenced by the volcanic eruptions (denoted by triangles above x axis) of Fuego in the early 1970s and Pinatubo in the early 1990s, but not Agung in the early 1960s or El Chichon in the early 1980s[23].

(horizontal dashed line, see Methods). Most 3- to 7-year hindcasts are skillful by that measure and exceed persistence (defined as the 5-year average anomalies before the initial year and persisted to forecast years three to seven) except for two periods, one in the mid-1970s and the other in the mid-1990s. These two periods have been shown to have reduced hindcast skill due to the effects of the Fuego and Pinatubo volcanic eruptions, respectively[23], when the post-eruption sequence of tropical Pacific SSTs does not match the model ensemble-average forced response. The previously documented skill in IPO transitions from negative to positive in the late 1970s and from positive to negative in the late 1990s is denoted by red dots above 95% significance for changes in shading in Fig. 2. Model skill in predicting transitions is indicated by the combination of having a good pattern correlation and poor performance of a persistence forecast (blue dashed line). It is noteworthy that the persistence predictions have no skill in predicting across major decadal transitions (actually having negative values in Fig. 2 at the time of the transitions), while the initialized predictions have significant skill across the transitions. In addition, the CCSM4 shows skill in predicting the correct state of the IPO even for years when there were significant El Niño (for example, no IPO transition for the 1982–83 El Niño) or La Niña (for example, no IPO transition for the 1988–89 La Niña) events.

Figure 2 also addresses the concern that an initialized hindcast could either trend to model climatology and always 'transition' away from its initial state or simply persist the initial state and never 'transition' to another IPO state. The performance of the model in the hindcasts shows that, with the exception of the volcano-related drops in skill noted above (with the caveat that the volcanoes affect 11 out of the 48 five-year average prediction periods in Fig. 2), the model can capture IPO transitions as well as persist existing initialized IPO states.

ENSO as a trigger for decadal transitions. As the previous two IPO transitions were simulated with some success as noted above and if ENSO events act as a trigger for the western tropical Pacific off-equatorial heat content anomalies in the ocean initial states noted in Fig. 1 as in previous studies[6], then ENSO evolution for those IPO transitions should have been simulated as well. There have been previous indications that ENSO events can be predicted with initialized climate models for lead times of at least 2 years[24]. Here we show the time evolution of the predictions for all ensemble members and the ensemble average for the first 5 years after initialization. Clearly, there will be spread in the predictions, but the hindcast verification in Fig. 2 uses the ensemble average and that is of primary interest here.

Owing to the limitations of the ocean initialization scheme that affect ocean initial states more in the pre-satellite (before the late-1970s) than post-satellite era (see Methods), we follow the conservative approach previously used to analyse predictions of the mid-1970s climate shift[1] and show results for the model initialized in January 1976. The ensemble average for the Niño3.4 index in the eastern equatorial Pacific (5°N–5°S, 170°W–120°W,

a standard ENSO index) tracks the observations for the onset of the El Niño in mid-1976 and then the ensemble average remains anomalously warm through 1978, in fact longer than what was observed (Fig. 3a). Although there is spread in the initialized predictions after 1977, the 1976–1977 period of above-normal eastern equatorial SSTs is captured by all ten ensemble members and was associated with the IPO transition from negative to positive for 3- to 7-year predictions during this time period[1].

For the IPO transition from positive to negative in the late 1990s, there is greater confidence in the ocean initial states (see Methods); thus, we follow a previous analysis[4] and show the predictions initialized in January 1996 that have more ensemble spread than for the 1976–77 case, but the ensemble average shows above-normal Niño3.4 SSTs for the 1997–98 event of lower amplitude than what was observed. Of more significance for the IPO transition, the ensemble average model prediction shows below-normal Niño3.4 SSTs for the observed 2-year La Niña from 1998 to 2000, with most ensemble members having negative values for that time period in qualitative agreement with the observations. This La Niña simulation was associated with the

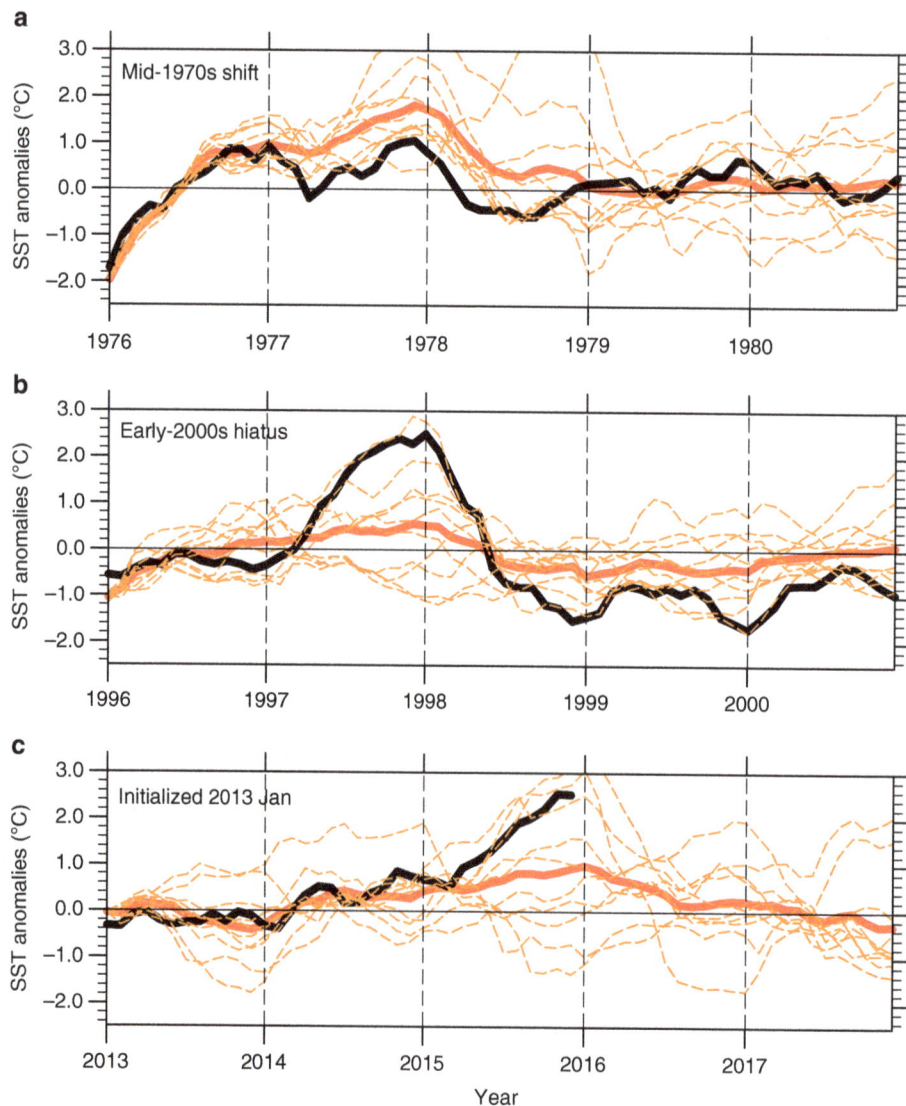

Figure 3 | Initialized predictions of eastern equatorial Pacific sea surface temperatures. Monthly mean time series of initialized predictions for the Niño3.4 area (5°S–5°N, 120°W–170°W), calculated as anomalies from a 1975 to 2014 climatology; observations (black lines[34]), individual model ensemble members are dashed orange lines; ensemble average is solid red line; (**a**) prediction for the mid-1970s shift, initialized 1 January 1976 (ref. 1); (**b**) prediction for early-2000s hiatus, initialized 1 January 1996 (ref. 4); (**c**) prediction initialized 1 January 2013.

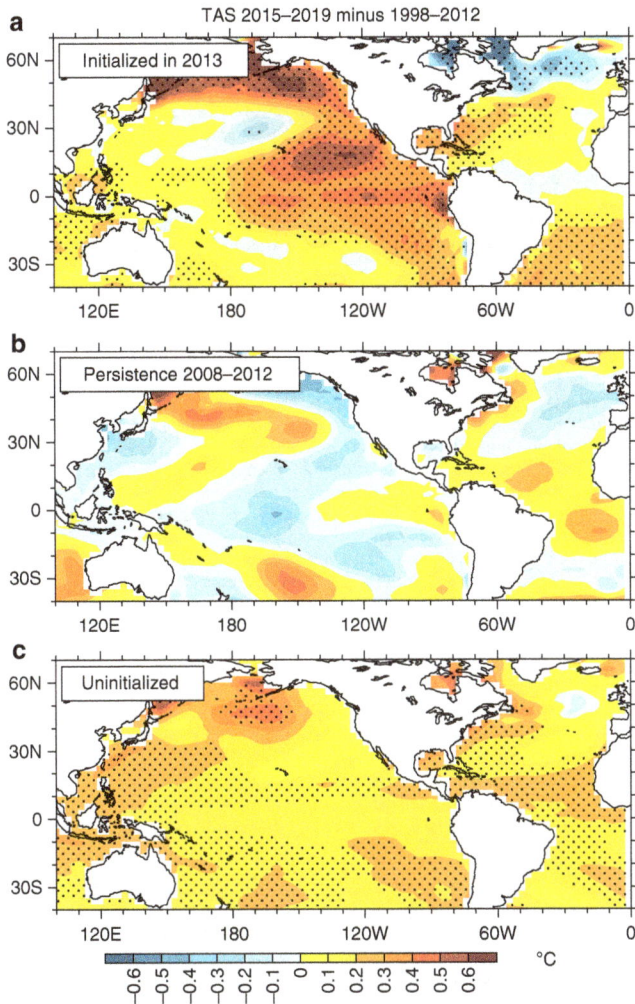

Figure 4 | Predicted patterns of surface temperature. (a) Global surface air temperature anomaly patterns (°C) for prediction initialized on 1 January 2013 for the 5-year average 2015–2019 minus observed reference period 1998–2012 from NCEP/NCAR reanalyses[34] (see Methods); (b) persistence prediction defined as the average of years three to seven in the future having the same SST anomalies as the 5-year average before the start of the prediction as in Fig. 2, for the 2015–2019 average having the same SST anomalies as the 2008–2012 average, and (c) uninitialized prediction for years 2015–2019 minus the model data averaged over the period 1998–2012; stippling indicates 95% significance level from one-sided t-test.

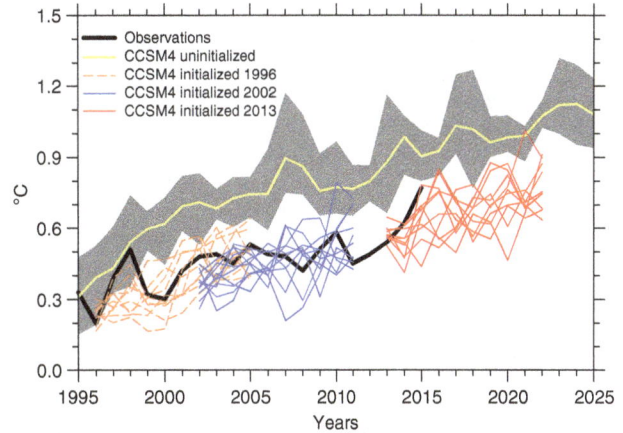

Figure 5 | Time series of global surface air temperature anomalies. Annual mean surface temperature anomalies (°C) relative to observed climatology of 1960–1990 for the CCSM4 uninitialized six member ensemble average (solid yellow line) and the envelope of individual ensemble simulations (grey shading); annual mean values from observations is solid black line[25]; individual ensemble member time series for 10-year predictions initialized 1 January 1996 (dashed orange lines) to capture the late-1990s transition from positive to negative IPO[4], an example of predictions for the hiatus initialized 1 January 2002 (blue lines) and predictions initialized on 1 January 2013 (red lines), the latter showing larger predicted warming rates than the former during the hiatus.

prediction of the transition of the IPO from positive to negative in the model for the 3- to 7-year predictions initialized in 1996 (ref. 1).

Prediction for 2015–2019. Thus, there are indications of some model skill in simulating ENSO events that could act as triggers for the off-equatorial ocean heat content in the western tropical Pacific to contribute to IPO transitions (Fig. 1). Therefore, it is of interest to investigate simulations initialized in 2013 for the 3- to 7-year predictions (2015–2019) of the IPO and globally averaged surface air temperature. We choose the initial year 2013 so that we have 2 years of subsequent observations to see whether there has been any skill for predicted Niño3.4 SSTs that could indicate a role in an IPO transition. As noted above, for a credible prediction of an IPO transition, there must not only be heat content build-up in the off-equatorial western Pacific over a period of ~15 years, but also a reasonable simulation of ENSO

for the first few years after initialization, for the 3- to 7-year prediction to have skill.

For predictions initialized in January 2013 (Fig. 3c), the observations started to warm in 2014 and the model-simulated Niño3.4 SSTs also warmed. In 2015, the observed Niño3.4 SSTs continued to increase, to approach the magnitude of the warmest predicted ensemble members (Fig. 3c). The ensemble average predictions from the model show above-normal Niño3.4 SSTs for 2015 continuing through 2016. The associated SST anomalies for the 3- to 7-year prediction (2015–2019 minus a 15-year average before the initial year, 1998–2012, used as a reference climatology with which to compare the initialized prediction) initialized in 2013 show a positive phase of the IPO, with above-normal SSTs predicted over the eastern tropical Pacific extending into the northeast Pacific, with low amplitude-negative SST anomalies in the northwest and southwest Pacific (Fig. 4a). This pattern is nearly opposite to the persistence prediction (persisting 2008–2012 average anomalies forward to 2015–2019; Fig. 4b) and is also considerably different from the uninitialized simulation that shows warming almost everywhere (Fig. 4c).

As noted above, there is a strong association between the IPO tendency and globally averaged surface air temperature trends. CCSM4 hindcasts initialized in 1996 simulated the onset of the reduced rate of global surface warming with the transition of the IPO from positive to negative in the late 1990s[4]. The predicted linear trend from 1996 to 2005 of $+0.30 \pm 0.09$ °C per decade (error bars are ± 1 s.d.) is smaller than the uninitialized value of $+0.39 \pm 0.08$ °C per decade and closer to the observed value of $+0.24$ °C per decade (Fig. 5). Meanwhile, a hindcast initialized in 2002 during the hiatus has a predicted linear trend from 2002 to 2011 of $+0.20 \pm 0.11$ °C per decade, roughly 30% smaller than the prediction initialized in 1996 before the hiatus.

Prediction of 2013–2022 global surface temperature trend. With a transition of the IPO from negative to positive in a prediction initialized in 2013, it could be expected from previous studies[14–18] that there should be a resumption of higher rates of globally averaged surface temperature increase. For the prediction initialized in 2013, the linear trend from 2013 to 2022 is $+0.22 \pm 0.13\,^{\circ}\mathrm{C}$ per decade, nearly 60% larger than the uninitialized projection for that time period of $+0.14 \pm 0.12\,^{\circ}\mathrm{C}$ per decade (Fig. 5) and nearly three times larger than the observed trend[25] of $+0.08\,^{\circ}\mathrm{C}$ per decade over the period 2001 to 2014 during the early twenty-first century slowdown[26]. This indicates an acceleration of the global surface temperature warming trend over the prediction period from 2013 to 2022 associated with the positive phase of the IPO, compared with that from external forcing alone and compared with the observed warming trend during previous negative phase of the IPO that was associated with a slowdown of the rate of increase of global surface temperature[25].

Discussion

The 3- to 7-year hindcasts with CCSM4 initialized every year show skill in simulating patterns of SST in the Pacific associated with the IPO, except for periods when there are disruptions from volcanic eruptions with a post-eruption sequence of Pacific SST patterns that do not agree with the ensemble average forced response in the model[23]. Other studies have shown less skill for the smaller North Pacific area defined for the Pacific Decadal Oscillation (PDO)[27]. Although the PDO and IPO are closely related[28], the smaller area strictly defined for the PDO in the North Pacific has more spatial noise and less predictive skill for that small area than the much larger Pacific-wide SST pattern defined for the IPO. In addition, the STC in the Southern Hemisphere shows larger variability in some models, affecting the STC transport convergence anomaly that could affect the North Pacific STC and thus the PDO index, which is defined for the North Pacific.

A proposed mechanism for the IPO indicates that a sustained build-up of off-equatorial ocean heat content in the western tropical Pacific for at least 15 years, due to tropical–mid-latitude interaction and ocean Rossby waves associated with the phase of the IPO, is required for an ENSO event to then trigger a transition of the IPO to its opposite phase. Initial state heat content used for the hindcasts with CCSM4 indicates that there was indeed such a build-up before the simulated IPO transition in the mid-1970s that was associated with the prediction of the 1976–77 El Niño event, with possible contributions from the 1972 to 1973 El Niño event. For the late-1990s IPO transition from positive to negative, a heat content deficit in the off-equatorial western tropical Pacific was associated with the simulated 1998–2000 La Niña event. Other El Niño or La Niña events did not trigger an IPO transition, owing to the requisite time for heat content to build up in the off-equatorial western Pacific. Since the late 1990s, heat content has built up in the initial state used for the predictions, thus suggesting that even a moderate-sized El Niño event could trigger a transition to the positive phase of the IPO. Predictions initialized in 2013 show that the simulated Niño3.4 SSTs have tracked the observations with low-amplitude warming in 2014 and larger warming in 2015. Such qualitative success in Niño3.4 simulations for past IPO transitions suggests that an IPO transition probably started in 2014. Indeed, the 3- to 7-year predictions for 2015–2019 that were initialized in 2013 indicate such an IPO transition has occurred, with a resumption of accelerated rates of global warming above those in the uninitialized model simulations. If such predictions were to become operational, the next step (beyond the scope of this study) would be to provide probabilistic climate change information for various time frames in the near-term future.

Methods

Initialization procedure. As described in a previous study[21], the hindcasts in CCSM4 are initialized using a forced ocean–sea ice simulation designed to reproduce the evolution of the ocean and sea ice states from the start of 1948 through the end of 2013. The CCSM4 ocean and sea ice models are coupled and forced at the surface with Coordinated Ocean-ice Reference Experiment (CORE-II) version 2-observed historical atmospheric data[29,30]. The forcing data include a complete set of air–sea momentum, heat and freshwater fluxes based on the National Centers for Environmental Prediction/National Center for Atmospheric Research (NCEP/NCAR) reanalyses. There is a weak restoring of model surface salinity to observed climatology and no assimilation of subsurface observations. An integration is run through four consecutive 66-year cycles of 1948–2013 forcing. The ocean and sea ice models in hindcast experiments are initialized with 1 January restart files for a particular year from the last (fourth) cycle of the spin-up simulation. Initial conditions for the atmosphere and land surface are taken from corresponding years of a six-member ensemble of uninitialized twentieth-century runs. The ten-member ensembles run for each initial year starting in 1960 are generated by randomly selecting atmosphere and land initial conditions from different uninitialized twentieth-century runs[31] and/or from different days in the month of January. Results from these uninitialized twentieth-century experiments[31] are shown in Figs 4c and 5. The initialized hindcasts and the uninitialized twentieth-century experiments include the same external forcings of solar irradiance, greenhouse gases, aerosols and volcanic activity.

As atmospheric model reanalysis data are used in the CORE-II forcing of the spin-up integrations used for the ocean initial states in the hindcasts, it is likely to be that the quality of the forcing data becomes better in the more recent period, whereas there are issues with reanalysis quality earlier in the record, in particular with precipitation and radiation fluxes over oceans (for example, before the 1970s)[32,33]. Therefore, most of the emphasis in this study is on initialized hindcasts after 1975.

Bias adjustment procedure. The initialized hindcast experiments are bias adjusted to remove model systematic errors[1,21]. Climatological monthly mean differences from observations (surface air temperature from the NCEP/NCAR reanalyses)[34] for the 10-year hindcasts are computed and composited by month following the initial dates of all the hindcast simulations. Next, these average time-evolving monthly mean differences are subtracted from each member of each 10-year hindcast to remove the model systematic error, leaving the model signal from external forcing and internally generated decadal climate variability.

Assessing hindcast skill. Anomalies for the 3- to 7-year bias-adjusted predictions are calculated as the year 3–7 average minus the average of the observations for the previous 15 years before the initial year of the hindcast[1,4]. Other metrics show similar results[2]. To test whether the pattern correlation coefficient between the prediction and observations is distinguishable from chance associations in the large-scale pattern, a Monte Carlo test is performed that consists of 10,000 randomly constructed patterns based on detrended twentieth-century simulations from a multi-model ensemble of CMIP5 models[4]. The 95th percentile of the pattern correlation coefficient of the random pattern is 0.47 in Fig. 2. Persistence is defined as the observed 5-year average before the initial year of the prediction minus a 15-year average for 6–20 years before the initial year[4].

Observations. The Nino3.4 index is available from http://www.esrl.noaa.gov/psd/gcos_wgsp/Timeseries/. Observed global mean surface temperature data available from http://www.ncdc.noaa.gov/cag/time-series/global. The NCEP/NCAR reanalyses are available from http://www.esrl.noaa.gov/psd/data/gridded/data.ncep.reanalysis.html.

References

1. Meehl, G. A. & Teng, H. Case studies for initialized decadal hindcasts and predictions for the Pacific region. *Geophys. Res. Lett.* **39,** L22705 (2012).
2. Meehl, G. A. & Teng, H. Regional precipitation simulations for the mid-1970s shift and early-2000s hiatus. *Geophys. Res. Lett.* **41,** 7658–7665 (2014).
3. Meehl, G. A. & Teng, H. CMIP5 multi-model initialized decadal hindcasts for the mid-1970s shift and early-2000s hiatus and predictions for 2016–2035. *Geophys. Res. Lett.* **41,** 1711–1716 (2014).
4. Meehl, G. A., Teng, H. & Arblaster, J. M. Climate model simulations of the observed early-2000s hiatus of global warming. *Nat. Clim. Change* **4,** 898–902 (2014).
5. Ding, H., Greatbatch, R. J., Latif, M., Park, W. & Gerdes, R. Hindcast of the 1976/77 and 1998/99 climate shifts in the Pacific. *J. Climate* **26,** 7650–7661 (2013).
6. Meehl, G. A. & Hu, A. Megadroughts in the Indian monsoon region and southwest North America and a mechanism for associated

multi-decadal Pacific sea surface temperature anomalies. *J. Climate* **19**, 1605–1623 (2006).

7. Farneti, R., Molteni, F. & Kucharski, F. Pacific interdecadal variability driven by tropical-extratropical interactions. *Clim. Dyn.* **42**, 3337–3355 (2014).

8. Giese, B. S., Urizar, S. C. & Fučkar, N. S. Southern Hemisphere origins of the 1976 climate shift. *Geophys. Res. Lett.* **29**, 1-1-1-4 (2002).

9. White, W. B., Tourre, Y. M., Barlow, M. & Dettinger, M. A delayed action oscillator shared by biennial, interannual, and decadal signals in the Pacific basin. *J. Geophys. Res.* **108**, 3070 (2003).

10. Wang, X., Jin, F.-F. & Wang, Y. A tropical ocean recharge mechanism for climate variability. Part I: equatorial heat content changes induced by the off-equatorial wind. *J. Climate* **16**, 3585–3598 (2003).

11. Wang, X., Jin, F.-F. & Wang, Y. A tropical ocean recharge mechanism for climate variability. Part II: a unified theory for decadal and ENSO modes. *J. Climate* **16**, 3599–3616 (2003).

12. Farneti, R., Dwivedi, S., Kucharski, F., Molteni, F. & Griffies, S. M. On Pacific subtropical cell variability over the second half of the twentieth century. *J. Climate* **27**, 7102–7112 (2014).

13. Jin, F. Low frequency modes of the tropical ocean dynamics. *J. Climate* **14**, 3872–3881 (2001).

14. Meehl, G. A. *et al.* Model-based evidence of deep-ocean heat uptake during surface-temperature hiatus periods. *Nat. Clim. Change* **1**, 360–364 (2011).

15. Kosaka, Y. & Xie, S.-P. Recent global-warming hiatus tied to equatorial Pacific surface cooling. *Nature* **501**, 403–407 (2013).

16. England, M. H. *et al.* Slowdown of surface greenhouse warming due to recent Pacific trade wind acceleration. *Nat. Clim. Change* **4**, 222–227 (2014).

17. Watanabe, M. *et al.* Contribution of natural decadal variability to global warming acceleration and hiatus. *Nat. Clim. Change* **4**, 893–897 (2014).

18. Meehl, G. A., Hu, A., Arblaster, J. M., Fasullo, J. & Trenberth, K. E. Externally forced and internally generated decadal climate variability associated with the Interdecadal Pacific Oscillation. *J. Climate* **26**, 7298–7310 (2013).

19. Meehl, G. A., Hu, A. & Santer, B. D. The mid-1970s climate shift in the Pacific and the relative roles of forced versus inherent decadal variability. *J. Climate* **22**, 780–792 (2009).

20. Thoma, M., Greatbatch, R. J., Kadow, C. & Gerdes, R. Decadal hindcasts initialized using observed surface wind stress: evaluation and prediction out to 2024. *Geophys. Res. Lett.* **42**, 6454–6461 (2015).

21. Yeager, S., Karspeck, A., Danabasoglu, G., Tribbia, J. & Teng H., H. A decadal prediction case study: late twentieth-century North Atlantic ocean heat content. *J. Climate* **25**, 5173–5189 (2012).

22. Meehl, G. A. *et al.* Decadal climate prediction: an update from the trenches. *Bull. Am. Meteorol. Soc.* **95**, 243–267 (2014).

23. Meehl, G. A., Teng, H., Maher, N. & England, M. H. Effects of the Mount Pinatubo eruption on decadal climate prediction skill of Pacific sea surface temperatures. *Geophys, Res. Lett.* **42**, 10,840–10,846 (2015).

24. Luo, J.-J., Masson, S., Behera, S. & Yamagata, T. Extended ENSO predictions using a fully coupled ocean-atmosphere model. *J. Climate* **21**, 84–93 (2008).

25. Karl, T. R. *et al.* Possible artifacts of data biases in the recent global surface warming hiatus. *Science.* **348**, 1469–1472 (2015).

26. Fyfe, J. C. *et al.* Making sense of the early-2000s warming slowdown. *Nat. Clim. Change* **6**, 224–228 (2016).

27. Kim, H.-M., Webster, P. J. & Curry, J. A. Evaluation of short-term climate change prediction in multi-model CMIP5 decadal hindcasts. *Geophys. Res. Lett.* **39**, L10701 (2012).

28. Han, W. *et al.* Intensification of decadal and multi-decadal sea level variability in the western tropical Pacific during recent decades. *Clim. Dyn.* **43**, 1357–1379 (2014).

29. Large, W. G. & Yeager, S. G. The global climatology of an interannually varying air-sea flux data set. *Clim. Dyn.* **33**, 341–364 (2008).

30. Griffies, S. M. *et al.* Coordinated ocean-ice reference experiments (COREs). *Ocean Model.* **26**, 1–46 (2009).

31. Meehl, G. A. *et al.* Climate system response to external forcings and climate change projections in CCSM4. *J. Climate* **25**, 3661–3683 (2012).

32. Trenberth, K. E. & Guillemot, C. J. Evaluation of the atmospheric moisture and hydrological cycle in the NCEP/NCAR reanalyses. *Clim. Dyn.* **14**, 213–231 (1998).

33. Trenberth, K. E., Fasullo, J. T. & Mackaro, J. Atmospheric moisture transports from ocean to land and global energy flows in reanalyses. *J. Climate* **24**, 4907–4924 (2011).

34. Kalnay, E. *et al.* The NCEP/NCAR 40-year reanalysis project. *Bull. Amer. Meteorol. Soc.* **77**, 437–471 (1996).

Acknowledgements

The authors thank Mark Cane for encouraging us to publish this work. Portions of this study were supported by the Regional and Global Climate Modeling Program (RGCM) of the U.S. Department of Energy's Office of Biological and Environmental Research (BER) Cooperative Agreement DE-FC02-97ER62402 and the National Science Foundation. The National Center for Atmospheric Research is sponsored by the National Science Foundation. The CCSM4 decadal prediction runs used resources of the National Energy Research Scientific Computing Center, which is supported by the Office of Science of the U.S. Department of Energy under contract DE-AC02-05CH11231, and the Oak Ridge Leadership Computing Facility located in the Oak Ridge National Laboratory, which is supported by the Office of Science of the Department of Energy under contract DE-AC05-00OR22725. An award of computer time was provided by the Innovative and Novel Computational Impact on Theory and Experiment (INCITE) program.

Author contributions

G.M. directed and wrote this work with contributions from all authors. G.M., A.H. and H.T. performed the analyses. H.T. ran the model ensemble hindcasts and predictions, and with A.H. conducted analyses. All the authors discussed the results.

Additional information

Competing financial interests: The authors declare no competing financial interests.

Global marine protected areas do not secure the evolutionary history of tropical corals and fishes

D. Mouillot[1,2], V. Parravicini[3], D.R. Bellwood[2], F. Leprieur[1], D. Huang[4], P.F. Cowman[5], C. Albouy[6], T.P. Hughes[2], W. Thuiller[7,8] & F. Guilhaumon[1]

Although coral reefs support the largest concentrations of marine biodiversity worldwide, the extent to which the global system of marine-protected areas (MPAs) represents individual species and the breadth of evolutionary history across the Tree of Life has never been quantified. Here we show that only 5.7% of scleractinian coral species and 21.7% of labrid fish species reach the minimum protection target of 10% of their geographic ranges within MPAs. We also estimate that the current global MPA system secures only 1.7% of the Tree of Life for corals, and 17.6% for fishes. Regionally, the Atlantic and Eastern Pacific show the greatest deficit of protection for corals while for fishes this deficit is located primarily in the Western Indian Ocean and in the Central Pacific. Our results call for a global coordinated expansion of current conservation efforts to fully secure the Tree of Life on coral reefs.

[1] UMR 9190 MARBEC, IRD-CNRS-IFREMER-UM, Université de Montpellier, Montpellier 34095, France. [2] Australian Research Council Centre of Excellence for Coral Reef Studies, James Cook University, Townsville, Queensland 4811, Australia. [3] CRIOBE, USR 3278 CNRS-EPHE-UPVD, Labex 'Corail', University of Perpignan, Perpignan 66860, France. [4] Department of Biological Sciences and Tropical Marine Science Institute, National University of Singapore, Singapore 117543, Singapore. [5] Department of Ecology & Evolutionary Biology, Yale University, 21 Sachem St, New Haven, Connecticut 06511 USA. [6] Département de biologie, chimie et géographie, Université du Québec à Rimouski, 300 Allée des Ursulines, Rimouski, Canada G5L 3A1. [7] Laboratoire d'Écologie Alpine (LECA), Univ. Grenoble Alpes, Grenoble F-38000, France. [8] Laboratoire d'Écologie Alpine (LECA), CNRS, Grenoble F-38000, France. Correspondence and requests for materials should be addressed to D.M. (email: david.mouillot@univ-montp2.fr).

Human activities are altering ecosystems worldwide, changing their biodiversity and composition, and imperilling their capacity to deliver ecosystem services[1]. In this context, protected areas are indisputably the flagship tool for protecting both ecosystems and biodiversity by limiting direct human impacts[2]. Conservation strategies have traditionally focused on vulnerable components of taxonomic diversity such as endemic, rare or threatened species[3,4]. However, phylogenetic diversity, represented by the Tree of Life, is becoming an increasingly important component of conservation science[5,6] since it represents the breadth of evolutionary history[7] and supports biodiversity benefits and uses, often unanticipated, for future generations[8,9]. Phylogenetically related species tend to have similar functional traits, environmental niches and ecological interactions[10,11], although numerous counter examples exist[12,13]. Therefore, species that are more phylogenetically distinct may have greater functional complementarity. In turn, species assemblages that are more phylogenetically diverse may promote greater biomass production within[14] and across[15] trophic levels even though a universal relationship between phylogenetic diversity and ecosystem functioning remains questionable[9,16]. Yet, few studies have quantitatively assessed the extent to which protected areas encompass phylogenetic diversity[17,18] and none have focused on marine taxa at a global scale.

Here, we tackle this critical issue for the iconic but threatened coral reefs of the world that support one of the largest concentrations of biodiversity, around 830,000 multi-cellular species[19], and provide vital ecosystem services to half a billion people including food security[20], financial incomes[21] and protection against natural hazards[22]. There is overwhelming evidence that human activities, particularly fishing pressure and pollution, affect coral reef ecosystem state[23], functioning[24] and resilience[25]. Thus, to counteract human impacts and maintain the integrity of coral reefs, thousands of marine-protected areas (MPAs) have been created worldwide[26]. However, the spatial design of the global MPA system is largely contingent on local socioeconomic conditions and history rather than regional or global considerations[27,28]. Furthermore, given the limited resources dedicated to conservation efforts[29] and the need to maintain coastal fisheries for people's livelihoods[21,30], MPAs cannot be extended to all coral reefs. Guiding future conservation strategies thus remains a key challenge, particularly at a global scale where deficits of protection must be identified and addressed to achieve effective protection of evolutionary history on coral reefs. Here we assessed the extent to which the global system of MPAs represents individual species and phylogenetic diversity for two major components of coral reef ecosystems, shallow-water corals in the order Scleractinia (805 species) and fishes in the family of Labridae (452 species). These groups contribute to the high biodiversity of tropical seas[31] and help maintaining productive and resilient reefs[32,33]. We show that the current global MPA system, covering 5.9% of the world's coral reef area, does not meet the minimum conservation targets considered necessary to adequately secure the branches of the Tree of Life for corals or fishes, particularly the longest branches that represent the greatest amount of evolutionary history.

Results and Discussion

Lag behind minimum conservation targets. Using global distribution maps of each scleractinian coral and labrid fish species (Methods), we reveal that only 5.7% of coral species and 21.7% of fish species meet a minimum protection target of 10% potential coverage of their geographic range by the global system of MPAs (Fig. 1). Regionally, the situation is even more contrasted. For example, coral species that occur exclusively in the Tropical Eastern Pacific all fall below the critical 10% coverage

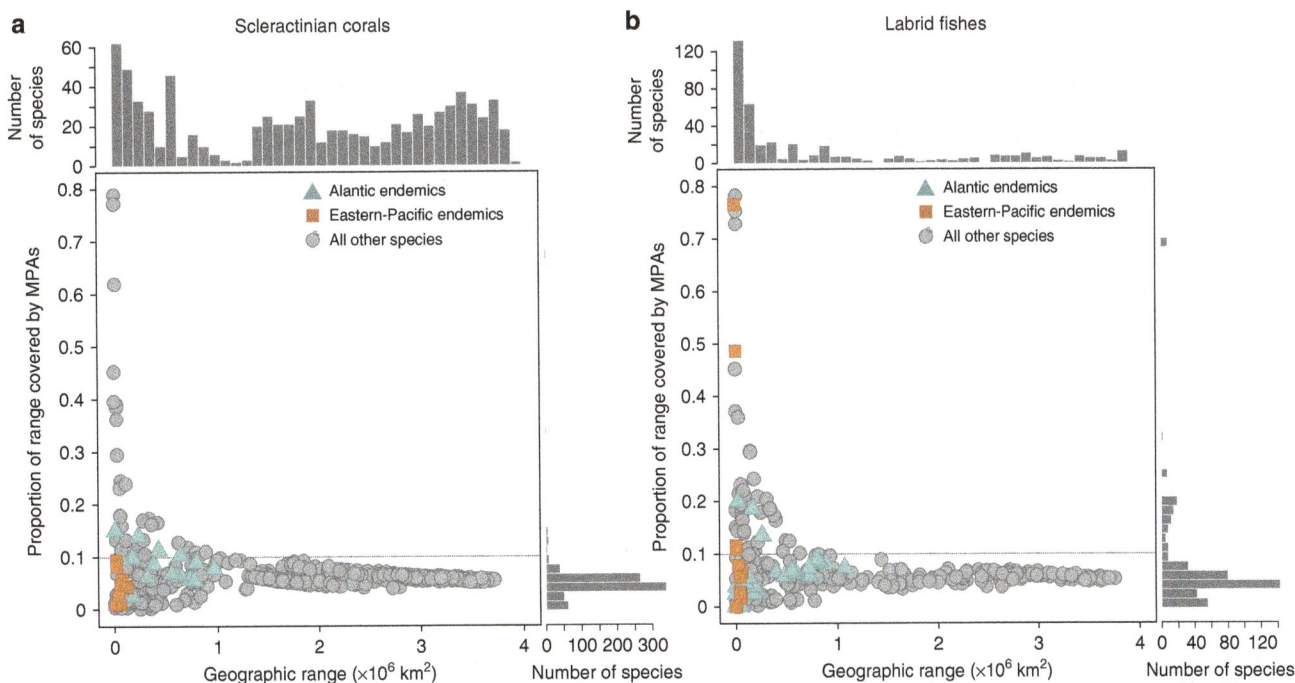

Figure 1 | Relationship between the total geographic range of species and the proportion of that range covered by the global system of MPAs.
(a) Scleractinian coral species and (b) fish species of the family Labridae. Histograms on top and to the right represent the distributions of total ranges and proportion of protection among species respectively. Coloured squares and triangles represent endemic species, that is, only present in one of the two biogeographic realms: Atlantic and Eastern Pacific, respectively. Dotted lines represent the 10% threshold corresponding to the minimum representation target for sustaining species persistence.

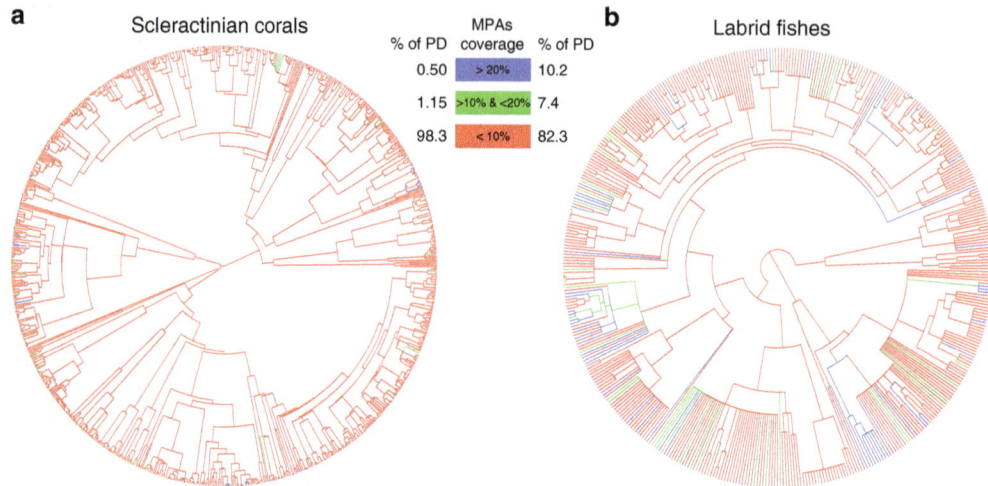

Figure 2 | Percentages of geographic ranges covered by the global system of MPAs for species and internal branches across the Tree of Life.
(**a**) Scleractinian coral species and (**b**) fish species of the family Labridae. Species or branches in red do not meet the minimum 10% representation threshold, that is, <10% of their geographic range is covered by MPAs, while green and blue colours indicate 10–20% and more than 20% coverage respectively. The corresponding percentage of total phylogenetic diversity (PD) is indicated for each coverage category.

threshold. Similarly, all coral and fish species found only in the Atlantic have <20% coverage (Fig. 1). This 10% threshold has been specifically advocated for wide-ranging species ($>250,000\,km^2$) and is regarded as a conservative target of coverage by protected areas for sustaining species persistence[3,34]. This conservative cut-off takes into account commission errors, that is, the potential absence of a given species from protected areas that lie within its geographic distribution due to chance or unsuitable habitats[34].

By applying the same reasoning to the internal branches of the phylogenetic trees (Methods), we show that only 1.7% ($\pm$0.2 s.d.) of the Tree of Life of corals and 17.6% ($\pm$0.6 s.d.) of fishes attain the minimum 10% coverage (Fig. 2). Thus 7,160 Myr of the evolutionary history of corals and 3,586 Myr of fishes are inadequately represented by the global MPA system, far more than for many other threatened taxonomic groups[8]. Globally, the amount of evolutionary history potentially covered by MPAs, that is, the proportion of the geographic range of evolutionary branches overlapping with the global MPA system, is only 6.0% ($\pm$0.1 s.d.) and 8.7% ($\pm$0.2 s.d.) for corals and fishes, respectively. Coral evolutionary history receives significantly less coverage than expected under a random distribution of species geographic ranges across the Tree of Life ($P<0.001$, $n=999$, randomization test) while fishes receive significantly more protection than expected by chance ($P<0.001$, $n=999$, randomization test) (Methods). The greatest amount of evolutionary history is supported by the longest branches on the Tree of Life. In our case, the top 10% longest extant and internal branches, corresponding to >8.68 Myr ($\pm$0.5 s.d.) for corals and >10.7 Myr ($\pm$0.25 s.d.) for fishes, support a disproportional amount of evolutionary history, with 62% ($\pm$0.9% s.d.) and 34% ($\pm$0.5% s.d.) for corals and fishes, respectively. These longest branches are overwhelmingly under-represented within the global MPA system (Fig. 2). Only 1.3% ($\pm$0.6% s.d.) of the longest branches in corals and 20.2% ($\pm$2.3% s.d.) in fishes are adequately protected by the minimum threshold of 10% geographic coverage by MPAs. If those poorly protected longest branches support endangered species we may expect large and abrupt changes in ecosystem functioning following extinctions. This situation already exists for the world's primates, where the most endangered species are both evolutionarily and ecologically distinct[35]. In the sea global

extinctions remain scarce, partly due to limited assessment[36], but the functionally most distinctive fish species on coral reefs tend to be rare either in their geographic extent or their local abundance[37]. We may thus anticipate a disproportional local loss of functional diversity within coral reef communities if the longest evolutionary branches are under threat and inadequately protected[38]. For instance long-branched lineages include relatively specialized forms, such as the large invertivore *Lachnolaimus* and the world's largest excavating parrotfsh *Bolbometopon* which are severely overexploited, suggesting that the loss of long branches may result in the loss of unique and functionally important groups[39].

Global distribution of protection deficits. To highlight the critical gaps in protecting the Tree of Life on coral reefs, we mapped the locations where the longest evolutionary branches that receive <10% coverage are concentrated using a regular grid of $5° \times 5°$ cells (Methods). For corals, the longest evolutionary branches with low protection are predominantly in the Atlantic, Eastern Pacific and, to a lesser extent, the North Indian Ocean (Fig. 3b). These deficits of protection are only marginally correlated with the heterogeneous MPA coverage at the global scale ($r=0.045$, $n=304$ $5° \times 5°$ grid cells, $P>0.05$, Fig. 3a). Instead, the high proportion of longest branches, and their unique evolutionary history, in the Atlantic and Eastern Pacific primarily drives this pattern[40] (Fig. 4a,b). For fishes, the highest concentrations of poorly protected long branches are located in the Western Indian, Central Pacific and, to a lesser extent, the Eastern Atlantic (Fig. 3c). As in corals, these deficits of protection are not correlated with the heterogeneous distribution of MPA coverage ($r=0.025$, $n=287$ grid cells, $P>0.05$, Fig. 3a). Instead, the pattern is driven by the relatively high proportion of long evolutionary branches of fishes at the periphery of the Indo-Pacific[41] (Fig. 4c,d). The correlation between the proportion of poorly protected longest evolutionary branches for corals and fishes within assemblages is negative ($r=-0.15$; $n=287$ grid cells, $P=0.30$) suggesting that there is a global spatial mismatch, albeit weak, of conservation needs for these two taxa. The Atlantic and Eastern Pacific tend to concentrate many long and poorly protected branches for corals but substantially less for fishes (Fig. 5). This most likely reflects the biogeographic history of the

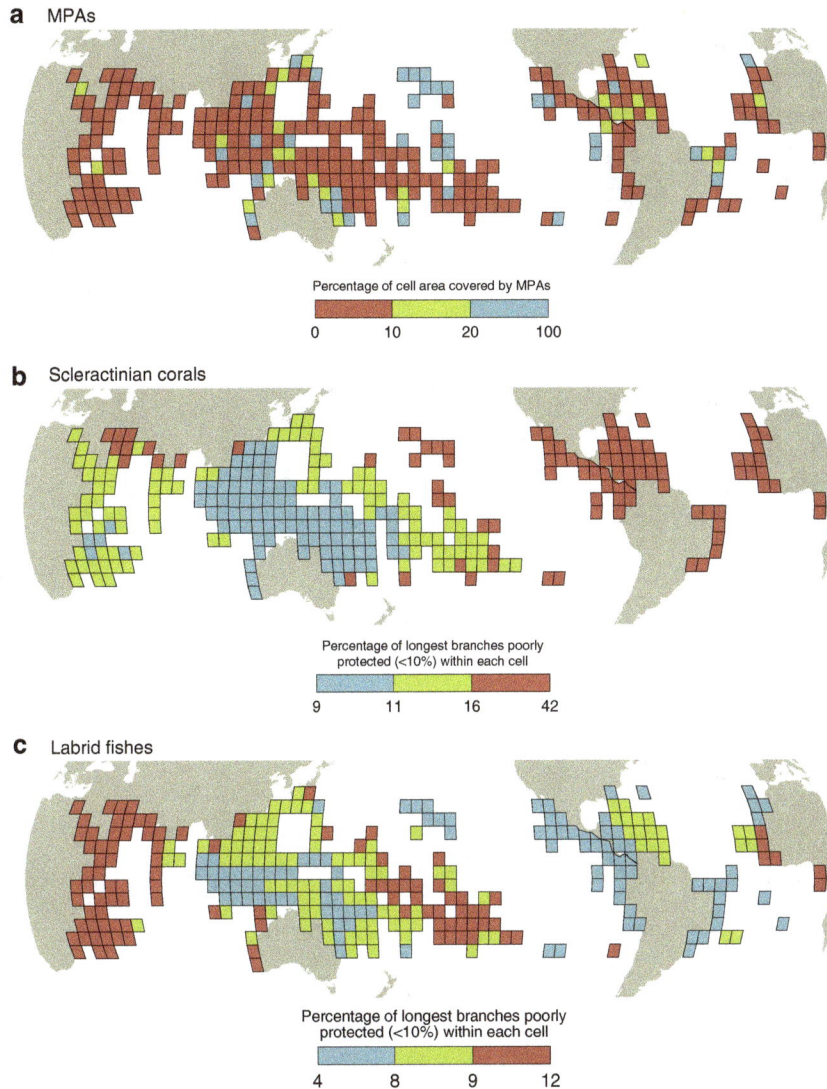

Figure 3 | Global distribution of protection deficits to secure the Tree of Life on coral reefs. Global maps representing, for each cell (5° × 5°), the percentage of coral reef habitat covered by MPAs (**a**), and the proportion of the longest evolutionary branches (top 10%) that receive less than the critical 10% coverage by the MPA system within coral (**b**) and fish (**c**) local assemblages. Colours correspond to three categories of values based on percentage of coverage for MPAs and on tertiles for corals and fishes.

tropical Atlantic which has been characterized by isolation, thus maintaining old coral lineages[40] in contrast to the recent diversification in younger fish lineages[42], especially along the Brazilian coast where there is extensive evidence of recent colonization[43]. In the Atlantic, therefore, there is a logical priority to emphasize the protection of older coral lineages. For fishes, the Atlantic hosts younger labrid lineages than the Indo-Pacific particularly in the Caribbean following cryptic speciation[42] and in the North Eastern Atlantic with subsequent diversification of Mediterranean lineages following the Messinian Salinity Crisis at 6 Myr (ref. 44). By contrast, the Coral Triangle, at the centre of the Indo-Pacific region, harboured most of the coral reef refugia during the Quaternary glaciations, hence acting as a 'museum' for the older labrid lineages[45].

Globally, the proportion of poorly protected longest branches in corals ranges from 9 to 42% compared with 4 to 12% in fishes (Fig. 3b,c), suggesting that conservation efforts should initially be focused on the Atlantic to better preserve the coral Tree of Life where it is most at risk. West African and, to a lesser extent, South American countries that border each side of the Atlantic, show

the slowest rate of MPA establishment worldwide although positive outliers in environmental governance also occur at both national and local levels[28]. For example, the Dominican Republic has already reached the target of 10% coverage. Similarly an increase in conservation investment has promoted MPA establishment in Eastern Africa[46]. Other countries of Western Africa and Eastern America remain far below the 10% coverage and should be priority areas to better protect the evolutionary history of corals. For fishes, conservation investment are primarily needed in the Western Indian Ocean where poorly protected longest branches are concentrated.

Limitations and less conservative protection assessment. Overall, our results show that the Tree of Life on coral reefs is inadequately represented by the current global MPA system, with most evolutionary branches, particularly the longest ones, receiving <10% protection. Despite the magnitude of this shortfall, our estimates are highly conservative because they are based on the assumption that all MPAs are able to protect every

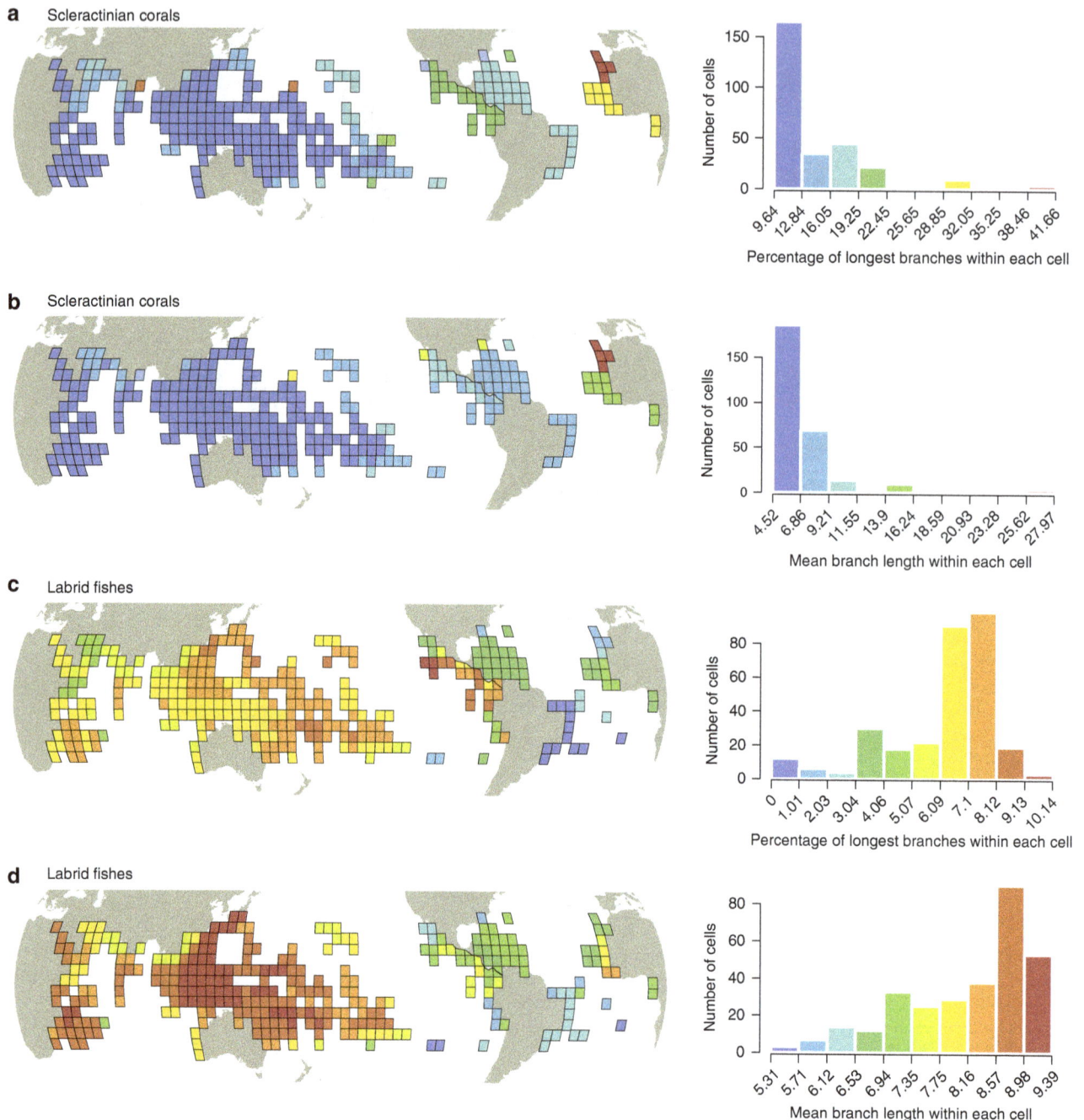

Figure 4 | Global distribution of the amount of evolutionary history on coral reefs. Global maps representing, for each grid cell (5° × 5°), the percentage of the longest evolutionary branches (top 10%) and the mean evolutionary branch length within coral (**a,b**) and fish (**c,d**) local assemblages, respectively. Colours correspond to classes of the histograms representing the distribution of values across the cells.

coral and fish species that geographically overlaps with them. It thus assumes that coral and fish species are present in all MPAs within their geographic ranges, and that all MPAs are effective in their protection. These assumptions may not be valid. First, we have no proof of individual species presence within MPAs. These commission errors are inevitable given the coarse grain of species geographic distributions and the small size of most MPAs. We therefore assess maximum potential protection while the conservation target of 10% is partly set to compensate for this limitation[34]. Second, although there is overwhelming evidence that MPAs can maintain or increase fish diversity, size and biomass[47,48], and strong evidence that the presence of intact fish communities can enhance coral persistence

and recovery[49–51], the extent of these benefits may vary among MPAs. Not all MPAs are able to ensure that fish and coral communities are protected, due to poor compliance and enforcement[52]. Furthermore, MPAs cannot prevent pulses of coral mortality from cyclones or coral bleaching[53], or from chronic declines in coral recruitment and growth due to degraded water quality[54,55]. MPAs in the Atlantic should better focus on coral lineages while those in the Western Indian Ocean should primarily limit fish overexploitation to protect the amount of evolutionary history on coral reefs. If we exclude MPAs that are not specifically designed to protect species and habitats and have a reduced capacity to protect fish diversity and biomass[48], that is, if only IUCN categories I to IV are considered (Methods), the

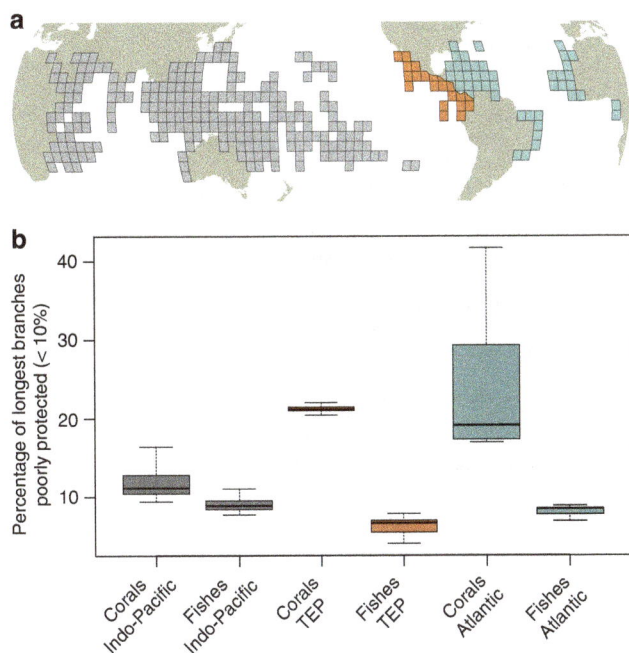

Figure 5 | Representation in MPAs for branches of the Tree of Life on coral reefs across marine realms. (**a**) Global map representing the three marine realms: Indo-Pacific (grey), Tropical Eastern Pacific (orange), and Atlantic (green). (**b**) Boxplots (median and quartiles) representing the percentage of the longest evolutionary branches (top 10%) that receive less than the critical 10% coverage by the MPA system within coral and fish local assemblages (in 5° × 5° grid cells) of the three marine realms.

proportion of the Tree of Life attaining the minimum target of 10% coverage by MPAs drops to 0.9% (± 0.2 s.d.) and 14.9% (± 2.0 s.d.) for corals and fishes, respectively.

Conclusions

Phylogenetic diversity is one of the key components of biodiversity[5,14]. However, the existing global system of MPAs does not meet the minimum levels considered necessary to adequately protect the Tree of Life for corals or fishes. If MPAs are to protect the Tree of Life, we need to carefully consider their features and future placement. Geographic variation in evolutionary history, and variable susceptibility to human impacts differs among fish and corals. The most notable example is in the Atlantic where there is a predominance of old coral lineages but a larger proportion of younger fish lineages. This mismatch brings to the fore the potential limitations of MPAs, and the differing needs of fishes, corals and other taxa. For corals, many of the major ongoing threats are not mitigated by MPAs. For effective protection we may need to look beyond traditional MPAs and develop new strategies that can encompass the full range of threats to reef biodiversity. A broader approach could include the protection of herbivorous fishes that promote local recovery of corals[50], management to control terrestrial influences and water quality[56] and effective action to mitigate climate change[57]. For future conservation efforts, we need to adequately secure greater amounts of evolutionary history on coral reefs in the Atlantic, Eastern Pacific and in the Western Indian Ocean.

Methods

Data. We restricted our database to shallow reef habitats (<50 m) showing a minimum monthly sea surface temperature (hereafter SST) of at least 17 °C to define tropical marine waters[58]. We built the geographic distribution of 452 tropical reef fish species from the Family Labridae by compiling 455 references

from 169 locations worldwide[58]. From these distributional data we obtained a range map for each species, defined as the convex polygon shaping the area where each species is present[58]. These were individually checked by expert to avoid the combination of disjointed ranges, for example, anti-tropical species.

We focused on labrid fishes since they (i) represent an exceptionally rich and diverse reef associated family, (ii) live in shallow waters, (iii) benefit from MPAs[59] as a common fisheries target[24] and (iv) have a well resolved phylogeny[41]. To incorporate unsampled taxa, new tips were grafted onto a backbone phylogeny based on other published phylogenies for the group[60,61], supplemented by species accounts from fish identification guides and FishBase (www.fishbase.org). Where information allowed, new tips representing unsampled species were added to direct sister species or to the base of the clade representing its genus. The full list of labrid fishes is provided as Supplementary Data 1.

We selected 805 coral species for which global range maps were downloaded at http://www.iucnredlist.org/technical-documents/spatial-data#corals. We considered only hard corals in shallow habitats. We used the supertree method to reconstruct the phylogeny of the scleractinian clade, comprising a total of 842 reef and 705 non-reef species[40]. The source trees were derived from a molecular phylogeny of 474 species (based on seven mitochondrial DNA markers), 13 morphological trees and 1 taxonomic tree. These were combined via the SuperFine-boosted Matrix Representation with Parsimony[62] and Matrix Representation with Likelihood[63]. The full list of coral species is provided as Supplementary Data 2.

We collected spatial information on MPAs from the WDPA (World Database on Protected Areas) database available at: http://protectedplanet.net/. The original database included 9,600 PAs covering a total surface of 17,633,881 km². We eliminated PAs on land, those that did not involve coastal habitat, defined as the portion of sea bottom from 0 to 200 m depth, and MPAs designated to protect species not considered in the present study (for example, birds). The latter were discarded after evaluating the description of the 'Designation' field in the original IUCN-WDPA database. MPAs for which IUCN criteria were either 'not applicable' or 'unknown' (for example, not communicated by the Authority), and are likely to be unreliable, were also removed. The final database included 3,625 MPAs covering a total surface of 942,568 km² (IUCN categories I–VI). We also used another restricted data set where we eliminated MPAs that are not specifically designed to protect species or habitats. We retained the 2,224 MPAs belonging to IUCN categories I to IV covering a total surface of 575,806 km² with a relatively higher degree of protection.

We then used a 5° × 5° grid cell corresponding to ~550 × 550 km at the equator to collate the presence of species, the area of tropical reef habitat, and the area of reef habitat protected within MPAs[64].

Analyses. Fossil records show that species extinction risk is primarily determined by geographic range size in the marine realm[65,66] with restricted ranged species being less buffered against demographic variability under changing environments. However, having at least 'one foot' in the MPA system does not ensure persistence[67]. We thus examined the proportion of the geographic range of species overlapping with the global MPA system. This represents a potential overlap since the presence of species within MPAs overlapping with their geographic range was not measured directly. We adopted a threshold of 10% spatial coverage by MPAs corresponding to a minimum (and conservative) target for effective protection[3,34]. This minimum threshold is based on the rational that some MPAs may be unsuitable for a given species, that protection is not effective in all MPAs and that the coarse grain of species distribution maps may induce commission errors by which species can be absent from protected areas that overlap their geographical ranges[34].

We applied the same reasoning to the internal branches of phylogenetic trees. The coverage by MPAs of the evolutionary history of a branch is therefore defined as the relative coverage by MPAs of the combined geographic ranges of the species subtending this branch. To evaluate the effectiveness with which MPAs protect the overall Tree of Life we measured the amount of evolutionary history represented by branches that pass the coverage threshold of 10%.

By grafting species we create polytomies on the phylogenetic trees that may bias the results since many species have artificially identical branch lengths. This may ultimately inflate the amount of evolutionary history supported by the tips and the level of phylogenetic conservatism[68]. To limit this bias and estimate the uncertainty of our results linked to the unresolved recent diversification events, polytomies were randomly resolved by a birth–death model[69] using BEAST[70]. Using 100 resolved trees for both corals and fishes, we provided the mean value and s.d. (±) for each result.

We also tested whether the current global system of MPAs is effective given the topology of the phylogenetic tree and thus the evolutionary constraints that have shaped species geographic ranges across history. To do so we performed a null model analysis where species labels were shuffled across the tips of the two phylogenies. By so doing the null model breaks the relationships between species ranges and their position on the phylogenetic tree while maintaining the amount of species coverage by MPAs. This procedure was applied 999 times for each of the 100 resolved trees to obtain a null frequency distribution for the overall amount of evolutionary history covered by MPAs. From this distribution, we extracted a P value for each resolved tree by assessing the positions of the observed in the null frequency distribution. These 100 P values were combined using the Fisher's combined probability test to provide a global P value quantifying whether the current global system of MPAs is more or less effective for the observed distribution of species geographic ranges across the phylogenies when compared to a random distribution.

To highlight the critical geographical gaps in protecting the Tree of Life on coral reefs, we mapped, at the grid cell level, the proportion of the longest evolutionary branches that receive <10% coverage. The longest branches are the top 10% for each Tree of Life. We also mapped the proportion of the longest evolutionary branches that are poorly protected, and the mean length of evolutionary history in each grid cell, that is, at the species assemblage level, and in each realm for both corals and fishes.

References

1. Cardinale, B. J. *et al.* Biodiversity loss and its impact on humanity. *Nature* **486**, 59–67 (2012).
2. Watson, J. E. M., Dudley, N., Segan, D. B. & Hockings, M. The performance and potential of protected areas. *Nature* **515**, 67–73 (2014).
3. Venter, O. *et al.* Targeting global protected area expansion for imperiled biodiversity. *PLoS Biol.* **12**, e1001891 (2014).
4. Selig, E. R. *et al.* Global priorities for marine biodiversity conservation. *PLoS ONE* **9**, e82898 (2014).
5. Winter, M., Devictor, V. & Schweiger, O. Phylogenetic diversity and nature conservation: where are we? *Trends Ecol. Evol.* **28**, 199–204 (2013).
6. Jetz, W. *et al.* Global distribution and conservation of evolutionary distinctness in birds. *Curr. Biol.* **24**, 919–930 (2014).
7. Faith, D. P. Conservation evaluation and phylogenetic diversity. *Biol. Conserv.* **61**, 1–10 (1992).
8. Veron, S., Davies, T. J., Cadotte, M. W., Clergeau, P. & Pavoine, S. Predicting loss of evolutionary history: where are we? *Biol. Rev. Camb. Philos. Soc.* in press (2015).
9. Srivastava, D. S., Cadotte, M. W., MacDonald, A. A. M., Marushia, R. G. & Mirotchnick, N. Phylogenetic diversity and the functioning of ecosystems. *Ecol. Lett.* **15**, 637–648 (2012).
10. Wiens, J. J. *et al.* Niche conservatism as an emerging principle in ecology and conservation biology. *Ecol. Lett.* **13**, 1310–1324 (2010).
11. Gomez, J. M., Verdu, M. & Perfectti, F. Ecological interactions are evolutionarily conserved across the entire tree of life. *Nature* **465**, 918–921 (2010).
12. Prinzing, A. *et al.* Less lineages—more trait variation: phylogenetically clustered plant communities are functionally more diverse. *Ecol. Lett.* **11**, 809–819 (2008).
13. Kelly, S., Grenyer, R. & Scotland, R. W. Phylogenetic trees do not reliably predict feature diversity. *Diversity Distrib.* **20**, 600–612 (2014).
14. Cadotte, M. W. Experimental evidence that evolutionarily diverse assemblages result in higher productivity. *Proc. Natl. Acad. Sci. USA* **110**, 8996–9000 (2013).
15. Thompson, P. L., Davies, T. J. & Gonzalez, A. Ecosystem functions across trophic levels are linked to functional and phylogenetic diversity. *PLoS ONE* **10**, e0117595 (2015).
16. Venail, P. *et al.* Species richness, but not phylogenetic diversity, influences community biomass production and temporal stability in a re-examination of 16 grassland biodiversity studies. *Funct. Ecol.* **29**, 615–626 (2015).
17. Thuiller, W. *et al.* Conserving the functional and phylogenetic trees of life of European tetrapods. *Philos. Trans. R. Soc. B Biol. Sci.* **370**, 20140005 (2015).
18. Guilhaumon, F. *et al.* Representing taxonomic, phylogenetic and functional diversity: new challenges for Mediterranean marine-protected areas. *Diversity Distrib.* **21**, 175–187 (2015).
19. Fisher, R. *et al.* Species richness on coral reefs and the pursuit of convergent global estimates. *Curr. Biol.* **25**, 500–505 (2015).
20. Hughes, S. *et al.* A framework to assess national level vulnerability from the perspective of food security: the case of coral reef fisheries. *Environ. Sci. Policy* **23**, 95–108 (2012).
21. Teh, L. S. L., Teh, L. C. L. & Sumaila, U. R. A global estimate of the number of coral reef fishers. *PLoS ONE* **8**, e65397 (2013).
22. Ferrario, F. *et al.* The effectiveness of coral reefs for coastal hazard risk reduction and adaptation. *Nat. Commun.* **5**, 3794 (2014).
23. McClanahan, T. R. *et al.* Critical thresholds and tangible targets for ecosystem-based management of coral reef fisheries. *Proc. Natl Acad. Sci. USA* **108**, 17230–17233 (2011).
24. Bellwood, D. R., Hoey, A. S. & Hughes, T. P. Human activity selectively impacts the ecosystem roles of parrotfishes on coral reefs. *Proc. R. Soc. B Biol. Sci.* **279**, 1621–1629 (2012).
25. Hughes, T. P., Graham, N. A. J., Jackson, J. B. C., Mumby, P. J. & Steneck, R. S. Rising to the challenge of sustaining coral reef resilience. *Trends Ecol. Evol.* **25**, 633–642 (2010).
26. Mora, C. *et al.* Coral reefs and the global network of marine protected areas. *Science* **312**, 1750–1751 (2006).
27. Marinesque, S., Kaplan, D. M. & Rodwell, L. D. Global implementation of marine protected areas: is the developing world being left behind? *Mar. Policy* **36**, 727–737 (2012).
28. Fox, H. E. *et al.* Explaining global patterns and trends in marine protected area (MPA) development. *Mar. Policy* **36**, 1131–1138 (2012).
29. Joppa, L. N. & Pfaff, A. Global protected area impacts. *Proc. R. Soc. B Biol. Sci.* **278**, 1633–1638 (2011).
30. Cinner, J. Coral reef livelihoods. *Curr. Opin. Environ. Sustain.* **7**, 65–71 (2014).
31. Tittensor, D. P. *et al.* Global patterns and predictors of marine biodiversity across taxa. *Nature* **466**, 1098–U1107 (2010).
32. Rogers, A., Blanchard, J. L. & Mumby, P. J. Vulnerability of coral reef fisheries to a loss of structural complexity. *Curr. Biol.* **24**, 1000–1005 (2014).
33. Bellwood, D. R., Hughes, T. P., Folke, C. & Nystrom, M. Confronting the coral reef crisis. *Nature* **429**, 827–833 (2004).
34. Rodrigues, A. S. L. *et al.* Global gap analysis: Priority regions for expanding the global protected-area network. *Bioscience* **54**, 1092–1100 (2004).
35. Redding, D. W., DeWolff, C. V. & Mooers, A. O. Evolutionary distinctiveness, threat status, and ecological oddity in primates. *Conserv. Biol.* **24**, 1052–1058 (2010).
36. Webb, T. J. & Mindel, B. L. Global patterns of extinction risk in marine and non-marine systems. *Curr. Biol.* **25**, 506–511 (2015).
37. Mouillot, D. *et al.* Rare species support vulnerable functions in high-diversity ecosystems. *PLoS Biol.* **11**, e1001569 (2013).
38. Mora, C. *et al.* Global human footprint on the linkage between biodiversity and ecosystem functioning in reef fishes. *PLoS Biol.* **9**, e1000606 (2011).
39. D'Agata, S. *et al.* Human-mediated loss of phylogenetic and functional diversity in coral reef fishes. *Curr. Biol.* **24**, 555–560 (2014).
40. Huang, D. W. & Roy, K. The future of evolutionary diversity in reef corals. *Philos. Trans. R. Soc. B Biol. Sci.* **370**, 20140010 (2015).
41. Cowman, P. F. & Bellwood, D. R. Coral reefs as drivers of cladogenesis: expanding coral reefs, cryptic extinction events, and the development of biodiversity hotspots. *J. Evol. Biol.* **24**, 2543–2562 (2011).
42. Floeter, S. R. *et al.* Atlantic reef fish biogeography and evolution. *J. Biogeogr.* **35**, 22–47 (2008).
43. Rocha, L. A. Patterns of distribution and processes of speciation in Brazilian reef fishes. *J. Biogeogr.* **30**, 1161–1171 (2003).
44. Meynard, C. N., Mouillot, D., Mouquet, N. & Douzery, E. J. P. A phylogenetic perspective on the evolution of mediterranean teleost fishes. *PLoS ONE 7*, e36443 (2012).
45. Pellissier, L. *et al.* Quaternary coral reef refugia preserved fish diversity. *Science* **344**, 1016–1019 (2014).
46. Wells, S., Burgess, N. & Ngusaru, A. Towards the 2012 marine protected area targets in Eastern Africa. *Ocean Coast. Manage.* **50**, 67–83 (2007).
47. Lester, S. E. *et al.* Biological effects within no-take marine reserves: a global synthesis. *Mar. Ecol. Prog. Ser.* **384**, 33–46 (2009).
48. Edgar, G. J. *et al.* Global conservation outcomes depend on marine protected areas with five key features. *Nature* **506**, 216–220 (2014).
49. Mumby, P. J. *et al.* Fishing, trophic cascades, and the process of grazing on coral reefs. *Science* **311**, 98–101 (2006).
50. Hughes, T. P. *et al.* Phase shifts, herbivory, and the resilience of coral reefs to climate change. *Curr. Biol.* **17**, 360–365 (2007).
51. Stockwell, B., Jadloc, C. R. L., Abesamis, R. A., Alcala, A. C. & Russ, G. R. Trophic and benthic responses to no-take marine reserve protection in the Philippines. *Mar. Ecol. Prog. Ser.* **389**, 1–15 (2009).
52. McCook, L. J. *et al.* Adaptive management of the Great Barrier Reef: a globally significant demonstration of the benefits of networks of marine reserves. *Proc. Natl Acad. Sci. U. S. A.* **107**, 18278–18285 (2010).
53. Graham, N. A. J., Jennings, S., MacNeil, M. A., Mouillot, D. & Wilson, S. K. Predicting climate-driven regime shifts versus rebound potential in coral reefs. *Nature* **518**, 94–97 (2015).
54. Hughes, T. P. *et al.* Shifting base-lines, declining coral cover, and the erosion of reef resilience: comment on Sweatman *et al.* (2011). *Coral Reefs* **30**, 653–660 (2011).
55. McClanahan, T. R., Ateweberhan, M., Darling, E. S., Graham, N. A. J. & Muthiga, N. A. Biogeography and change among regional coral communities across the western indian ocean. *PLoS ONE* **9**, e93385 (2014).
56. Weber, M. *et al.* Mechanisms of damage to corals exposed to sedimentation. *Proc. Natl Acad. Sci. USA* **109**, E1558–E1567 (2012).
57. Rogers, A. *et al.* Anticipative management for coral reef ecosystem services in the 21st century. *Glob. Chang. Biol.* **21**, 504–514 (2015).
58. Parravicini, V. *et al.* Global patterns and predictors of tropical reef fish species richness. *Ecography* **36**, 1254–1262 (2013).
59. McClanahan, T. R. & Arthur, R. The effect of marine reserves and habitat on populations of east African coral reef fishes. *Ecol. Appl.* **11**, 559–569 (2001).
60. Choat, J. H., Klanten, O. S., Van Herwerden, L., Robertson, D. R. & Clements, K. D. Patterns and processes in the evolutionary history of parrotfishes (Family Labridae). *Biol. J. Linn. Soc.* **107**, 529–557 (2012).
61. Alfaro, M. E., Brock, C. D., Banbury, B. L. & Wainwright, P. C. Does evolutionary innovation in pharyngeal jaws lead to rapid lineage diversification in labrid fishes? *BMC. Evol. Biol.* **9**, 255 (2009).

62. Swenson, M. S., Suri, R., Linder, C. R. & Warnow, T. SuperFine: fast and accurate supertree estimation. *Syst. Biol.* **61**, 214–227 (2012).

63. Nguyen, N., Mirarab, S. & Warnow, T. MRL and SuperFine plus MRL: new supertree methods. *Algorithms Mol. Biol.* **7**, 3 (2012).

64. Parravicini, V. *et al.* Global mismatch between species richness and vulnerability of reef fish assemblages. *Ecol. Lett.* **17**, 1101–1110 (2014).

65. Harnik, P. G., Simpson, C. & Payne, J. L. Long-term differences in extinction risk among the seven forms of rarity. *Proc. R. Soc. B Biol. Sci.* **279**, 4969–4976 (2012).

66. Payne, J. L. & Finnegan, S. The effect of geographic range on extinction risk during background and mass extinction. *Proc. Natl Acad. Sci. USA* **104**, 10506–10511 (2007).

67. Boyd, C. *et al.* Spatial scale and the conservation of threatened species. *Conserv. Lett.* **1**, 37–43 (2008).

68. Davies, T. J., Kraft, N. J. B., Salamin, N. & Wolkovich, E. M. Incompletely resolved phylogenetic trees inflate estimates of phylogenetic conservatism. *Ecology* **93**, 242–247 (2012).

69. Kuhn, T. S., Mooers, A. O. & Thomas, G. H. A simple polytomy resolver for dated phylogenies. *Methods Ecol. Evol.* **2**, 427–436 (2011).

70. Drummond, A. J., Suchard, M. A., Xie, D. & Rambaut, A. Bayesian Phylogenetics with BEAUti and the BEAST 1.7. *Mol. Biol. Evol.* **29**, 1969–1973 (2012).

Acknowledgements

The FRB CESAB-GASPAR project is thanked for providing fish geographical data. D.H. is supported by NUS Start-up Grant R-154-000-671-133.

Author contributions

D.M., D.R.B., W.T. and F.G. conceived the project; all authors designed the study; V.P., F.L., D.H., P.F.C, C.A. and F.G. collected the data and performed the analyses; D.R.B, F.L., T.P.H., C.A. and F.G. drew the figures, D.M. and F.G. wrote the first draft and all authors contributed substantially to revisions.

Additional information

Ocean currents generate large footprints in marine palaeoclimate proxies

Erik van Sebille[1,†], Paolo Scussolini[2,†], Jonathan V. Durgadoo[3], Frank J. C. Peeters[2], Arne Biastoch[3], Wilbert Weijer[4], Chris Turney[1], Claire B. Paris[5] & Rainer Zahn[6,7]

Fossils of marine microorganisms such as planktic foraminifera are among the cornerstones of palaeoclimatological studies. It is often assumed that the proxies derived from their shells represent ocean conditions above the location where they were deposited. Planktic foraminifera, however, are carried by ocean currents and, depending on the life traits of the species, potentially incorporate distant ocean conditions. Here we use high-resolution ocean models to assess the footprint of planktic foraminifera and validate our method with proxy analyses from two locations. Results show that foraminifera, and thus recorded palaeoclimatic conditions, may originate from areas up to several thousands of kilometres away, reflecting an ocean state significantly different from the core site. In the eastern equatorial regions and the western boundary current extensions, the offset may reach 1.5 °C for species living for a month and 3.0 °C for longer-living species. Oceanic transport hence appears to be a crucial aspect in the interpretation of proxy signals.

[1] ARC Centre of Excellence for Climate System Science and Climate Change Research Centre, School of Biological, Earth and Environmental Sciences, University of New South Wales, Sydney, New South Wales 2010, Australia. [2] Earth and Climate Cluster, Faculty of Earth and Life Sciences, VU University, 1081 HV Amsterdam, The Netherlands. [3] GEOMAR Helmholtz Centre for Ocean Research Kiel 24148 Kiel, Germany. [4] Los Alamos National Laboratory, Los Alamos, New Mexico 87545. [5] Rosenstiel School for Marine and Atmospheric Science, University of Miami, Miami, FL 33149 Florida, USA. [6] Institució Catalana de Recerca i Estudis Avançats (ICREA) 08010 Barcelona, Spain [7] Departament de Física, Institut de Ciència i Tecnologia Ambientals (ICTA), Universitat Autònoma de Barcelona, 08193 Bellaterra (Cerdanyola), Spain. † Present addresses: Grantham Institute & Department of Physics, Imperial College London, London SW7 2AZ, UK (E.v.S.); Institute for Environmental Studies (IVM), VU University Amsterdam, The Netherlands (P.S.). Correspondence and requests for materials should be addressed to E.v.S. (email: e.van-sebille@imperial.ac.uk).

Marine sediment archives have been paramount in forming our understanding of centennial- to orbital-scale climate and environmental change[1-5]. Much of the palaeoclimatic information has been obtained from the geochemistry of fossil planktic foraminifer shells and from their species assemblage composition. It has been known for a long time that the drift of planktic foraminifera may mean they record water conditions different from conditions at the core site[6]. The influence of the provenance of foraminifera on proxy signals during their life cycle, however, has not been assessed and quantified in a rigorous manner, using high-resolution ocean models.

Besides the fact that planktic foraminifera employ a mechanism to control their depth habitat[7], they can be considered as passive particles, sensitive to advective processes by ocean currents. As they grow their calcite shell during their lifespan, foraminifera may drift across different climate zones and ocean regimes. At the end of their life cycle—during the phase of gametogenesis—foraminifera lose their ability to stay buoyant in the upper ocean and their shells sink to the ocean floor to become part of the sedimentary geological archive[1-5,8]. Although the horizontal advection distance for post-mortem sinking foraminifera has been estimated at a few hundred kilometres[6,9-12], there is a remarkable dearth of information on the geographical footprint of foraminifera during their lifespan.

Here we quantify the lateral distance that planktic foraminifera can cover during their lifespan and quantify the impact of the ambient temperature along their trajectory on the signal incorporated into their shells. We show that this impact is potentially highly significant in regions of fast-flowing surface currents such as western boundary currents. To illustrate the impact of the trajectory integrated temperature signal during life and transport on the proxy, we focus on the Agulhas region, where planktic foraminifera have been extensively used to study variations and global influence of the amount of warm tropical Indian Ocean water flowing into the Atlantic Ocean[5,13,14].

Results

Foraminiferal traits and their relation to drift. We use two ocean models of contemporary circulation, which both include mesoscale eddies, to study the advection during the life span and the post-mortem sinking of foraminifera. Both models have a $1/10°$ horizontal resolution, but their domains differ: the INALT01 model[15] is focused around southern Africa in the Agulhas system and is among the best-performing models in that region[13,15-17], while the Ocean model For the Earth Simulator (OFES) model[18] is global in extent, allowing us to place these results in a wider context. In both models, we advect the virtual foraminifera as passive Lagrangian particles using the Connectivity Modeling System (CMS)[19], simulating both their trajectories during their lifetime, as well as their post-mortem sinking. The local *in situ* temperature from the hydrodynamic models is interpolated onto the particle trajectories and used to reconstruct the incorporation of the temperature signal during the virtual foraminifera's lifetime. We compare the model results to combined single-shell $\delta^{18}O$ and multiple-shell Mg/Ca temperature reconstructions from *Globigerinoides ruber* from core tops at two locations in the Agulhas region[5]: (1) site CD154-18-13P below the Agulhas Current and (2) site MD02-2594 below the Agulhas leakage area.

Foraminifer traits such as depth habitat, lifespan, seasonality, post-mortem sinking speed and rate of growth (which is related to rate of calcification) vary widely between species and are often poorly constrained[6,8,20,21]. Focusing on surface-dwelling foraminifera, we therefore undertook a sensitivity assessment of these different traits. Values for sinking speed employed in the models were 100, 200 and 500 m per day and depth habitats were 30 and 50 m. Lifespans were related to the synodic lunar cycle[8,20], with 15 days for *G. ruber* and 30 days for other surface-dwelling foraminifera. However, as some studies report even longer life spans for upper water column dwelling foraminifera[6,20], we also investigated extended lifespans of 45 days within the INALT01 model and 180 days within the global OFES model. Two growth rates were used to simulate different calcification scenarios. One was a linear growth scenario, where the recorded calcification temperature of a virtual foraminifer is the mean temperature along its trajectory during its lifespan. The other was an exponential growth scenario, with a growth rate[7,22] of 0.1 per day, so that the later life stages of the foraminifera weigh more heavily in the final calcification temperature[8,23]. Finally, we studied the effect of a seasonal growth cycle on the recorded

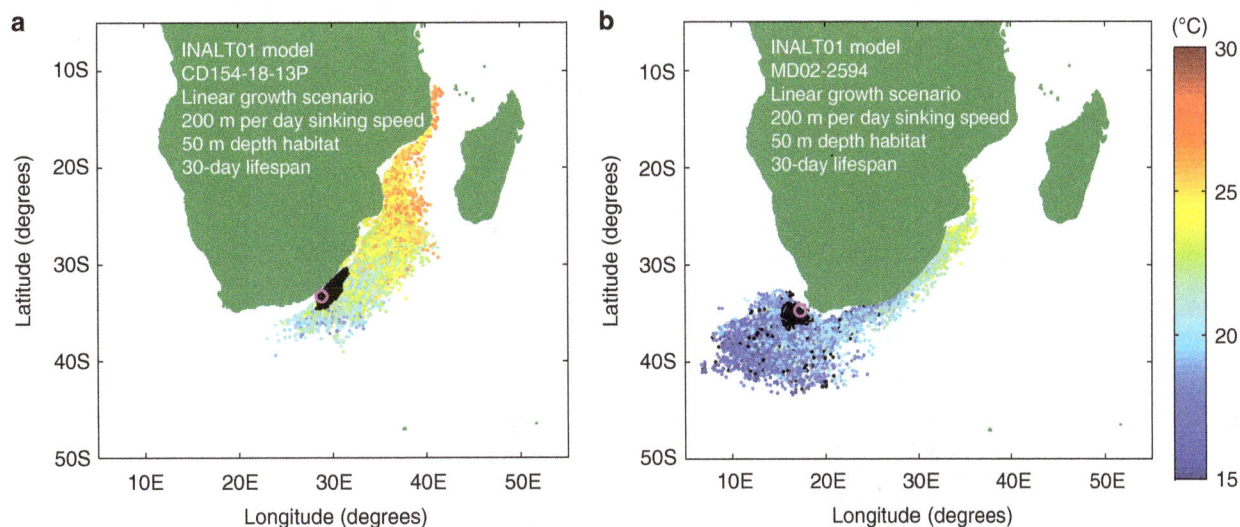

Figure 1 | Foraminifera footprints in the Agulhas region. Maps of the footprint for two core sites in (**a**) the Agulhas Current and (**b**) the Agulhas leakage region. Each coloured dot represents the location where a virtual foraminifer starts its 30-day life, colour-coded for the recorded temperature along its trajectory. Black dots represent where foraminifera die and start sinking to the bottom of the ocean (at 200 m per day) to end up at the core location (indicated by the purple circle).

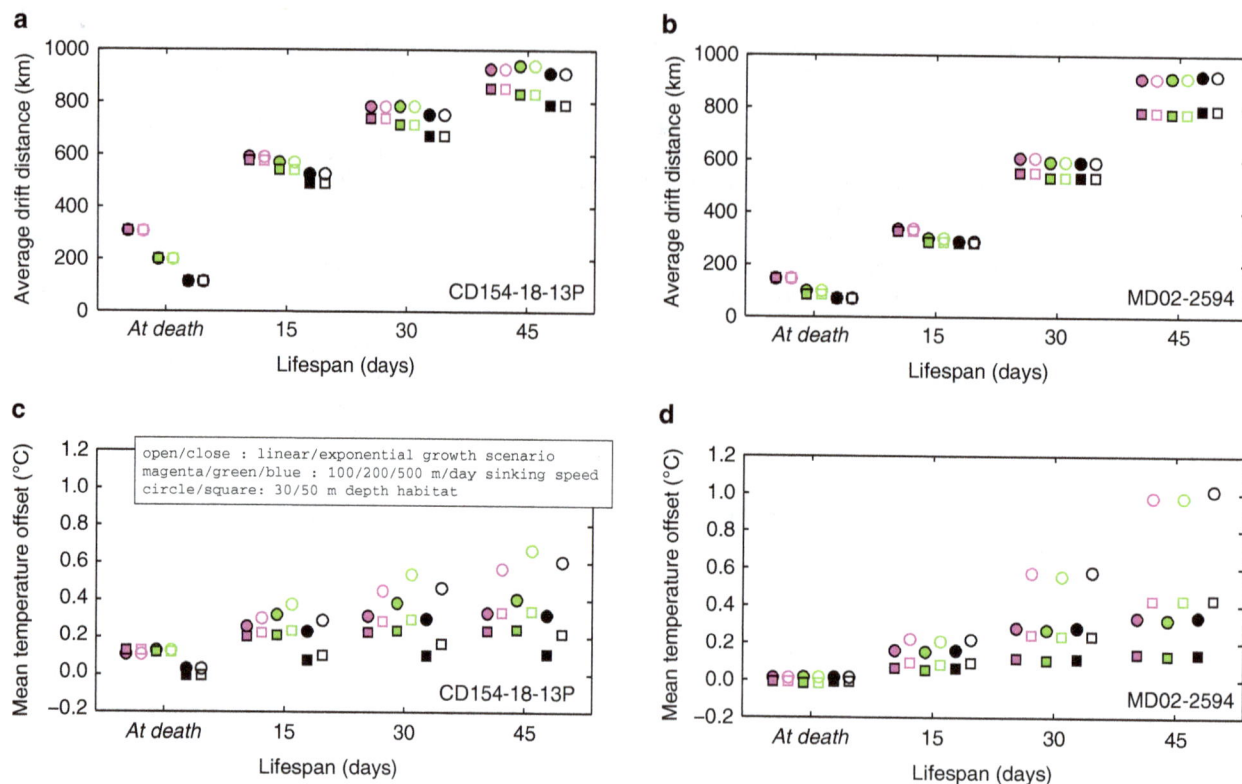

Figure 2 | The dependence of foraminifera footprint on life traits. The sensitivity of the chosen foraminiferal traits on (**a,b**) the average distance between spawning and core location, and on (**c,d**) the offset between the mean recorded temperature and the local temperature at the two core sites depicted in Fig. 1. Lifespan is on the x axis, with 'at death' the assumption that foraminifera record the temperature of the location in the last day before they die and start sinking. The results depend noticeably on the traits, except for the sinking speed (colours), which seems to have little effect on mean recorded temperature.

temperatures. See Methods section for further methodological information.

Foraminifera drift in the Agulhas region. A substantial fraction of the particles incorporated in the cores from both the Agulhas Current and the Agulhas leakage region appears to originate from hundreds of kilometres away (Fig. 1). Using a depth habitat of 50 m, a lifespan of 30 days and a sinking speed of 200 m per day, the average drift distances, which are defined as the average shortest distance from spawning location to the core site for all virtual foraminifera, are 713 and 533 km in the Agulhas Current and Agulhas leakage, respectively. These distances are more than four times larger than the drift distances during their post-mortem sinking (which are 166 and 71 km for the Agulhas Current and Agulhas leakage, respectively, Fig. 2a,b), highlighting the impact of drift during the virtual foraminifer's life.

This surface drift has implications for the recorded temperature. In the case of the Agulhas Current core (Fig. 1a), some of the virtual foraminifera start their life in the Mozambique Channel and the temperature recorded by these specimens along their 30-day life is up to 5 °C warmer than at the core site. Such an offset is much larger than the uncertainty of 1.5 °C (2σ) that is associated with foraminifera proxy-based temperature reconstructions[9–11,24]. In the core at the Agulhas leakage region (Fig. 1b), some particles arrive from warmer subtropical temperature regimes of the northern Agulhas Current, whereas others—in our model—originate from the sub-Antarctic cold waters of the Southern Ocean in the south.

Both the average drift distances as well as the recorded temperatures are strongly dependent on the values chosen for the foraminifer traits (Fig. 2). The dependence is nonlinear and different for the two sites, although general patterns emerge: sinking speed is the least important trait; growth scenario becomes more important for longer-living foraminifera; depth habitat has far less effect on drift distance than on recorded temperature (Supplementary Figs 1–4). There are also differences between the INALT01 and the OFES models, particularly in the amount of virtual foraminifera originating far upstream in the Agulhas Current, which show the dependency of the results on the circulation state (Supplementary Fig. 5). However, there does not seem to be a seasonal variation in the temperature offsets (Fig. 3).

The distribution of the calcification temperatures of the virtual foraminifera can be compared with proxy temperature distributions derived from the *G. ruber* from the core tops (see Methods). The mean ± 1 s.d. of the INALT01, OFES and proxy distributions overlap (Fig. 4). The spread in temperatures is larger than the typical sensitivities to the choice of life trait values (which is < 1 °C, Fig. 2c,d). According to a two-sample Kolmogorov–Smirnov test, the *G. ruber* proxy data in the Agulhas Current core is most closely matched by the virtual foraminifera within OFES with a depth habitat of 30 m ($P = 0.47$, which means the OFES and proxy distributions are statistically indistinguishable). The *G. ruber* proxy data in the Agulhas leakage core is most closely matched by the virtual foraminifera within INALT01, with a depth habitat of 50 m ($P = 0.06$). All other virtual foraminifera distributions are statistically different from the *G. ruber* proxy data ($P < 0.05$), even though in all cases means and s.d. are within 1.5 °C of the *G. ruber* proxy data.

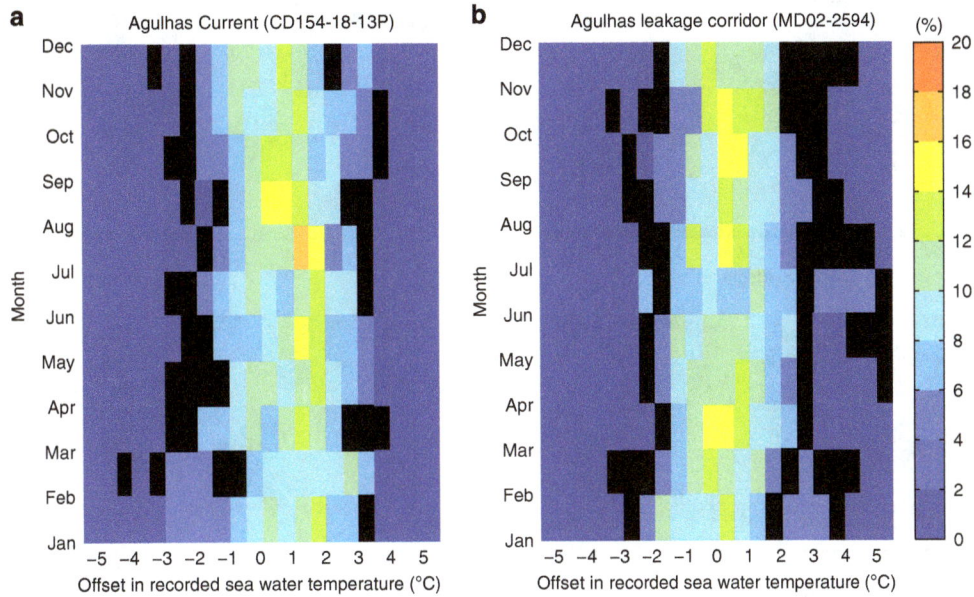

Figure 3 | The effect of seasonality on the temperature offsets. Seasonal cycle of the offset of recorded temperature for the virtual foraminifera with respect to the local *in-situ* temperature in (**a**) the Agulhas Current core and (**b**) the Agulhas leakage core. For each month, the difference between the recorded temperatures and the instantaneous temperatures at the core is plotted with a 0.5 °C interval, as a percentage of the total number of virtual foraminifera that reach the core in that month. The virtual foraminifera have a lifespan of 30 days, a depth habitat of 50 m, a linear growth scenario and a sinking speed of 200 m per day. There is no clear seasonal variation in offset of recorded temperatures with time of year.

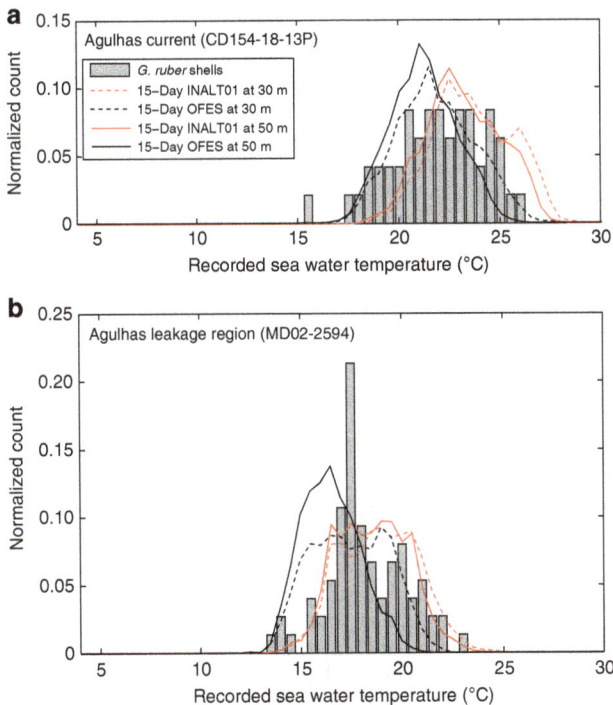

Figure 4 | Distributions of temperature at two cores in the Agulhas region. The observed proxy temperatures (grey bars) at (**a**) the Agulhas Current core and (**b**) the Agulhas leakage core are compared with the temperature distributions for the virtual foraminifera experiments in the INALT01 model (red) and the OFES model (blue). Traits used are the *G. ruber* lifespan[20] of 15 days, a depth habitat of 30 m (dashed) or 50 m (solid), a sinking speed of 200 m per day and a linear growth scenario. Note that the spread in recorded temperature is larger than the sensitivity of the means with foraminiferal trait choices (Fig. 2c,d).

A global estimate of foraminifera drift. A global analysis (Fig. 5), using the OFES model, of virtual foraminifera released on a 5° × 5° global grid reveals that the virtual foraminifera can drift for up to a thousand kilometres during an assumed 30-day lifespan (Fig. 5c). This is one order of magnitude larger than the lateral drift, which dead virtual foraminifera experience during the 200 m per day sinking (Fig. 5a). Drifts are largest in regions with largest horizontal velocities such as along the equator, in the western boundary currents and their extensions, and in the Southern Ocean, while drift distances are smaller in the centres of the gyres.

This horizontal drift can introduce large offsets when foraminiferal records are interpreted as representative of local conditions: for example, in the reconstruction of temperatures, the discrepancy with the local temperatures varies greatly with region (Fig. 5b,d,f). If it is assumed that the foraminifera document the local temperature at the location where they die and start sinking, the offsets are smaller than 0.1 °C almost everywhere (Fig. 5b). However, for lifespans of 30 days[6,20], offsets can be as large as 1.5 °C (Fig. 5d), which is equal to the uncertainty associated with proxy-based palaeotemperature estimates[9–11,24]. Virtual *G. ruber*, with lifespans of 15 days, have similar offsets (Supplementary Fig. 6). For virtual foraminifera with more extended lifespans of 180 days (Fig. 5e,f), average drift distances can reach 3,000 km and the associated offsets in average recorded temperature can be >3 °C. In the case of virtual foraminifera with depth habitats of 30 m, these temperature offsets are up to 4 °C (Supplementary Fig. 7), while they are up to 2 °C in the case of an exponential growth scenario (Supplementary Fig. 8).

Discussion

We have shown that ocean currents affect the signals incorporated in foraminiferal proxies. There appears to be a clear global pattern in the global temperature offsets, which are positive along the equator and within the western boundary currents, and

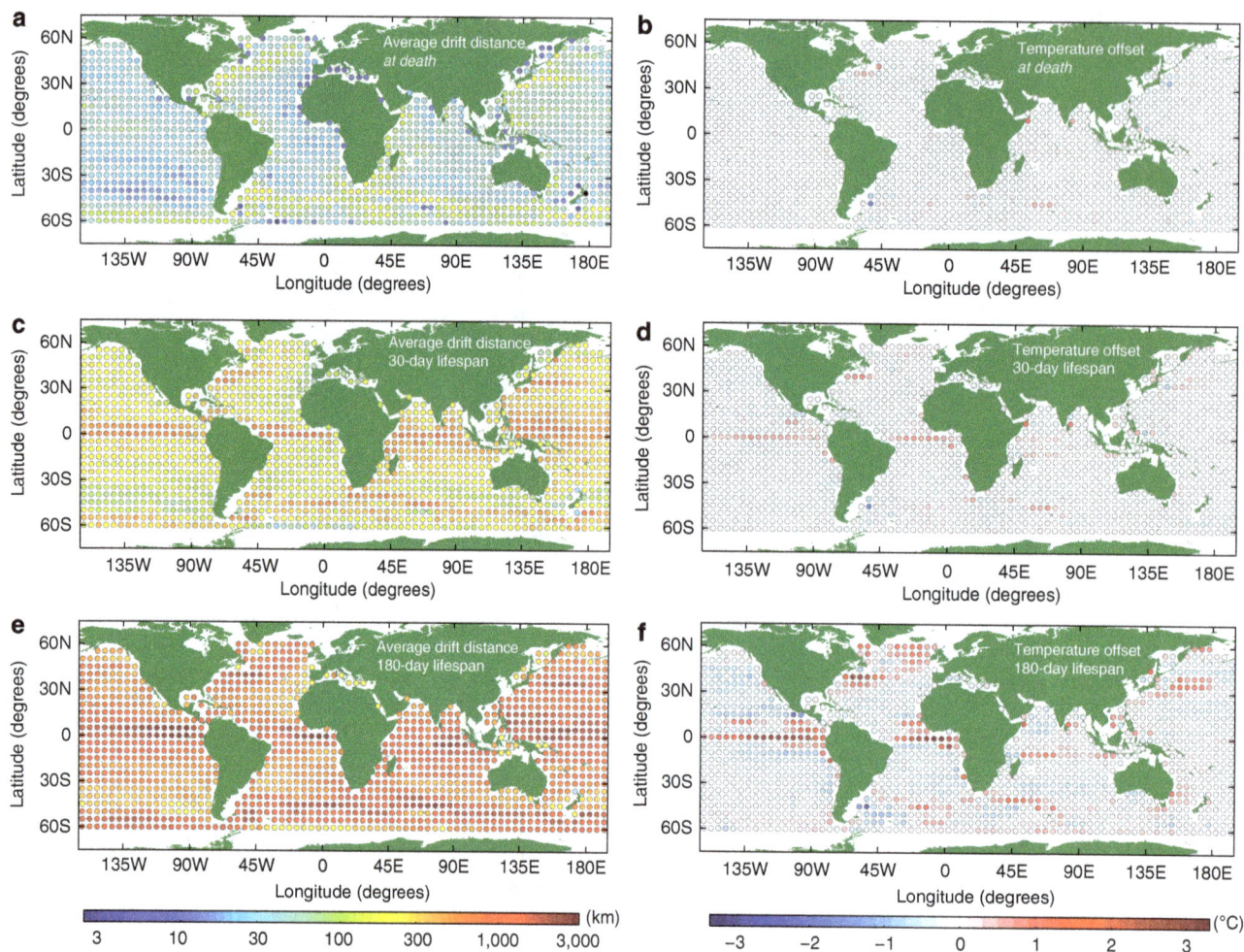

Figure 5 | Global analysis of drift distances and temperature offsets. (a,c,e) The average distance between spawning location and the core site for virtual foraminifera in the OFES model that record the temperature (**a**) in the last day before they die and start sinking and for virtual foraminifera with lifespans of (**c**) 30 days and (**e**) 180 days. In all cases, a depth habitat of 50 m, a linear growth scenario and a sinking speed of 200 m per day were used. Note that the colour scale is logarithmic. (**b,d,f**) Offsets, defined as the difference between along-trajectory recorded temperatures and local temperatures at the core site. Offsets reach up to 1.5 °C for 30-day lifespans and up to 3 °C for 180-day lifespans, when the virtual foraminifera travel more than 1,000 km.

negative in the centres of the subtropical gyres. The regions with largest temperature offsets are those closely related to regions of high ocean surface velocity and consequently lateral drift: in the eastern Tropical Pacific and Atlantic Ocean, and in the extensions of the western boundary currents such as the Gulf Stream, Kuroshio and Agulhas currents. However, there are also regions of high lateral drift where temperature offsets are much smaller such as the Southern Ocean and the Tropical Indian Ocean. The difference is that the regions of high offsets are also the regions of some of the largest lateral temperature gradients (often related to large ocean–atmosphere heat fluxes). Larger temperature changes experienced by the foraminifera along their pathway result in larger offsets with respect to the temperature above the core site. The implication is that, depending on species traits and locations, the temperature offsets can be significant if the shells in the core are interpreted as representative of the conditions right above the core location.

An analysis such as the one presented here could also be used *a priori* to identify the amount of advective bias at a potential drilling site. Another tantalizing application could be to 'invert' the problem and use our approach to determine where different fossil specimens most probably grew their shell, so that the temperatures recorded by the fossils could be geolocated to the location where the microorganism actually grew its shell, rather

than where it reached the ocean floor. This would allow disentanglement of proxy data from microorganisms with different traits and a better spatial interpretation of the signal around the location of the sediment core site. Coccolithophores, for example, are also paleoclimatological proxy carriers of primary importance, with life traits and settling dynamics that differ notably from planktic foraminifera[25]. With an approach similar to ours, coccolithophoric footprints could be calculated and compared with the foraminiferal ones, potentially vastly increasing the amount of information that can be obtained from a single sediment core. A vital prerequisite to this application, however, is a better understanding and quantification of the organism's ecology[20,26], including species-specific lifespans, depth habitats, calcification rates and sinking speeds.

Methods

Ocean model data. We used data from two ocean circulation models. The first is the INALT01 model configuration[15], which is based on the NEMO ocean model[27], extending an earlier setup[16]. The model was specifically set up to study the dynamics of the Agulhas region and includes a 1/10° high-resolution region with 46 vertical levels that spans the entire South Atlantic and western part of the Southern Indian (between 70°W–70°E and 50°S–8°N), which is nested in a 1/2° global model. We used 28 years (1980–2007) of the hindcast experiment, a period for which the dynamics of the model has been shown to agree well with observations[15,16]. The atmospheric forcing builds on the CORE reanalysis

Table 1 | Values of sinking speeds retrieved from laboratory studies.

Species	Size fraction	Speed (m per day)	Reference	Notes
General planktic foraminifera		30-480	39	
General planktic foraminifera	~400 μm	1,210	40	
General planktic foraminifera	177-250 μm	864	41	Combusted (empty) shells, in freshwater
General planktic foraminifera	>250 μm	1,987	41	Combusted (empty) shells, in freshwater
G. ruber	~300-400 μm	838 ± 441	42	From sea floor sediment; combusted; in freshwater
G. sacculifer	~400-600 μm	1,396 ± 652	42	From sea floor sediment; combusted; in freshwater
G. ruber	**314 ± 49 μm**	**198 ± 94**	21	**Non-ashed shells from plankton tows**
G. ruber	289 ± 82 μm	723 ± 321	21	Ashed shells from plankton tows
Orbulina universa	573 ± 74 μm	277 ± 144	21	Non-ashed shells from plankton tows
O. universa	521 ± 52 μm	701 ± 219	21	Ashed shells from plankton tows
G. sacculifer	328 ± 99 μm	274 ± 143	21	Non-ashed shells from plankton tows
G. sacculifer	340 ± 170 μm	1054 ± 531	21	Ashed shells from plankton tows
G. bulloides	299 ± 44 μm	328 ± 174	21	Non-ashed shells from plankton tows
G. bulloides	211 ± 28 μm	208 ± 46	21	Ashed shells from plankton tows
Neogloboquadrina pachyderma	200-300 μm	358 ± 67	11	Empty shells; from sediment traps
Turborotalita quinqueloba	~200 μm	180	11	Empty shells; from sediment traps
G. ruber	~550 μm	2,000 ± 270	26	Sea floor sediment, uncleaned
G. sacculifer	~690 μm	2,600 ± 310	26	Sea floor sediment, uncleaned
O. universa	~640 μm	2,760 ± 890	26	Sea floor sediment, uncleaned

Values are ordered chronologically by publishing date. The species most commonly used for palaeo-reconstructions are reported. Bold indicates the values we considered the most appropriate for our study. Average values for multiple one-shell experiments are reported, along with the s.d.. In addition, the size of the shells is specified.

products[28] and is applied via bulk air–sea flux formulae. We used the same 28 years of data from the Japanese OFES[18], which is also 1/10° horizontal resolution and has a near-global coverage between 75°S and 75°N, and 53 vertical levels. The model is forced using National Centers for Environmental Prediction (NCEP) wind and flux fields. Results from both models have been shown to be consistent with important observed features of the modern ocean circulation, including among others the trajectories of surface buoys[29] and the deep currents in the North Atlantic[30], the South Atlantic[31] and the Agulhas region[17,32].

Virtual foraminifera trajectory calculations. The virtual foraminifera were advected within the INALT01 and OFES velocity data using the CMS version 1.1b[19]. The CMS employs a fourth-order Runge–Kutta method and can output along-trajectory temperature and salinity.

For each core, we computed Lagrangian particle trajectories in reverse time. We started one particle every day at the core site itself, near the ocean floor, for a total of almost 10,000 particles per site (amounting to 27 simulated years). These particles were then integrated backwards in time by reversing the sign of the velocity components. A sinking velocity was added to the particles. Once near the surface, the particles were advected for another 45 days (180 days in the global case) at their prescribed depth habitat, using only the horizontal velocity fields and without any explicit diffusivity (see below). During this part of their trajectory, representing the lifespan of the foraminifera, the location as well as the *in situ* temperature of the particle was stored every day for further analysis. These along-trajectory temperatures were then used to offline calculate the recorded temperature based on growth scenario. The temperature distributions along the trajectory paths were then compared with *in situ* conditions at the core location.

For sites poleward of 40°N and 40°S in the global experiment, we used only those virtual foraminifera that lived for their full life in the warmer months (April to September for the Northern Hemisphere and October to March for the Southern Hemisphere). In all other cases, including those of the Agulhas region cores, we used virtual foraminifera throughout the year and have not observed a bias in the results that would be associated with seasonality (Fig. 3).

Sensitivity to the addition of diffusion in foraminiferal transport. The particles in this study have been computed using the three-dimensional model velocity fields, without any additional diffusion due to sub-grid scale processes. Here we show that the effect of diffusion is an order of magnitude smaller than that of advection with the currents (Supplementary Fig. 9).

In these simulations, we used the turbulent diffusion module of the CMS (equation 3 in ref. 19) with $K_h = 50\,\mathrm{m}^2\,\mathrm{s}^{-1}$ for the MD02-2594 core and with $K_h = 250\,\mathrm{m}^2\,\mathrm{s}^{-1}$ for the CD154-18-13P core. We chose the first of these values for diffusion ($K_h = 50\,\mathrm{m}^2\,\mathrm{s}^{-1}$) as the most appropriate for the INALT01 and OFES models, which both have a resolved scale of 10 km (Fig. 2 of ref. 33). The second of these values ($K_h = 250\,\mathrm{m}^2\,\mathrm{s}^{-1}$) was chosen to study the effect of an extremely high diffusivity.

The experiments revealed that for both cores, the effect of diffusion on the core footprints is minimal. In the case of core MD02-2594, the average shortest distance between spawning location and core site changed by only 10 km. In the case of core

CD154-18-13P, which had the much higher diffusivity, the average distance changed only by 18 km.

This finding is in agreement with previous results where it was shown (Fig. 1 of ref. 33) that diffusion on time scales of months is <50 km. It is also in agreement with the theoretical estimate of dispersion in the absence of advective flow. A Brownian motion process gives for the spread of particles std(X) = sqrt(2 K_h T), where std(X) is the s.d. of distance (that is, the spread due to diffusion) and T is the length of integration. For $T = 30$ days and $K_h = 50\,\mathrm{m}^2\,\mathrm{s}^{-1}$ this leads to std(X) = 16 km, whereas for the longer OFES runs with $T = 180$ days and $K_h = 50\,\mathrm{m}^2\,\mathrm{s}^{-1}$ this leads to std(X) = 40 km.

In summary, diffusivity in the 1/10° resolution OFES and INALT01 models is at least an order of magnitude smaller than the advective spread we find in our study, and hence diffusion will not affect our main conclusions.

Literature review of the sinking speed of planktic foraminifera. We consider a set of surface-dwelling planktic foraminifer species, widely used to reconstruct sea surface conditions such as temperatures. The depth habitat of these species can be confidently constrained to the mixed layer, therefore warranting the assumption that no significant vertical migration during living time occurs[8,20].

We reviewed the specialized literature for the most accurate quantification of the sinking speed of foraminifera shells (Table 1). The results of previous studies (see references in Table 1) confirm that the sinking speed of planktic foraminifera depends mainly on the shell weight (in turn related to the shell size, that is, diameter) and the presence of spines. From the same studies, it appears that the shell morphology, which is characteristic of each species, is also determinant for the sinking speed. Shell thickening is also important and it is related to the life stage of the specimen, which in turn is arguably proportional to the shell size.

Therefore, following ref. 21, we chose to use a sinking speed of 200 m per day for non-ashed G. ruber with a common size of ~300 μm. This was based on four considerations: first, G. ruber, Globigerinoides sacculifer and Globigerinoides bulloides are among the most used surface foraminifer species in palaeo-reconstructions; second, foraminifera in the size fraction between 200 and 350 μm are the most used; third, even though foraminifera might undergo partial post-mortem degradation of their plasma content, and although before sinking they normally release their gametes, which constitute a large part of their organic composition, the non-ashed, plankton-tow sample probably resembles the form in which a foraminifer sinks just after death; and finally, seawater (as opposed to freshwater) experiments more closely mime the dynamics of foraminifera sinking.

Empirical data from G. ruber shells. Shells of planktic foraminifer G. ruber, white variety, were picked from the top centimetre of cores MD02-2594 (Agulhas leakage region, 34° 42.6′ S, 17° 20.3′ E, 2,440 m depth) and CD154-18-13P (Agulhas Current, 33° 18.3′ S, 28° 50.8′ E, 3,090 m depth), from the size fraction 250–355 μm. Both core tops represent contemporary climate (see below). Stable isotope ($\delta^{18}O$) analyses were conducted on the single shells with a Thermo Finnigan Delta Plus mass spectrometer at the VU University Amsterdam, with the method outlined in ref. 13. We analysed 79 G. ruber shells from core MD02-2594 and 48 shells from core CD154-18-13P.

From core MD02-2594, we also analysed the Mg/Ca value of a group of 20 shells of *G. ruber*, using an inductively coupled plasma/optical emission spectrometry, after rigorous cleaning of the sample, following a standard procedure[34]. Analysis was performed at the Trace Elements Laboratory of Uni Research, Bergen. As for core CD154-18-13P the amount of shells did not allow carrying out Mg/Ca analysis, we used the Mg/Ca value of the core top of adjacent core CD154-17-17K (33° 16.1′ S, 29° 7.3′ E, 3,330 m depth)[14], which is located 26 km to the SE.

Temperature reconstructions from *G. ruber* geochemistry. The Mg/Ca values were converted to calcification temperatures using a species-specific calibration[24]. We used a previously established approach to assign calcification temperatures to individual foraminiferal shell[2], which consists of first anchoring the mean temperature of the foraminiferal population using the Mg/Ca-derived temperature of a group of shells; then calculating the offset of each shell $\delta^{18}O$ value from the mean of all measurements; and finally converting each $\delta^{18}O$ offset into a temperature offset by dividing it by a factor of -0.22, which approximates the dependency of equilibrium calcite $\delta^{18}O_{eq}$ on temperature[35]. This method necessarily assumes that only temperature determines the foraminiferal $\delta^{18}O$ ($\delta^{18}O_f$), thus ignoring a potential effect of changes in seawater $\delta^{18}O$ ($\delta^{18}O_w$) that can be measurable near ocean fronts[36] such as the subtropical front near 40°S south of Africa. Given the northerly location of our Agulhas leakage core at 34°S, this is not a major concern for our study and we consider this approach to yield a reasonable first-order approximation of palaeo upper water column temperature variability from a foraminiferal population as previously shown[2].

Radiocarbon dating of the core tops. One assumption in the comparison between palaeo proxy data and INALT01 model (Fig. 4) is that the two core tops are representative of the same contemporary circulation as the model. We support the validity of this assumption in the following.

Core MD02-2594 in the Agulhas leakage area has been dated at a depth of 50–51 cm, to be 2,815 ± 57 years before present[37]. Therefore, the core top itself will be younger than that. Core CD154-18-13P in the Agulhas Current has not been radiocarbon dated, but the core top of core CD154-17-17K, <50 km away, has been dated at a calibrated age of between 1,760 and 1,849 years before present[38]. As a further confirmation that the core top material of CD154-18-13P is representative at least of the Holocene, we verify that the average $\delta^{18}O$ value of the core top *G. ruber* specimens we analysed (-1.29 ± 0.5‰; error is s.d. of 48 measurements) is comparable—if not more negative—to that of CD154-17-17K core top (-1.13 ± 0.1‰; error is instrument precision[38]). A radiocarbon date on CD154-18-13P core top should be obtained to certify this, but this was not possible due to scarcity of material.

In summary, both core tops are of at least Late Holocene age, which suggests that our foraminiferal analyses should reflect the dynamics and ocean properties of the modern Agulhas System.

References

1. Katz, M. E. *et al.* Traditional and emerging geochemical proxies in foraminifera. *J. Foraminiferal Res.* **40**, 165–192 (2010).
2. Ganssen, G. M. *et al.* Quantifying sea surface temperature ranges of the Arabian Sea for the past 20 000 years. *Clim. Past* **7**, 1337–1349 (2011).
3. Emiliani, C. Pleistocene temperatures. *J. Geol.* **63**, 538–578 (1955).
4. Mix, A. C., Ruddiman, W. F. & McIntyre, A. Late quaternary paleoceanography of the Tropical Atlantic, 1: spatial variability of annual mean sea-surface temperatures, 0–20,000 years B.P. *Paleoceanography* **1**, 43–66 (1986).
5. Peeters, F. J. C. *et al.* Vigorous exchange between the Indian and Atlantic oceans at the end of the past five glacial periods. *Nature* **430**, 661–665 (2004).
6. Weyl, P. K. Micropaleontology and ocean surface climate. *Science* **202**, 475–481 (1978).
7. Furbish, D. J. & Arnold, A. J. Hydrodynamic strategies in the morphological evolution of spinose planktonic foraminifera. *Geol. Soc. Am. Bull.* **109**, 1055–1072 (1997).
8. Hemleben, C., Spindler, M. & Erson, O. R. *Modern Planktonic Foraminifera* (Springer, 1989).
9. Waniek, J., Koeve, W. & Prien, R. D. Trajectories of sinking particles and the catchment areas above sediment traps in the northeast Atlantic. *J. Mar. Res.* **58**, 983–1006 (2000).
10. Siegel, D. A. & Deuser, W. G. Trajectories of sinking particles in the Sargasso Sea: modeling of statistical funnels above deep-ocean sediment traps. *Deep Sea Res. I* **44**, 1519–1541 (1997).
11. Gyldenfeldt, von, A. B., Carstens, J. & Meincke, J. Estimation of the catchment area of a sediment trap by means of current meters and foraminiferal tests. *Deep Sea Res. II* **47**, 1701–1717 (2000).
12. Qiu, Z., Doglioli, A. M. & Carlotti, F. Using a Lagrangian model to estimate source regions of particles in sediment traps. *Sci. China Earth Sci.* **57**, 2447–2456 (2014).
13. Scussolini, P., van Sebille, E. & Durgadoo, J. V. Paleo Agulhas rings enter the subtropical gyre during the penultimate deglaciation. *Clim. Past.* **9**, 2631–2639 (2013).
14. Simon, M. H. *et al.* Millennial-scale Agulhas Current variability and its implications for salt-leakage through the Indian–Atlantic Ocean Gateway. *Earth Planet Sci. Lett.* **383**, 101–112 (2013).
15. Durgadoo, J. V., Loveday, B. R., Reason, C. J. C., Penven, P. & Biastoch, A. Agulhas leakage predominantly responds to the Southern Hemisphere westerlies. *J. Phys. Oceanogr.* **43**, 2113–2131 (2013).
16. Biastoch, A., Böning, C. W., Schwarzkopf, F. U. & Lutjeharms, J. R. E. Increase in Agulhas leakage due to poleward shift of Southern Hemisphere westerlies. *Nature* **462**, 495–U188 (2009).
17. Cronin, M. F., Tozuka, T., Biastoch, A., Durgadoo, J. V. & Beal, L. M. Prevalence of strong bottom currents in the greater Agulhas system. *Geophys. Res. Lett.* **40**, 1772–1776 (2013).
18. Masumoto, Y. *et al.* A fifty-year eddy-resolving simulation of the world ocean—preliminary outcomes of OFES (OGCM for the Earth Simulator). *J. Earth Simul.* **1** 35–56 (2004).
19. Paris, C. B., Helgers, J., van Sebille, E. & Srinivasan, A. Connectivity Modeling System: A probabilistic modeling tool for the multi-scale tracking of biotic and abiotic variability in the ocean. *Environ. Modell. Software* **42**, 47–54 (2013).
20. Schiebel, R. & Hemleben, C. Modern planktic foraminifera. *Paläontol. Zeitschr.* **79**, 135–148 (2005).
21. Takahashi, K. & Be, A. W. H. Planktonic foraminifera: factors controlling sinking speeds. *Deep Sea Res. I* **31**, 1477–1500 (1984).
22. Lombard, F., Labeyrie, L., Michel, E., Spero, H. J. & Lea, D. W. Modelling the temperature dependent growth rates of planktic foraminifera. *Marine Micropaleontol.* **70**, 1–7 (2009).
23. Erez, J. The source of ions for biomineralization in foraminifera and their implications for paleoceanographic proxies. *Rev. Mineral. Geochem.* **54**, 115–149 (2003).
24. Anand, P., Elderfield, H. & Conte, M. H. Calibration of Mg/Ca thermometry in planktonic foraminifera from a sediment trap time series. *Paleoceanography* **18**, 28 (2003).
25. Bach, L. T. *et al.* An approach for particle sinking velocity measurements in the 3–400 μm size range and considerations on the effect of temperature on sinking rates. *Mar. Biol.* **159**, 1853–1864 (2012).
26. Caromel, A. G. M., Schmidt, D. N., Phillips, J. C. & Rayfield, E. J. Hydrodynamic constraints on the evolution and ecology of planktic foraminifera. *Marine Micropaleontol.* **106**, 69–78 (2014).
27. Madec, G. *NEMO Ocean Engine* (Institut Pierre-Simon Laplace (IPSL), 2006).
28. Large, W. G. & Yeager, S. G. The global climatology of an interannually varying air–sea flux data set. *Clim. Dyn.* **33**, 341–364 (2009).
29. van Sebille, E., van Leeuwen, P. J., Biastoch, A., Barron, C. N. & de Ruijter, W. P. M. Lagrangian validation of numerical drifter trajectories using drifting buoys: application to the Agulhas system. *Ocean Modelling* **29**, 269–276 (2009).
30. van Sebille, E. *et al.* Propagation pathways of classical Labrador Sea water from its source region to 26°N. *J. Geophys. Res.* **116**, C12027 (2011).
31. van Sebille, E., Johns, W. E. & Beal, L. M. Does the vorticity flux from Agulhas rings control the zonal pathway of NADW across the South Atlantic? *J. Geophys. Res.* **117**, C05037 (2012).
32. Biastoch, A., Beal, L. M., Lutjeharms, J. R. E. & Casal, T. G. D. Variability and coherence of the Agulhas undercurrent in a high-resolution ocean general circulation model. *J. Phys. Oceanogr.* **39**, 2417–2435 (2009).
33. Okubo, A. Oceanic diffusion diagrams. *Deep Sea Res.* **18**, 789–802 (1971).
34. Barker, S., Greaves, M. & Elderfield, H. A study of cleaning procedures used for foraminiferal Mg/Ca paleothermometry. *Geochem. Geophys. Geosyst.* **4**, 9 (2003).
35. Kim, S. T. & ONeil, J. R. Equilibrium and nonequilibrium oxygen isotope effects in synthetic carbonates. *Geochim. Cosmochim. Acta* **61**, 3461–3475 (1997).
36. LeGrande, A. N. & Schmidt, G. A. Global gridded data set of the oxygen isotopic composition in seawater. *Geophys. Res. Lett.* **33**, L12604 (2006).
37. Martínez-Méndez, G. *et al.* Contrasting multiproxy reconstructions of surface ocean hydrography in the Agulhas Corridor and implications for the Agulhas Leakage during the last 345,000 years. *Paleoceanography* **25**, PA4227 (2010).
38. Ziegler, M. *et al.* Development of Middle Stone Age innovation linked to rapid climate change. *Nat. Commun.* **4**, 1905 (2013).
39. Kuenen, P. H. *Marine Geology* (John Wiley & Sons, 1950).
40. Berthois, L. & Le Calvez, Y. Etude de la vitesse de chute des coquilles de foraminifères planctoniques dans un fluide comparativement à celle des grains de quartz. *Revue des Travaux de l'Institut des Pêches Maritimes* **24**, 294–301 (1960).
41. Berger, W. H. & Piper, D. J. W. Planktonic foraminifera—differential settling, dissolution, and redeposition. *Limnol. Oceanogr.* **17**, 275–287 (1972).
42. Fok-Pun, L. & Komar, P. D. Settling velocities of planktonic-foraminifera—density variations and shape effects. *J. Foraminiferal Res.* **13**, 60–68 (1983).

Acknowledgements

E.v.S. was supported by the Australian Research Council via grants DE130101336 and CE110001028. W.W. was supported by the Regional and Global Climate Modeling

Program of the United States Department of Energy's Office of Science. J.V.D. and A.B. acknowledge funding by the Bundesministerium für Bildung und Forschung project SPACES 03G0835A. C.T. acknowledges the support of an ARC Laureate Fellowship (FL100100195). C.B.P. is funded by GOMRI through C-IMAGE Consortium. P.S., J.V.D., F.P., A.B. and R.Z. acknowledge funding by the European Community's Seventh Framework Programme (FP7) Marie-Curie Initial Training Network 'GATEWAYS' under Grant Agreement 238512. We thank Kelsey Dyez for providing core top material of MD02-2524, Ian Hall and Margit Simon for providing core top material of CD154-18-13P, and the Trace Element Laboratory of Uni Research Bergen for enabling the Mg/Ca measurements. INALT01 CMS experiments were performed at the super-computer at Kiel University. The OFES simulation was conducted on the Earth Simulator under the support of JAMSTEC.

Author contributions

E.v.S. lead the model analysis and wrote the manuscript. P.S. and F.P. performed the proxy analysis and reviewed foraminiferal traits. J.V.D. and A.B. are custodian of the INALT01 model; its Lagrangian simulations were performed by J.V.D. C.B.P. developed code for biotic particle movement. All authors contributed to the planning of the experiments, the writing of the manuscript, and the discussion and interpretation of the results.

Additional information

The exposure of the Great Barrier Reef to ocean acidification

Mathieu Mongin[1], Mark E. Baird[1], Bronte Tilbrook[1,2], Richard J. Matear[1], Andrew Lenton[1], Mike Herzfeld[1], Karen Wild-Allen[1], Jenny Skerratt[1], Nugzar Margvelashvili[1], Barbara J. Robson[3], Carlos M. Duarte[4], Malin S.M. Gustafsson[5], Peter J. Ralph[5] & Andrew D.L. Steven[1]

The Great Barrier Reef (GBR) is founded on reef-building corals. Corals build their exoskeleton with aragonite, but ocean acidification is lowering the aragonite saturation state of seawater (Ω_a). The downscaling of ocean acidification projections from global to GBR scales requires the set of regional drivers controlling Ω_a to be resolved. Here we use a regional coupled circulation–biogeochemical model and observations to estimate the Ω_a experienced by the 3,581 reefs of the GBR, and to apportion the contributions of the hydrological cycle, regional hydrodynamics and metabolism on Ω_a variability. We find more detail, and a greater range (1.43), than previously compiled coarse maps of Ω_a of the region (0.4), or in observations (1.0). Most of the variability in Ω_a is due to processes upstream of the reef in question. As a result, future decline in Ω_a is likely to be steeper on the GBR than currently projected by the IPCC assessment report.

[1] CSIRO Oceans and Atmosphere, Hobart, Tasmania 7000, Australia. [2] Antarctic Climate and Ecosystems Co-operative Research Centre, Hobart, Tasmania 7000, Australia. [3] CSIRO Land and Water, Canberra, Australian Capital Territory 2601, Australia. [4] Red Sea Research Center, King Abdullah University of Science and Technology, Thuval 23955-6900, Kingdom of Saudi Arabia. [5] Plant Functional Biology and Climate Change Cluster (C3), Faculty of Science, University of Technology Sydney, Sydney, New South Wales 2007, Australia. Correspondence and requests for materials should be addressed to M.M. (email: Mathieu.Mongin@csiro.au).

The Great Barrier Reef (GBR) ecosystem, described as one of the seven natural wonders of the world, is under increasing pressure from local and global anthropogenic stressors[1]. Coral calcification has continued to decline over the last few decades at rates similar to less well-managed reefs[2], due to damage from cyclones, disease, invasive species (crown-of-thorns starfish), coral bleaching and possibly ocean acidification. In the long term, ocean acidification is expected to become an increasing threat to the GBR ecosystem[3].

Ocean acidification is the decrease in pH and altered carbonate chemistry of ocean waters, including a decrease in aragonite saturation state of seawater (Ω_a) due to the uptake of anthropogenic carbon dioxide that will impact the ability of many reef-building corals to precipitate $CaCO_3$ (ref. 4). The Ω_a of tropical surface waters is predicted to decline by about 0.1 per decade over this century[5]. Informed decision making for the future management of the GBR in the face of global ocean acidification will necessitate that the current mean state, variability and drivers of Ω_a are known for individual reefs[6,7].

The task to measure Ω_a at all individual reefs in the GBR (Fig. 1) is impossible. Earth System Models assessing current and future Ω_a (ref. 8) have spatial and temporal resolutions that are too coarse to resolve the variability in Ω_a in coastal and shelf environments, where vulnerable reefs, and a diversity of processes that contribute to regulating Ω_a, exist[9]. Therefore, the current state of the carbon chemistry of the GBR system at the scale of individual reefs remains unknown.

The variability in Ω_a in coastal ecosystems is driven by complex interactions between forcing from the open-ocean carbon dynamics, the delivery of freshwater and carbon from coastal watersheds, and metabolic effects within the ecosystem[9]. Due to spatially and temporally variable processes affecting carbon chemistry, and the complex circulation of the GBR, the relative contribution of the drivers of Ω_a has not been well quantified for the region.

Observational studies on a small number of GBR reefs have shown that Ω_a is modified by processes such as calcification/dissolution and production/respiration on a reef[10-13], and by the broader-scale impact of regional changes in carbon chemistry flowing onto each reef[8,14,15]. The relative contribution of the differing processes affecting Ω_a needs to be resolved to establish the vulnerability and likely impact on coral reef ecosystems to ocean acidification, and to incorporate ocean acidification into conservation and management strategies.

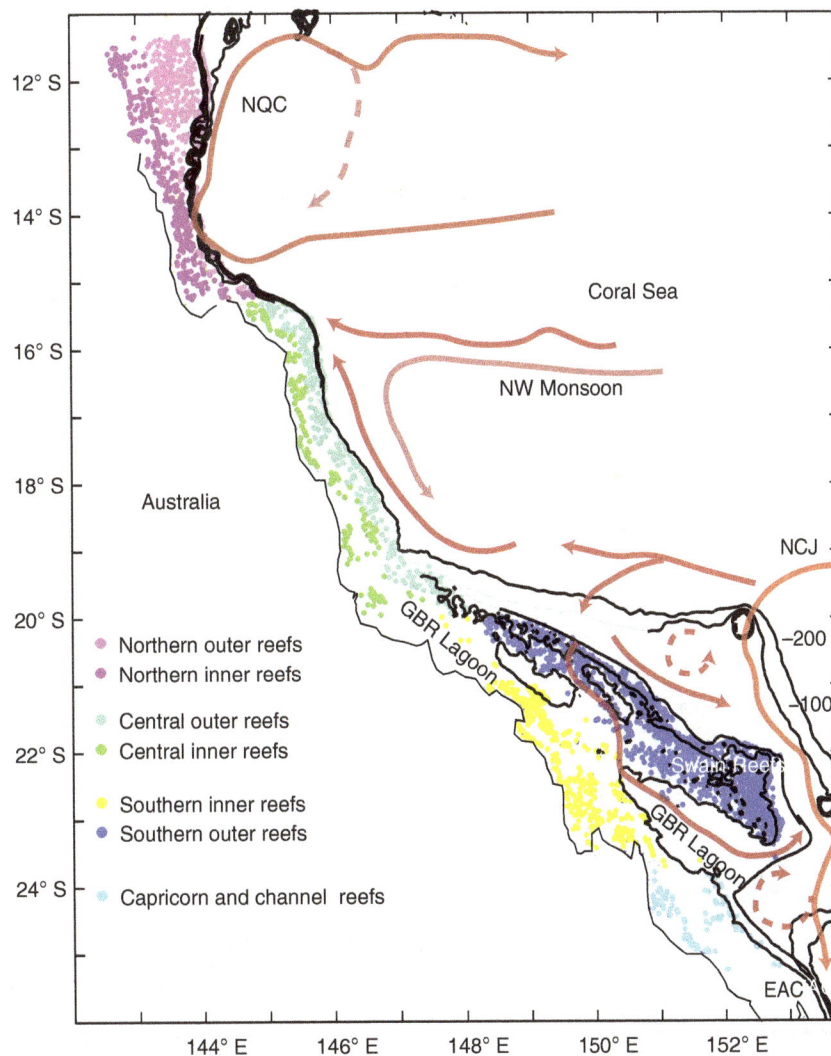

Figure 1 | Great Barrier Reef living coral distribution. Black lines shows the 60, 100 and 200 m depth contours. Red lines show major surface currents. NQC: North Queensland Current; EAC: East Australian current; NCJ: North Caledonia Jet. Dashed orange lines show transient currents. The coloured dots show the individual reefs as identified by the Great Barrier Reef Marine Park Authority (GBRMPA) shelf features e-atlas separated into different geographical zones.

The novelty of the approach applied here allows us to estimate Ω_a and its drivers at over 3,000 reefs, a capability that is not possible using conventional approaches. A combination of total alkalinity (A_T), dissolved inorganic carbon (C_T), salinity, and temperature observations from 22 sites located inshore of the GBR (Supplementary Fig. 1), combined with a coupled catchment, hydrodynamic, sediment and biogeochemical model[16–18] was used to estimate for the first time the Ω_a, and the processes driving its variability, for the 22 coastal observations sites and 3,581 individual reefs (see Methods for model description). The observations are used to estimate Ω_a variability and to evaluate the approach and model performance.

The new maps quantifying the exposure of individual reefs to ocean acidification demonstrate a greater spatial variability than previous studies could resolve. The coupled model shows a clear spatial structure in Ω_a, with large gradients across and along the GBR shelf, created by the interaction of reef processes and ocean circulation.

Results

Simulated ocean properties. In the biogeochemical model, the photosynthesis–respiration balance for planktonic and benthic (including coral) organisms is parameterized as a function of temperature, nutrients supply, light availability and predation control. Calcification–dissolution of calcium carbonate is a function of temperature, light, Ω_a and the type of benthic substrate. The model captures the heterogeneity of primary production on the GBR, with episodic riverine and upwelled shelf nutrient inputs driving local phytoplankton blooms along the shore and the continental shelf break[19,20].

The modelled A_T, C_T, S, and Ω_a, has temporal mean errors (root mean square errors, Supplementary Tables 1 and 2) at least a factor of 5 smaller than the observed spatial and temporal

variability at the 22 observation sites, thereby providing confidence in the use of the model predictions (see Methods).

Aragonite saturation-state variability. Across the 3,581 reefs, the model predicts an Ω_a that varies between 2.51 and 3.94 (Fig. 2a), (observations at 22 sites varied between 3.04 and 3.53, Table 1, Supplementary Fig. 1). Total alkalinity was generally below that of the Coral Sea mean, with values at the 3,581 reefs spreading from the mean open ocean (Fig. 2a). The southern GBR reefs showed a larger spread in C_T and A_T compared with the northern and central reefs regions (Fig. 2a).

A mean value of $\Omega_a = 3.61 \pm 0.19$ (mean ± s.e.) was estimated for the offshore Coral Sea source waters. The simulated Ω_a decreased towards the coast and was much lower in inshore reef waters (in areas <20 km from land, $\Delta\Omega_{reef-ocean} = -0.64$, range = 0.03 to -1.37, Fig. 2b), due primarily to higher C_T. The outer reefs of the GBR (on the eastern side of the GBR lagoon) typically had Ω_a values above those of the mean Coral Sea ($\Delta\Omega_{reef-ocean} = 0.12$, range = -0.1 to 1.21), with the exception of the southern region, south of 20 °S (around the Swain Reefs).

Drivers of aragonite saturation state variability. We considered three groups of processes driving change in the Ω_a: freshwater fluxes from river flow, evaporation and precipitation and ocean circulation, referred to as the hydrological cycle component, $\Delta\Omega_{fresh}$; net calcification, or calcification minus dissolution, $\Delta\Omega_{cd}$; and the sum of photosynthesis minus respiration and net air–sea gas exchange, $\Delta\Omega_{pra}$. The unique combination of impacts that each of these drivers have on temperature, A_T, C_T, and salinity (see Fig. 3 and Methods) enabled us to quantify the relative contribution of each driver to the change in Ω_a as waters flowed from the Coral Sea through the GBR shelf.

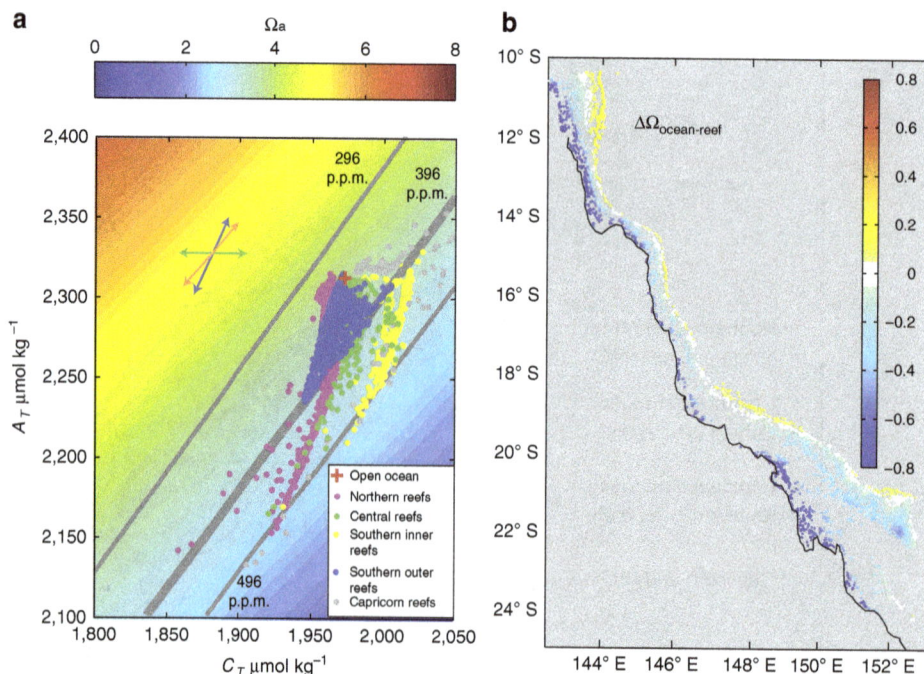

Figure 2 | Aragonite saturation state for the 3581 coral reefs. (**a**) Simulated aragonite saturation state (as background shading) versus dissolved inorganic carbon, C_T, and total alkalinity, A_T, for the individual coral reefs (average from daily September 2010–September 2012 values). Grey lines show the surface pCO$_2$ of 396 and 296 p.p.m. The Ω_a was calculated at a temperature of 25 °C and salinity of 35. If the Ω_a used as the background shading in **a** is calculated at 21 and then 27 °C, the range observed in the GBR surface waters, Ω_a at a constant C_T and A_T changes by < ± 0.07 from the shaded values. The process arrows are approximations that are further discussed in Fig. 3. (**b**) For the individual reefs the mean difference in aragonite saturation state between the open ocean value and the value simulated at the reef ($\Delta\Omega_{ocean-reef} = (\Omega_{ocean} - \Omega_{reef})$).

Table 1 | Mean change in aragonite saturation state.

	Model mean (range)	Observations mean (range)	Model - Observations
Ω_{reef}	3.14 (2.46-3.47)	3.21 (3.04-3.53)	− 0.07
$\Delta\Omega_{reef-ocean}$	− 0.59 (− 1.26, − 0.28)	− 0.52 (− 0.71, − 0.20)	− 0.07
$\Delta\Omega_{fresh}$	0.01 (− 0.05, 0.12)	0.05 (− 0.03, 0.14)	− 0.04
$\Delta\Omega_{cd}$	− 0.20 (− 0.57, 0.07)	− 0.07 (− 0.14, 0.17)	− 0.13
$\Delta\Omega_{pra}$	− 0.40 (− 0.82, − 0.20)	− 0.48 (− 0.93, -0.11)	0.08

Average $\Delta\Omega_{reef-ocean}$ and the drivers of this change, at the 22 observation sites (15 locations—some locations have multiple depths), as determined from model output and observations[5] using A_T, C_T, S, and temperature. The number of observation points in time at each site ranges from 6 to 41. We compared model outputs of a 4-year hindcast (2010–2014), taken the day and hour of the observations, with the observation. Discrepancies between model and observations are not unexpected as the model represents the mean properties in a 4 by 4-km region, compared to a single observation point, in a highly dynamic coastal environment.

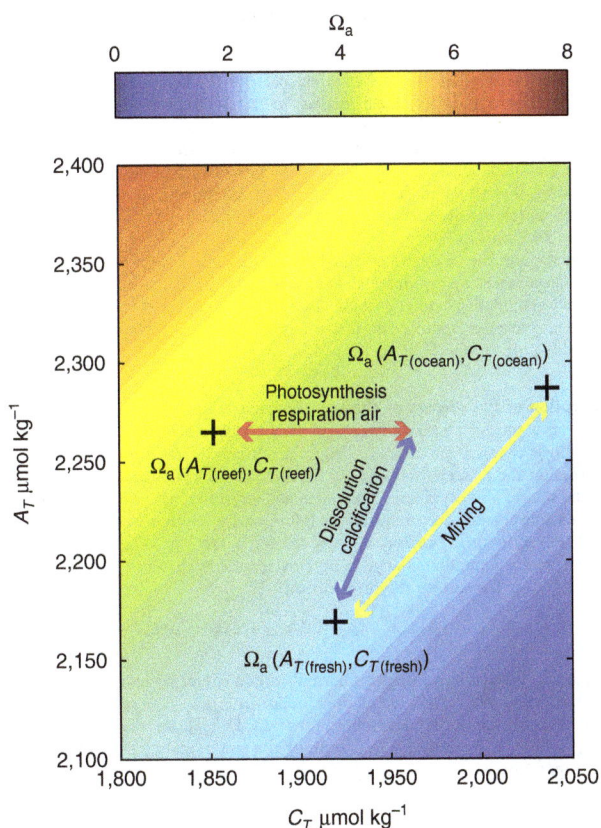

Figure 3 | Driver of aragonite saturation-state change schematic. The change from the open ocean value is separated into three drivers: (1) hydrological cycle calculated considering dilution/mixing associated with a change in salinity (yellow arrow); (2) Net calcification is calculated as a shift in alkalinity and dissolved inorganic carbon at a ratio of 2:1 (blue arrow); (3) net carbon uptake due photosynthesis, respiration and air-sea flux (red arrow). Based on these values, the change in aragonite saturation from the open-ocean state is calculated. Background shows aragonite saturation-state versus dissolved inorganic carbon, C_T, and total alkalinity, A_T, at a temperature of 25 °C and salinity of 35.

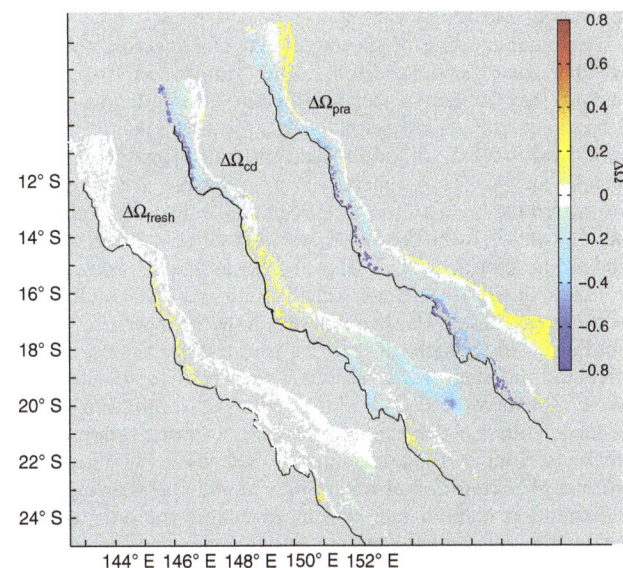

Figure 4 | Drivers of change in aragonite saturation. The changes are separated into the affect due to hydrological cycle ($\Delta\Omega_{fresh}$; left), net calcification ($\Delta\Omega_{cd}$; centre), and due to the combination of photosynthesis, respiration and air-sea exchange ($\Delta\Omega_{pra}$; right), average from daily September 2010–September 2012 values.

We found that freshwater fluxes ($\Delta\Omega_{fresh}$) resulting from river flows, evaporation/precipitation, and oceanic circulation had limited influence on $\Delta\Omega_{reef-ocean}$ ($\Delta\Omega_{fresh} < 0.1$ of $\Delta\Omega_{ocean-reef}$) for most of the GBR reefs. The largest freshwater fluxes were observed in the southern outerreef region (near the Swain Reefs), south of 20° S ($\Delta\Omega_{fresh} = 0.2$), where ocean circulation alters the salinity of the region (Supplementary Fig. 2), with a corresponding change in Ω_a. Net calcification and metabolic processes had a much greater impact on $\Delta\Omega_{reef-ocean}$.

The change in Ω_a due to net calcification, $\Delta\Omega_{cd}$, was generally negative, illustrating the importance of upstream net calcification in depleting carbonate ion concentration, thus reducing Ω_a at downstream reefs within the GBR (Fig. 4). The large negative $\Delta\Omega_{cd}$ in the northern inner and southern GBR waters ($ − 0.8 < \Delta\Omega_{cd} < − 0.4$ excluding the Capricorn Group) resulted from a strong net calcification signal in the outer reefs being transported into the northern and southern regions of the mid to inner shelf. In contrast, the outer reefs were flushed with relatively high Ω_a Coral Sea water.

The net balance of photosynthesis and respiration was a strong contributor to a downward shift in Ω_a (Fig. 4) as a consequence of respiration exceeding uptake of carbon due to photosynthesis, and a CO_2 influx through air–sea gas exchange (Fig. 3). Metabolic effects were strongest ($\Delta\Omega_{pra} = − 0.9$) for the inner reefs, particularly for the central (14° S to 19° S) GBR ($\Delta\Omega_{pra} \sim − 0.4$), compared with the southern and northern GBR reefs ($\Delta\Omega_{pra} \sim − 0.2$, Fig. 4). Positive $\Delta\Omega_{pra}$ observed for most of the outer reefs suggested photosynthesis exceeds respiration in the water upstream of these reefs.

Discussion

The results show Ω_a for an individual reef is strongly modulated by the circulation of waters and the regional balance of production/respiration, and dissolution/calcification processes. These processes cause the mean Ω_a value of reefs of the GBR to be 0.51 less than that of the open ocean. Further, the mean spatial variability in Ω_a within the GBR is large and is approximately equal to the temporal trend expected over the twenty first century under high carbon emission scenarios. As a result, we find that the projected Ω_a under all emissions scenarios will be lower on large parts of the GBR than estimated in the IPCC AR5 report[5]. This, combined with the projected changes in thermal stress[21,22], suggests that even low future emissions scenarios may not be sufficient to avoid significant impacts of low Ω_a values of coral reef calcification with potential losses in coral cover, ecosystem biodiversity and resilience[23].

The understanding of the regulation of Ω_a across the GBR provides new insights into how future environmental changes may impact ocean acidification on individual reefs. Changes in the hydrological cycle (for example, freshwater inputs) will have a limited impact, particularly as few reefs are located in regions where river plumes are frequent. The outermost reefs on the shelf edge are flushed with relatively high Ω_a waters from the Coral Sea, and currently net primary production counteracts and in some locations exceeds the decreases in Ω_a due to net calcification. As net calcification decreases on these reefs the influence of net primary production could result in Ω_a increases relative to Coral Sea values. For the inner to mid-shelf reefs, Ω_a decreases relative to upstream values. A source of C_T associated with net respiration or gas exchange causes a decline in Ω_a, while net calcification in the North and South GBR causes further declines in Ω_a. The model indicates that central reef waters may already be showing signs of dissolution in inter-reef regions exceeding the net calcification in the reefs and this region is likely to expand north and south with time. The balance of net calcification and primary production will be influenced by multiple factors that are difficult to predict.

Thus the future Ω_a regime of the GBR is dependent on the balance of calcification and primary production, which have different drivers and cannot be extrapolated directly from forecasts for the open ocean. Accurate forecasts of the exposure and vulnerability of GBR reefs to ocean acidification will require the future trajectories of regional carbon cycle and calcification/dissolution processes, both likely to also be impacted by ocean warming, and other anthropogenic pressures[14].

The present methodology represents a step forward in our capacity to assess the contribution of various processes to Ω_a regulation in coastal ecosystems like the GBR. The lack of observations in carbon chemistry space still makes the validation of the model for the entire GBR difficult and this remains an important issue that needs resolution.

The quantitative understanding of the drivers of Ω_a provides insights that can be used to inform local management practices. The identification of regions with higher Ω_a, combined with estimates of how the dominant drivers of Ω_a at these locations will likely be impacted by climate change, provides the best opportunity to identify zones on the GBR that will be least impacted by global acidification in this century. The approach developed here provides a pathway to advance beyond coarse ocean acidification models that focused on open ocean surface waters, to address the complex regional and local variability characteristic of the coastal ecosystems. In addition, the study offers a methodology and approach, based on a simple conceptual model (Fig. 3) that can be applied to other coastal ecosystems to elucidate key drivers of Ω_a variability.

Methods

Carbon chemistry calculations. Calcifying organisms utilize dissolved carbonate (CO_3^{2-}) and calcium (Ca^{2+}) ions, to produce calcium carbonate ($CaCO_3$) skeletal materials and shells. Aragonite is a metastable form of calcium carbonate that is precipitated biogenically by many reef forming corals and other species. The aragonite saturation state, Ω_a, is commonly used to describe the ability of corals to calcify and is given by:

$$\Omega_a = \frac{[CO_3^{2-}][Ca^{2+}]}{K_{sp}} \quad (1)$$

where Ksp is the solubility product[24]. A fall in CO_3^{2-} ion concentration is expected to cause a decrease in coral growth[25].

The Ocean-Carbon Cycle Model Intercomparison Project (OCMIP) numerical methods are used to quantify air–sea carbon fluxes and the carbon dioxide system equilibria in seawater. The OCMIP procedures quantify the state of the CO_2 system using two prognostic variables, the concentration of dissolved inorganic carbon, C_T, and total alkalinity, A_T. The value of these prognostic variables, along with salinity and temperature, are used to calculate the partial pressure of carbon dioxide, pCO_2, in the surface waters using a set of governing chemical equations that are solved using a Newton–Raphson method[26] with literature-derived constants[24,27-29]. One alteration from the typical implementation of the OCMIP algorithm is that we increased the search space for the iterative scheme from ± 0.5 pH units (appropriate for global models) to ± 2.5. This altered OCMIP scheme converges over the broader range of C_T and A_T values[30], and is essential in our biogeochemical model that includes sources and sinks from tropical rivers, shallow coastal areas and offshore reefs. The air–sea flux of carbon dioxide is calculated based on 2-h wind speed data from the Australian Bureau of Meteorology operational forcing products, and the air–sea gradient in pCO_2, using an empirical net flux relationship[31].

Throughout the study, we assumed that in the region of interest, total alkalinity was approximated by carbonate alkalinity, hence neglecting the effect of biological nitrogen assimilation and release on A_T. Wolf-Gladrow et al.[32], showed that biological processes could be an important sink of A_T in some regions, but is a negligible term on the GBR shelf[33].

Separation of the drivers of change. The change in Ω_a over an individual reef from that of the open ocean, $\Delta\Omega_{reef-ocean}$, can be broken down into the change due to three biogeochemical processes or drivers: freshwater fluxes from river flow, evaporation and precipitation, $\Delta\Omega_{fresh}$; the sum of calcification and dissolution, $\Delta\Omega_{cd}$, and the sum of photosynthesis, respiration and air–sea exchange of CO_2, $\Delta\Omega_{pra}$. This separation is possible due to the unique influence the drivers have on total alkalinity, A_T, dissolved inorganic carbon, C_T, and salinity, S, see Fig. 3 for details

Thus, we assume that change in Ω_a between the open ocean and the local reef, $\Delta\Omega_{reef-ocean}$, is the sum of the three drivers:

$$\Delta\Omega_{reef-ocean} = \Delta\Omega_{fresh} + \Delta\Omega_{cd} + \Delta\Omega_{pra} \quad (2)$$

$\Delta\Omega_{reef-ocean}$ is calculated using equilibrium carbon chemistry and the alkalinity and dissolved inorganic carbon in the open ocean ($A_{T(ocean)}$, $C_{T(ocean)}$, respectively) to determine the open-ocean aragonite saturation (Ω_{ocean}), and that above the reef ($A_{T(reef)}$, $C_{T(reef)}$, respectively) to determine the reef aragonite saturation Ω_{reef} (at temperature 25 °C and salinity 35):

$$\Delta\Omega_{reef-ocean} = \Omega_{reef}\left(A_{T(reef)}, C_{T(reef)}\right) - \Omega_{ocean}\left(A_{T(ocean)}, C_{T(ocean)}\right) \quad (3)$$

where $C_{T(ocean)} = 1{,}966 \,\mu mol\,kg^{-1}$ and $A_{T(ocean)} = 2{,}300 \,\mu mol\,kg^{-1}$ ($\Omega_a = 3.8$) are climatological values from the Coral Sea[34].

The model is forced with current, temperature and salinity at the ocean boundary by an ocean global circulation model[35]. C_T and A_T values at the ocean boundary are computed using existing salinity relationships[36,37].

$$C_{T(ocean\ boundary)} = 1{,}966.66 + 83.3\left(S_{ocean\ boundary} - 35\right) \quad (4)$$

$$A_{T(ocean\ boundary)} = 2{,}300.0 + 63.3\left(S_{ocean\ boundary} - 35\right) \quad (5)$$

The salinity at the boundary varies temporally and spatially from 34.5 to 35.5; leading to a change in A_T from 2,278 to 2,331 $\mu mol\,kg^{-1}$ and C_T from 1,880.3 to 2,008.4 $\mu mol\,kg^{-1}$.

The variables A_T, C_T and S are concentrations, independent of temperature. In the absence of biological processes and air–sea exchange, each is a conservative tracer. This property has been used to determine a climatology of C_T and A_T for the region:

$$C_{T(fresh)} = 1{,}966.66 + (83.3(S_{reef} - 35)) \text{ if } S_{reef} < 25 C_{T(fresh)}$$
$$= 1{,}133 \,\mu mol\,kg^{-1} \quad (6)$$

$$A_{T(fresh)} = 2{,}300.0 + (63.3(S_{reef} - 35)) \text{ if } S_{reef} < 44 A_{T(fresh)} = 900 \,\mu mol\,kg^{-1} \quad (7)$$

Using these values, the changes in Ω_a due to freshwater fluxes is given by:

$$\Delta\Omega_{fresh} = \Omega\left(A_{T(fresh)}, C_{T(fresh)}\right) - \Omega\left(A_{T(ocean)}, C_{T(ocean)}\right) \quad (8)$$

The salinity thresholds for C_T and A_T are different as we used the same value of 900 µmol kg^{-1} for freshwater C_T and A_T.

Calcification and dissolution consume/produce twice as much A_T as C_T. The remaining processes (photosynthesis, respiration and air–sea exchange) only change C_T. Therefore, the change in the alkalinity from that of the reef, $A_{T(reef)}$, from the value obtained considering only conservative fluxes, $A_{T(fresh)}$, is due to calcification/dissolution only. Further, C_T must change by half this amount. Thus, the change in Ω_a from open ocean due to calcification and dissolution processes is:

$$\Delta\Omega_{cd} = \Omega\left(A_{T(fresh)}, C_{T(fresh)} + \frac{A_{T(reef)} - A_{T(fresh)}}{2}\right) - \Omega\left(A_{T(fresh)}, C_{T(fresh)}\right) \tag{9}$$

Finally, we can calculate the remaining driver:

$$\Delta\Omega_{pra} = \Delta\Omega_{reef-ocean} - \Delta\Omega_{fresh} - \Delta\Omega_{cd} \tag{10}$$

Where $\Delta\Omega_{pra}$ is the change in Ω_a from open ocean due to photosynthesis minus respiration (by benthic and pelagic organisms), and includes net air–sea gas exchange driven by changes in the solubility of CO_2 (a function of water temperature, salinity and pressure) and biological processes.

As A_T, C_T and S are also available for the 22 observations sites, we use the same methodology to derive $\Delta\Omega_{inshore,obs-ocean}$ and the corresponding drivers (presented in Supplementary Fig. 1).

To remove any bias from seasonal cycles, we averaged the model over a 2-year period (1 dry and 1 wet year) and validated the model using a 4-year hindcast. The spatial gradient $\Delta\Omega_{reef-ocean}$ did not change seasonally (Supplementary Fig. 3), which indicates that the annual mean of this value is a good representation of the spatial structure.

Skill of model predictions of carbon chemistry and validation. The skill of the model assessed at the 22 inshore observational sites provided a measure of the uncertainty of the Ω_a predictions at the 3,581 reef locations. The root mean square (r.m.s.) error of time series of A_T, C_T, S and temperature at each site was calculated (Supplementary Figs 4 and 5, Supplementary Tables 1 and 2) . The mean (and range) of r.m.s. errors across the 22 sites were: A_T: 39.90 mmol m^{-3}(8.5, 91.5 mmol m^{-3}); C_T: 35.9 mmol m^{-3} (12.5, 63.97); S: 0.47 (0.15, 0.93); temperature: 0.87 °C (0.63, 1.24), resulting in an error in the calculated Ω_a of 0.23 (0.09, 0.54).

Due to their geographical locations, the observational sites are strongly influenced by freshwater plumes from rivers. The model is too coarse to resolve some of the small-scale water circulation features, such as internal waves, filament and small freshwater plumes. Freshwater footprints are difficult to accurately be represented in a 4-km resolution ocean model. For example, the real freshwater plumes could be thinner than the model grid cell, or could be offset in space and time, which make a comparison with observations at a point in space deceptive. Despite these problems, we are confident that the amount and frequency of freshwater inputs is correct (and has been validated in our model by many salinity observations), but their spatial footprint and timing could be offset.

Supplementary Fig. 6 shows, as an example, a point-by-point time series comparison at one of the inshore sites. Most of the variability at this location, for both the simulated and observed quantities, follows the seasonal cycle of temperature and the episodic flooding events (illustrated as freshwater input flow from the nearest river, as a green line on the salinity panel). The model reproduces baseline evolution of temperature, salinity, C_T, A_T and Ω_a . When a mismatch between observed and simulated entities occurs (like in April 2011), it usually follows a discrepancy in simulated salinity (most likely due to a different plume dynamics in the model compared to the reality, as mentioned above). Consequently, during these periods, simulated C_T, A_T and Ω_a (representing a mean over a 4-km grid) are failing to get as low as the single-point observations.

This does not mean the model is incorrect, it just shows that we cannot always use a single point observation to validate the model. As long as the model is not used to predict the dynamic of a single point, the approach remains valid.

However, given the available observations, we show that our mechanistic model does simulate the Ω_a variability and its drivers at the 22 sites where we have sufficient carbon data to make the comparisons. Elsewhere, the assessment of other key features of the model, like the circulation, salinity and primary production show we do capture the spatial mean variability evident in the observations.

eReefs hydrodynamic and biogeochemical model configuration. The model output is generated by the eReefs coupled hydrodynamic, sediment and biogeochemical modelling system[17–19]. The hydrodynamic model is nested within a global circulation model, to provide accurate forcing data along the boundary within the Coral Sea, and forced by atmospheric winds and radiation and 22 rivers flows. We used a 4-km resolution eReef grid[18], and output from a hindcast from September 2010 to July 2014. The hydrodynamic model produces hourly fluxes between grid cells, which are used to determine the advection of the sediment and biogeochemical tracers.

The model is designed to simulate water chemistry around reefs and throughout the GBR reef matrix (4 km scale), but does not resolve processes inside the lagoon of the individual reefs (100 m scale). We use a complex biogeochemical model[38]

that simulates optical, nutrient, plankton, benthic organisms (seagrass macroalgae and coral), detritus, chemical and sediment dynamics across the whole GBR region, spanning estuarine systems to oligotrophic offshore reefs (Supplementary Fig. 7). Ocean acidification is quantified by the A_T and C_T (in turn determining pH and Ω_a), that are altered by processes of calcification, carbonate dissolution, primary production, remineralization and gas exchange processes. The model also simulates the nutrient, sediment and freshwater inputs from the rivers system along the GBR.

The eReefs coral growth model considers corals and zooxanthellae as two separate entities, allowing corals to grow through both autotrophic and heterotrophic means[39]. Coral calcification rates are based on the biomass of corals and bottom light levels[40]. The calcification and reef dissolution rates were adapted from[10,11], and are similar to those used in ref. 40. Sediment dissolution for grid cells on the continental shelf (depth < 200 m) and between reefs was set to 0.0001 mmol C m^{-2} s^{-1}. This constant rate of sediment dissolution was required to keep A_T from drifting below observations in coastal regions. For the purposes of predicting Ω_a, the sediment dissolution rate was the least well-constrained parameter, and is likely to vary spatially and temporally. The current model does not consider calcite as another form of $CaCO_3$, which are specific to the *Halimeda* beds, particularly in the northern GBR.

The GBRMPA Great Barrier Reef and Coral Sea Geomorphic Features data set identifies 3,860 individual reefs[41]. This database was used to set the initial distribution of corals in the model domain (3,581 reefs resolved). Due to the 4-km resolution of the model, in some instances > 1 identified reef occurs within a single grid cell. Thus the 4-km model contains only 1,725 cells with reefs.

Hydrodynamic-simulated GBR ocean circulation. The general circulation of the GBR region is dominated by the westward flow of the South Equatorial Current flowing through the Coral Sea, as a number of water jets controlled by the complex plateau, seamount and ridge topography. On approaching the western boundary of the Coral Sea, these multiple jets are steered by the continental shelf to form the southward flowing East Australian Current and the northward flowing North Queensland Current (Fig. 1). Flow within the relatively shallow GBR lagoon is determined by the interaction of the above-mentioned basin-scale forcing, seasonal winds patterns, and the restricted flow through GBR outer reefs.

Quantification of local processes using an age tracer. We defined local processes affecting the Ω_a of the waters overlying a reef to be those biogeochemical processes occurring on each individual reef with no contribution from upstream reefs. To quantify the relative importance of upstream versus local processes, we focused on the net calcification driver, assuming that the other carbon cycle processes and the hydrological cycle had an intrinsically regional impact. We computed the component of the change in Ω_a from the open-ocean values due to net calcification at the reef site ($\Delta\Omega_{cd,local}$), using a simulated age tracer as a proxy for residence time above a reef[40], and the average change in Ω_a due to net calcification on a well-studied GBR reef.

We carried out an age tracer experiment[42,43] to diagnose the residence time of water on different reefs in the model. The age tracer concentration, τ, was initialized at 0 days within each reef. The age tracer was advected and diffused by the hydrodynamic model in a similar manner to salinity.

We allowed the age to decay off the reef so that water from one reef did not unduly impact on the age of adjacent reefs. This gave the local age. When on a reef grid cell, the age increased at the rate of :

$$\frac{d\tau}{dt} = 1\,d\,d^{-1} \tag{11}$$

When the age tracer was not on a reef, age reduced at a rate proportional to its present age:

$$\frac{d\tau}{dt} = \tau\,d\,d^{-1} \tag{12}$$

On the Heron Island reef, with an average depth of 2 m, a numerical simulation showed the local rate of change of Ω_a with age, is equal to 0.43 per day, similar to an observed value of 0.38 per day for the nearby One Tree Island reef[44]. For reefs with a depth of h, the local rate of change in Ω_a due to local net calcification in a model cell, using the Heron Island value, was given by:

$$\Omega_{cd,local} = \frac{2}{h}\left(\frac{d\Omega_a}{d\tau}\right)_{Heron} \tau \tag{13}$$

For the purposes of comparing local and upstream processes, we assumed all reefs had a mean depth of 2 m. Under this assumption, we overestimated the influence of local processes on reefs deeper than 2 m, and underestimated local processes for reefs shallower than 2 m. Local processes were shown at the scale of a reef (16 km^2 in the model) to be small relative to the regional processes (Supplementary Fig. 8).

Results showed that the change in Ω_a due to local processes was almost negligible for the vast majority of the GBR, and only reached − 0.05 in the inner northern GBR and Swain Reef regions where the largest reefs were present (Supplementary Fig. 1). While large variability in Ω_a is possible on sections of a reef, and will be significant for sub-reef scale community processes[45,15], the mean Ω_a of an entire reef is generally forced by upstream processes.

eReefs model description. The hydrodynamic model SHOC (Sparse Hydro-dynamic Ocean Code[46]) was used in this study and is a general purpose model that is applicable on spatial scales ranging from estuaries to regional ocean domains. It is a three-dimensional finite-difference hydrodynamic model, based on primitive equations. Inputs required by the model include forcing due to wind, atmospheric pressure gradients, surface heat and water fluxes, and open-boundary conditions such as tides and low frequency ocean currents. The 4-km model is forced with OceanMAPS (http://www.bom.gov.au/bluelink/products/prod_oceanmaps.html) data on the open boundaries. The tide is introduced through 22 constituents derived from the global CSR tide model. The surface fluxes are obtained from ACCESS-R (http://www.bom.gov.au/nwp/doc/access/NWPData.shtml). Surface fluxes comprise momentum, heat and freshwater sources. Bathymetry has been be sourced from the Digital Elevation Model of the GBR produced at 100 m spatial resolution[41]. The model uses a curvilinear orthogonal grid in the horizontal and fixed 'z' coordinates in the vertical.

The sediment transport model adds a multilayer sediment bed to the hydrodynamic model grid and simulates sinking, deposition and resuspension of multiple size-classes of suspended sediment[47]. The model solves advection-diffusion equations of the mass conservation of suspended and bottom sediments and is particularly suitable for representing fine sediment dynamics, including resuspension and transport of biogeochemical particles.

The model is forced with freshwater inputs at 21 rivers along the GBR and the Fly River. River flows input into the model are obtained from the Department of Environment and Resource Management gauging network (http://www.derm.qld.gov.au/water/monitoring/current_data). Relationships are used to account for nutrient and sediment inputs from rivers into the model (statistical relationships between river flow and nutrient concentrations)[18,19,48].

The biogeochemical model is organized into three zones: pelagic; epibenthic; and sediment. The epibenthic zone overlaps with the lowest pelagic layer and the top sediment layer, sharing the same dissolved and suspended particulate material fields. The sediment is modelled in multiple layers with a thin layer of easily resuspendable material overlying thicker layers of more consolidated sediment. Dissolved and particulate biogeochemical tracers are advected and diffused throughout the model domain in an identical fashion to temperature and salinity. In addition, biogeochemical particulate substances sink and are resuspended in the same way as sediment particles. Biogeochemical processes are organized into pelagic processes of phytoplankton and zooplankton growth and mortality, detritus remineralization and fluxes of dissolved oxygen, nitrogen and phosphorus; epibenthic processes of growth and mortality of macroalgae, seagrass and corals; and sediment based processes of phytoplankton mortality, microphytobenthos growth, detrital remineralisation and fluxes of dissolved substances.

The biogeochemical model considers four groups of microalgae (small and large phytoplankton, *Trichodesmium* and microphytobenthos), three macrophytes types (seagrass species *Zostera* and *Halophila*, macroalgae) and coral communities. Photosynthetic growth is determined by concentrations of dissolved nutrients (nitrogen and phosphate) and photosynthetically active radiation. Autotrophs take up dissolved ammonium, nitrate, phosphate and inorganic carbon. Microalgae incorporate carbon (C), nitrogen (N) and phosphorus (P) at the Redfield ratio (106C:16N:1P), while macrophytes do so at the Atkinson ratio (550C:30N:1P). Microalgae contain two pigments (chlorophyll a and an accessory pigment), and have variable carbon:pigment ratios determined using a photoadaptation model.

Micro- and mesozooplankton graze on small and large phytoplankton, respectively, at rates determined by particle encounter rates and maximum ingestion rates. Half of the grazed material is released as dissolved and particulate carbon, nitrogen and phosphate, with the remainder forming detritus. Additional detritus accumulates by mortality. Detritus and dissolved organic substances are remineralized into inorganic carbon, nitrogen and phosphate with labile detritus transformed most rapidly (days), refractory detritus slower (months) and dissolved organic material transformed over the longest timescales (years). The production (by photosynthesis) and consumption (by respiration and remineralisation) of dissolved oxygen is also included in the model and depending on prevailing concentrations, facilitates or inhibits the oxidation of ammonia to nitrate and its subsequent denitrification to di-nitrogen gas, which is then lost from the system.

References

1. Death, D. G., Lough, J. M. & Fabricius, K. E. Declining coral calcification on the Great Barrier Reef. *Science* **323**, 116–119 (2009).
2. Brodie, J. & Waterhouse, J. A critical review of environmental management of the 'not so Great' Barrier Reef. *Estuar. Coast. Shelf. Sci.* **104**, 1–22 (2012).
3. Johnson, J. E. & Marshall, P. A. (eds). *Climate Change and the Great Barrier Reef* (Great Barrier Marine Park Authority and Australian Greenhouse Office, 2007).
4. Orr, J. C. et al. Anthropogenic ocean acidification over the twenty-first century and its impact on calcifying organisms. *Nature* **437**, 681–686 (2005).
5. Gattuso, J.-P. et al. Cross-chapter box on ocean acidification. in *IPCC, Climate Change 2014: Impacts, Adaptation, and Vulnerability. Part A: Global and Sectoral Aspects Contribution of Working Group II to the Fifth Assessment Report of the Intergovernmental Panel on Climate Change* (eds Pörtner, H.-O. et al.) (Cambridge Univ. Press, 2015).
6. Hughes, T. P., Day, J. C. & Brodie, J. Securing the future of the Great Barrier Reef. *Nat. Clim. Change* **5**, 508–511 (2015).
7. Strong, A. L., Kroeker, K. J., Teneva, L. T., Mease, L. A. & Kelly, R. P. Ocean acidification 2.0: managing our changing coastal ocean chemistry. *Bioscience* **64**, 581–592 (2014).
8. Ricke, K. L., Orr, J. C., Schneider, K. & Caldeira, K. Risks to coral reefs from ocean carbonate chemistry changes in recent earth system model projections. *Environ. Res. Lett.* **8**, 034003 (2013).
9. Duarte, C. M. et al. Is ocean acidification an open-ocean syndrome? Understanding anthropogenic impacts on seawater pH. *Estuar Coast* **36**, 221–236 (2013).
10. Anthony, K. R. N., Kleypas, J. A. & Gattuso, J.-P. Coral reefs modify their seawater carbon chemistry–implications for impacts of ocean acidification. *Glob. Change Biol.* **17**, 3655–3666 (2011).
11. Kleypas, J. A., Anthony, K. R. N. & Gattuso, J.-P. Coral reefs modify their seawater carbon chemistry–case study from a barrier reef (Moorea, French Polynesia). *Glob. Change Biol.* **17**, 3667–3678 (2011).
12. Shaw, E. C., McNeil, B., Tilbrook, B., Matear, R. & Bates, M. Anthropogenic changes to seawater buffer capacity combined with natural reef metabolism induce extreme future coral reef CO_2 conditions. *Glob. Change Biol.* **19**, 1632–1641 (2013).
13. Albright, R., Langdon, C. & Anthony, K. R. N. Dynamics of seawater carbonate chemistry, production, and calcification of a coral reef flat, central Great Barrier Reef. *Biogeosciences* **10**, 6747–6758 (2013).
14. Uthicke, S., Furnas, M. & Lonborg, C. Coral reefs on the edge? Carbon chemistry on inshore reefs of the Great Barrier Reef. *PLoS ONE* **9**, e109092 (2014).
15. Eyre, B. D., Andersson, A. J. & Cyronak, T. Benthic coral reef calcium carbonate dissolution in an acidifying ocean. *Nat. Clim. Change* **4**, 969–976 (2014).
16. Schiller, A., Herzfeld, M., Brinkman, R. & Stuart, G. Monitoring, predicting, and managing one of the seven natural wonders of the world. *Bull. Am. Meteorol. Soc.* **95**, 23–30 (2014).
17. Herzfeld, M. & Gillibrand, P. Active open boundary forcing using dual relaxation time-scales in downscaled ocean models. *Ocean Model.* **89**, 71–83 (2015).
18. Furnas, M. *Catchments and Corals: Terrestrial Runoff to the Great Barrier Reef* 34 (Australian Institute of Marine Science, 2003).
19. Furnas, M., Alongi, D., McKinnon, D., Trott, L. & Skuza, M. Regional-scale nitrogen and phosphorus budget for the northern (14S) and central (17S) Great Barrier Reef shelf ecosystem. *Cont. Shelf Res.* **31**, 1967–1990 (2011).
20. Baird, M. E. et al. Remote-sensing reflectance and true colour produced by a coupled hydrodynamic, optical, sediment, biogeochemical model of the Great Barrier Reef, Australia: comparison with remotely-sensed data. *Environ. Model. Software* **78**, 79–96 (2015).
21. Frieler, K. et al. Limiting global warming to 2 °C is unlikely to save most coral reefs. *Nat. Clim. Change.* **3**, 165–170 (2013).
22. Gattuso, J.-P. et al. Contrasting futures for ocean and society from different anthropogenic CO_2 emissions scenarios. *Science* **349**, aac4722 (2015).
23. Fabricius, K. E. et al. Losers and winners in coral reefs acclimatized to elevated carbon dioxide concentrations. *Nat. Clim. Change.* **1**, 165–169 (2011).
24. Mucci, A. The solubility of calcite and aragonite in seawater at various salinities, temperatures, and one atmosphere total pressure. *Am. J. Sci.* **283**, 780–799 (1983).
25. Atkinson, M. J. & Cuet, P. Possible effects of ocean acidification on coral reef biogeochemistry: topics for research. *Mar. Ecol. Prog. Ser.* **373**, 249–256 (2008).
26. Orr, J., Najjar, R., Sabine, C. & Joos, F. *Abiotic-HOWTO Internal OCMIP Report* 25pp (Tech. Rep., LSCE/CEA Saclay, Gif- sur-Yvette, France, 1999).
27. Dickson, A. G. & Millero, F. J. A comparison of the equilibrium constants for the dissociation of carbonic acid in seawater media. *Deep Sea Res.* **34**, 1733–1743 (1989).
28. Dickson, A. G. Standard potential of the reaction - $AgCl(s) + 1/2$ $H_2(g) = Ag(s) + HCl(aq)$ and the standard acidity constant of the ion $HSO4^-$ in synthetic sea-water from 273.15 K to 318.15 K. *J. Chem. Thermodyn.* **22**, 113–127 (1990).
29. Mehrbach, C., Culberso, C. H., Hawley, J. E. & Pytkowic, R. M. Measurement of apparent dissociation-constants of carbonic-acid in seawater at atmospheric-pressure. *Limnol. Oceanogr.* **18**, 897–907 (1973).
30. Munhoven, G. Mathematics of the total alkalinity-pH equation - pathway to robust and universal solution algorithms: the SolveSAPHE package v1.0.1. *Geosci. Model. Dev.* **6**, 1367–1388 (2013).
31. Wanninkhof, R. Relationship Between wind-speed and gas-exchange over the ocean. *J. Geophys. Res.* **97**, 7373–7382 (1992).
32. Wolf-Gladrow, D. A., Zeebe, R. E., Klaas, C., Kortzinger, A. & Dickson, A. G. Total alkalinity: The explicit conservative expression and its application to biogeochemical processes. *Mar. Chem.* **106**, 287–300 (2007).

33. Alongi, D. & McKinnon, A. The cycling and fate of terrestrially-derived sediments and nutrients in the coastal zone of the Great Barrier Reef shelf. *Mar. Poll. Bull.* **51**, 239–252 (2005).

34. Key, R. *et al.* A global ocean carbon climatology: Results from Global Data Analysis Project (GLODAP). *Global Biogeochem. Cy* **18**, 1–23 (2004).

35. Oke, P. R., Brassington, G. B., Griffin, D. & Schiller, A. The Bluelink ocean data assimilation system (BODAS). *Ocean Model.* **21**, 46–70 (2008).

36. Lenton, A., Tilbrook, B., Matear, R. J., sasse, T. & Nojiri, Y. Historical reconstruction of ocean acidification in the Australian region. *Biogeosci. Disc.* **12**, 8265–8297 (2015).

37. Kuchinke, M., Tilbrook, B. & Lenton, A. Seasonal variability of aragonite saturation state in the Western Pacific. *Mar. Chem.* **161**, 1–13 (2014).

38. Baird, M. E., Ralph, P. J., Rizwi, F., Wild-Allen, K. & Steven, A. D. L. A dynamic model of the cellular carbon to chlorophyll ratio applied to a batch culture and a continental shelf ecosystem. *Limnol. Oceanogr.* **58**, 1215–1226 (2013).

39. Gustafsson, M. S. M., Baird, M. E. & Ralph, P. J. The interchangeability of autotrophic and heterotrophic nitrogen sources in Scleractinian coral symbiotic relationships: A numerical study. *Ecol. Model.* **250**, 183–194 (2013).

40. Mongin, M. & Baird, M. The interacting effects of photosynthesis, calcification and water circulation on carbon chemistry variability on a coral reef flat: A modelling study. *Ecol. Model.* **284**, 19–34 (2014).

41. Beaman, R. *School of Earth and Environmental Sciences* (James Cook University, 2012).

42. Monsen, N., Cloern, J., Lucas, L. & Monismith, S. G. A comment on the use of  flushing time, residence time, and age as transport time scales. *Limnol. Oceanogr.* **47**, 1545–1553 (2002).

43. Hall, T. M. & Haine, T. W. N. On ocean transport diagnostics: The idealized age  tracer and the age spectrum. *J. Geophys. Res.* **32**, 1987–1991 (2002).

44. Silverman, J. *et al.* Carbon turnover rates in the One Tree Island reef: a 40-year  298 perspective. *J. Geophys. Res.* **117**, G03023 (2012).

45. Shaw, E. C., Mcneil, B. I. & Tilbrook, B. Impacts of ocean acidification in naturally variable coral reef flat ecosystems. *J. Geophys. Res. Ocean.* **117**, C03038 (2012).

46. Herzfeld, M. An alternative coordinate system for solving finite difference ocean models. *Ocean Model.* **14**, 174–196 (2006).

47. Margvelashvili, N., Saint-Cast, F. & Condie, S. Numerical modelling of the suspended sediment transport in Torres Strait. *Cont. Shelf Res.* **28**, 2241–2256 (2008).

48. Herzfeld, M. Methods for freshwater riverine input into regional ocean models. *Ocean Model.* **90**, 1–15 (2015).

Acknowledgements

This study was undertaken with the support of CSIRO Oceans and Atmosphere, and through resources made available through the eReefs project, the CSIRO Marine and Coastal Carbon Biogeochemistry Cluster, and the Australian Climate Change Science Program. The observations at Yongala were sourced as part of the Integrated Marine Observing System (IMOS)—IMOS is supported by the Australian Government through the National Collaborative Research Infrastructure Strategy and the Super Science Initiative. We thank Sven Uthicke for making his observations publically available, Professorss Tom Trull and Sabina Belli for their constructive comments on the manuscript, Miles Furnas who willingly shared his knowledge of the region and provided observations and three anonymous reviewers.

Author contributions

M.M. and M.E.B. conceived the study, implemented the analysis, and carried out the experiments; M.M., M.E.B., B.T., R.J.M., A.L., C.M.D. and A.D.L.S. developed the analysis; J.S. compiled the observational data and error statistics; M.H. developed the hydrodynamic model; K.W.-A. and B.J.R. implemented the catchment–biogeochemical model; N.M. developed the sediment model; M.E.B., M.S.M.G. and P.J.R. developed the coral growth model; M.M. and M.E.B. led the manuscript following writing contributions from all co-authors.

Additional information

Morphological plasticity of the coral skeleton under CO_2-driven seawater acidification

E. Tambutté[1,2,*], A.A. Venn[1,2,*], M. Holcomb[1,†], N. Segonds[1,2], N. Techer[1,2], D. Zoccola[1,2], D. Allemand[1,2] & S. Tambutté[1,2]

Ocean acidification causes corals to calcify at reduced rates, but current understanding of the underlying processes is limited. Here, we conduct a mechanistic study into how seawater acidification alters skeletal growth of the coral Stylophora pistillata. Reductions in colony calcification rates are manifested as increases in skeletal porosity at lower pH, while linear extension of skeletons remains unchanged. Inspection of the microstructure of skeletons and measurements of pH at the site of calcification indicate that dissolution is not responsible for changes in skeletal porosity. Instead, changes occur by enlargement of corallite-calyxes and thinning of associated skeletal elements, constituting a modification in skeleton architecture. We also detect increases in the organic matrix protein content of skeletons formed under lower pH. Overall, our study reveals that seawater acidification not only causes decreases in calcification, but can also cause morphological change of the coral skeleton to a more porous and potentially fragile phenotype.

[1] Marine Biology Department, Centre Scientifique de Monaco, 8 Quai Antoine 1er, Monaco 98000, Monaco. [2] Laboratoire Européen Associé 647 « BIOSENSIB » Centre Scientifique de Monaco- Centre National de la Recherche Scientifique, 8 Quai Antoine 1er, Monaco 98000, Monaco. * These authors contributed equally to the work. † Present address: ARC Centre of Excellence in Coral Reef Studies, School of Earth and Environment & Oceans Institute, The University of Western Australia, Crawley, WA 6009, Australia. Correspondence and requests for materials should be addressed to A.V. (email: avenn@centrescientifique.mc) or to S.T. (email: stambutte@centrescientifique.mc).

Calcification by scleractinian corals is a major contributor to the structural foundation of coral-reef ecosystems, which harbour an estimated one-quarter to a third of biodiversity in the ocean[1]. Climate change-related phenomena that impinge on the capacity of corals to calcify are therefore considered to be major threats to these marine resources. Declines in seawater pH and associated decreases in $CaCO_3$ mineral saturation states (Ω) driven by seawater uptake of CO_2 (ocean acidification) are predicted to be unfavourable to marine calcification[2,3]. An ever-growing body of laboratory and field-based research supports this prediction, and meta-analysis of data derived from such studies indicates declines in coral calcification of 15 to 22 % at levels of ocean acidification predicted to occur under business-as-usual scenarios of CO_2 emissions by the end of this century[4].

Nevertheless, there remains uncertainty about the extent of the threat of ocean acidification to coral calcification[5]. Some of this uncertainty stems from the lack of understanding of how and why ocean acidification affects coral skeleton formation[6]. This is partly due to the limited number of mechanistic studies that have attempted to dissect the effects of ocean acidification on multiple aspects of coral growth[7]. Calcification rate (mass addition) is a product of skeletal density (mass unit per volume) and extension (linear growth)[8], but ocean acidification studies rarely consider both of the latter two parameters. Furthermore, these parameters may vary due to morphological plasticity of the coral skeleton under different environmental regimes[9]. An integrated understanding of how ocean acidification affects coral skeleton formation is crucially needed to form a clearer perspective on physiological tradeoffs or processes of acclimatization that corals could undertake in an ocean with reduced pH and $\Omega_{aragonite}$.

Here we aim to provide a holistic picture of the impact of ocean acidification on coral skeleton formation, against a backdrop of coral physiology, to provide mechanistic insight into the vulnerability of the coral calcification process to reduced pH and $\Omega_{aragonite}$. We conducted a controlled laboratory study on *Stylophora pistillata*, which has been used extensively for investigations into mechanisms of coral resilience to ocean acidification[10–13] and for which molecular resources and physiological data are available[14,15]. We exposed colonies of *S. pistillata* for more than 1 year to four pH treatments; pH 8, 7.8, 7.4 and 7.2 with corresponding $\Omega_{aragonite}$ values of 3.2, 2.3, 1.1 and 0.7, respectively. Although these acidification treatments extend well beyond ocean acidification scenarios predicted by the International Panel on Climate Change (IPCC) in the coming century, they are similar to treatments in previous investigations, for example[10,11,16–18], that have proved useful in identifying clear trends in coral physiology under decreased pH/higher pCO_2. We characterized the response of a broad suite of interrelated physiological and calcification parameters to these conditions. Our findings reveal that changes in coral calcification rate during CO_2-driven seawater acidification can occur, not by physico-chemical dissolution of the skeleton, but rather by a change in skeleton morphology.

Results

Changes in skeletal growth parameters. Throughout the investigation, coral colonies remained visibly healthy (polyps remained extended and corals unbleached), and calcified in all experimental pH treatments, including treatment pH 7.2 ($\Omega_{aragonite} < 1$) (carbonate chemistry of treatments given in Table 1). Respiration and photosynthetic rates of coral colonies remained unchanged across experimental treatments, with respiration rates higher than net photosynthetic rates in all treatments (Table 2). As anticipated, rates of calcification declined under seawater acidification, with significant decreases measured at pH 7.4 and 7.2 relative to calcification at pH 8 (Fig. 1a). By contrast, rates of linear extension of corals did not change significantly across pH treatments (Fig. 1b). Instead, skeletal bulk density decreased significantly at low pH (Fig. 1c) and quantitative x-ray micro-computed tomography CT (micro-CT) analysis revealed an increase in skeletal porosity at pH 7.4 and 7.2 relative to pH 8 (Fig. 1d). Together, these data show that declines in calcification rate are manifested in a change in skeletal density and porosity, and not in extension rates of the skeleton.

Causes of increases in skeletal porosity. We endeavoured to determine the cause of increasing skeletal porosity, one possibility being that dissolution is responsible for such an effect. To investigate whether dissolution could occur in the coral's internal calcifying fluid where skeletal aragonite crystals form, we used confocal microscopy to measure calcifying fluid pH [11,19,20]. Our calcifying fluid pH measurements were consistent with our earlier work on *S. pistillata*[11], showing that, although seawater acidification depresses pH in the calcifying fluid to a certain extent, calcifying fluid pH remains high relative to the surrounding seawater (Fig. 2). Here we show for the first time, that calcifying fluid pH remains elevated in darkness as well as in light under conditions of seawater acidification (Fig. 2). These data are entirely consistent with our published estimates of $\Omega_{aragonite}$ of the calcifying fluid at similar pH values[11], and indicate that the calcifying environment in *S. pistillata* is supersaturated with respect to aragonite in corals from pH 8 and 7.2 treatments in light and darkness. Dissolution due to decreases in pH at the site of calcification would therefore not be expected to be responsible for increases in skeletal porosity.

We also inspected aragonite crystal morphology in coral skeletons taken from the four pH treatments using scanning electron microscopy (SEM). We compared aragonite crystal morphology across treatments, on the skeleton surface (Fig. 2) and within pores inside the skeleton. In addition to daytime sampling, we also examined samples taken at the end of the night time period, to check whether dissolution was occurring in darkness (Supplementary Fig. 1). In all cases aragonite crystal morphology was very similar, comprising bundles of fibres that ranged between flat blades and rhomb-shaped structures, as observed previously for corals[21,22]. No signs of damage or

Table 1 | Carbonate chemistry parameters in the four experimental pH treatments.

Treatment name	pH_T	Total alkalinity ($\mu mol\,kg^{-1}$-SW)	pCO_2 (μatm)	HCO_3^- ($\mu mol\,kg^{-1}$-SW)	CO_3^{2-} ($\mu mol\,kg^{-1}$-SW)	Total carbon ($\mu mol\,kg^{-1}$-SW)	Ω_{ar}
pH 7.2	7.2 ± 0.01	2530.96 ± 7.15	3792.57 ± 52.54	2423.67 ± 5.22	44.23 ± 0.81	2573.68 ± 4.57	0.69 ± 0.01
pH 7.4	7.40 ± 0.01	2490.00 ± 6.06	2256.88 ± 32.98	2323.49 ± 3.20	68.38 ± 1.21	2454.76 ± 3.48	1.06 ± 0.02
pH 7.8	7.79 ± 0.01	2485.12 ± 5.25	856.17 ± 13.63	2119.11 ± 1.40	149.94 ± 2.37	2292.90 ± 1.58	2.33 ± 0.04
pH 8.0	7.95 ± 0.00	2461.73 ± 6.31	538.41 ± 4.85	1961.44 ± 1.48	204.27 ± 2.06	2180.71 ± 3.42	3.17 ± 0.03

Data are means ± s.d. Parameters of carbonate seawater chemistry were calculated from total scale pH, TA, temperature and salinity using the free-access CO2SYS package (ref. 10) using constants from Mehrbach et al.[39] as refit by Dickson and Millero[40].

Table 2 | Physiological parameters measured in the four pH treatments.

	Units	7.20	7.40	7.80	8.0	F	P-value
Normalized by surface area							
Photosynthesis	$\mu mol\ O_2\ h^{-1} cm^{-2}$	1.14 ± 0.18	0.82 ± 0.15	1.00 ± 0.18	1.01 ± 0.23	$F_{3,36}=0.988$	0.409
Respiration	$\mu mol\ O_2\ h^{-1} cm^{-2}$	1.39 ± 0.24	1.31 ± 0.23	1.33 ± 0.20	1.25 ± 0.18	$F_{3,36}=0.072$	0.974
Protein	$mg\ cm^{-2}$	1.18 ± 0.12	0.91 ± 0.11	0.95 ± 0.11	0.97 ± 0.10	$F_{3,36}=2,493$	0.76
Symbiont density	$cells\ cm^{-2}$	$2.11 \times 10^6 \pm 1.68 \times 10^5$	$1.78 \times 10^6 \pm 1.68 \times 10^5$	$1.50 \times 10^6 \pm 1.78 \times 10^5$	$2.14 \times 10^6 \pm 2.03 \times 10^5$	$F_{3,16}=2.110$	0.139
Chla	$\mu g\ cm^{-2}$	9.77 ± 0.58	11.20 ± 0.59	9.35 ± 0.58	7.80 ± 1.05	$F_{3,16}=1.710$	0.205
Chlc$_2$	$\mu g\ cm^{-2}$	2.01 ± 0.16	2.27 ± 0.16	1.63 ± 0.16	1.74 ± 0.12	$F_{3,16}=2.458$	0.100
Normalized by protein							
Photosynthesis	$\mu mol\ O_2\ h^{-1} mg^{-1}$	0.95 ± 0.11	0.82 ± 0.18	0.95 ± 0.18	0.92 ± 0.14	$F_{3,48}=0.444$	0.723
Respiration	$\mu mol\ O_2\ h^{-1} mg^{-1}$	1.03 ± 0.21	1.40 ± 0.36	1.45 ± 0.21	1.24 ± 0.12	$F_{3,48}=0.554$	0.648
Symbiont density	$Cell\ mg^{-1}$	$2.12 \times 10^6 \pm 1.53 \times 10^5$	$1.82 \times 10^6 \pm 1.72 \times 10^5$	$1.71 \times 10^6 \pm 1.54 \times 10^5$	$2.56 \times 10^6 \pm 1.58 \times 10^5$	$F_{3,16}=0.699$	0.566
Chla	$\mu g\ mg^{-1}$	9.82 ± 0.51	11.44 ± 0.59	10.39 ± 0.72	9.41 ± 1.16	$F_{3,16}=0.851$	0.486
Chlc$_2$	$\mu g\ mg^{-1}$	2.01 ± 0.14	2.02 ± 0.16	2.20 ± 0.18	2.34 ± 0.24	$F_{3,16}=1.933$	0.165
Normalized per symbiont cell							
Photosynthesis	$\mu mol\ O_2\ h^{-1} cell^{-1}$	$4.15 \times 10^{-7} \pm 2.26 \times 10^{-8}$	$3.97 \times 10^{-7} \pm 2.50 \times 10^{-7}$	$3.66 \times 10^{-7} \pm 2.23 \times 10^{-8}$	$3.89 \times 10^{-7} \pm 3.13 \times 10^{-8}$	$F_{3,16}=2.493$	0.967
Chla	$pg\ cell^{-1}$	$4.69^* \pm 0.30$	$4.95^* \pm 0.28$	$5.03^* \pm 0.10$	5.42 ± 0.14	$F_{3,16}=6,173$	0.005
Chlc$_2$	$pg\ cell^{-1}$	0.96 ± 0.06	0.99 ± 0.05	1.04 ± 0.05	1.09 ± 0.05	$F_{3,16}=3,152$	0.052

Data are mean ± s.e. Data were analysed by 1 way ANOVA.
*Denotes mean values significantly different than the mean value from treatment with pH 8 ($P<0.05$).

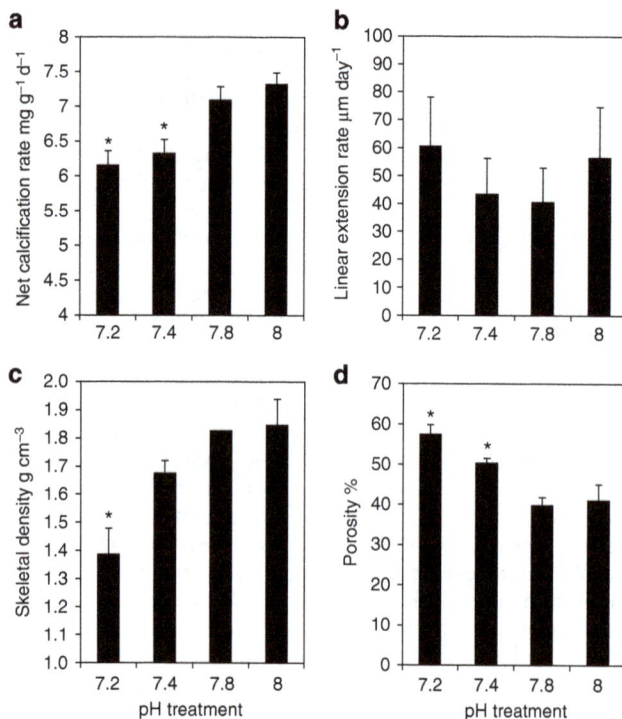

Figure 1 | Skeletal growth parameters in the four pH treatments.
(a) Net calcification rate (one way ANOVA, $n=12$, $F_{3,44}=4.11$, $P<0.05$).
(b) Linear extension (one way ANOVA, $n=15$, $F_{3,56}=0.62$, $P>0.05$).
(c) Bulk skeletal density (one way ANOVA, $n=7$, $F_{3,24}=16.44$, $P<0.001$).
(d) Skeletal porosity (one way ANOVA, $n=3$, $F_{3,8}=11.05$, $P<0.05$).
Data are means ± s.e.m. Asterisk (*) indicates values that are significantly different for treatment with pH 8 ($P<0.05$).

dissolution were detected to the fibre-bundle arrangement nor to the smooth faces of the fibres. The crystal morphology we observed is unlike coral aragonite crystals that have undergone dissolution, which are irregular and corroded by comparison, as shown in previously published SEM images of coral skeleton dissolution[18].

An alternative explanation for the increase in skeletal porosity under acidification was found at higher levels of organization in the skeleton's architecture. We measured the size of corallite calyxes (the interior diameter of the cup-shaped openings in the skeleton that house the polyps[23] (see Supplementary Fig. 2 for nomenclature of coral skeletal elements)). The size of corallite calyxes significantly increased with decreasing pH (Fig. 3a,b) and as larger corallite calyxes formed under acidification, there was a corresponding decrease in the number of corallites per unit surface area (Supplementary Fig. 3). Corallite calyx size from the four pH treatments was significantly correlated with both porosity and bulk density, indicating that the acidification-induced enlargement of corallite calyx size is a factor explaining the decrease in skeletal density and increase in porosity (Spearman's rank-order tests: $r_s=0.783$, $P<0.01$; $r_s=0.744$, $P<0.01$). In addition, micro-CT images showed that enlargement of corallite calyxes at the surface of the skeleton corresponded to larger corallites extending throughout the interior of the skeleton, making it more porous (Fig. 4). Micro-CT images also revealed that corallites formed at lower pH are characterized by thinner septae, thecae and areas of coenosteum (Fig. 4), indicating that changes in the development of these features also contributed to changes in skeletal porosity. Taken together, these observations constitute an important finding: changes in calcification rate can be explained, not by physico-chemical dissolution of the skeleton, but rather by a change in skeleton phenotype.

Organic matrix protein incorporation in the skeleton. Changes in skeleton morphology were accompanied by a significant increase in organic matrix (OM) proteins per gram of skeleton in corals from pH 7.2 versus pH 8 (Fig. 5). The increase in the ratio of OM proteins to gram of skeleton is not driven solely by the decrease in calcification, because, when OM per gram is converted to OM per cm^3 using mean bulk density values (Fig. 1c), an increase in OM is still observed ($30.60\ \mu g\ cm^{-3}$ at pH 8 versus $34.16\ \mu g\ cm^{-3}$ at pH 7.2). This finding is the first indication that ocean acidification causes an enrichment in OM incorporation in the coral skeleton and is also consistent with transcriptomic data from other coral species showing that the expression of certain OM protein genes are upregulated during acidification in coral juveniles[24] and adults[17].

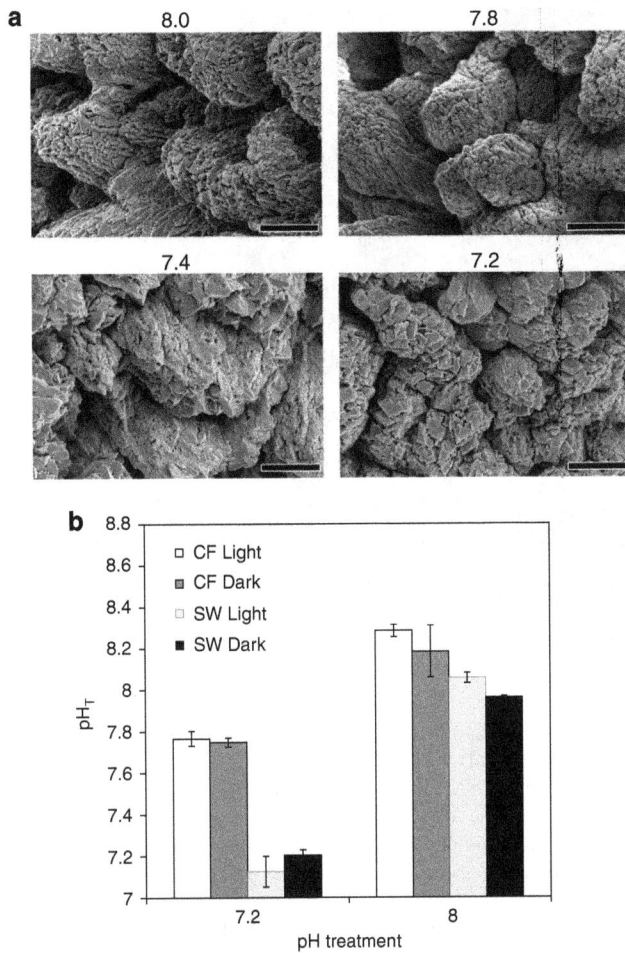

Figure 2 | Indicators of physio-chemical conditions at the site of calcification in corals in the four pH treatments. (a) Aragonite crystal morphology imaged by scanning electron microscopy. Scale bar, 5 μm. pH treatment indicated above each image. (b) Calcifying fluid pH in corals under light and dark conditions at seawater pH 7.2 and pH 8.0. Data are means ± s.e.m. CF = calcifying fluid; SW = seawater surrounding the colonies.

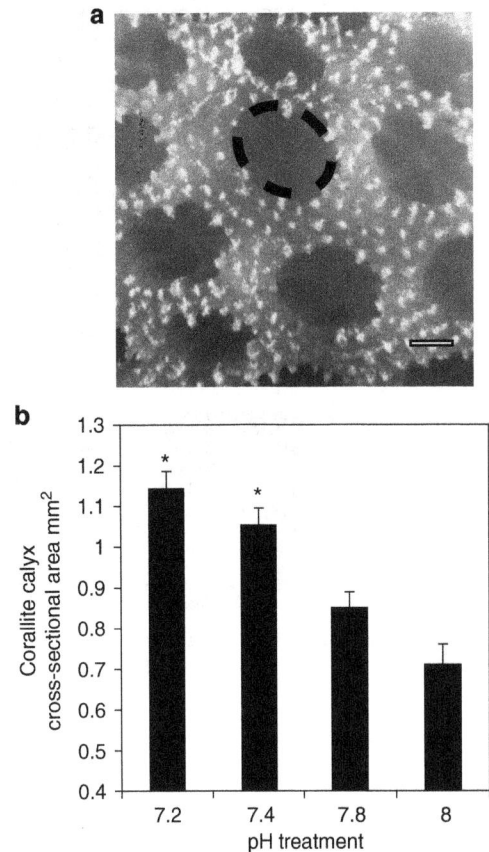

Figure 3 | Corallite calyx size in the four pH treatments. (a) Image of corallite calyxes in the skeleton of S. pistillata. Dotted line shows the extent of cross-sectional area of a representative corallite calyx. Scale bar, 0.5 mm. (b) Corallite calyx size (cross-sectional area) (one way ANOVA, $n = 9$, $F_{3,32} = 21.60$, $P < 0.001$). Data are means ± s.e.m. Asterisk (*) indicates values that are significantly different for treatment with pH 8 ($P < 0.05$).

Discussion

The vulnerability of the coral calcification process to ocean acidification is well documented, but not well understood from a mechanistic point of view. As such, the current study aimed to investigate mechanisms underlying the calcification response of the model coral species S. pistillata, by analysing several interrelated calcification and physiological parameters. The principle finding of our study is that declines in coral calcification induced by seawater acidification were accompanied by a change in the morphology of the coral skeleton. At lower pH, corals were characterized by larger corallite calyxes and thinner associated skeletal structures including septae and thecae, resulting in more porous skeletons.

Recent field-based research has also noted increases in skeletal porosity associated with decreased pH and $\Omega_{aragonite}$, but with the caveat that changes in skeletal growth may also have been affected by other environmental factors that co-vary in the field[25]. Here, in our laboratory experiments we were able to ensure that only pH and carbonate chemistry varied between our treatments, whereas other environmental factors were kept constant, enabling us to directly assign a role of seawater acidification in causing changes in skeletal formation. In addition, variation in rates of

photosynthesis and respiration can be difficult to control in the field, but here under our experimental set-up, rates of these parameters were not different between the four treatments. This result is informative from a mechanistic perspective because it indicates that the calcification response we observed can be attributed to a direct effect of seawater acidification, rather than indirect effects via changes in photosynthetic or respiration rates.

In considering the mechanisms that contribute to changes in skeletal formation, dissolution could be considered as a potential factor, but our findings do not support this idea. Indeed SEM imaging of coral skeletons revealed similar aragonite crystal morphology in all the treatments, which is not consistent with aragonite crystals dissolving at lower pH treatments. This finding is in agreement with a previous study that demonstrated how living coral tissue protects the underlying skeleton from dissolution, even when corals are exposed for extended periods to seawater, which is under-saturated with respect to aragonite[18]. Given that our samples were entirely covered by tissue, dissolution would not be expected to be a factor contributing to changes in skeleton formation as long as pH and $\Omega_{aragonite}$ at the site of calcification is controlled.

Indeed, calcifying fluid pH measurements indicate that $\Omega_{aragonite}$ states were maintained at elevated levels in the calcifying environment in the lowest pH treatment, in both light and darkness. These data in the current study agree with a previous study that this species has a capacity to regulate calcifying fluid pH above seawater pH in the absence of photosynthesis[26]. This

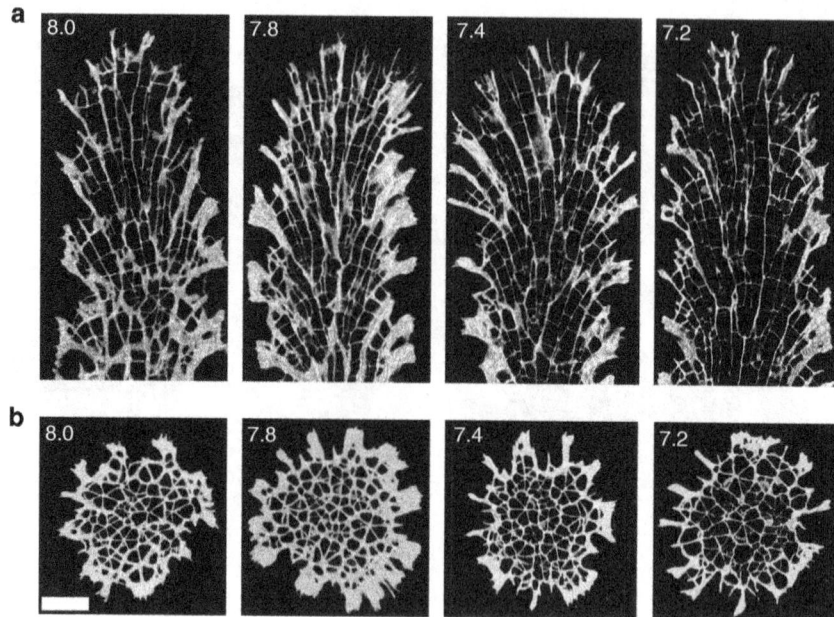

Figure 4 | Coral skeleton morphology in the four pH treatments imaged by micro-CT. (**a**) Representative longitudinal sections; (**b**) transverse sections. pH treatment is indicated in the top left corner of each image. Scale bar, 1 mm.

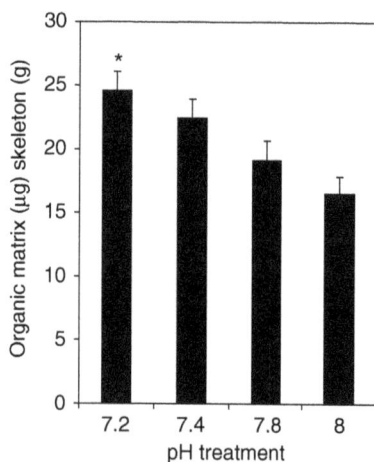

Figure 5 | Organic matrix protein content of the coral skeletons in the four pH treatments. (one way ANOVA, $n = 6$, $F_{3,20} = 3.384$, $P < 0.05$). Data are means ± s.e.m. Asterisk (*) indicates values that are significantly different for treatment with pH 8 ($P < 0.05$).

conditions indicate that dissolution was not a factor leading to increasing porosity of coral skeletons.

Interestingly, we observed increased incorporation of OM proteins into the coral skeleton in corals grown under seawater acidification. This observation is intriguing considering that one of the primary roles of the OM in most biomineralizing systems is to reduce the free energy required for crystal nucleation, thereby facilitating calcification at lower saturation state values[29]. Recent research suggests that OM proteins do indeed perform this function in corals, as molecular characterization of OM proteins in *S. pistillata* and *in vitro* studies of their properties suggest that OM proteins catalyse precipitation of new aragonite crystals[12]. Given that corals exposed to reduced seawater pH experienced slight declines in the calcifying fluid pH and therefore $\Omega_{aragonite}$, it is possible that corals in the current study could have increased levels of OM proteins under acidification to promote calcification under less favourable calcifying fluid $\Omega_{aragonite}$. This is only a hypothesis, but future research into the role of the OM in coral biomineralization under conditions of seawater acidification could explore this possibility. Furthermore, in other marine calcifiers OM proteins have been found to direct orientation of aragonite crystals and determinate skeletal morphology[30,31]. These functions have not yet been demonstrated in corals, but the association of changes in skeletal morphology with the increase in OM proteins is intriguing and is also an area for future research.

We synthesized the results of our study to describe the phenotypic response of *S. pistillata* to seawater acidification (Fig. 6). Exposure of corals to seawater acidification results in increased rates of proton removal from the calcifying fluid to maintain pH elevated at the site of calcification[11]. Nevertheless, as the proton gradient with the surrounding seawater becomes less favourable, calcifying fluid pH (and therefore $\Omega_{aragonite}$) decreases slightly[11]. At the same time, *S. pistillata* increases production of OM proteins per unit mass of $CaCO_3$. In these conditions, corals continue to calcify and dissolution of the skeleton does not occur, even when seawater $\Omega_{aragonite} < 1$. However, pH upregulation and OM production are energy-requiring processes; thus, calcification becomes more difficult and energetically more costly under seawater acidification. Rates of $CaCO_3$ production decrease and

trait is also found in non-photosynthetic deep water coral species that have the capacity to maintain elevated calcifying fluid pH under decreased seawater pH[27]. The mechanisms by which corals achieve pH regulation of the calcifying fluid are not fully understood, but upregulation of calcifying fluid pH is believed to be an energy-requiring process that involves removal of protons from the calcifying fluid by Ca^{2+} ATPases (reviewed in ref. 15). Available energy for pH upregulation is anticipated to be limited in darkness, as research with microsensors indicates that tissues near the skeleton are hypoxic in dark conditions in some corals[28]. Further research must better characterize pH regulation by calcifying (calicoblastic) cells and its energetics to reach a clearer understanding of this aspect of calcification physiology in a range of coral species. In any case, when determining the causes of changes in skeletal growth in the context of the current study, the key interpretation of the pH data presented here is that elevated calcifying pH values in the low pH treatment in dark or light

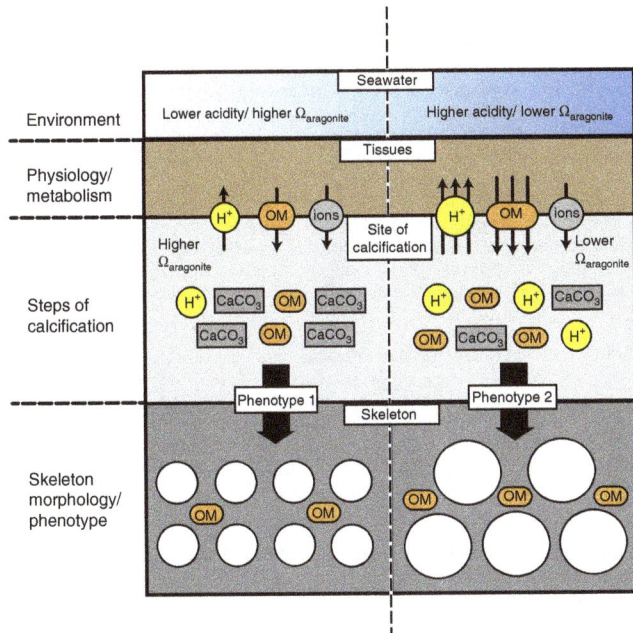

Figure 6 | Schematic summary of the impact of ocean acidification on skeletal growth in *Stylophora pistillata*. Environmental change in the form of seawater acidification depresses pH and $\Omega_{aragonite}$ at the site of calcification. Coral physiology responds by increasing proton removal from the calcifying fluid to maintain elevated pH and $\Omega_{aragonite}$ that favours calcification. *S. pistillata* also increases production of organic matrix proteins (OM = organic matrix) per unit mass of $CaCO_3$. In these conditions, corals continue to calcify, and dissolution of the skeleton does not occur, even when seawater $\Omega_{aragonite} < 1$. However, with lower saturation states in the calcifying fluid and increased energy expenditure for calcification, *S. pistillata* changes its skeleton phenotype to a morphology characterized by larger corallite calyxes. The resulting skeleton is more porous. 'OM' = organic matrix. 'Ions' represent both transcellular and paracellular transport of ions needed for the steps of calcification. White circles in skeleton represent corallites. 'Steps of calcification' encompass steps of skeleton precipitation and assembly outlined in Tambutté *et al.*[15].

S. pistillata changes the morphology of the skeleton, shifting to a skeletal phenotype characterized by larger corallite calyxes and thinner septae and thecae. The resulting skeleton is more porous and less dense, but rates of linear extension are maintained.

The potential ecological consequences of the trade-off of maintaining linear extension while producing a more porous skeleton are unknown, but it can be speculated that this may be a useful strategy for benthic organisms such as corals, where competition for space and light is an intense selective pressure[32]. However, as increasing skeletal porosity has been shown previously to result in more fragile skeletons[33], this trade-off could present significant disadvantages. Fragile, porous skeletons are more susceptible to bioerosion[34] and could be more easily damaged in high waveenergy environments or during events such as hurricanes[33]. Interestingly, our model species *S. pistillata* has been previously categorized as a species that is tolerant to ocean acidification over the same pH range tested here[10,11,16]. This is largely based on the fact this species effectively mitigates the effect of decreasing seawater pH and $\Omega_{aragonite}$ at the site of calcification by upregulating pH[10,11]. However, the present study demonstrates that a more comprehensive, multi-parameter assessment of skeleton formation under seawater acidification can reveal previously unseen vulnerabilities in the calcification process; in this case, a morphological shift to more porous and potentially more fragile skeletal architecture.

In any case, extrapolation of our laboratory-based study to predictions of how corals will behave in the field in a future high-CO_2 world should be made with caution. While informative from a mechanistic point of view, our study was conducted inside narrowly controlled environmental parameters, which do not simulate the dynamic and changing environment on the reef. For example, corals living in shallow water can experience greater midday irradiances than levels provided in the current experiments. It is possible that the combination of higher irradiances with seawater acidification may have resulted in additional effects on coral physiology, notably bleaching (loss of algal symbionts), as observed in previous work[35]. Furthermore our mechanistic study has relied on a single, model species that is tractable for laboratory studies, but the calcification and associated physiology of other coral species may respond differently to seawater acidification. Notable examples include previous accounts of where ocean acidification has caused some coral species to change from colonial forms to non-calcifying single polyp forms[36].

In conclusion, our study on *S. pistillata* reveals how CO_2-driven seawater acidification can cause a shift in coral skeleton morphology that occurs in concert with a decrease in coral calcification rates. The novelty of these findings lies in the demonstration that seawater acidification not only decreases the rate at which coral skeletons grow, but it can also affect assembly of the overall skeletal architecture. The porous-phenotype that we describe here may potentially lead to more fragile coral skeletons under conditions of acidification making corals more vulnerable to damage and bioerosion. More generally, our findings underline the importance of using a multi-parameter analysis of skeletal growth when investigating the vulnerability of coral species to ocean acidification.

Methods

Biological material and treatments. Colonies of the tropical coral *Stylophora pistillata* were exposed to long-term seawater acidification as described previously[11]. Corals were kept in aquaria supplied with Mediterranean seawater (exchange rate 70%/h) at a salinity of 38, temperature 25 °C and irradiance of 170 μmol photons $m^{-2} s^{-1}$ on a 12-h/12-h photoperiod provided by HQI-10000K metal halide lamps (BLV Nepturion). Carbonate chemistry was manipulated by bubbling with CO_2 to reduce pH to the target values of pH 7.2, pH 7.4 and pH 7.8 (Table 1). A fourth treatment (pH 8) was not bubbled with CO_2 (Table 1). pH and temperature probes (Ponsel-Mesure, France) were installed in the four treatment aquaria and connected to a custom-made monitoring system (Enoleo, Monaco), which controlled CO_2 bubbling. On a daily basis, pH checks were carried with a Digital ODEON pH meter (Ponsel-Mesure, France). All probes were calibrated to pH total scale. In addition, biweekly pH measurements were made using the indicator dye m-cresol purple (Acros 199250050) adapted from Dickson *et al.*[37]; the absorbance was measured using a spectrophotometer (UVmc2; Safas, Monaco).

Measurements of total alkalinity (TA) were made weekly according to protocols described in Dickson *et al.*[37]. TA was measured via titration with 0.03 N HCl containing 40.7 g NaCl per litre using a Metrohm Titrando 888 Dosimat controlled by Tiamo software to perform automated titrations of 4-ml samples, and alkalinity was calculated using a regression routine based on Department of Energy guidelines[38]. For each sample run, certified seawater reference material supplied by the laboratory of A. G. Dickson (Scripps Institution of Oceanography, La Jolla, CA) was used to verify acid normality. Parameters of carbonate seawater chemistry were calculated from total scale pH, TA, temperature and salinity using the free-access CO2SYS package (10) using constants from Mehrbach *et al.*[39] as refit by Dickson and Millero[40]. Parameters of carbonate seawater chemistry in each treatment are given in Table 1.

In each of the four aquariums, submersible pumps ensured high water circulation. Corals were fed twice a week with *Artemia salina* nauplii. To avoid 'tank' effects, aquariums were rigorously cleaned every week to prevent the growth of epiphytic algae and fouling communities or the accumulation of detritus. Biomass in the aquariums was kept relatively constant by sampling and by addition of new corals (coral colonies were identified by labelling colonies with a radio frequency identification microchip). This maintenance, the high seawater renewal, and the regular monitoring of seawater chemistry (described below), temperature and irradiance ensured that similar conditions prevailed in each aquarium, except for the carbonate chemistry.

Experimental design and statistical analysis. Analysis of physiological and calcification parameters was conducted on coral colonies that had already experienced 1 year of exposure to the pH treatments. Analysis of each parameter was carried out in sequential experimental trials. Each trial included replicates from each of the four treatments. Sample sizes for each parameter maximized replication while keeping experiments feasible. The number of replicates and trials for each parameter are given in the following text.

For each experimental parameter, results of the experimental trials were compared using two-way analysis of variance (ANOVA) using trial and pH treatment as factors. If no significant difference was found between the experimental trials or if the trend of the response to acidification was identical in trials, then the results of the trials were pooled and analysed together using one-way ANOVA, with Student-Newman-Keuls *post hoc* analysis to identify significant differences between treatments. Correlation analysis was carried out by using Spearman rank-order tests. Where necessary, the data were logarithmically or square-root transformed to adhere to the assumptions necessary for parametric analysis (normal distributions with homogenous variances). Percentage data were arcin-square root transformed for normalization. Data were analysed with SPSS Statistics 21 software (IBM, France).

Calcification rate.
Net calcification rate was measured by buoyant weight analysis[41]. Working with microcolonies grown suspended on nylon threads, corals were suspended below a microbalance (XP205 delta range, Mettler Toledo, France) in seawater maintained at 25 °C. Corals were buoyant weighed in this manner weekly for 20 weeks. Net calcification rate was determined by normalizing the change in buoyant weight of each sample by the initial buoyant weight of the sample and then dividing by the number of days between sampling times. Data were thus expressed as $mg\,g^{-1}\,d^{-1}$.

The analysis was carried out in two experimental trials using six microcolonies per treatment each trial.

Linear extension.
Linear extension was assessed by measuring the increase in height of branches attached to glass slides. Branch apexes $\sim 1\,cm$ high were attached to glass slides and allowed to recover. Calipers were then used to measure the height of the colony, and measurements repeated 2–3 months later. The analysis was carried out in five experimental trials with three replicates per treatment each trial.

Skeletal bulk density.
Bulk density was obtained by methods adapted from Bucher et al.[42]. Branches of similar size were sampled from seven colonies in each pH treatment and placed in 10% NaClO to remove tissues. Branches were then rinsed and oven-dried to constant weight and weighed to obtain clean dry weight (DW_{clean}). Samples were then coated with paraffin wax and the dry weight (DW_{wax}) measured again. The buoyant weight (BW_{wax}) was obtained in dH_2O at 20 °C. Total enclosed volume was calculated by the following formula:

$$V_{enclosed} = (DW_{wax} - BW_{wax})\,/\,\partial m$$

Where ∂m is the density of the medium (that is $1\,g\,cm^3$). The bulk density was determined by the following equation:

$$Bulk\ density = DW_{clean}\,/\,V_{enclosed}$$

Bulk density was analysed in one experimental trial, with seven replicates per treatment.

Skeletal porosity.
X-Ray micro-computed tomography (Micro-CT) is a non-invasive high-resolution imaging method for visualizing the external and internal structure of objects that has recently been validated to examine porosity in coral skeletons[43,44]. Micro-CT analysis was carried out at the Polyclinique St Jean, Cagnes sur Mer, France, with an Skyscan 1173 compact micro-CT (SkyScan, Antwerp, Belgium). A microfocus X-ray tube with a focal spot of $10\,\mu m$ was used as a source (80 kV, 100 μA). The sample was rotated 360° between the X-Ray source and the camera. Rotation steps of 1.5° were used. At each angle an X-Ray exposure was recorded on the distortion-free flat-panel sensor (resolution $2,240 \times 2,240$ pixels). The resulting slice is made of voxels, the three-dimensional equivalent of pixels. Each voxel is assigned a grey value derived from a linear attenuation coefficient that relates to the density of materials being scanned. All specimens were scanned at the same voxel size. The radial projections were reconstructed into a three-dimensional matrix of isotropic voxels ranging from 5 to $10\,\mu m$, depending on the exact height of the coral tip.

X-Ray images were transformed by NRecon software (Skyscan) to reconstruct 2-D images for quantitative analysis. From these 2-D images evaluation of the morphometric parameters was performed using CTann® CT analysis software (SkyScan). A manual greyscale threshold was implemented manually on the first set of images and then applied to all specimens.

For each sample, a digital region of interest was created to extend through $100\,\mu m$ of skeleton at 7 mm distance from the apex corresponding to about 15 slices. Then the percentage negative space versus skeleton was determined, providing the measure of porosity.

For each treatment condition, three branches of similar size were taken from the apical part of a colony. Porosity was analysed in one experimental trial with three replicates per treatment.

Calyx size (interior diameter of corallites).
Branches of similar size were sampled from colonies in each pH treatment, and placed in a 10% NaClO solution for 30 min to remove tissues. Skeletons were then rinsed several times in ultrapure H_2O and dried at room temperature. Samples were observed under a macroscope at $\times 15$ magnification (Z16APO, Leica Microsystems) using a digital camera (Tri-CCD JAI AT200 GE 2MP) to determine calyx size and density. The interior diameter of corallites (which we call calyx according to the terminology of (Johnson 1981)) was determined by measuring the cross-sectional area of all calyxes per branch and obtaining the mean. Corallite calyx density was calculated by normalizing the total number of calyxes on the branch to its surface area. Cross-sectional area measurements and counting of calyxes were achieved using SAISAMsoftware (Microvision Instruments, France). Calyx size and density were assessed in three experimental trials with three replicates per trial.

SEM of skeleton microstructure.
Skeleton microstructure (for example, crystal morphology) was examined in three colonies from each treatment at the end of daytime and night time periods by SEM as described previously in Tambutté et al.[45]. In brief, the soft tissues were removed with NaClO 10%, rinsed with distilled water then oven-dried. All samples were coated with gold and observed at 5 kV in a JEOL JSM-6010LV. Analysis of skeleton microstructure was carried out in one experimental trial.

Confocal microscope measurements of calcifying fluid pH.
Analysis was carried out as described previously[11] on samples grown laterally on glass coverslips fitted in semi-closed perfusion chambers (PeCon, Germany) that were mounted on the confocal microscope and supplied with seawater drawn from the desired acidification treatment. The seawater pH and carbonate chemistry of the perfused seawater was checked by measuring TA and pH in the inflowing and outflowing seawater to check that carbonate chemistry did not drift away from the target values in treatment aquariums during the period of measurement. We used higher flow rates than in previous work (for example, ref. 11) to achieve stable pH and carbonate chemistry in both light and dark conditions (renewal rate of 50% per min of a 2.5-ml vol). Sample sizes were restricted to $1\,cm^2$ in surface area, irradiance provided at $170\,\mu mol$ photons $m^{-2}\,s^{-1}$ (Philips 21V 150-W halogen bulb) in light treatments, and temperature maintained at 25 °C. Measurement of oxygen in seawater in the perfusion chamber with a needle-type microsensor (PreSens, Germany) in light and darkness indicated oxygen levels also remained stable between values of $265–280\,\mu mol\,l^{-1}$ under these conditions.

Dye loading and pH measurements.
Measurements of pH of the calcifying fluid ($=$ subcalicoblastic medium) in the light and dark were made by inverted confocal microscopy (Leica SP5, Germany) and the ratiometric dye SNARF-1 (Invitrogen) according to methods we published previously (refs 11,26).

Before pH measurements, samples were first perfused with seawater from the desired pH treatment for 10 min in either light or darkness. Samples were then loaded with pH-sensitive dye by perfusion of seawater containing $45\,\mu M$ cell-impermeable SNARF-1 for an additional 5 min under the same conditions. Perfusion of samples with seawater at the desired pH treatment containing SNARF-1 continued for an additional 10 min in light or darkness, during which five measurements of calcifying fluid pH were made 2 min apart to ensure a stable pH value was obtained.

Calcifying fluid pH measurements were carried out at $\times 40$ magnification by excitation of SNARF-1 at 543 nm at 30% laser intensity and fluorescence captured at emission wavelengths of $585 \pm 10\,nm$ and $640 \pm 10\,nm$. For each measurement, several optical sections were captured in a Z-stack, with an acquisition time of $\sim 10\,s$, during which the part of the colony under analysis is exposed to the laser. The ratio of fluorescence at the emission wavelengths was calibrated to pH by methods previously published[26], except that the standard curves were produced using seawater adjusted to the range pH 7–9 on total scale rather than National Bureau of Standards (NBS) scale.

Calcifying fluid pH was measured in light and dark conditions in three samples from pH treatments pH 7.2 and pH 8.

OM protein content.
Branches were sampled and incubated in 10% sodium hypochlorite (NaClO) to eliminate soft tissues and separate skeletons. Skeletons were thoroughly rinsed with ultrapure water and cryo-ground (Spex SamplePrep 6770 apparatus) into powder of homogeneous granulometry (about $30\,\mu m$ diameter). The powder was incubated with 10% NaClO at 4 °C for 24 h to remove potential contaminants such as endoliths. The resulting solution was centrifuged (3,500 g, 5 min, 4 °C). The pellets of skeleton powders were rinsed several times with ultrapure water and freeze dried. To separate the organic fraction from the mineral fraction, 15 g of powder of skeletons were demineralized in 0.25 M EDTA (pH 7.8, 23 h, 4 °C), the solution was prefiltered on $0.2\,\mu m$ of polyethersulfone filters and then filtered in tandem through two Sep-Pak Plus C18 cartridges (Waters, 5 kDa) according to the protocol of Rahman et al., 2013. The eluted macromolecules were frozen at $-80\,°C$ and subsequently lyophilized. Protein content was determined using the bicinchoninic acid assay kit (BC Protein Assay, Interchim). The standard curve was established with bovine serum albumin and

the absorbance was measured with a microplate reader (EpochTM, Bioteck, US) at 562 nm.

The analysis was carried out in two experimental trials with three replicates per treatment for each trial.

Photosynthesis and respiration rates and biomass parameters. Five microcolonies of similar size were used for measuring photosynthetic rates and respiration, after which they were frozen at $-80\,°C$. Protein content, chlorophyll (Chl) content and zooxanthella density were determined as described below.

Each microcolony was placed on a nylon net in a closed beaker agitated using a magnetic stirrer. Incubations were performed in the same conditions of temperature ($25\,°C$) and seawater chemistry as in experimental tanks but either under light conditions (170 μmol photons m^{-2} s^{-1}) for photosynthesis, or dark conditions for respiration (seawater pH and alkalinity were checked at the end of each experiment). An oxygen optode sensor system (oxy-4 mini, PreSens, Regensburg, Germany) was used to quantify oxygen flux. Data were recorded with OXY4v2_11FB software (PreSens). Before each measurement, the oxygen sensor was calibrated against air-saturated seawater (100% oxygen) and a saturated solution of sodium sulfite (zero oxygen). Rates of photosynthesis and respiration were estimated by regressing oxygen data against time and normalized to surface area, protein content or symbiont cell. Photosynthesis and respiration rates were measured in three experimental trials, with five replicates per treatment each trial.

Frozen samples were placed in filtered seawater and tissues removed with a jet of pressurized nitrogen. The skeleton was collected, rinsed in distilled water, oven-dried for 24 h at $100\,°C$ and used for measuring surface area. The tissues were homogenized with a Potter grinder and divided into three aliquots, one for measuring protein content, one for measuring chlorophyll content and one for measuring symbiont density.

For protein content, homogenized tissues in filtered seawater were centrifuged and the pellet suspended in 1 M NaOH at $90\,°C$ for 10 min. Protein content was determined using the bicinchoninic acid assay kit (BC Protein Assay, Interchim). The standard curve was established with bovine serum albumin and the absorbance was measured with a microplate reader (EpochTM, Bioteck, US) at 562 nm. Data were normalized to skeleton surface area. For chlorophyll pigment content, homogenized tissues in filtered seawater were centrifuged and the pellet suspended in 100% acetone at $4\,°C$ for 24 h. The extraction was repeated and the extracts were pooled and centrifuged at 10,000 g for 15 min. Absorbance of extracts were measured at 630 and 663 nm in a spectrophotometer (UVMC2, Safas, Monaco) against an acetone blank. Concentrations of chlorophyll a and c$_2$ were calculated using the equations of Jeffrey and Humphrey[46].

For symbiont density, counting was performed using the HistolabH 5.2.3 image analysis software (Microvision Intruments, rance) (ref. 47).

Surface area of colonies was measured utilizing one of the common methods currently used, the paraffin wax method, Stimson and Kinzie 1991 (ref. 48). In brief, coral skeletons were coated in paraffin wax (Paraplastwax—Sigma, France) at $65\,°C$. Surface area of the specimens was obtained by referring the weight of the paraffin wax coated on the specimen to the standard curve of paraffin wax versus surface area. The standard curve was generated by regressing weight of the paraffin wax to known surface area density blocks.

References

1. Knowlton, N. et al.in Life in the World's Oceans. (ed. Mcintyre, A.) Ch. 4 (Wiley-Blackwell, 2010).
2. Caldeira, K. & Wickett, M. E. Oceanography: anthropogenic carbon and ocean pH. Nature **425**, 365 (2003).
3. Orr, J. C. et al. Anthropogenic ocean acidification over the twenty-first century and its impact on calcifying organisms. Nature **437**, 681–686 (2005).
4. Chan, N. C. & Connolly, S. R. Sensitivity of coral calcification to ocean acidification: a meta-analysis. Glob. Chang. Biol. **19**, 282–290 (2013).
5. Mumby, P. J. & van Woesik, R. Consequences of ecological, evolutionary and biogeochemical uncertainty for coral reef responses to climatic stress. Curr. Biol. **24**, R413–R423 (2014).
6. Weis, V. M. & Allemand, D. What determines coral health? Science **324**, 1153–1155 (2009).
7. Goffredo, S. et al. Biomineralization control related to population density under ocean acidification. Nat. Clim. Change **4**, 593–597 (2014).
8. Dodge, R. E. & Brass, G. W. Skeletal extension, density and calcification of the reef coral, Montastrea annularis: St Croix, U.S. Virgin Islands. Bull. Mar. Sci. **34**, 288–307 (1984).
9. Todd, P. A. Morphological plasticity in scleractinian corals. Biol. Rev. Camb. Philos. Soc. **83**, 315–337 (2008).
10. McCulloch, M., Falter, J., Trotter, J. & Montagna, P. Coral resilience to ocean acidification and global warming through pH up-regulation. Nat. Clim. Change **2**, 623–627 (2012).
11. Venn, A. A. et al. Impact of seawater acidification on pH at the tissue-skeleton interface and calcification in reef corals. Proc. Natl Acad. Sci. USA **110**, 1634–1639 (2013).
12. Mass, T. et al. Cloning and characterization of four novel coral acid-rich proteins that precipitate carbonates in vitro. Curr. Biol. **23**, 1126–1131 (2013).
13. Holcomb, M. et al. Coral calcifying fluid pH dictates response to ocean acidification. Sci. Rep. **4**, 5207 (2014).
14. Karako-Lampert, S. et al. Transcriptome analysis of the scleractinian coral Stylophora pistillata. PLoS ONE **9**, e88615 (2014).
15. Tambutté, S. et al. Coral Biomineralization: from the gene to the environment. J. Exp. Mar. Biol. Ecol. **408**, 58–78 (2011).
16. Krief, S. et al. Physiological and isotopic responses of scleractinian corals to ocean acidification. Geochim. Cosmochim. Acta **74**, 4988–5001 (2010).
17. Vidal-Dupiol, J. et al. Genes related to ion-transport and energy production are upregulated in response to CO$_2$-driven pH decrease in corals: new insights from transcriptome analysis. PLoS ONE **8**, e58652 (2013).
18. Rodolfo-Metalpa, R. et al. Coral and mollusc resistance to ocean acidification adversely affected by warming. Nat. Clim. Change **1**, 308–312 (2011).
19. Ries, J. B. A physicochemical framework for interpreting the biological calcification response to CO2-induced ocean acidification. Geochim. Cosmo. Acta **75**, 4053–4064 (2011).
20. Gagnon, A. C. Coral calcification feels the acid. Proc. Natl Acad. Sci. USA **110**, 1567–1568 (2013).
21. Cohen, A. L., McCorkle, D. C., De Putron, S., Gaetani, G. A. & Rose, K. A. Morphological and compositional changes in the skeletons of new coral recruits reared in acidified seawater: insights into the biomineralization response to ocean acidification. Geochem. Geophys. Geosyst. **10**, 1–12 (2009).
22. Holcomb, M., Cohen, A. L., Gabitov, R. I. & Hutter, J. L. Compositional and morphological features of aragonite precipitated experimentally from seawater and biogenically by corals. Geochim. Cosmochim. Acta **73**, 4166–4179 (2009).
23. Johnston, I. The ultrastructure of skeletogenesis in zooxanthellate corals. Int. Rev. Cytol. **67**, 171–214 (1980).
24. Moya, A. et al. Whole transcriptome analysis of the coral Acropora millepora reveals complex responses to CO(2)-driven acidification during the initiation of calcification. Mol. Ecol. **21**, 2440–2454 (2012).
25. Crook, E. D., Cohen, A. L., Rebolledo-Vieyra, M., Hernandez, L. & Paytan, A. Reduced calcification and lack of acclimatization by coral colonies growing in areas of persistent natural acidification. Proc. Natl Acad. Sci. USA **110**, 11044–11049 (2013).
26. Venn, A., Tambutté, E., Holcomb, M., Allemand, D. & Tambutté, S. Live tissue imaging shows reef corals elevate pH under their calcifying tissue relative to seawater. PLoS ONE **6**, e20013 (2011).
27. McCulloch, M. et al. Resilience of cold-water scleractinian corals to ocean acidification: Boron isotopic systematics of pH and saturation state up-regulation. Geochim. Cosmo Acta **87**, 21–34 (2012).
28. Kuhl, M., Cohen, Y., Dalsgaard, T., Jorgensen, B. & Revsbech, N. Microenvironment and photosynthesis of zooxanthellae in scleractinian corals studied with microsensors for O2, pH and light. Mar. Ecol. Prog. Ser. **117**, 159–172 (1995).
29. Mann, S. Biomineralization, Principles And Concepts in Bioinorganic Materials Chemistry (Oxford University Press, 2001).
30. Sollner, C. et al. Control of crystal size and lattice formation by starmaker in otolith biomineralization. Science **302**, 282–286 (2003).
31. Suzuki, M. et al. An acidic matrix protein, Pif, is a key macromolecule for nacre formation. Science **325**, 1388–1390 (2009).
32. Chadwick, N. & Morrow, K. in Coral Reefs: An Ecosystem in Transition. (eds Dubinsky, Z. & Stambler, N) (Springer, 2011).
33. Chamberlain, J. R. Mechanical properties of the coral skeleton: compressive strength and its adaptive significance. Paleobiology **4**, 419–435 (1978).
34. Reyes-Nivia, C., Diaz-Pulido, G., Kline, D., Guldberg, O. H. & Dove, S. Ocean acidification and warming scenarios increase microbioerosion of coral skeletons. Glob. Chang. Biol. **19**, 1919–1929 (2013).
35. Anthony, K. R., Kline, D. I., Diaz-Pulido, G., Dove, S. & Hoegh-Guldberg, O. Ocean acidification causes bleaching and productivity loss in coral reef builders. Proc. Natl Acad. Sci. USA **105**, 17442–17446 (2008).
36. Fine, M. & Tchernov, D. Scleractinian coral species survive and recover from decalcification. Science **315**, 1811 (2007).
37. Dickson, A. G., Sabine, C. L. & Christian, J. R. Guide to best practices for ocean CO$_2$ measurements. PICES Special Publication **3**, 191 (2007).
38. DOE. Handbook of methods for the analysis of the various parameters of the carbon dioxide system in sea water (version 2). (eds Dickson, AG & Goyet, C.) (ORNL/CDIAC-74, 1994).
39. Mehrbach, C., Culberso, C., Hawley, J. & Pytkowic, R. Measurement of apparent dissociation constants of carbonic-acid in seawater at atmospheric pressure. Limnol. Oceanogr. **18**, 897–907 (1973).
40. Dickson, A. G. & Millero, F. J. A comparison of the equilibrium-constants for the dissociation of carbonic-acid in seawater media. Deep Sea Res. A **34**, 1733–1743 (1987).
41. Jokiel, P. L., Maragos, J. E. & Franzisket, L. Coral growth: buoyant weight technique (UNESCO, 1978).

42. Bucher, D. J., Harriott, V. J. & Roberts, L. G. Skeletal bulk density, micro-density and porosity of acroporid corals. *J. Exp. Mar. Biol. Ecol.* **228**, 117–135 (1998).

43. Roche, R. C., Abel, R. A., Johnson, K. G. & Perry, C. T. Quantification of porosity in *Acropora pulchra* (Brook 1891) using X-ray micro-computed tomography techniques. *J. Exp. Mar. Biol. Ecol.* **396**, 1–9 (2010).

44. Roche, R. C., Abel, R. L., Johnson, K. G. & Perry, C. T. Spatial variation in porosity and skeletal element characteristics in apical tips of the branching coral *Acropora pulchra* (Brook 1891). *Coral Reefs* **30**, 195–201 (2011).

45. Tambutte, E. *et al.* Observations of the tissue-skeleton interface in the scleractinian coral *Stylophora pistillata*. *Coral Reefs* **26**, 517–529 (2007).

46. Jeffrey, S. W. & Humphrey, G. F. New spectrophotometric equations for determining chlorophylls a, b, c_1 and c_2 in higher plants, algae and natural phytoplankton. *Biochem. Physiol. Pflanzen* **167**, 191–194 (1975).

47. Rodolfo-Metalpa, R., Richard, C., Allemand, D. & Ferrier-Pages, C. Growth and photosynthesis of two Mediterranean corals, *Cladocora caespitosa* and *Oculina patagonica*, under normal and elevated temperatures. *J. Exp. Biol.* **209**, 4546–4556 (2006).

48. Stimson, J. & Kinzie, R. A. The temporal pattern and rate of release of zooxanthellae from the reef coral Pocillopora under nitrogen enrichment and control conditions. *J. Exp. Mar. Biol. Ecol.* **153**, 63–74 (1991).

Acknowledgements

We thank Dr Pierre Alemanno and Dr Christophe Sattonnet (Polyclinique St Jean, Cagnes sur Mer, France) for access to the micro-CT and Dr Nathalie Rochet and Ivana Starzic for technical assistance. We also thank Dominique Desgre for assistance with coral culture and Philippe Ganot for useful comments on the manuscript. We thank three anonymous reviewers for useful comments. M.H. was supported by a NSF International Research Fellowship. We acknowledge the Fondation Paul Hamel for the purchase of the Scanning Electron Microscope. This research was funded by the Government of the Principality of Monaco.

Author contributions

E.T., A.V., M.H. and S.T. conceived and designed research. E.T., A.V., M.H., N.S. and N.T. performed laboratory experiments. E.T., A.V., M.H., N.S., N.T. and S.T. analysed data. E.T., A.V., M.H., N.S., N.T., D.Z., D.A., and S.T. contributed to the manuscript and participated in scientific discussion of the findings.

Additional information

Bidecadal North Atlantic ocean circulation variability controlled by timing of volcanic eruptions

Didier Swingedouw[1], Pablo Ortega[2], Juliette Mignot[2,3,4], Eric Guilyardi[2,5], Valérie Masson-Delmotte[6], Paul G. Butler[7], Myriam Khodri[2] & Roland Séférian[8]

While bidecadal climate variability has been evidenced in several North Atlantic paleoclimate records, its drivers remain poorly understood. Here we show that the subset of CMIP5 historical climate simulations that produce such bidecadal variability exhibits a robust synchronization, with a maximum in Atlantic Meridional Overturning Circulation (AMOC) 15 years after the 1963 Agung eruption. The mechanisms at play involve salinity advection from the Arctic and explain the timing of Great Salinity Anomalies observed in the 1970s and the 1990s. Simulations, as well as Greenland and Iceland paleoclimate records, indicate that coherent bidecadal cycles were excited following five Agung-like volcanic eruptions of the last millennium. Climate simulations and a conceptual model reveal that destructive interference caused by the Pinatubo 1991 eruption may have damped the observed decreasing trend of the AMOC in the 2000s. Our results imply a long-lasting climatic impact and predictability following the next Agung-like eruption.

[1] Environnements et Paléoenvironnements Océaniques et Continentaux (EPOC), UMR CNRS 5805 EPOC—OASU—Université de Bordeaux, Allée Geoffroy Saint-Hilaire, Pessac 33615, France. [2] LOCEAN/IPSL Sorbonne Universités (UPMC, Univ Paris 06)-CNRS-IRD-MNHN, 4 place Jussieu, Paris F-75005, France. [3] Climate and Environmental Physics, Physics Institute, University of Bern, Falkenplatz 16, 3012 Bern, Switzerland. [4] Oeschger Centre of. Climate Change Research, University of Bern, Falkenplatz 16, Bern 3012, Switzerland. [5] NCAS-Climate, Univeristy of Reading, Reading RG6 6BB, UK. [6] Laboratoire des Sciences du Climat et de l'Environnement (Institut Pierre Simon Laplace, CEA-CNRS-UVSQ, UMR8212), 91191 Gif-sur-Yvette, France. [7] School of Ocean Sciences, Bangor University, Menai Bridge, Anglesey LL59 5AB, UK. [8] Centre National de Recherches Météorologiques-Groupe d'Etude de l'Atmosphère Météorologique/Groupe de Météorologie de Grande Echelle et Climat, Toulouse 31100, France. Correspondence and requests for materials should be addressed to D.S. (email: didier.swingedouw@u-bordeaux1.fr).

The Atlantic Meridional Overturning Circulation (AMOC) plays a key role in the meridional heat transport, and in heat and carbon storage in the ocean[1]. Changes in the AMOC affect surface oceanic conditions in the North Atlantic (salinity, temperature and sea level), with impacts on marine ecosystems and regional climate (for example, Greenland glaciers[2]). Understanding the mechanisms driving AMOC decadal variability is therefore critical for climate predictability, particularly in the Northern Hemisphere[3-5]. In response to increased greenhouse gas concentrations, climate models project a gradual AMOC slowdown during the 21st century[6,7]. AMOC strength has only recently been monitored through observation networks. Existing data sets do not reveal a strong trend in the 2000s[8,9], although a declining tendency appears from the early 2010s[10,11]. This is also supported by ocean reanalyses[12,13], which depict an AMOC maximum in the 1990s at subpolar latitudes, followed by a decrease and stabilization in the 2000s. Beyond direct information on AMOC, North Atlantic observations[14,15] and proxy records indicate a 20-year preferential variability in this region in the atmosphere[16], sea ice[17] and the ocean[18,19]. Such variability can be associated with the dynamics of subpolar gyre, whose characteristic decadal timescales are associated with advection processes and the size of the gyre[20,21].

Indeed, such advective processes have been observed during Great Salinity Anomaly (GSA) events[22-24]. For instance, a salinity anomaly was first identified in the Nordic Seas in 1968, and then detected in the Labrador Sea around 1971. This anomaly was monitored along its propagation within the subpolar gyre during the following 7 years, and it reached the eastern part of the Nordic Seas in 1978 (ref. 22).

Explosive volcanic eruptions have a short-lived but strong radiative impact through the loading of a large amount of sulphate aerosols into the stratosphere, which leads to a cooling of the Earth's surface during the 2–3 years following the onset of the eruption[25,26]. Moreover, the volcanic sulphate aerosol injection causes a significant stratospheric warming in the tropical band due to long-wave heat absorption. The subsequent strengthening of the meridional atmospheric temperature gradient leads to intensified zonal winds and jets through thermal wind balance. As a result, observations show that volcanic eruptions trigger dynamical changes[27,28], with a tendency towards a positive phase of the winter North Atlantic Oscillation (NAO, first mode of atmospheric variability in the north Atlantic sector[29]) during the few years following the eruption[27,30,31]. This dynamical mechanism is not well represented in most climate simulations, at least partly due to a too coarse vertical resolution of their atmospheric model component[27]. In addition, climate model analysis suggests that the volcanic-driven, short-lived cooling of the upper ocean can induce longer-lived changes on North Atlantic climate, notably through its influence on the AMOC[32-34]. The robustness of the mechanisms at play have, however, been challenged by differences in the simulated timing and response, suggesting a sensitivity of the results to the model used or to the prescribed volcanic forcing. Oceanographic data have not yet been used to evaluate the exact oceanic processes at play.

Here we evaluate the potential impact of moderate explosive volcanic eruptions (similar to Agung or Pinatubo) on the North Atlantic bidecadal preferential variability. For this purpose, we use available outputs from different climate models using the Coupled Model Intercomparison Project Phase 5 (CMIP5) database[35] complemented by additional simulations performed using one model, and *in situ* recent oceanic observations as well as longer paleoclimate proxy records of the last millennium. We find that moderate volcanic eruptions may reset a 20-year intrinsic variability mode in the North Atlantic both in model

simulations as well as in the data analyzed, leading to interference patterns over the recent period and in the near future.

Results

Analysis of the CMIP5 database.
We investigate the mechanisms involved in bidecadal North Atlantic and AMOC variations and reduced AMOC variability around the trend during the 2000s using CMIP5 simulations and sensitivity tests conducted specifically with the IPSL-CM5A-LR model[36]. Historical climate simulations (1870–2005) driven by natural and anthropogenic forcing archived in the CMIP5 database exhibit a large spread in the simulated AMOC at 48°N (Fig. 1a). Out of 19 available model simulations, we select the subset of 9 (8 + IPSL-CM5A-LR called 'Bi-Dec' ensemble) which exhibit peaks of spectral energy in the

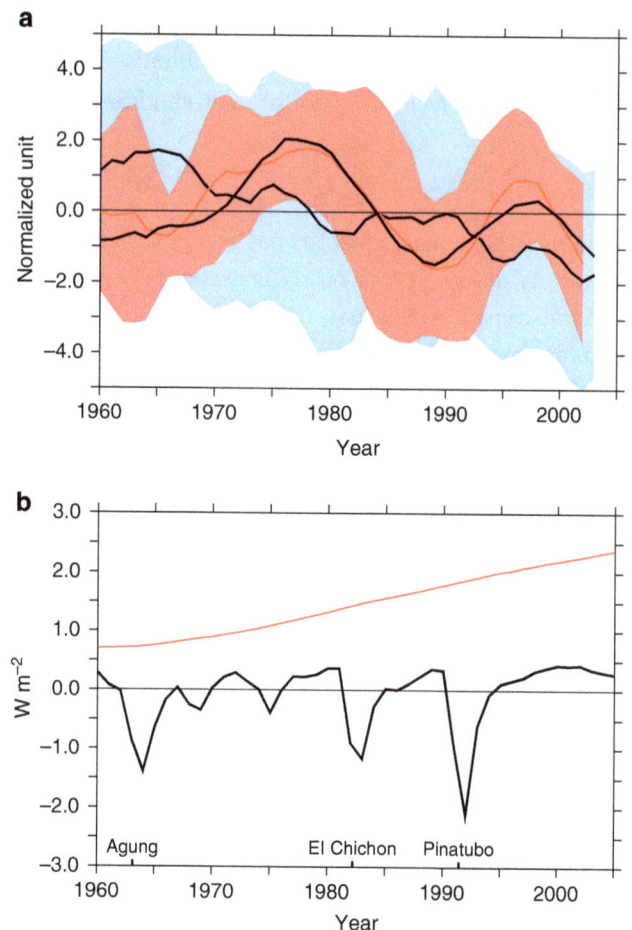

Figure 1 | Simulated AMOC changes and radiative external forcing.
(**a**) Variations of the AMOC maximum at 48°N over the period 1960-2005 for the ensemble of five historical simulations performed with the IPSL-CM5A-LR model (black); for the Bi-Dec ensemble excluding the IPSL-CM5A-LR simulation (subset of eight CMIP5 models, which also exhibit variability in the 10-30 years spectral band, in red, see Methods, Supplementary Table 1 and Supplementary Fig. 1); for the ensemble mean of the 10 other CMIP5 historical simulations (in blue). The standard deviation (s.d.) of the two CMIP5 ensembles is shown with the red and blue envelopes. All the AMOC indices have been normalized with the s.d. of the detrended time series over the period 1850-2005. (**b**) External radiative forcing (in $W m^{-2}$) computed in the IPSL-CM5A-LR historical simulation. The black curve is the natural forcing including solar and volcanic eruptions and the red curve represents the anthropogenic forcing including greenhouse gas changes and the anthropogenic aerosols effects. A 5-year running mean has been applied to all time series from **a**.

10–30 years band for either the historical or the pre-industrial simulations (Supplementary Table 1, Supplementary Figs 1 and 2). Because this frequency peak is clearly evidenced in North Atlantic proxy records[16–18], we expect these selected models to potentially capture the actual variability pattern. In the IPSL-CM5A-LR model, strong 20-year variability in the North Atlantic is related to the time scale of temperature and salinity advection in the subpolar gyre, with salinity anomalies driving convection and deep ocean circulation[20]. The Bi-Dec ensemble shows coherent AMOC variability at 48°N (with an ensemble correlation coefficient of 0.64, significant at the 99% level), not seen in the ensemble of the remaining CMIP5 historical runs (Fig. 1a and Supplementary Fig. 2). More precisely, the Bi-Dec members systematically exhibit an AMOC maximum in the late 1970s. This consistent timing among them strongly suggests the response to a common external forcing. Earlier modelling studies have indeed shown intensified AMOC about a decade after strong volcanic eruptions[26,32,33], the lag reflecting the time required for the advection of cold anomalies from the subtropics to high latitudes, and for the response of the AMOC to changes in atmospheric momentum forcing. Here we suggest that the 1963 Agung eruption (Fig. 1b) has reset bidecadal variability in the eight CMIP5 simulations, leading to a first AMOC maximum about 15 years later (late 1970s). A second (weaker) simulated maximum occurs in the 1990s, corresponding to a secondary peak from the 20-year mode (Fig. 1). The same features are reproduced in a five-member ensemble of historical simulations from the IPSL-CM5A-LR model[13].

We do not expect such coherent variability to arise from the gradual long-term anthropogenic forcing. While solar forcing has been shown to influence ocean circulation variability in a climate model[37], solar forcing had a weak radiative amplitude of variability compared with volcanic eruptions, especially during the second half of the 20th century. We therefore propose that the coherent signal identified in the simulations has been triggered by the 1963 Agung volcanic eruption. Such large volcanic eruptions are indeed known to strongly cool the surface of the North Atlantic subpolar gyre and the Nordic Seas[32,34,38] (see for example, Supplementary Fig. 3). The bidecadal time scale of the subsequent dynamical adjustment to temperature anomalies involves the advection of temperature and salinity anomalies[15] (for example, GSAs) or baroclinic instabilities leading to large-scale baroclinic Rossby waves[21,39,40], strongly constrained by the size of the subpolar basin, and interactions with sea ice and atmospheric circulation. The specific processes accounting for the bidecadal variability may differ in the different climate models[21]. In the following, we will focus on one specific climate model (IPSL-CM5A-LR), for which the mechanisms underlying this variability[20,41] as well as its response to recent volcanic eruptions have been recently investigated in depth[15]. Simulations over the historical period, the last millennium, as well as additional sensitivity simulations performed with this model are discussed and compared with *in situ* oceanic data as well as proxy records.

Comparison with recent oceanic data. Recent AMOC variability cannot be assessed in the short 2004-to-present observational record[8,11]. An alternative strategy is to use salinity observations, which exhibit large and well-documented variability in the North Atlantic, most notably as GSA. These GSAs occurred in the 1970, 1980 and 1990s (Fig. 2b). The current understanding is that the 1980s event was forced by the atmospheric circulation, while the 1970 and 1990s events are associated with Arctic Ocean variability via the East Greenland Current (EGC)[22,24]. Changes in salinity play a crucial role in the stability of the water column. They can drive changes in oceanic convection, which is directly

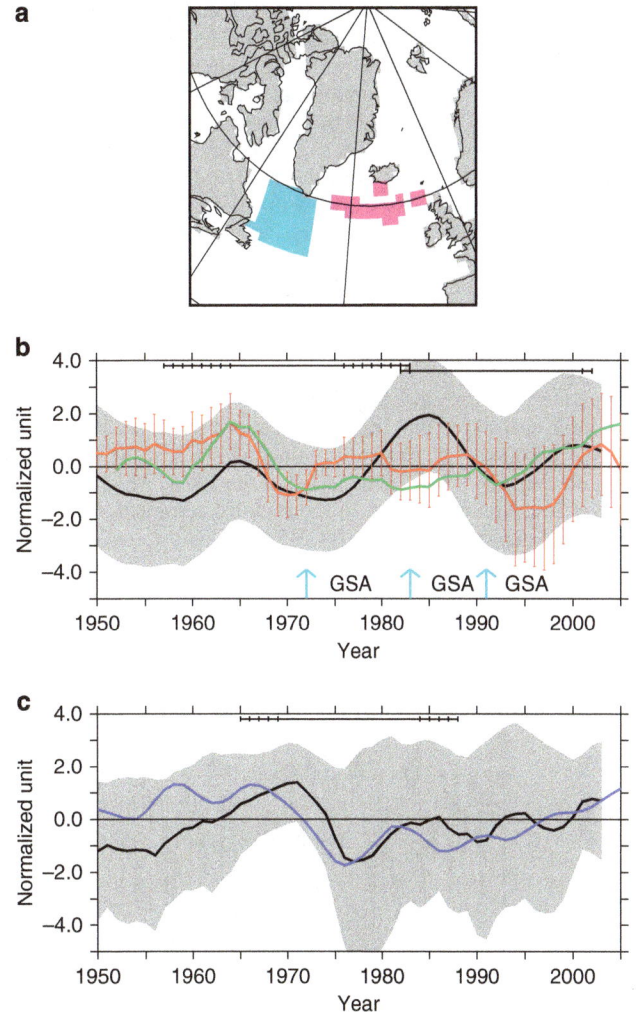

Figure 2 | Observed and simulated changes in salinity. (a) Location of the two key regions (Labrador Sea and subpolar gyre) studied here. **(b)** *In situ* salinity data from the Labrador Sea averaged over the cyan region down to 300 m compiled from the Bedford Institute (see for example, Methods, red line, with vertical line representing 2-s.d. errors) and from EN3 (ref. 43) (green) observational data sets. **(c)** Sea Surface Salinity from *in situ* data[42] (in blue) averaged over the eastern subpolar gyre (magenta region). In **b,c**, the 5-member ensemble mean outputs from historical simulations using IPSL-CM5A-LR are also displayed (black, mean value; grey shaded, s.d. envelope). The horizontal black lines at the top of **b,c** delimit the 20-year sliding windows for which significant correlation is detected (at the 95% confidence level) between the simulations and the observations (using the Bedford Institute data only in **b**. A 5-year running mean has been applied to all time series.

related to AMOC variability. Here we use a combination of North Atlantic surface and subsurface salinity observational data sets (spanning years 1949–2010): a reconstruction of surface salinity in the eastern subpolar gyre[42], a new compilation of salinity data over a region enclosing large part of the Labrador Sea and reaching down to 300 m (compiled from the Bedford Institute of Oceanography website, see for example, Methods) and the global EN3 (ref. 43) data set (Fig. 2b). These data sets are compared with the IPSL-CM5A-LR ensemble of historical simulations.

Figure 2 shows that the IPSL-CM5A-LR simulations capture the observed 1970 and 1990s GSAs in the Labrador Sea and Eastern subpolar gyre. In this model, an increase in the EGC leads to a decreased salinity in the Labrador Sea, followed by AMOC variations about 10 years later[20]. Detailed analyses[15] have shown that changes in salinity simulated by the IPSL-CM5A-LR in the 1970s are a dynamical response to the Agung eruption. Here we conclude that the Agung eruption has excited the bidecadal mode of the North Atlantic in both observations and simulations, explaining not only the GSA of the 1970s but also the subsequent GSA of the 1990s. During both the 1970 and 1990s GSA events, salinity anomalies are associated with local temperature minima in the observations (Supplementary Fig. 4), consistently with the mechanism at play in the IPSL-CM5A-LR model[20]. This agreement between model and observations can be extended to the Bi-Dec ensemble. Indeed similar GSAs as the observed ones in the 1970 and 1990s are found in this ensemble mean as in the observations in the Labrador Sea both for salinity and temperature (Supplementary Fig. 5). The large spread in the ensemble also shows that the exact locations of the anomalies and of the processes at play are model dependent. It has been suggested that the GSA encountered in the 1980s, which is clearly weaker than the two others in the Bedford compilation (Fig. 2), and is not depicted by the IPSL-CM5A-LR historical simulations,

may have a different origin[22]. This specific GSA was indeed associated with extremely severe winters of the early 1980s, with a possible contribution of the Arctic freshwater outflow via the Canadian Archipelago[22]. If this GSA was indeed driven by extreme stochastic atmospheric processes, this might explain the fact that it is not captured in the historical simulations, as well as the smaller amplitude and depth of this event as compared with the two other GSAs (Fig. 2b). We therefore hypothesize that this smaller GSA may not have strongly affected convection in the subpolar gyre, with therefore a very weak signature on the AMOC variability, in contrast with the two other GSA. This is however not fully supported by the fingerprint of the 1980s GSA in the EN3 data set, challenging this interpretation.

Last millennium perspective. To further assess the validity of the excitation of bidecadal variations by volcanic eruptions, we investigate two high-resolution proxy records and a new IPSL-CM5A-LR simulation spanning the last millennium, which includes several earlier Agung-like volcanic eruptions. This

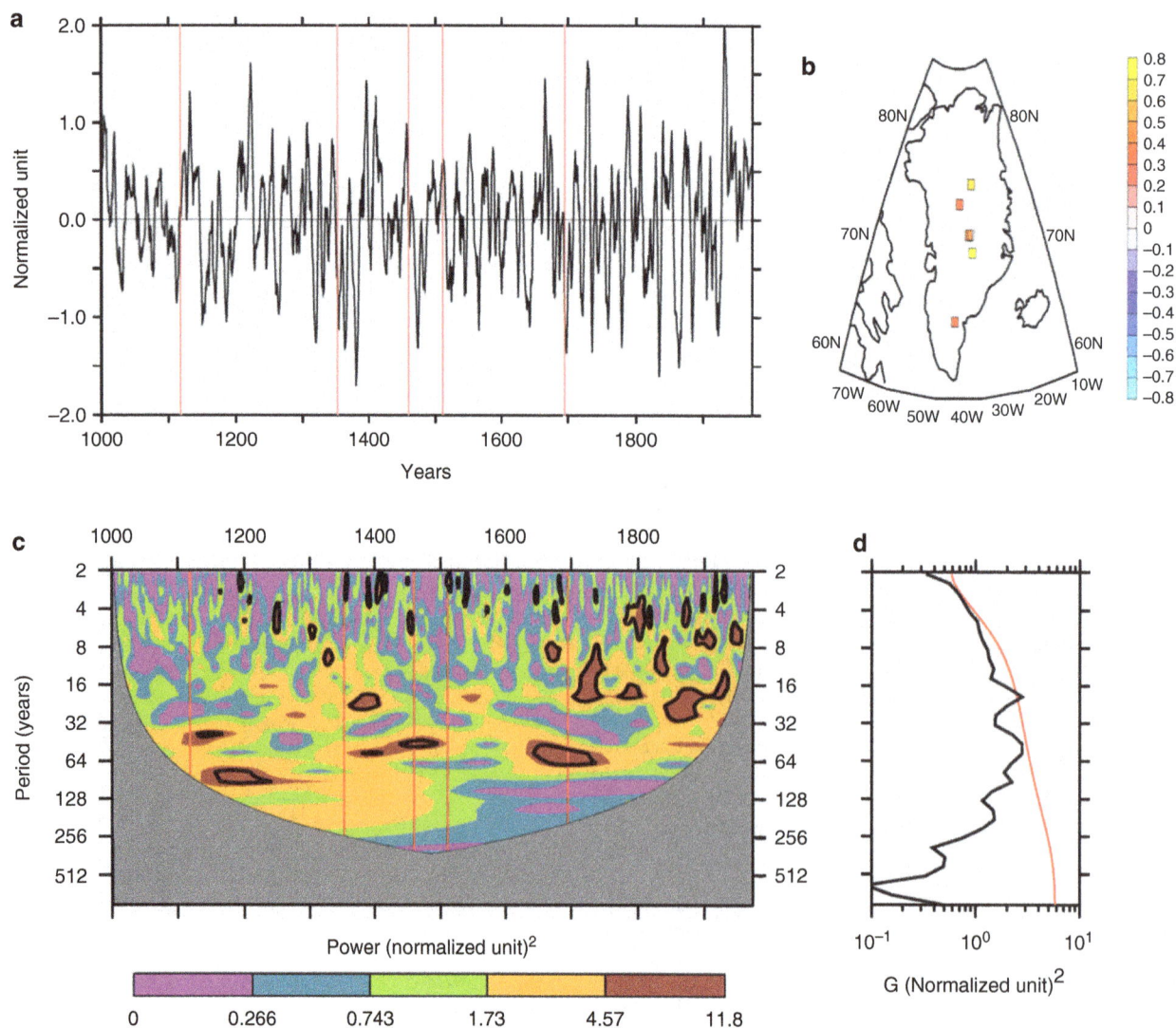

Figure 3 | Bidecadal variability in Greenland ice core data. (a) Principal component time series (with a 5-year running mean) associated with (b) the first EOF of the $\delta^{18}O$ data from six annually resolved Greenland ice cores, computed over the period 1000–1973 (see for example, Methods). (c) Wavelet analysis of the time series shown in a but without any filtering. The region significant at the 95% level are marked using bold contours. (d) Power spectrum of the time series in black compared with an AR1 signal shown in red. In a,c, the vertical red bars stand for the onset of the five eruptions selected over the last millennium.

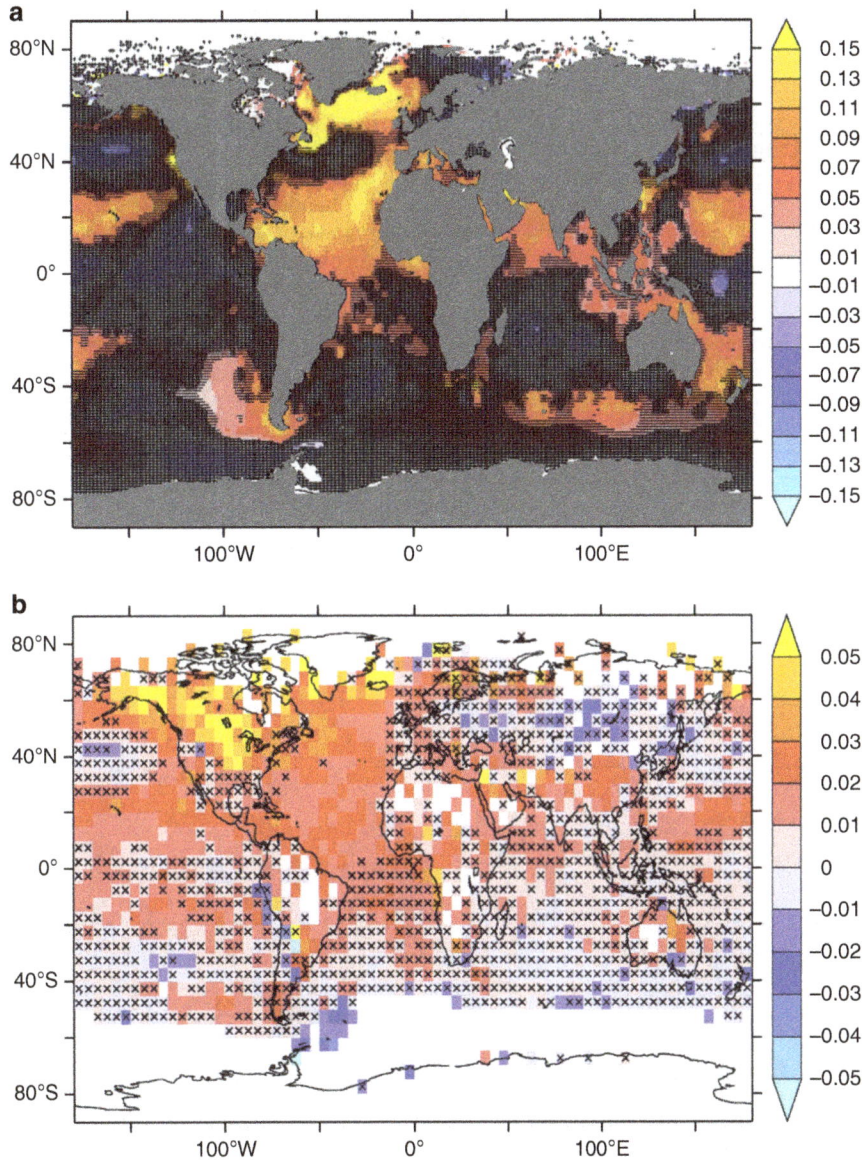

Figure 4 | Regression of the principal component of the Greenland ice core δ¹⁸O with sea surface temperature data sets. (**a**) For the period 1870-1974 with SST data from HadISST[44] and (**b**) over the period 1000-1850 with gridded surface temperature reconstructions[46]. The non-significant zones for the regression at 95% level are marked with a black cross.

simulation also exhibits a clear spectral peak at the 20-year time scale for the 48°N AMOC variability (Supplementary Fig. 6).

The first proxy record is an updated compilation of six annually resolved δ¹⁸O reconstructions from Greenland ice cores (see Methods and Supplementary Fig. 1). From an Empirical Orthogonal Function (EOF) analysis, their leading common signal is extracted, corresponding to a monopole that exhibits preferential variability at a 20-year time scale (Fig. 3), coherent with earlier ice core composites[3]. This data set is highly correlated with North Atlantic Sea Surface Temperature (SST) as shown by the regression of HadISST data[44] on the principal component (PC) of the first EOF from the ice cores for the period 1850–1973 (Fig. 4). Significant correlation is indeed found in the Atlantic, approximately from the equator to the Nordic Seas, a spatial pattern that is reminiscent of the Atlantic Multidecadal Oscillation[45] (AMO). The response is especially large in the North Atlantic (Fig. 4a). To test this relationship on a longer time scale, we use a gridded SST reconstruction[46] and perform the same regression analysis for the period 1000–1850. Again,

significant correlation is mainly found in the Atlantic basin north of the equator (Fig. 4b). Although this second result should be taken with caution, given that the SST reconstruction is based on a multi-proxy approach, also including Greenland ice core data, we argue that the ice core compilation provides an annually resolved record of bidecadal North Atlantic surface temperature variability. Such bidecadal timescales are usually filtered out in AMO reconstructions[46,47], which indeed act as low pass filters and stress multidecadal variability. We thus interpret this first EOF of Greenland ice cores as a proxy for North Atlantic SST variations in line with a former hypothesis linking a stack of Greenland ice cores with surrounding SST[48].

The second proxy record corresponds to growth increments from the bivalve *Arctica islandica* collected north of Iceland[49], also annually resolved. These bivalve data are not an unequivocal proxy for SST as they also depend on other factors, notably the nutrient supply[49]. The latter depends itself on vertical movements that can be related to AMOC variations, notably through convection processes which can help deep ocean nutrients to come to the surface[50]. For instance, in the isopycnal layered ESM2G Earth System model, strong inter-decadal changes in surface salinity associated with changes in AMOC produce spatially heterogeneous variability in convection, nutrients supply and thus diatom biomass[44]. As the IPSL-CM5A-LR model includes a bio-geochemical component[36,51], we use it to test this hypothesis through a pseudo-proxy approach[52]. In particular, we analyze the link between nutrient concentrations (PO_4, NO_3, Si and Fe) and the 48°N AMOC index in the last millennium simulation. In this case, silicate and iron in the Nordic Seas are the main limiting nutrients for phytoplankton growth (not shown) and therefore expected to be also the limiting nutrients for bivalve growth, motivating a focus on these nutrients. Around Iceland, the simulated nutrients supply is significantly correlated with the simulated variations in 48°N AMOC, the latter leading the former by 1–3 years (Fig. 5a). We argue here that this

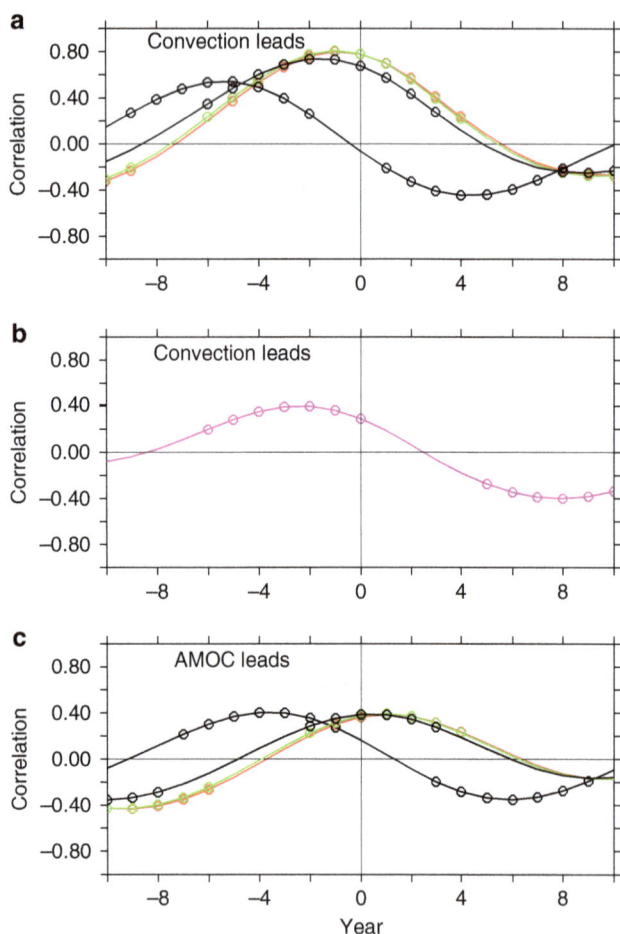

relationship arises from changes in convection in the Nordic Seas, which precedes AMOC changes by 1–4 years[20] and changes in nutrient supplies around Iceland by 2–5 years in our model (Fig. 5c). This last lag can be explained by the upwelling of nutrients stored at depth by vertical currents associated with convection, and a subsequent period of a few years to export these nutrients from the convection sites towards north Iceland. Based on the hypothesis that the nutrient supply to Iceland is the limiting factor for the growth of the bivalve *A. islandica*, we therefore use growth increment records as a proxy of the subpolar AMOC, with a response lag of 1–3 years.

Using estimates of past volcanic radiative forcing[53], we identify volcanic eruptions with a radiative forcing ranging between one

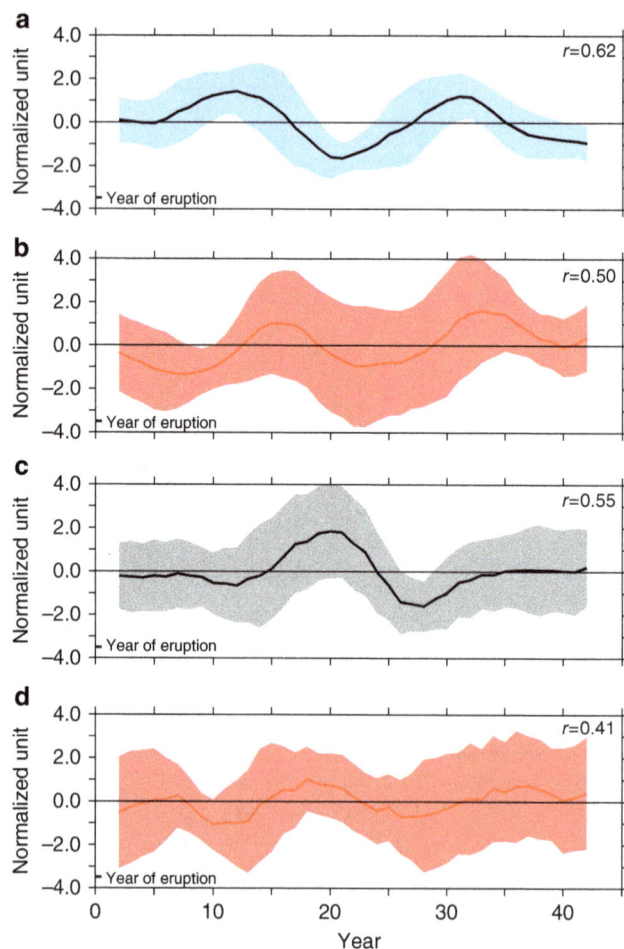

Figure 5 | Phase relationship between key variables of the last millennium IPSL-CM5A-LR simulation. Cross-correlation coefficients (y-axis) computed over the whole length of the last millennium simulation (850–1850) for different time lags (x-axis), (**a**) mixed layer depth in the Nordic Seas (an indicator of convection) and nutrient concentration north of Iceland, averaged between 50 and 150 m, where the bivalves are located. (**b**) Cross-correlation between mixed layer depth in the Nordic Seas and 48°N AMOC. (**c**) Same as **b** except that mixed layer depth in the Nordic Seas is replaced by 48°N AMOC. In **a**,**c**, different colours are used to represent different nutrients (red, NO_3; green, PO_4; blue, Si and black, Fe). Correlation coefficients which are significant using a 95% confidence level Student t-test (where degrees of freedom account for data autocorrelation[70]) are highlighted with circles.

Figure 6 | Response to five Agung-like eruptions of the last millennium in reconstructions and IPSL-CM5-LR simulations. Time series of the composite formed by the ensemble of the five members for five Agung-like eruptions (taking place in 1118, 1352, 1460, 1511, 1695) over the last millennium identified through the same volcanic forcing reconstruction used in last millennium simulation[53]: (**a**) Simulated AMOC maximum at 48°N, (**b**) Growth increments in the bivalve *Arctica islandica*[49], (**c**) Simulated North Atlantic SST between 25 and 55°N, (**d**) first principal component of six yearly Greenland $\delta^{18}O$ ice core reconstructions. The ensemble correlation between the five individual time series following the respective eruptions is shown for each index in top right of each panel. A 5-year running mean has been applied to all time series. Data have been standardized and expressed in normalized units. The average of the response following the five eruptions is shown with lines, and the envelope of the response is displayed through the shading around these lines.

to one and a half times the amplitude of the 1963 Agung eruption (Pinatubo being 1.4 times Agung in terms of global radiative forcing used in IPSL-CM5A-LR, see for example, Methods). To avoid any interference from successive events (see below), we retain only those not followed within 40 years by eruptions larger than 1.5 Agung. Only five of all the eruptions recorded over the last 1,000 years meet these two criteria (onset of the eruption in 1118, 1352, 1460, 1511, 1695, see for example, Supplementary Fig. 7).

We first compare the simulated AMOC variations following these five eruptions with the Icelandic bivalve data. They exhibit a maximum both around 15 and 35 years following the eruptions (Fig. 6a,b). The slight shift that is found between the 48°N AMOC from the model and the bivalve observations[49] is consistent with the 3-year lag found in Fig. 5. Indeed the 3-year lagged correlation between the model AMOC and the bivalve data ensemble means reaches 0.66 (p-value <0.1 see for example, Supplementary Fig. 8a).

We then compare the simulated SST variations in the North Atlantic to the Greenland ice core data. They both exhibit coherent warming around 20 and 40 years after the eruptions (Fig. 6c,d and Supplementary Fig. 8b). The lag between AMOC and North Atlantic SST changes is attributed to a lagged accumulation of warm waters in the North Atlantic when the circulation accelerates[41] (Supplementary Fig. 9) and is consistent with the results of other climate models[54]. Altogether, these results for five different Agung-type eruptions of the last millennium support the mechanism by which Agung-like volcanoes reset the 20-year AMOC variability. Figure 7 summarizes the different processes proposed here as well as the time lags between them, including the proxy records analyzed.

Interference pattern. The implications of this oceanic response to moderate volcanoes like Agung or Pinatubo over the most recent 30-year period and the very near future are now investigated. Indeed, the Agung eruption was followed by the El Chichon eruption 19 years after and the Pinatubo eruption 28 years later. While the timing of El Chichon may have led to a constructive interference with the Agung-excited 20-year cycle, we make the hypothesis that the Pinatubo eruption interfered destructively; thereby explaining the damped observed and simulated AMOC variability in the 2000s.

This hypothesis is tested using a five-member ensemble of IPSL-CM5A-LR starting in 1991 and using the same forcings as in the historical simulations but without the Pinatubo eruption. This 'no Pinatubo' sensitivity ensemble deviates from the historical ensemble (Fig. 8a) and exhibits much stronger practically bidecadal variability, with a minimum around 2010, a maximum in 2015 and a new minimum in 2025–2030. By contrast, the historical ensemble does not exhibit any remarkable decadal AMOC excursions from 2005 to 2030. Although the signal-to-noise ratio remains small and leads to an overlap of the s.d. envelopes of the two ensembles (Supplementary Fig. 10), this sensitivity test confirms that the Pinatubo eruption induced a suppression of the bidecadal AMOC variability in the model. The simulated regular AMOC weakening in the 2010s after the Pinatubo eruption is consistent with the RAPID array observations[8,11] and with oceanic reanalyses[12] (Fig. 8).

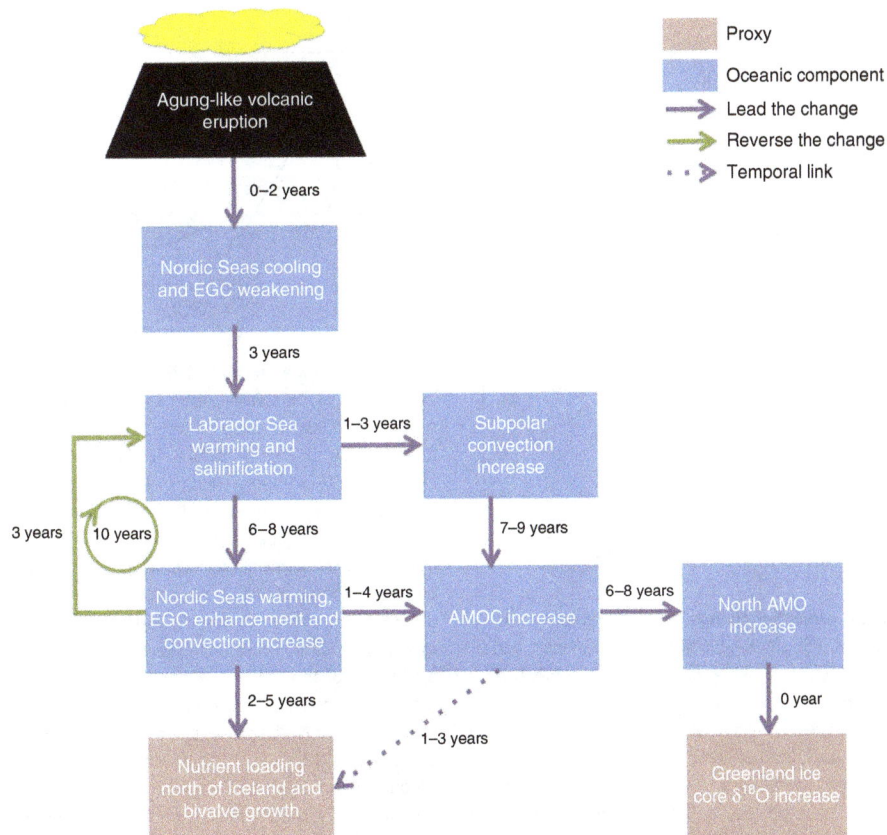

Figure 7 | Scheme summarizing the main processes at play in the North Atlantic response to volcanic eruptions. Blue boxes depict oceanic processes and red boxes depict the proxies considered here. Purple arrows represent causality links together with the associated time lags in years. Dotted arrows only depict a temporal link, with no direct interaction. The green arrow represents the delayed negative feedback that triggers the reversal phase of the 20-year cycle.

This interference pattern from successive volcanoes can be captured by a simple conceptual model (Fig. 9b), which takes into account the timing of large volcanic eruptions that excited bidecadal oscillations over the last 60 years, assuming a constant response time, together with a long-term weakening trend[6] (see Methods). This conceptual model successfully reproduces the main features of the interference theory developed here, and closely follows the IPSL-CM5A-LR simulations with and without Pinatubo (Fig. 8). However, neither the forced simulations nor this conceptual model are able to produce the rapid AMOC increase identified in the early 1990s in a few AMOC reconstructions[12,13] (Fig. 8b). It has been suggested that this specific shift is linked with internal atmospheric variability and 5-year-earlier changes in the NAO, which affected North Atlantic SST and therefore the decadal variability of the AMOC[3,55,56]. Indeed, climate simulations nudged towards observed SST anomalies performed using IPSL-CM5A-LR do produce a strong AMOC enhancement in the late 1990s that can be attributed to a NAO positive phase a few years earlier[15] (Fig. 8). Similarly, implementing additional NAO variations in the conceptual model with a 5-year lagged response of the ocean (see Methods) reproduces most of the observed variability (Fig. 8b) as well as the very recent weakening over the last few years[10].

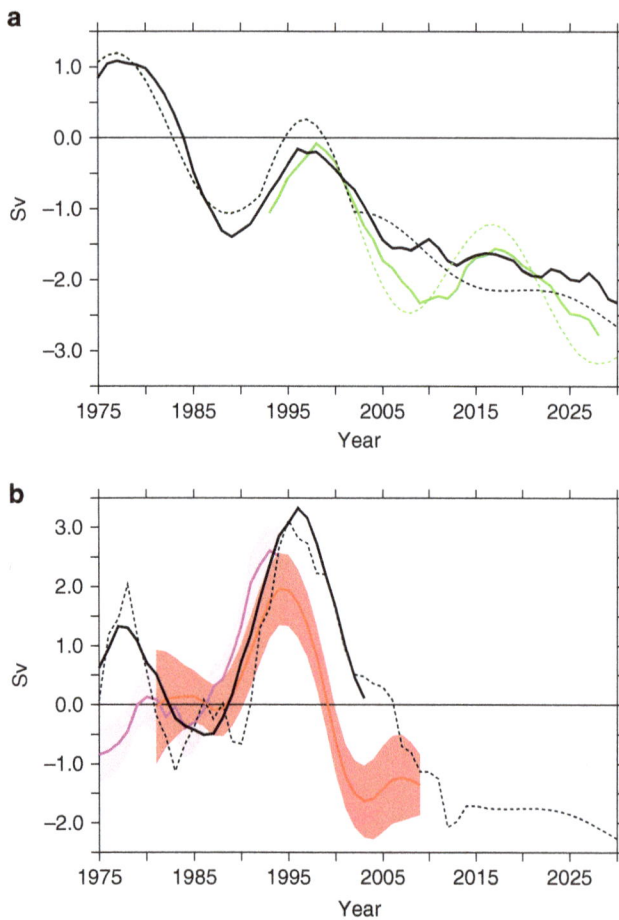

Figure 8 | Recent AMOC and projected change. (a) Simulated AMOC anomalies at 48°N in historical IPSL-CM5A-LR simulations (black), and in the sensitivity tests performed without taking into account the Pinatubo eruption (green). Results of the corresponding conceptual model (SI) are shown in dotted lines. **(b)** Estimates of AMOC at 45°N (1979–1988) based on hydrographic data[13] (pink curve) and ensemble mean of the AMOC maximum at 45°N from 12 ocean reanalyses[12] (red). This ensemble mean has been rescaled by a factor $\sqrt{12}$, to correct for the loss of variance caused by averaging different ocean models with different data assimilation. The five-member ensemble average of IPSL-CM5A-LR simulations nudged towards observed SST anomalies[14] is shown in blue. The dashed blue curve is the associated conceptual model including NAO variation (Methods). A 5-year running mean has been applied to all time series.

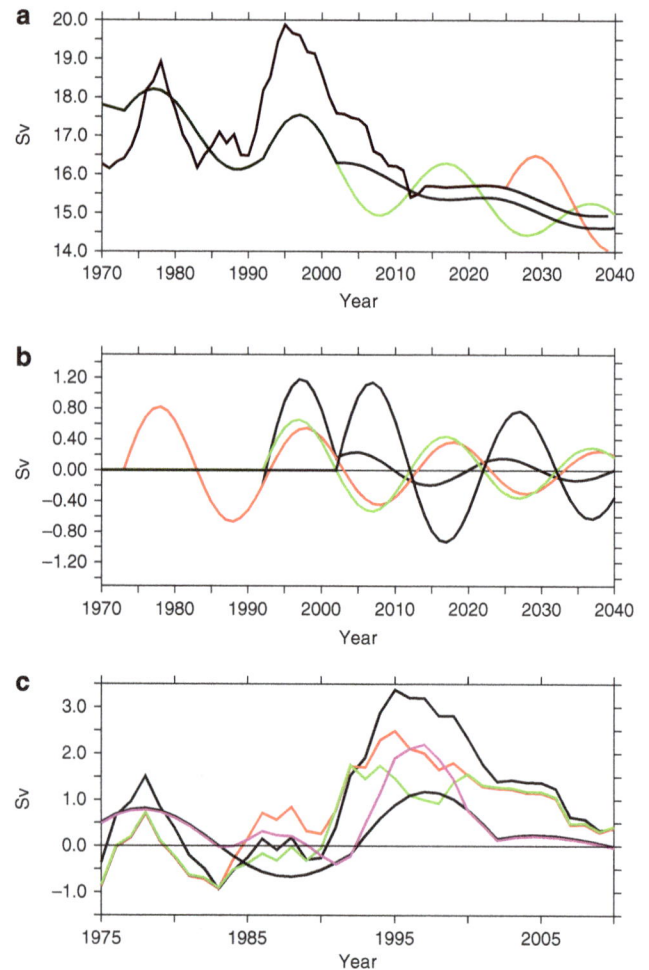

Figure 9 | Time evolution of the AMOC as computed by different conceptual models (described in Methods). (a) Different conceptual models with two (green) or three (black) volcanoes and with the NAO included (blue). For the future, when no NAO information is available, we take a null NAO phase. This allows illustration of the impact of an eruption in 2015 as shown in red. All the conceptual models include a long-term decreasing trend due to global warming (GW). **(b)** Decomposition of the impact of each volcano, Agung (in red), El Chichon (in green), Pinatubo (in blue) and the sum of the three (in black). **(c)** AMOC at 48°N from the different conceptual models, including or not including NAO variations following volcanic eruptions (see Methods). This decomposition allows estimation of the part of variance of the AMOC resulting from volcanic forcing and/or NAO, as compared with the total variance induced by volcanoes and NAO (black curve, 100% of the total variance). The red curve shows the calculation over the years 1975–2010 result when accounting for NAO (57% of the total variance), while the calculation accounting for NAO with the exception of the 3 years following each volcanic eruption is shown in green (44% of the total variance). The blue curve shows the bidecadal mode excitation by the volcanoes only (20% of the total variance), while the magenta curve shows the calculation when accounting for the impact of volcanoes plus the effect of the NAO in the 3 years following each eruption (30% of the total variance).

For the 1975–2010 period, the conceptual model allows the relative importance of the response to volcanoes and to the NAO to be distinguished (see Methods). The NAO-induced AMOC variations account for 57% of the total AMOC variance, while 20% of the AMOC variance is due to excitation of the AMOC 15 years after volcanic eruptions, as a damped eigen mode (see for example, Fig. 9c). The sum of the variance from this decomposition into two components does not equal 100%, indicating that there exists covariance between the components.

This simple approach does not account for the potential impact of volcanoes on the NAO. While a strong positive NAO occurred in the 2–3 years following the Pinatubo eruption[28,30], this is not a systematic response to all eruptions[57]. For instance, the winter following the Agung eruption had a negative NAO phase. This different response may arise from the intrinsic chaotic nature of the NAO, for which internal oscillations produce a large amount of noise, which can therefore mask a volcanic-forced signal. To address this issue, we repeat our analysis removing the NAO signature of the three winters following the three main volcanic eruptions considered. In the conceptual model, the fraction of AMOC variance explained by the response to volcanic forcing in 1975–2010 now increases to 30%, while NAO-only (without considering the three winters following eruptions) accounts for 44% of the variance (see for example, Fig. 9c and Methods). Once again, we note some covariance within this decomposition leading to a sum lower than 100%.

Discussion

We have shown here multiple lines of evidence supporting the long-lasting impact of moderate (Agung-size) volcanic eruptions on ocean circulation in the North Atlantic. The timing of subsequent volcanic eruptions can lead to constructive (for example, Agung and El Chichon) or destructive interferences (for example, Agung and Pinatubo) to force internal modes of ocean variability. In the IPSL-CM5A-LR model, the underlying mechanism is related to the build-up and advection of salinity anomalies near the Labrador region, and its subsequent effect on North Atlantic convection. Moderate volcanoes increase the sea-ice cover in the Nordic Seas, thus reducing the export of freshwater through the Denmark Strait and leading to anomalous positive anomalies in the Labrador region, explaining the observed GSAs of the 1970 and 1990s. While the exact processes at play may be model dependent, the coherence of the Bi-Dec model responses suggest a key role for the propagation of salinity and temperature anomalies as observed during GSA events. The exact path and pattern for this propagation may vary among the models that exhibit bidecadal AMOC variability, including, for example, the strength of the amplifying mechanisms through sea ice and atmosphere interactions, clearly depicted by the IPSL-CM5A-LR model. Further intercomparisons of the precise mechanisms at play within the different models are needed to understand their potential diversity. Nevertheless, the present results provide a good working hypothesis based on the data that is currently available. Indeed, this impact of volcanic eruptions explains several aspects of recent AMOC variations, and associated patterns (for example, GSAs). Moreover recent analysis of the observed variability mode of bidecadal sea level reveals a strong regime shift since the 1970s[14], which can be related to the ocean circulation changes described here in response to the Agung eruption.

Nevertheless, processes other than volcanic eruptions may have shaped variability in the North Atlantic over the last six decades, such as the NAO (Fig. 8) or the 1980s GSA, which is not attributed here to volcanic eruptions. We also stress the fact that the proposed mechanisms identified in the IPSL-CM5A-LR simulations are not at play for eruptions with a radiative forcing

50% larger than that of Agung (see for example, Supplementary Fig. 11). Different ranges of the sea-ice response[34] can indeed either increase or decrease convection, depending on the size of the eruption. This sensitivity to volcanic forcing is therefore expected to be model dependent and other models from our Bi-Dec ensemble may exhibit different thresholds and sensitivity to volcanic eruption size. Further analysis of this sensitivity is necessary to improve our understanding of the response of the climate system to volcanic eruptions, within a coordinated intercomparison framework. This will hopefully be implemented within the CMIP6 framework in the near future.

The present findings also imply a significant predictability of the North Atlantic dynamics if an Agung- or Pinatubo-like eruption occurs in the future (Fig. 9a). Excited bidecadal variability would be superimposed on the long-term decreasing trend, which is driven by global warming[6]. Given the potential influence of AMOC and the North Atlantic Ocean on hurricane activity[58], African Sahel drought[59], Greenland melt[2], regional sea level[60] and marine ecosystems, such decadal predictability is of key interest to a diversity of stakeholders, including the agriculture, energy, fisheries, insurance industries, water supply and management agencies.

Methods

CMIP5 analysis. We use the CMIP5 database of historical simulations for the period 1850–2005 (http://cmip-pcmdi.llnl.gov/cmip5/), performed using solar, volcanic and anthropogenic forcings, and pre-industrial control simulations, performed without changes in forcing, to explore the variability of the simulated subpolar AMOC. This analysis is restricted to the 19 models for which the meridional stream function output is available (Supplementary Table 1). Based on the information from several paleoclimate archives from the North Atlantic, which consistently show bidecadal variability (see main text), we select the subset of models for which a peak of energy significantly larger than a red noise process is present in the spectral band of 10 to 30 years. A power spectrum analysis (Supplementary Fig. 1) is used to identify this subset of models for which spectral power is expressed in the 10–30-year period band, and passes a significance test (that is, the one that refutes the red noise null hypothesis) in either the pre-industrial or historical simulations. Almost half of the CMIP5 climate models (9 among 19 models) do satisfy this criterion: CCSM4, CESM1-BGC, MRI-ESM, NorESM1-ME, NorESM1-M, CanESM2, GFDL-ESM2M, GFDL-CM3 and IPSL-CM5A-LR. Supplementary Table 1 summarizes the main findings in terms of spectral characteristics of the AMOC at 48°N in the different models and simulations. Supplementary Figure 2 shows the AMOC evolution from some of the selected models over the period 1950–2005. These models show relatively well-phased variability close to harmonic variations with a maximum in the late 1970s and another one in the late 1990s. This extends the findings of Swingedouw et al.[15] using the IPSL-CM5A-LR model. Supplementary Figure 5 illustrates the phasing in terms of salinity and temperature, averaged in the upper Labrador Sea, found in these models

IPSL-CM5A-LR simulations. The ocean-atmosphere coupled model mainly used in this study is the IPSL-CM5A (ref. 36) in its low-resolution (LR) version as developed for CMIP5. The atmospheric model is LMDZ5 (ref. 61) with a 96 × 96 × L39 regular grid and the oceanic model is NEMO[62] with an 182 × 149 × L31 non-regular grid, in version 3.2, including the LIM-2 sea ice model[63] and the PISCES[51] module for oceanic biogeochemistry. The internal variability of the AMOC has been described in the study by Escudier et al.[20] It has a 20-year preferential variability in the whole subpolar gyre. The mechanism explaining such a variability was shown to be a cycle beginning (for instance) with an intensification of the EGC that brings more cold and fresh water to the Labrador Sea where it accumulates and gives rise to negative SST and sea surface salinity (SSS) anomalies. These anomalies are advected along the subpolar gyre; they affect the convection all along their path up to the Nordics Seas, where the negative SST anomalies increase the sea-ice cover, which in turn induces a positive sea-level pressure anomaly and a localized anticyclonic atmospheric circulation. This leads to a decrease in the wind stress along the eastern coast of Greenland and thus a weakening of the EGC, leading to an opposite phase of the cycle around 10 years after its onset. This cycle impacts the AMOC variations through the contribution of SSS anomalies to the convective activity in the subpolar North Atlantic. The list of simulations performed with this model and used in this study is summarized in the Supplementary Table 2 and described below.

The five-member ensemble of historical simulations use a prescribed external radiative forcing from the observed increase in greenhouse gases and aerosol concentrations as well as the ozone changes (Fig. 1) and the land-use modifications (not shown, see the study by Dufresne et al.[36]). They also include estimates of solar irradiance variations and of volcanic eruptions over the historical period represented as a decrease in the total solar irradiance (depending on the intensity of the volcanic eruption, Fig. 1). These historical simulations start from year 1850. Their initial conditions come from different dates of the 1000-year control simulation under pre-industrial conditions, each separated by 10 years. This control pre-industrial simulation is itself starting after thousands of years of spin-up procedure.

To evaluate the hypothesis of a destructive interference due to the Pinatubo eruption, we prepared an additional five-member ensemble using the IPSL-CM5A-LR model. The simulations start on 1 January 1991 from each historical simulation described earlier. They use rigorously the same external forcing except for the effect of the Pinatubo eruption, which was removed from the total solar irradiance. The AMOC response with the s.d. of the ensemble is shown in Supplementary Fig. 10.

We also consider a five-member ensemble of nudged simulations, for which each simulation includes a nudging term towards observed anomalous monthly SST (ref. 15). Each simulation starts on 1 January 1949 from one of the historical simulations presented above. These nudged simulations are using the same external forcing as the historical ones (Fig. 1). The nudging technique consists in adding a heat flux term Q to the SST equation under the form $Q = -\kappa(SST'_{mod} - SST'_{obs})$ where SST'_{mod} stands for the modelled anomalous SST at each time step and grid point, and SST'_{obs} the anomalous observed SST (Reynolds et al. (2007)). Anomalies are computed with respect to the average SST over the period 1949–2005 in the corresponding historical simulation and in the observations. We use a restoring coefficient κ of $40\,W\,m^{-2}\,K^{-1}$ corresponding to a relaxing time scale of ~60 days (for a mixed layer of 50 m depth). See the study by Swingedouw et al.[15] for further details.

The CMIP5 last millennium simulation starts in 850. It includes solar[64] and volcanic forcing[53]. The implementation of the volcanic forcing in this simulation is slightly different from that used for the historical simulations. In the latter, the effect of stratospheric injection of sulphate volcanic aerosols is only considered as a modulation in the total solar irradiance variations, while in this simulation the volcanic aerosols in the stratosphere are transported latitudinally in the model following Grieser and Schonwiese[65] parameterization. We consider 48 latitude bands for this transport. The model uses prescribed changes in aerosols optical depth and interactively computes the perturbed (longwave and shortwave) radiative budgets. The reconstruction of aerosols optical depth uses sulphate data from ice cores records from Greenland and Antarctica[53]. In this simulation, the AMOC at 48°N shows very strong variability in the bidecadal band (Supplementary Fig. 6), as in pre-industrial simulations analyzed in the study by Escudier et al.[20] and in the historical simulations (Supplementary Fig. 1).

For the analysis of the Agung-like eruptions, we consider the 40 years following each selected event as an individual sensitivity experiment, like the member of an ensemble. The selected volcanoes are shown with respect to the full time series of volcanic forcing over the last millennium in Supplementary Fig. 7. The ensemble mean and the s.d. of the five members are shown in Fig. 3. We only consider eruptions with a magnitude comparable to Agung (1963), since larger eruptions lead to very strong cooling resulting in a different regime response for the AMOC[34]. Note that although the selection of the individual volcanic events includes a condition on the absence of significant eruption during the 40 years after the event, no condition is imposed regarding the years preceding the selected eruption. It is nevertheless possible that volcanic eruptions occurring a few years earlier could also impact North Atlantic variability, and indeed it is possible that the 1460 eruption is affected in this way, although none of the other volcanoes selected. Since removing this eruption does not modify the main conclusions, we keep it in our pool.

Sensitivity of IPSL simulations to the size of eruptions. Our analyses are focused on the model response to moderate eruptions. Indeed, a composite computation of key oceanic variables in the North Atlantic for all volcanoes lower than one and a half Agung as shown in Supplementary Fig. 11 for the IPSL-CM5A-LR model clearly supports our results: this shows again that moderate-size volcanic eruptions excite the 20-year variability through a cooling of the Nordic Seas at lag 1 year, when the subpolar region is less affected. This leads in the following years to the build-up of SST positive anomalies (as well as SSS, not shown) in the Labrador Sea, which are then advected along the subpolar gyre affecting the mixed layer depth around lag 5–6 years, explaining the 15-year lag for maximum AMOC following the simple scheme from the study by Escudier et al.[20]

However, volcanic eruptions larger than one and half Agung have a direct negative impact on SST in the convection region south of Iceland, leading almost immediately to local negative mixed layer depth anomalies for lags 2–6 years (Supplementary Fig. 11) following the eruption onset, which will lead to AMOC weakening around year 9–13, consistent with the study by Mignot et al.[34]

Such sensitivity could clearly be model dependent, so that further analysis of different models will be necessary to exactly evaluate the volcanic eruption size that lead to different regimes in different models and in reality.

Salinity data. Observational salinity data have been downloaded from the Bedford Institute of Oceanography website (http://www.bio.gc.ca/science/data-donnees/base/start-commencer-eng.php). Their analysis is focused on the Labrador Sea area, which is a key for the processes involved in the 20-year variability[20]. Since

changes in deep convection driving the AMOC can be triggered by salinity anomalies in the Labrador Sea as deep as 300 m, we also consider subsurface data to this depth[20]. To evaluate the observational uncertainty, we also use the version 2a of data compilation from EN3 (http://www.metoffice.gov.uk/hadobs/en3), and a SSS data compilation from ships of opportunity and oceanographic cruises[42] for the eastern part of the subpolar gyre.

***In situ* temperature anomalies.** In the IPSL-CM5A-LR climate model, the salinity anomalies are associated with temperature anomalies in the Labrador Sea and all along the subpolar gyre and Nordic Seas[20]. This is related to variations of the EGC, which brings cold and fresh water from the Arctic. We check in Supplementary Fig. 4 that, in the data analyzed, the salinity anomalies shown in Fig. 2 have a temperature counterpart (using the same source of data, that is, Bedford Institute of Oceanography and EN3). It is clearly the case in the Labrador Sea, with once again an agreement between historical simulations and temperature data, following the GSAs of the 1970 and 1990s. We note that the GSA of the 1980s is not clear in the Bedford Institute of Oceanography data set and absent in the model. We argue that this GSA is primarily driven by exceptional atmospheric forcing[22]. Given the lack of strong signals in the eastern subpolar gyre, the correspondence between temperature and salinity anomalies is weaker in this area.

Greenland ice core composite. The $\delta^{18}O$ Greenland records used herein correspond to six annually resolved millennial-long ice core records covering the period 1000–1973 (B18, Crete, DYE-3, GISP2, GRIP, NGRIP, see for example, Supplementary Table 3). During the instrumental period, the first PC associated with the leading EOF of these proxies (accounting for 25% of the variance of the six ice core records) is closely related to precipitation-weighted Greenland surface temperature changes[66], ultimately driven by changes in the four major weather regimes in the North Atlantic, namely the two phases of the NAO, the Atlantic Ridge and the Scandinavian Blocking[66]. Our PC is closely related, during the overlapping period, ($R = 0.67$, $P < 0.05$) to the average of five Arctic and Greenland $\delta^{18}O$ ice core records previously used to characterize the bidecadal SST variability in the North Atlantic during the past 700 years[16].

Bivalve proxy records. The bivalve data used are described in the study by Butler et al.[49] This proxy is a unique record of bivalve growth north of Iceland, annually resolved over a period of 1,357 years. Radiocarbon measurements taken on absolutely dated shell material have been used to show the co-varying influence of Atlantic and Arctic waters on the North Icelandic Shelf over the past millennium[67]. The correlation between growth and observed temperature is positive and weak (0.217, $P < 0.05$ using HadISST1 gridded data), suggesting that the bivalves are responding to a nutrient signal that is only partially temperature driven.

Link between last millennium proxies and model simulation. The AMO[45] represents the low-frequency variations of SST averaged over the Atlantic from the equator to 60°N. These multidecadal variations influence surrounding regions. In particular, Greenland climate is considered to be very sensitive to these variations. For instance, surface air temperature at Nuuk, West Greenland and the AMO Index are highly correlated ($r = 0.49$, $P = 0.05$) over the last 150 years[68].

Last, we show that in IPSL-CM5A-LR (and other models) the AMOC is leading the AMO by a few years[41]. Here we focus on the link between the AMOC and the SST in the North Atlantic and investigate the phase lag in Supplementary Fig. 8 using 1,000 years of the last millennium simulation. We show that AMOC index at 48°N is leading the SST in the Atlantic between 25 and 55°N by around 5–10 years. This phasing also clearly appears in the 40 years following selected volcanic eruptions from Fig. 3. Strong correlations are also identified for the control simulation under pre-industrial conditions, showing that this behaviour is not triggered by external forcing but is linked to the internal mechanisms relating AMOC and North Atlantic SST (not shown).

Conceptual models. We build here a family of simple conceptual models to capture the essence of the main mechanisms at play in our climate model and hopefully in reality.

Our first assumption, suggested by our results from the different CMIP5 screened models and data sets, is that Pinatubo-like eruptions lead to an AMOC maximum around 15 years after the onset of volcanic eruption. This delay is related to the characteristic time required for the North Atlantic to adjust to the volcanic signal, through induced changes in convection. It accounts for the lag between cooling of Nordic Seas, reductions in EGC, positive anomalies advected along the subpolar gyre, changes in convection sites and, a few years later, changes in the AMOC[15]. This AMOC response corresponds to the excitation of damped oscillatory variations with a typical time scale of 20 years[69].

We call τ the time delay for a particular volcanic eruption to have an impact on the AMOC; it is expected to depend on the typical 15-year delay as well as on the exact timing of the eruption. The subsequent damped oscillation response excited by the volcano can then be described through the following function of time $f(t)$:

$$f(t) = H(t-\tau) \sin\left(\frac{2\pi}{20}t\right) e^{\frac{t-\tau}{\Delta}} \qquad (1)$$

where H stands for the Heaviside function, t is the time and Δ is a characteristic damping time scale.

Over the time period 1960–2010, there have been three main eruptions (Agung, El Chichon and Pinatubo) whose radiative forcing in the IPSL-CM5A-LR historical simulations are $1.5\,W\,m^{-2}$, $1.2\,W\,m^{-2}$ and $2.1\,W\,m^{-2}$, respectively (Fig. 1). To account for the size of the eruption, we introduce a scaling factor λ_i for each volcano (equal to its radiative forcing). We finally use a factor α to scale globally the effect from all the volcanoes.

For comparison with historical simulations and projections, we must also take into account the AMOC decreasing trend produced in response to anthropogenic forcing by CMIP5 models[6]. This effect of global warming is implemented through the $GW(t)$ function, which corresponds to the anthropogenic radiative forcing multiplied by a scaling factor β.

$$f(t) = \alpha \sum_{i=1}^{3} \lambda_i H(t-\tau_i) \sin\left(\frac{2\pi}{20}t\right) e^{\frac{t-\tau_i}{\Delta}} - \beta \times GW(t) \qquad (2)$$

This conceptual model can be tested with and without the Pinatubo eruption. When excluding the Pinatubo eruption, the sum in equation (2) only accounts for the two other volcanoes, and the rest stays the same.

In reality, AMOC is also affected by the internal variability of the atmospheric circulation and is particularly sensitive to the NAO[3,55,56] with a 5-year lag. To include this aspect in the conceptual model, an additional term is introduced. It corresponds to the winter NAO index from NOAA (http://www.cpc.ncep. noaa.gov/products/precip/CWlink/pna/JFM_season_nao_index.shtml), with a lag of 5 years, due to oceanic adjustment to changes in surface fluxes and scaled by a coefficient γ.

$$f(t) = \alpha \sum_{i=1}^{3} \lambda_i H(t-\tau_i) \sin\left(\frac{2\pi}{20}t\right) e^{\frac{t-\tau_i}{\Delta}} - \beta \times GW(t) + \gamma \times NAO(t-5) \qquad (3)$$

The tuning of the adjustable parameters of this conceptual model (α, β, Δ and γ) is performed with a minimization technique using IPSL-CM5A-LR simulations as targets. The global warming scaling β is first adjusted using all IPSL-CM5A-LR available projections. A linear regression performed between anthropogenic forcing and the simulated 48°N AMOC index leads to a scaling factor β of $1.2\,Sv\,m^2\,W^{-1}$. We then define a cost function to be minimized, as the mean squared error between the conceptual model results and the reference simulation (historical and sensitivity experiments without Pinatubo eruption) over the period 1975–2030 (Supplementary Fig. 12). The best choice for the scaling factor α for the volcanoes and the damping factor Δ for the oscillations is unequivocally 0.5 Sv and 60 years, respectively. Finally, the scaling factor γ for the NAO-forced AMOC variations is adjusted using the IPSL-CM5A-LR nudged simulations towards observed SST and optimized for a value of 2.5 Sv.

We compare the results of two different conceptual models, which account for volcanic-induced variability and NAO forcing. Indeed, between 1975 and 2010, the volcanic excitations of the bidecadal mode acts on the timing of decadal variations; however, NAO is the main driver of the magnitude of AMOC variations in the conceptual model scaled on IPSL-CM5A-LR simulations (Fig. 9c). Indeed, if we compute the ratio of variance between the different conceptual models (Fig. 9c), including different pieces of forcing of the AMOC, we find that NAO alone accounts for 57% of the variance of the model including all processes, while the excitation of the bidecadal model accounts for 20%. Given that for any X and Y random variables, we have: $Var(X+Y) = Var(X) + Var(Y) + Cov(X,Y)$ where $Var()$ is the variance operator and $Cov()$ the covariance one. The sum of the variance explained by NAO and by volcanic-induced variability is not 100%, due to covariance.

To avoid accounting twice for the role of volcanic eruption (through the radiative forcing and through the potential consequence on the subsequent winter NAO phase[27]), we consider an alternative conceptual model, in which we remove the NAO forcing in the three winters following each eruption. In this case, the variance ratio between the variance of AMOC volcanically induced and the variance of AMOC, which is purely induced by the NAO (removing the three years following volcanic eruptions), is 30% and only 44% for the pure NAO-forced AMOC with volcanic years removed, respectively (Fig. 9c).

The conceptual model is finally used to explore the impact of a hypothetical Pinatubo-like eruption starting in 2015 (Fig. 9a). Without such an event, and without large NAO changes, the conceptual model simulates only the gradual decreasing trend driven by global warming. If a large eruption were to occur in 2015, then it would excite large AMOC decadal oscillations with maximum strength in 2030 and minimum strength in 2040, potentially complicating the detection of long-term trends associated with global warming.

References

1. Wunsch, C. Oceanography. What is the thermohaline circulation? *Science* **298**, 1179–1181 (2002).
2. Straneo, F. & Heimbach, P. North Atlantic warming and the retreat of Greenland's outlet glaciers. *Nature* **504**, 36–43 (2013).
3. Latif, M., Collins, M., Pohlmann, H. & Keenlyside, N. A review of predictability studies of Atlantic sector climate on decadal time scales. *J. Clim.* **19**, 5971–5987 (2006).
4. Smith, D. M. *et al.* Improved surface temperature prediction for the coming decade from a global climate model. *Science* **317**, 796–799 (2007).
5. Doblas-Reyes, F. J. *et al.* Initialized near-term regional climate change prediction. *Nat. Commun.* **4**, 1715 (2013).
6. Weaver, A. J. *et al.* Stability of the Atlantic meridional overturning circulation: a model intercomparison. *Geophys. Res. Lett.* **39**, L70209. doi: 10.1029/2012GL053763 (2012).
7. Swingedouw, D. *et al.* On the reduced sensitivity of the Atlantic overturning to Greenland ice sheet melting in projections: a multi-model assessment. *Clim. Dyn.* doi: 10.1007/s00382-014-2184-7 (2015).
8. McCarthy, G. *et al.* Observed interannual variability of the Atlantic meridional overturning circulation at 26.5°N. *Geophys. Res. Lett.* **39**, L19609. doi: 10.1029/2012GL052933 (2012).
9. Mercier, H. *et al.* Variability of the meridional overturning circulation at the Greenland–Portugal OVIDE section from 1993 to 2010. *Prog. Oceanogr.* (in the press).
10. Robson, J., Hodson, D., Hawkins, E. & Sutton, R. Atlantic overturning in decline? *Nat. Geosci.* **7**, 2–3 (2013).
11. Smeed, D. a. *et al.* Observed decline of the Atlantic Meridional Overturning Circulation 2004 to 2012. *Ocean Sci.* **10**, 1619–1645 (2013).
12. Reichler, T., Kim, J., Manzini, E. & Kröger, J. A stratospheric connection to Atlantic climate variability. *Nat. Geosci.* **5**, 783–787 (2012).
13. Huck, T., de Verdière, A. C., Estrade, P. & Schopp, R. Low-frequency variations of the large-scale ocean circulation and heat transport in the North Atlantic from 1955–1998 in situ temperature and salinity data. *Geophys. Res. Lett.* **35**, 1–5 (2008).
14. Vianna, M. L. & Menezes, V. V. Bidecadal sea level modes in the North and South Atlantic Oceans. *Geophys. Res. Lett.* **40**, 5926–5931 (2013).
15. Swingedouw, D., Mignot, J., Labetoulle, S., Guilyardi, E. & Madec, G. Initialisation and predictability of the AMOC over the last 50 years in a climate model. *Clim. Dyn.* **40**, 2381–2399 (2013).
16. Chylek, P., Folland, C. K., Dijkstra, H. A., Lesins, G. & Dubey, M. K. Ice-core data evidence for a prominent near 20 year time-scale of the Atlantic Multidecadal Oscillation. *Geophys. Res. Lett.* **38**, L13704. doi 10.1029/2011GL047501 (2011).
17. Divine, D. V. & Dick, C. Historical variability of sea ice edge position in the Nordic Seas. *J. Geophys. Res.* **111**, C01001 (2006).
18. Sicre, M.-A. *et al.* Decadal variability of sea surface temperatures off North Iceland over the last 2000 years. *Earth Planet. Sci. Lett.* **268**, 137–142 (2008).
19. Cronin, T., Farmer, J. & Marzen, R. Late holocene sea-level variability and Atlantic meridional overturning circulation. *Paleoceanography* **29**, 765–777 (2014).
20. Escudier, R., Mignot, J. & Swingedouw, D. A 20-year coupled ocean-sea ice-atmosphere variability mode in the North Atlantic in an AOGCM. *Clim. Dyn.* **40**, 619–636 (2013).
21. Frankcombe, L. M., von der Heydt, A. & Dijkstra, H. A. North Atlantic multidecadal climate variability: an investigation of dominant time scales and processes. *J. Clim.* **23**, 3626–3638 (2010).
22. Belkin, I. M., Levitus, S., Antonov, J. & Malmberg, S.-A. 'Great Salinity Anomalies' in the North Atlantic. *Prog. Oceanogr.* **41**, 1–68 (1998).
23. Belkin, I. M. Propagation of the 'Great Salinity Anomaly' of the 1990s around the northern North Atlantic. *Geophys. Res. Lett.* **31**, 5–8 (2004).
24. Sundby, S. & Drinkwater, K. On the mechanisms behind salinity anomaly signals of the northern North Atlantic. *Prog. Oceanogr.* **73**, 190–202 (2007).
25. Robock, A. Volcanic eruptions and climate. *Rev. Geophys.* **38**, 191–219 (2000).
26. Timmreck, C. Modeling the climatic effects of large explosive volcanic eruptions. *Wiley Interdiscip. Rev. Clim. Change* **3**, 545–564 (2012).
27. Driscoll, S., Bozzo, A., Gray, L. J., Robock, A. & Stenchikov, G. Coupled Model Intercomparison Project 5 (CMIP5) simulations of climate following volcanic eruptions. *J. Geophys. Res. Atmos.* **117**, D17105 doi 10.1029/2012JD017607 (2012).
28. Stenchikov, G. Arctic Oscillation response to the 1991 Mount Pinatubo eruption: effects of volcanic aerosols and ozone depletion. *J. Geophys. Res.* **107**, 4803 (2002).
29. Hurrell, J. W. Decadal trends in the north atlantic oscillation: regional temperatures and precipitation. *Science* **269**, 676–679 (1995).
30. Robock, A. & Mao, J. Winter warming from large volcanic eruptions. *Geophys. Res. Lett.* **19**, 2405–2408 (1992).
31. Otterå, O. H. Simulating the effects of the 1991 Mount Pinatubo volcanic eruption using the ARPEGE atmosphere general circulation model. *Adv. Atmos. Sci.* **25**, 213–226 (2008).
32. Zanchettin, D. *et al.* Bi-decadal variability excited in the coupled ocean-atmosphere system by strong tropical volcanic eruptions. *Clim. Dyn.* **39**, 419–444 (2012).

33. Otterå, O. H., Bentsen, M., Drange, H. & Suo, L. External forcing as a metronome for Atlantic multidecadal variability. *Nat. Geosci.* **3**, 688–694 (2010).

34. Mignot, J., Khodri, M., Frankignoul, C. & Servonnat, J. Volcanic impact on the Atlantic Ocean over the last millennium. *Clim. Past* **7**, 1439–1455 (2011).

35. Taylor, K. E., Stouffer, R. J. & Meehl, G. A. A. Summary of the CMIP5 experiment design. *Bull. Am. Meteorol. Soc.* **93**, 485–498 (2012).

36. Dufresne, J.-L. *et al.* Climate change projections using the IPSL-CM5 Earth System Model: from CMIP3 to CMIP5. *Clim. Dyn.* **40**, 2123–2165 (2013).

37. Menary, M. B. & Scaife, A. A. Naturally forced multidecadal variability of the Atlantic meridional overturning circulation. *Clim. Dyn.* **42**, 1347–1362 (2014).

38. Stenchikov, G. *et al.* Arctic Oscillation response to volcanic eruptions in the IPCC AR4 climate models. *J. Geophys. Res. D Atmos.* **107**, D244803 doi: 10.1029/2002JD002090 (2002).

39. Colin de Verdière, A. & Huck, T. Baroclinic instability: an oceanic wavemaker for interdecadal variability. *J. Phys. Oceanogr.* **29**, 893–910 (1999).

40. Dijkstra, H. A. & Ghil, M. Low-frequency variability of the large-scale ocean circulation: a dynamical systems approach. *Rev. Geophys.* **43**, 1–38 (2005).

41. Persechino, A., Mignot, J., Swingedouw, D., Labetoulle, S. & Guilyardi, E. Decadal predictability of the Atlantic meridional overturning circulation and climate in the IPSL-CM5A-LR model. *Clim. Dyn.* **40**, 2359–2380 (2013).

42. Reverdin, G. North Atlantic subpolar gyre surface variability (1895–2009). *J. Clim.* **23**, 4571–4584 (2010).

43. Ingleby, B. & Huddleston, M. Quality control of ocean temperature and salinity profiles—Historical and real-time data. *J. Mar. Syst.* **65**, 158–175 (2007).

44. Rayner, N. a. Global analyses of sea surface temperature, sea ice, and night marine air temperature since the late nineteenth century. *J. Geophys. Res.* **108**, D14 4407. doi: 10.1029/2002JD002670 (2003).

45. Kerr, R. A. A north atlantic climate pacemaker for the centuries. *Science* **288**, 1984–1985 (2000).

46. Mann, M. E. *et al.* Global signatures and dynamical origins of the Little Ice Age and Medieval Climate Anomaly. *Science* **326**, 1256–1260 (2009).

47. Gray, S. T., Graumlich, L. J., Betancourt, J. L. & Pederson, G. A tree-ring based reconstruction of the Atlantic Multidecadal Oscillation since 1567A.D. *Geophys. Res. Lett.* **31**, 2–5 (2004).

48. Chylek, P. *et al.* Greenland ice core evidence for spatial and temporal variability of the Atlantic Multidecadal Oscillation. *Geophys. Res. Lett.* **39**, L09705. doi: 10.1029/2012GL051241 (2012).

49. Butler, P. G., Wanamaker, A. D., Scourse, J. D., Richardson, C. a. & Reynolds, D. J. Variability of marine climate on the North Icelandic Shelf in a 1357-year proxy archive based on growth increments in the bivalve *Arctica islandica*. *Palaeogeogr. Palaeoclimatol. Palaeoecol.* **373**, 141–151 (2013).

50. Auger, P. a. *et al.* Interannual control of plankton communities by deep winter mixing and prey/predator interactions in the NW Mediterranean: results from a 30-year 3D modeling study. *Prog. Oceanogr.* **124**, 12–27 (2014).

51. Aumont, O. & Bopp, L. Globalizing results from ocean in situ iron fertilization studies. *Glob. Biogeochem. Cycles* **20**, GB2017. doi: 10.1029/2005GB002591 (2006).

52. Lehner, F., Raible, C. C. & Stocker, T. F. Testing the robustness of a precipitation proxy-based North Atlantic Oscillation reconstruction. *Quat. Sci. Rev.* **45**, 85–94 (2012).

53. Ammann, C. M., Joos, F., Schimel, D. S., Otto-Bliesner, B. L. & Tomas, R. A. Solar influence on climate during the past millennium: results from transient simulations with the NCAR Climate System Model. *Proc. Natl Acad. Sci. USA* **104**, 3713–3718 (2007).

54. Knight, J. R. A signature of persistent natural thermohaline circulation cycles in observed climate. *Geophys. Res. Lett.* **32**, L20728. doi: 10.1029/2005GL024233 (2005).

55. Eden, C. & Willebrand, J. Mechanism of interannual to decadal variability of the North Atlantic circulation. *J. Clim.* **14**, 2266–2280 (2001).

56. Böning, C. W., Scheinert, M., Dengg, J., Biastoch, A. & Funk, A. Decadal variability of subpolar gyre transport and its reverberation in the North Atlantic overturning. *Geophys. Res. Lett.* **33**, L21S01 (2006).

57. Graf, H.-F., Zanchettin, D., Timmreck, C. & Bittner, M. Observational constraints on the tropospheric and near-surface winter signature of the Northern Hemisphere stratospheric polar vortex. *Clim. Dyn.* **43**, 3245–3266 (2014).

58. Goldenberg, S. B., Landsea, C. W., Mestas-Nunez, A. M. & Gray, W. M. The recent increase in Atlantic hurricane activity: causes and implications. *Science* **293**, 474–479 (2001).

59. Hoerling, M., Hurrell, J., Eischeid, J. & Phillips, A. Detection and attribution of twentieth-century Northern and Southern African rainfall change. *J. Clim.* **19**, 3989–4008 (2006).

60. Levermann, A., Griesel, A., Hofmann, M., Montoya, M. & Rahmstorf, S. Dynamic sea level changes following changes in the thermohaline circulation. *Clim. Dyn.* **24**, 347–354 (2005).

61. Hourdin, F. *et al.* The LMDZ4 general circulation model: climate performance and sensitivity to parametrized physics with emphasis on tropical convection. *Clim. Dyn.* **27**, 787–813 (2006).

62. Madec, G. *NEMO Ocean Engine* (Institut Pierre-Simon Laplace (IPSL), 2008).

63. Fichefet, T. & Morales Maqueda, A. M. Sensitivity of a global sea ice model to the treatment of ice thermodynamics and dynamics. *J. Geophys. Res.* **102**, 2609–2612 (1997).

64. Vieira, L. E. A., Solanki, S. K., Krivova, N. A. & Usoskin, I. Evolution of the solar irradiance during the Holocene. *Astron. Astrophys.* **531**, A6 doi: 10.1051/0004-6361/201015843 (2011).

65. Grieser, J. & Schonwiese, C. D. Parameterization of spatio-temporal patterns of volcanic aerosol induced stratospheric optical depth and its climate radiative forcing. *Atmosfera* **12**, 111–133 (1999).

66. Ortega, P. *et al.* Characterizing atmospheric circulation signals in Greenland ice cores: insights from a weather regime approach. *Clim. Dyn.* (2014).

67. Wanamaker, A. D. *et al.* Surface changes in the North Atlantic meridional overturning circulation during the last millennium. *Nat. Commun.* **3**, 899 (2012).

68. Drinkwater, K. F. *et al.* The Atlantic Multidecadal Oscillation: its manifestations and impacts with special emphasis on the Atlantic region north of 60°N. *J. Mar. Syst.* **133**, 117–130 (2013).

69. Sévellec, F. & Fedorov, A. V. The leading, interdecadal Eigenmode of the Atlantic meridional overturning circulation in a realistic ocean model. *J. Clim.* **26**, 2160–2183 (2013).

70. Bretherton, C., Widmann, M., Dymnikov, V., Wallace, J. & Blade, I. The effective number of spatial degrees of freedom of a time-varying field. *J. Clim.* **12**, 1990–2009 (1999).

Acknowledgements

This research was partly funded by the ANR CEPS Green Greenland project and by ANR MORDICUS project (ANR-13-SENV-0002-02). It is also funded by the SPECS project funded by the European Commission's Seventh Framework Research Programme under the grant agreement 308378 and by the EMBRACE project with Research Number 282672. To analyze the CMIP5 data, this study benefited from the IPSL Prodiguer-Ciclad facility, which is supported by CNRS, UPMC, Labex L-IPSL, which is funded by the ANR (Grant #ANR-10-LABX-0018) and by the European FP7 IS-ENES2 project (Grant #312979). We are grateful to the Bedford Institute of Oceanography, EN3 and Gilles Reverdin for freely providing their ocean data. We also thank Laurent Terray and Aurélien Ribes for interesting discussions around this paper and Vincent Hanquiez and Patrick Brockmann for help with the figures.

Author contributions

D.S. conceived the study and led the writing of the paper. P.O. performed the Greenland ice core analysis. All authors contributed to data interpretation and writing of the manuscript.

Additional information

Competing financial interests: The authors declare no competing financial interests.

Both respiration and photosynthesis determine the scaling of plankton metabolism in the oligotrophic ocean

Pablo Serret[1,2], Carol Robinson[3], María Aranguren-Gassis[1,†], Enma Elena García-Martín[1,†], Niki Gist[4], Vassilis Kitidis[4], José Lozano[1,2], John Stephens[4], Carolyn Harris[4] & Rob Thomas[5]

Despite its importance to ocean–climate interactions, the metabolic state of the oligotrophic ocean has remained controversial for >15 years. Positions in the debate are that it is either hetero- or autotrophic, which suggests either substantial unaccounted for organic matter inputs, or that all available photosynthesis (P) estimations (including ^{14}C) are biased. Here we show the existence of systematic differences in the metabolic state of the North (heterotrophic) and South (autotrophic) Atlantic oligotrophic gyres, resulting from differences in both P and respiration (R). The oligotrophic ocean is neither auto- nor hetero-trophic, but functionally diverse. Our results show that the scaling of plankton metabolism by generalized P:R relationships that has sustained the debate is biased, and indicate that the variability of R, and not only of P, needs to be considered in regional estimations of the ocean's metabolic state.

[1]Departamento de Ecología y Biología animal, Universidad de Vigo, E36310 Vigo, Spain. [2]Estación de Ciencias Marinas de Toralla, Universidad de Vigo, Toralla island, E-36331 Vigo, Spain. [3]School of Environmental Sciences, University of East Anglia, Norwich Research Park, Norwich NR4 7TJ, UK. [4]Plymouth Marine Laboratory, Prospect Place, Plymouth PL1 3DH, UK. [5]British Oceanographic Data Centre, Joseph Proudman Building, 6 Brownlow Street, Liverpool L3 5DA, UK. †Present addresses: 4WK Kellogg Biological Station, Michigan State University, 3700 E. Gull Lake Drive, Hickory Corners MI 49060, USA (M.A.-G.); School of Environmental Sciences, University of East Anglia, Norwich Research Park, Norwich NR4 7TJ, UK (E.E.G.-M.). Correspondence and requests for materials should be addressed to P.S. (email: pserret@uvigo.es).

The microbial biota of the surface ocean is a prime determinant of the ocean's productivity and biogeochemical functioning. Marine phytoplankton is responsible for half of the Earth's oxygenic photosynthesis, which contributes to the redox state of the planet and the global cycles of C and O_2 in the ocean and atmosphere, affecting the global climate and creating habitable conditions for many organisms[1,2]. Part of the photosynthetically produced organic matter is respired in the surface ocean, either to support the maintenance costs of the phytoplankton or by the heterotrophic activity of other microbes. This oxidation counterbalances the fluxes of C and O_2 from photosynthesis, and reduces the net amount of organic matter available for consumption by metazoans. The difference between plankton photosynthesis (P) and auto- plus heterotrophic respiration (R) is net community production (NCP), the amount of photosynthetically fixed C available for sequestration to the deep ocean or for transfer up the marine food web[3]. Over significant scales, NCP summarizes the energy flow and metabolic state of a planktonic ecosystem, setting upper bounds to its contribution to O_2 and CO_2 fluxes, and hence to global climate.

The oligotrophic ocean is the most extensive biome of the world, occupying about half of the Earth's surface. Its areal productivity is limited because of the strong vertical stratification of the water column that reduces nutrient inputs to the illuminated zone. However, its extent makes its contribution to global productivity, C export and biologically driven ocean–atmosphere fluxes significant[4–6]. In addition, climate change is increasing both the area[7] and stratification of oligotrophic gyres[7,8], with feedbacks to the biological rates and the metabolic state. Despite its importance, the metabolic state of the oligotrophic ocean has remained controversial[3], since del Giorgio et al.[9] published a seminal paper on the relationship between primary production and bacterial respiration in the ocean. Direct measurements of community metabolism in low-production ecosystems show a prevalence of heterotrophy (P < R)[10], that leads to the 'intriguing, if not disquieting' idea that the extensive and remote oligotrophic ocean is heterotrophic[3]. These results are based on the in vitro incubation of small seawater samples (~125 ml) during short periods of time (~24 h), which are susceptible to biases from the bottle confinement of natural plankton communities[11]. An alternative to incubation-based methods is to study the indirect impact of plankton metabolism on inventories of biogenic or limiting compounds in large water bodies (km^2) during long periods of time (weeks to months)[12–14]. Contrary to direct measurements of planktonic activity, results from this approach show a prevalence of autotrophy (P > R) in the oligotrophic ocean[11]. These calculations involve a large uncertainty derived from the estimations of water mass mixing and air–sea gas exchange[10], and assume a constancy in gradients of stocks and water transport that conflicts with non-steady ocean dynamics[15].

Beyond methodological biases that have played a major role in the debate[3,10,11], these two approaches measure different processes over completely different time and spatial scales—snapshots of the NCP of single communities versus the integrative metabolism of a succession of communities in a water mass. Comparison hence requires scaling-up in vitro data to the large and long spatial and temporal extent of in situ estimates. To this aim, empirical P:R relationships have been used to predict NCP from the extensive global data set of P (mainly ^{14}C derived)[16–19]. All the extrapolations carried out during the last 15 years have found heterotrophy to prevail whenever P is low, leading some authors to conclude that the oligotrophic ocean is heterotrophic and the (autotrophic) in situ data are biased[10]. According to these authors, the input of organic matter to the open ocean is much higher than the observations of stocks and export from the euphotic zone, and non-oxygenic autotrophic processes (not included in P estimations) are significantly more relevant than currently believed[10]. On the contrary, others have concluded that the oligotrophic ocean is autotrophic, and that all the in vitro measurements of plankton production—including tens of thousands of ^{14}C measurements—could be biased because of the perturbation of the light conditions[11]. Such a widespread and constant state of autotrophy would require a supply of inorganic nutrients to the surface oligotrophic ocean at rates much higher than observations of stocks and fluxes indicate[15]. Both positions hence ultimately imply unresolved explanatory hypotheses that would force a reconsideration of our knowledge about the magnitude, variability and control of organic matter production and cycling in the open ocean.

Despite the lengthy discussion on methodological biases, very little attention has been paid to the scaling procedure that is essential for data comparisons. In all cases, the empirical P:R[16–19] or P/R:P[17,20] relationships are assumed universal[16–18,20] or constant for a certain latitudinal band (for example, 10–40°)[19], which relies on some important and untested assumptions. The idea that NCP (that is, P minus R) may be predicted in the oligotrophic ocean from P alone is based on the hypothesis that changes in P but not in R control the metabolic balance either globally[10,16–18,20] or at low latitudes[11,19]. This may occur because the variability of R is irrelevant or minor[20–22], or because it is tightly coupled to the variability of P[16–19]. In either case, the use of a single P:R relation implies that the influence of P-independent processes on heterotrophic R (for example, magnitude and quality of allochthonous organic matter inputs, consumption of organic matter accumulated during productive phases, composition, structure and activity of the heterotrophic communities) are uniform throughout at least the low latitude band of the global ocean. This is a very tenuous assumption, as even in the most extensive data sets[18,19] used both to calculate balances and to derive P:R relationships[10,11,18,19], the great majority of data from the oligotrophic ocean come from one province, the Eastern Atlantic gyre (NAST-E). However, a reduced number of observations in the South Atlantic gyre have suggested differences in community metabolism that would reflect distinct P:R relationships[23,24].

To test the assumption that no regional differences exist in either the P:R relationship or the metabolic balance throughout the oligotrophic ocean, that is, to assess the basis of the scaling procedure sustaining the debate, we have compiled 194 vertical profiles (median five depths) in the epipelagic zone (defined here as the layer between the surface and the 1% of photosynthetically active radiation—PAR—depth) of in vitro derived P, R and NCP measurements (ΔO_2 after 24-h light and dark incubations) made across the Atlantic (50° N–50° S) during 10 Atlantic Meridional Transect (AMT11-18, AMT21-22) cruises from 2000 to 2012. All the cruises except AMT17 and AMT18 included both the North and South gyres (NAST and SATL). This is a very consistent data set in terms of sampling, incubation and analytical methodology and precision, it is equivalent to ~76% of non-AMT Atlantic data in the latest published global NCP data set[18], and includes 46 profiles from the very undersampled SATL. The analysis of this data set confirms the prevalence of net heterotrophy in the NAST, but shows that the plankton metabolism in the similarly oligotrophic SATL was predominantly autotrophic. Our data reveal significant differences in R, P:R relationship and NCP between the two Atlantic ocean gyres, indicating that the oligotrophic ocean is functionally diverse. This would bias any regional to global prediction of NCP based on generalized P:R relationships, constraining the scaling of plankton metabolism to functionally coherent ecosystems.

Results

Plankton metabolic rates in the Atlantic gyres. The AMT cruises traversed several biogeographic provinces of the Atlantic Ocean (Fig. 1). Gyre stations were identified when the depth of the thermocline and chlorophyll a maxima were $>100\,m$ (~20–$38°$ N for the NAST; ~10–$32°$ S for the SATL).

Integrated NCP was clearly different between the NAST and SATL (t-test ($n=91$) $=-7.14$, $P=0.0001$; Fig. 2a, Table 1). Net heterotrophy prevailed in the NAST, agreeing with the literature[10,17–20,23,24]; 85% of our integrated NCP data were negative, the mean ($\pm$ s.e.m.) NCP was $-14.55\pm2.42\,mmolO_2\,m^{-2}$ per day and the 95% confidence interval was -19.31 to $-9.79\,mmolO_2\,m^{-2}$ per day. However, in the SATL, a balanced or net autotrophic metabolism prevailed (78% of the data), with a mean NCP of $9.78\pm2.34\,mmolO_2\,m^{-2}$ per day and a 95% confidence interval of 4.96 to $14.59\,mmolO_2\,m^{-2}$ per day. The composite vertical and latitudinal variability of volumetric rates during AMT11 to AMT22 shows that the significant differences in NCP between the NAST and SATL were consistent throughout the entire epipelagic zone (t-test ($n=433$) $=-6.63$, $P=0.0001$; Fig. 3). Sixty-two per cent of the 202 volumetric data from the NAST were net heterotrophic, with a mean ($\pm$ s.e.m.) of $-0.12\pm0.02\,mmolO_2\,m^{-3}$ per day, while 63% of the 233 data from the SATL were net autotrophic, with a mean ($\pm$ s.e.m.) of $0.06\pm0.02\,mmolO_2\,m^{-3}$ per day (Table 1). Although these percentages in the NAST concur with the literature (57% (ref. 10)), conversion of volumetric data into discrete categories with a cutoff NCP of zero (that is, auto- or heterotrophic) misrepresents the prevailing metabolism because of the relatively high variability of these very low rates (Fig. 4), especially in the upper 25 m ($\sim33\%$ light depth) and at the 1% light depth, where the mean NCP rates were very close to zero at both the NAST and SATL (Fig. 5). This explains the lower percentage of the volumetric data that were auto- or heterotrophic in the SATL and NAST (63 and 62%, respectively), compared with the depth integrated data (78–85%). Vertical profiles of the mean and median NCP through the epipelagic zone provide a more realistic account of the clear differences in the prevailing metabolism of the two gyres (Fig. 5).

Most interestingly, the regional differences observed in NCP were not caused only by differences in P, but fundamentally by the variation of R. The assumed relative constancy of R[16–22] is only found when we analyse the entire latitudinal range of the transect, that is, when the highly productive equatorial and temperate provinces are included together with the oligotrophic gyres (Figs 2b and 3). However, within the two gyres, we found similar ranges of variability for both volumetric and integrated P and R rates (Figs 2b and 3 and Table 1), and this variability of R has consequences for the differences in metabolic balance between the gyres. The mean integrated P was lower in the heterotrophic NAST than in the autotrophic SATL (57.51 ± 3.50 versus $68.32\pm3.13\,mmolO_2\,m^{-2}$ per day, respectively), consistent with the hypothesis that P alone controls NCP[10–20]. However, the mean integrated R was lower in the autotrophic SATL than in the heterotrophic NAST (58.53 ± 3.85 versus $72.09\pm3.47\,mmolO_2\,m^{-2}$ per day), where very few data lower than $\sim35\,mmolO_2\,m^{-2}$ per day were found (Fig. 2b). Volumetric rates emphasize the importance of R to the significant differences in NCP between the NAST and SATL. R was significantly higher in the heterotrophic NAST ($0.68\pm0.03\,mmolO_2\,m^{-3}\,d^{-1}$) than in the autotrophic SATL ($0.48\pm0.02\,mmolO_2\,m^{-3}$ per day; t-test ($n=439$) $=6.08$, $P=0.0001$), while P rates were not significantly different (0.56 ± 0.02 and $0.55\pm0.02\,mmol\,O_2\,m^{-3}$ per day in the NAST and SATL, respectively; t-test ($n=433$) $=0.56$, $P=0.57$; Table 1). Volumetric rates of P were mostly lower in both gyres than the latest published threshold P for net heterotrophy ($1.26\,mmolO_2\,m^{-3}$ per day)[18] (Fig. 3). However, the corresponding prediction of net heterotrophy was only fulfilled in the NAST, precisely the oligotrophic province that contributes the most to the global data set whose generalized P:R relationship sustains both this threshold[18] and the hypothesis of relative R constancy[10,18].

Scaling plankton metabolism in the Atlantic gyres. The difference in the relative importance of R, and its consequences for scaling the metabolic balance are clearly shown in a plot of the P:R relationships in the two gyres (Fig. 6). We focus on the integrated rates because they provide a more accurate representation of the interaction of auto- and heterotrophic processes in the epipelagic ocean[16,23]. The NAST P:R relationship

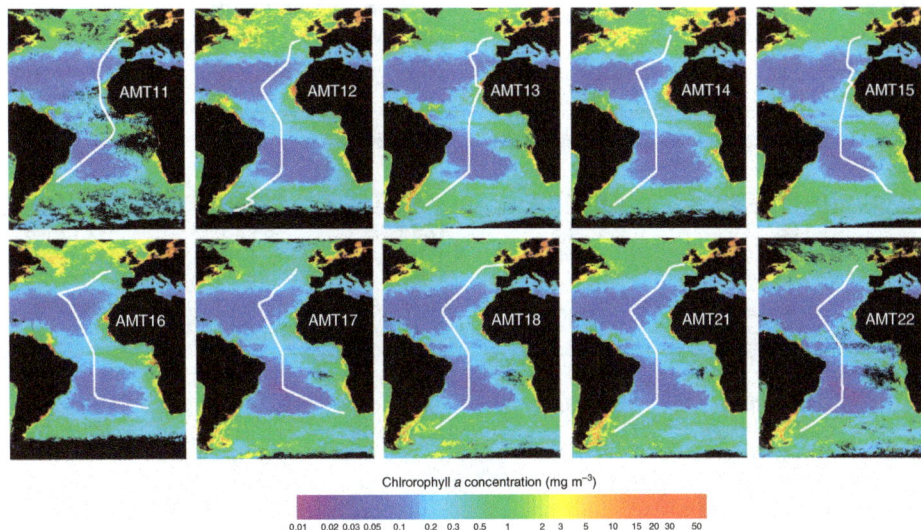

Figure 1 | The AMT cruises. Tracks of Atlantic Meridional Transect (AMT) cruises 11-18 and 21-22 overlaid on SeaWiFS chlorophyll a composites. After AMT11 (September–October 2000), cruises took place biannually from May 2003 (AMT12) until October 2005 (AMT17); AMT18, 21 and 22 took place in September–October 2008, 2011 and 2012, respectively. Earth observation data produced by the ESA Ocean Colour Climate change Initiative, courtesy of the NERC Earth Observation Data Acquisition and Analysis Service, Plymouth, the Ocean Biology Processing Group, NASA and European Space Agency.

resembles the generalized relationships that have been used to scale-up *in vitro* data for the last 15 years[16-19]. The degree of heterotrophy is inversely related to the magnitude of P, so that below a certain threshold P ($\sim 120 \, mmolO_2 \, m^{-2}$ per day), the net heterotrophy would always be predicted. As P in the oligotrophic ocean is typically below calculated thresholds, using this type of relationship to extrapolate NCP, together with the assumption of trophic invariability, has led to the conclusions that either the oligotrophic ocean is net heterotrophic

and requires substantial allochthonous inputs of organic matter[10], or *in vitro* data are biased[11]. However, the SATL P:R relationship is essentially a 1:1 relationship (*t*-test, $P < 0.01$), indicating a tight coupling of local auto and heterotrophic processes. This relationship would never predict net heterotrophy from P alone. Actually, neither the NAST nor SATL relationship is able to correctly predict the metabolic balance in the other oligotrophic gyre from P alone. Consequently, differences in R between the two gyres, and not only in P are critical for the prediction of metabolic balances in the oligotrophic Atlantic, which should therefore be based on distinct P:R relationships for (at least) each of the two gyres.

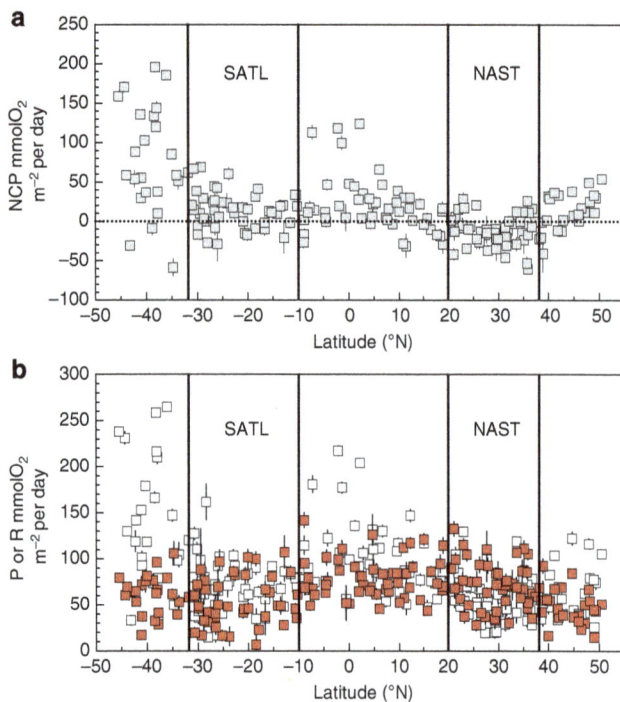

Potential seasonal bias. Most AMT cruises took place in boreal autumn, with only AMT12, 14 and 16 in spring. This implies that the seasonality of plankton metabolism in the gyres might bias our conclusions on regional variability. Seasonal differences in P, R and NCP in the NAST and SATL (Table 2) concur with a previous analysis of a partially overlapping data set[24] (note that gyre boundaries were identified in Gist *et al.*[24] from salinity/ density fronts and surface chlorophyll rather than the depth of the thermocline and DCM). To assess a potential seasonal bias, we have separated the entire AMT11–22 data set into astronomical season and repeated the analysis. Geographic patterns of P, R and NCP remain unaltered irrespective of the season, except for the fact that autumn NCP in the S gyre is not different from metabolic balance, which nonetheless sustains the significant difference between the N and S gyres (Table 2). Moreover, the slope of the P:R relationship in the SATL remains 1 in both spring and autumn, differing from the < 1 slopes of the N gyre relationships (Fig. 7). Although further empirical evidence is necessary to define the seasonality of plankton metabolism in the Atlantic gyres, our generalized analysis of regional differences appears unbiased.

Discussion

The debate about the metabolic state of the oligotrophic ocean has evolved into a polarized dispute between two contrasting positions that are very difficult to reconcile (that is, the oligotrophic ocean is heterotrophic or it is autotrophic) and require substantial unaccounted for allochthonous inputs of either organic or inorganic nutrients[10,11]. Both positions arise from the same fundamental view that 'the oligotrophic ocean' is a

Figure 2 | Integrated metabolism in the epipelagic zone. Latitudinal variation of integrated (**a**) NCP and (**b**) photosynthesis (P, open squares) and respiration (R, red filled squares) data during the AMT11 to AMT22 cruises. The error lines represent the propagated s.e. The dotted line on **a** is the NCP = 0 line, indicating the transition from auto- to heterotrophy. The approximate location of the South Atlantic gyre (SATL) and North Atlantic gyre (NAST) provinces is indicated.

Table 1 | Metabolic rates in the North and South Atlantic subtropical gyres.

	NCP NAST	NCP SATL	P NAST	P SATL	R NAST	R SATL
Integrated (± s.e.m.)	**− 14.5 ± 2.4**	**9.8 ± 2.3**	**57.5 ± 3.5**	**68.3 ± 3.1**	**72.1 ± 3.5**	**58.5 ± 3.8**
95% CI for the mean	− 19.3 to − 9.8	5.0 to 14.6	50.84 to 64.17	61.6 to 75.1	64.8 to 79.4	51.2 to 65.9
Median	− 14.5	11	54.8	68.5	73.4	54.8
Range	− 60.5 to 20.6	− 21.0 to 42.3	19.8 to 118.9	16.8 to 103.4	27.8 to 121.2	6.7 to 107.2
n	47	46	47	46	47	46
P value	0.0001		0.026		0.0001	
Volumetric (± s.e.m.)	**− 0.12 ± 0.02**	**0.06 ± 0.02**	0.56 ± 0.02	0.55 ± 0.02	**0.68 ± 0.03**	**0.48 ± 0.02**
95% CI for the mean	− 0.16 to − 0.08	0.03 to 0.1	0.52 to 0.61	0.50 to 0.59	0.63 to 0.73	0.43 to 0.52
Median	− 0.09	0.07	0.52	0.50	0.64	0.45
Range	− 1.06 to 0.48	− 1.30 to 0.75	0.00 to 2.38	0.00 to 1.58	0.00 to 2.75	0.00 to 2.03
n	202	233	202	233	204	237
P value	0.0001		0.57		0.0001	

CI, confidence interval; NCP, net community production; NAST, North Atlantic gyre; P, photosynthesis; R, respiration; SATL, South Atlantic gyre.
Mean (± s.e.), 95% confidence interval for the mean, median, range and number of observations of euphotic zone integrated ($mmolO_2 \, m^{-2}$ per day) and volumetric ($mmolO_2 \, m^{-3}$ per day) rates of NCP, P and R measured in the NAST and SATL provinces during AMT11–22 cruises. Numbers in bold indicate statistically significant differences (at $\alpha = 0.05$) between metabolic rates in the NAST and SATL. P value is the probability, assuming the null hypothesis, of the results of the Student's *t*-tests performed to compare the means of P, R and NCP rates in the NAST and SATL.

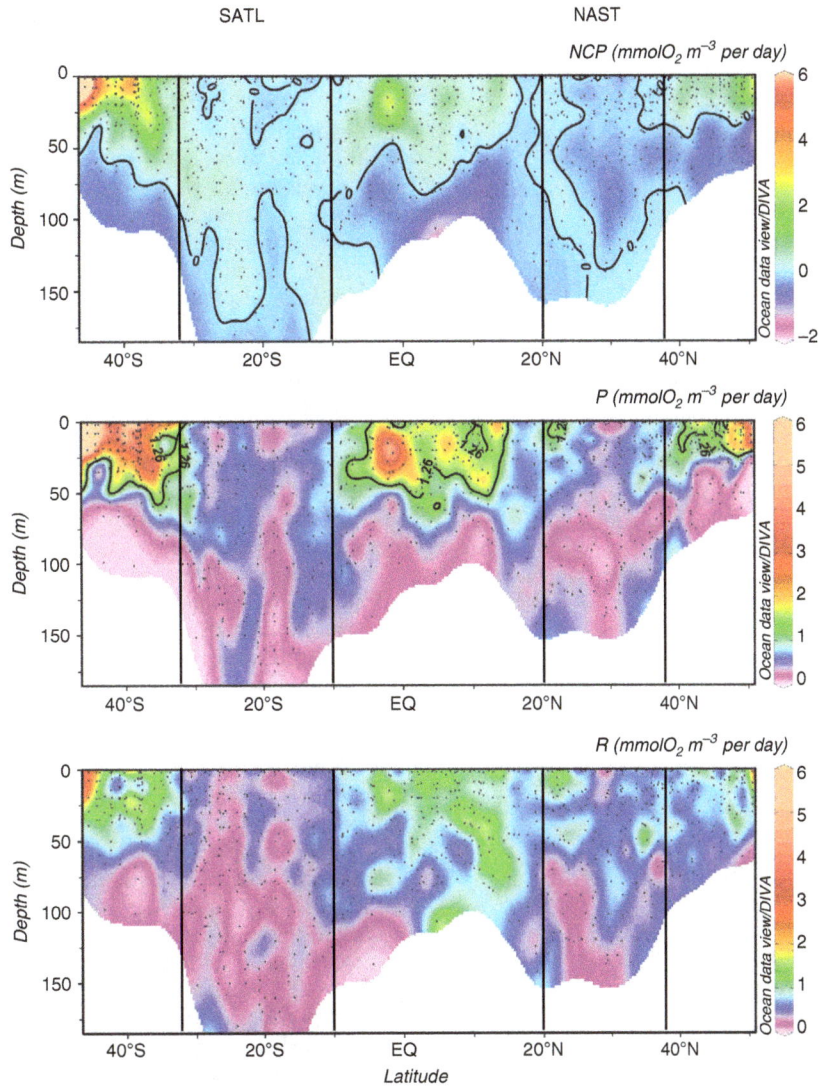

Figure 3 | Variation of plankton metabolism through the epipelagic waters of the Atlantic Ocean. Composite depth-latitudinal sections of plankton NCP, P and R (AMT11–22). Data from the NW Africa upwelling from AMT11, AMT13 and AMT15 are excluded. The 0 mmolO_2 m^{-3} per day NCP isoline marks the auto-heterotrophy transition. The 1.26-mmolO_2 m^{-3} per day P isoline marks the threshold for net heterotrophy in Regaudie-de-Gioux and Duarte[18]. The approximate location of the SATL and NAST provinces is indicated.

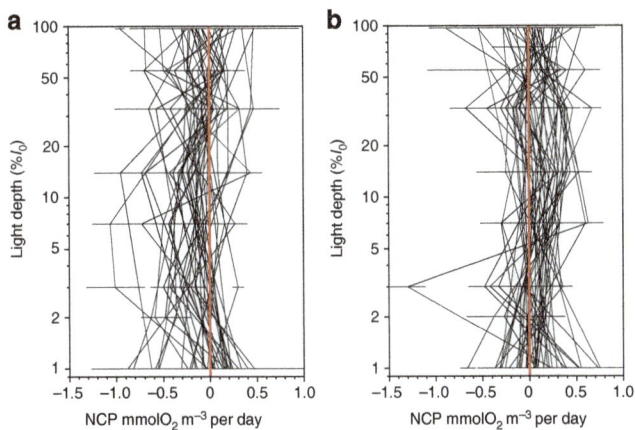

Figure 4 | Vertical variation of NCP in the Atlantic gyres. Profiles of NCP measurements (± s.e.m.) at different light depths in the epipelagic zone of the (**a**) NAST and (**b**) SATL stations in the AMT11–22 data set. The red lines mark the 0-mmolO_2 m^{-3} per day NCP.

single steady-state ecosystem whose balance of trophic processes is meaningful across scales. This perception is evident in the very discussion about its 'metabolic state', a property of either organisms or ecosystems whose calculation requires knowledge of the spatial and temporal scales of key trophic processes and connections in the system (production, accumulation, transport, consumption and oxidation of organic matter)[25]. However, the definition of the oligotrophic ocean is solely based on the level of productivity, controlled by the degree of water column stratification that determines nutrient limitation for the phytoplankton. Moreover, the oligotrophic ocean includes distant regions whose trophic connections are difficult to determine. Hence, estimating a single metabolic state for the global oligotrophic ocean from scattered measurements of P and R assumes scale independency in its trophic dynamics, that is, rests on the important assumptions of functional unity and steady state.

Our findings indicate that both this logic and the empirical evidence sustaining the debate are flawed. The systematic prevalence of autotrophy in the oligotrophic SATL challenges

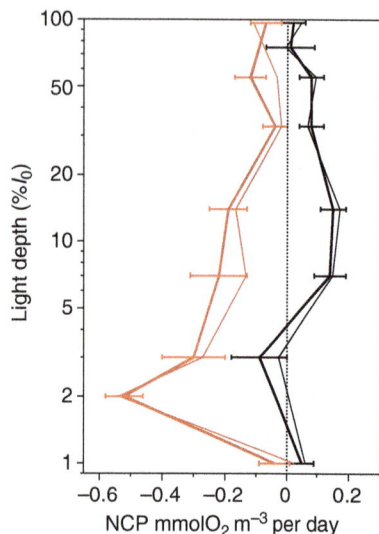

Figure 5 | Vertical distribution of NCP in the Atlantic gyres. Profiles of the average (± s.e.m.; thick lines) and median (thin lines) NCP rates at different light depths in the epipelagic zone of the NAST (red lines) and SATL (black lines) stations in the AMT11-22 data set. The dotted line marks the 0-$mmolO_2 m^{-3}$ per day NCP.

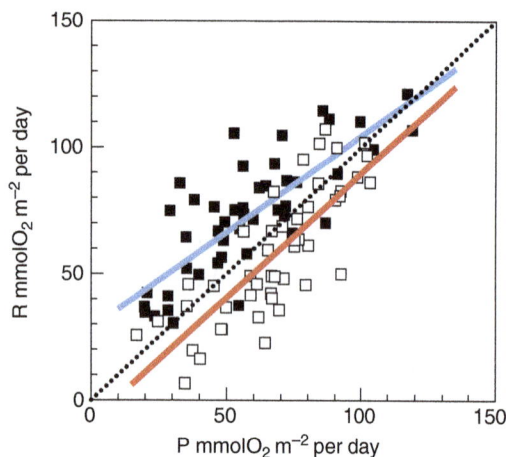

Figure 6 | Relationship between P and R. Relationship between integrated rates of photosynthesis (P) and respiration (R) in the NAST (filled squares) and SATL (open squares). The blue line is the regression line for the NAST data: $R = 0.76P + 28.25$, $R^2 = 0.59$, $n = 47$. The red line is the regression line for the SATL data: $R = 0.98P - 8.56$; $R^2 = 0.64$; $n = 46$. Both the slope and intercept of the two equations are significantly different (t-test ($n = 89$) = 12.89, $P = 0.0001$; t-test ($n = 89$) = 22.28, $P = 0.0001$, respectively). The dotted line indicates the 1:1 line.

the generally accepted view that net heterotrophy prevails in *in vitro* NCP measurements in the oligotrophic ocean[10,11], and the conclusion that such a prevalence is due to systematic methodological biases[11,19]. Moreover, the difference in NCP between the NAST and SATL refutes the hypothesis that a single type of balance prevails throughout the oligotrophic ocean[10,11]. The oligotrophic ocean is neither auto- nor heterotrophic, but functionally diverse. The debate has partially resulted from the use of universal P:R relationships where the heterotrophic NAST-E is over-represented.

The prevalence of R over P in the NAST implies carbon and energy deficits that need to be compensated through anoxygenic

primary production, inputs of allochthonous organic matter or a higher efficiency in the use of metabolic energy. The paucity of reducing substrates and feeble redox gradients in the epipelagic zone of the oligotrophic ocean constrain the efficiency of chemoautotrophy and anoxygenic photosynthesis, whose relative contributions to euphotic zone carbon fluxes are <1% compared to oxygenic photosynthesis[26,27], that is, too small to explain the metabolic differences between the NAST and SATL. Photoheterotrophic prokaryotes can use both light and organic matter for energy but require organic molecules as sources of carbon and electrons. Although proteorhodopsin-based and aerobic anoxygenic phototrophic bacteria are abundant and active in the oligotrophic ocean[28], the implications of photoheterotrophy for the metabolic balance in the upper ocean are unclear. On one hand, phototrophic energy production reduces the need for aerobic respiration to sustain the maintenance costs of heterotrophs. On the other hand, photoheterotrophy increases the survival, and on occasions the growth, of prokaryotes under very limiting conditions[29], possibly by supporting the energetic costs of processes such as active transmembrane transport, production of ectoenzymes, breakdown of complex organic molecules and cell motility[30], but also by light induced anaplerotic CO_2 fixation, which may provide up to 18% of cellular carbon demand[29]. This augmented survival, biomass production and the ability for organic matter utilization would increase the potential for heterotrophic respiration in the oligotrophic gyres. The consequences of photoheterotrophy for the metabolic cycling of carbon and the competitive advantage of planktonic bacteria[28-30] suggest an important, and yet unknown, role in energy and carbon balances in the oligotrophic ocean. However, it is difficult to infer the relationship between photoheterotrophy and the regional differences in R and NCP, and photoheterotrophy would not reduce the estimated deficit of organic carbon in the NAST *per se*.

Calculations of dissolved organic carbon import to the surface of the NAST are one to two orders of magnitude too small to support previous estimations of net heterotrophy[11]. However, the mean mixed layer (upper 50 m) *in vitro* NCP deficit in the NAST used in these calculations ($-1 mmolC m^{-3}$ per day)[11] is ~10 times higher than the $-0.12 mmolO_2 m^{-3}$ per day mean epipelagic NCP (average 112 m depth) estimated from our data set (Table 1; see also Fig. 5). In addition, our data show a depth-dependent distribution of NCP through the epipelagic layer, with lower heterotrophy near the 1% light depth (~100 m depth) and in the upper 25 m (~33% incident PAR; Figs 3–5). Although we are reluctant to calculate annual balances from our data set, assuming a respiratory quotient of 0.8[11] and a mean mixed layer of 50 m[11], 365 times our mean NCP in the upper 50 m of the NAST ($-0.07 ± 0.03 mmolO_2 m^{-3}$ per day) amounts to $-1.1 ± 0.4 molC m^{-2}$ per year, that is, less than twice the 0.7 $molC m^{-2}$ per year dissolved organic carbon input estimated 'with high uncertainty' because of the ill constrained inputs from the margins[14]. These same lateral inputs are also needed to balance phosphorous budgets in the NAST. The phytoplankton of the North and South Atlantic gyres is limited primarily by nitrogen availability[31]; however, a distinctive characteristic of the NAST is that the concentration of bioavailable phosphorous in the euphotic layer is the lowest of all the ocean gyres and limits phytoplankton production[13]. The similar rates of phytoplankton carbon fixation[32] and P (Table 1) in the NAST and SATL hence concur with the similar euphotic zone concentrations of bioavailable nitrogen, but are at odds with the chronic phosphorous limitation in the NAST. The lateral input of dissolved organic phosphorous (DOP) from the shelf region of NW Africa into the gyre interior helps to balance this discrepancy[33]. DOP imported by a combination of gyre and

Table 2 | Metabolic rates in spring and autumn in the Atlantic subtropical gyres.

	NCP NAST	NCP SATL	P NAST	P SATL	R NAST	R SATL
Astronomical spring ($\pm$ s.e.m.)	**−10.4 ± 2.2**	**12.3 ± 2.4**	61.7 ± 4.2	70.8 ± 2.6	72.4 ± 3.5	58.4 ± 3.7
95% CI for the mean	−18.9 to −1.9	7.1 to 17.6	50.1 to 73.3	63.6 to 77.9	56.0 to 85.8	50.13 to 66.6
Median	−9.12	14	66.6	70.4	75.1	50.1
Range	−45.9 to 12.6	−21.0 to 42.3	20.0 to 119.0	34.7 to 103.4	33.1 to 115.0	6.68 to 107.2
n	14	37	14	37	14	37
P value	0.0001		0.19		0.079	
Astronomical autumn ($\pm$ s.e.m.)	**−16.3 ± 2.5**	**−0.73 ± 1.5**	55.7 ± 3.2	58.2 ± 4.5	72.0 ± 3.5	59.2 ± 4.6
95% CI for the mean	−22.1 to −10.6	−11.7 to 10.3	47.2 to 64.3	41.9 to 74.6	62.8 to 81.2	41.6 to 76.8
Median	−16.9	−0.3	52.9	56.3	71.7	59.5
Range	−60.5 to 20.6	−16.7 to 14.4	19.8 to 117.0	16.8 to 101.0	27.8 to 121.2	19.6 to 102.0
n	33	9	33	9	33	9
P value	0.015		0.79		0.20	

CI, confidence interval; NCP, net community production; NAST, North Atlantic gyre; P, photosynthesis; R, respiration; SATL, South Atlantic gyre.
Mean ($\pm$ s.e.), 95% confidence interval for the mean, median, range and number of observations of integrated rates of NCP, P and R (mmolO$_2$ m^{-2} per day) measured in spring and autumn in the NAST and SATL provinces during AMT11–22 cruises. Numbers in bold indicate statistically significant differences (at $\alpha = 0.05$) between metabolic rates in the NAST and SATL. P value is the probability, assuming the null hypothesis, of the results of the Student's t-tests performed to compare the means of P, R and NCP rates in the NAST and SATL.

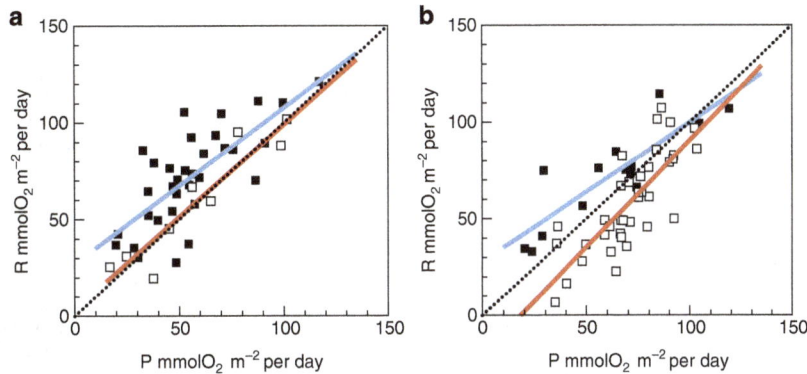

Figure 7 | Seasonality of P:R relationships. Relationship between the integrated rates of photosynthesis (P) and respiration (R) in the NAST (filled squares) and SATL (open squares) in (**a**) autumn and (**b**) spring. The blue lines are the regression lines for the NAST data: $R = 0.80P + 26.9$, $R^2 = 0.54$, $n = 33$ (autumn); $R = 0.72P + 28.01$, $R^2 = 0.74$, $n = 14$ (spring). The red lines are the regression lines for the SATL data: $R = 0.96P + 3.56$; $R^2 = 0.87$; $n = 9$ (autumn); $R = 1.10P − 19.79$, $R^2 = 0.62$, $n = 37$ (spring). The dotted lines indicate the 1:1 line.

eddy circulations is estimated to support up to 70% of the particle export over much of the gyre[33], and should therefore impact on the balance between P and R. DOP may be directly utilized by the phytoplankton or recycled by heterotrophic bacteria, but the latter have been found to easily outcompete *Prochlorococcus* (the most abundant phytoplankter in the oligotrophic ocean) for DOP (ATP) uptake[34]. Estimating the metabolic impact of DOP import into the NAST requires resolving the variability of group-specific differences in DOP uptake and utilization, which might also interact with the photoheterotrophic enhancement of dissolved organic matter bioavailability, and with an anticipated increase in the open ocean phosphorous limitation with global warming[35]. A mechanistic understanding of the regional differences in community metabolism requires not only solving the spatial and temporal variation of NCP and relative inputs of allochthonous inorganic and organic nutrients within the gyres, but also the study of the multiple biochemical strategies for nutrient stress[36] and energy and carbon acquisition across spatial and temporal scales[37].

Our results show that R heterogeneity in the oligotrophic ocean is important for NCP variability and prediction, and that global metabolic balances in the oligotrophic ocean represent the average of different metabolic states, rather than the metabolic state of a single global biome. While the potential of P:R

relationships for NCP scaling remains, extrapolation should be specific to provinces showing a coherent trophic functioning[38]. The different geographic extent of these provinces requires spatially explicit approaches, that is, a trophic biogeographic partitioning of the ocean based on the spatial and temporal variability not only of P but also of R[23,38]. This is a difficult task because only the substantial P data set and reliable predictive algorithms[39] allow for comprehensive regional and seasonal depictions[39–41], which is not enough to reveal the trophic diversity of the ocean. Besides, the scale dependency of ecosystem metabolism along the temporal axis is arguably as important for the highly dynamic, non-steady-state planktonic ecosystem, and particularly relevant in the context of anthropogenic change[42]. Although R is a slow variable compared with P, its temporal variability could be important for prediction of trophic dynamics over scales significant to environmental change[43].

The estimation of the metabolic balance of the oligotrophic ocean continues to be an important and urgent challenge for ocean biogeochemistry. Integrative *in situ* approaches are invaluable for constraining metabolic balances over long and large scales. However, only *in vitro* or other instantaneous methods may resolve the spatial heterogeneity and temporal dynamics of ecosystem processes, which is a prerequisite for prediction. Our findings call for an effort to improve the scope and resolution of R and NCP measurements in the ocean. While we are far from a mechanistic understanding of the variability of plankton trophic functioning, the scale difference between *in situ* and *in vitro* methods provides unique opportunities to derive and test system-dependent empirical models for NCP extrapolation. A better appreciation of the scaling of metabolic processes to biogeochemical fluxes is a priority to improve our prediction of the ocean's interaction with the changing climate.

Methods

Sampling. Ten latitudinal ($\sim 50°$ N–45° S) transects of ~ 70 stations across the Atlantic Ocean were conducted from 2000 to 2012 on RRS James Clark Ross, RRS Discovery and RRS James Cook between the United Kingdom and South America (Fig. 1). All the cruises except AMT17 and AMT18 included measurements of metabolism in both the North and South gyres (NAST and SATL). AMT cruises 11, 13, 15 and 17–22 took place in boreal autumn, while AMT12, 14 and 16 were in boreal spring. Details of sampling, incubation and analytical procedures for each cruise, including the complete cruise reports, may be found at http://www.amt-uk.org/ and Robinson *et al.*[44]

Chlorophyll *a* concentration. During each cruise, vertical profiles of temperature and fluorescence were obtained twice daily (at ~ 2 h before dawn and at solar noon) to a depth of ~ 300 m with SeaBird and WETLabs Wet star sensors fitted to a SeaBird 9/11 plus CTD system. Chlorophyll *a* concentration was derived from calibrated fluorescence readings. A rosette of 24×20 l Niskin bottles fitted with the CTD system was used to collect seawater samples from six to nine depths (including light depths sampled for photosynthesis and respiration incubations—see below) to calibrate the CTD fluorometer following Welschmeyer[45]. Samples of 250 ml were filtered through 47-mm 0.2-μm polycarbonate filters. The filters were placed in a vial with 10 ml 90% acetone and left in a freezer for 24 h. The samples were then analysed on a precalibrated Turner Designs Trilogy fluorometer with a non-acidified chl module (CHL NA #046) fitted.

Plankton metabolism. Gross photosynthesis (P), dark community respiration (R) and NCP were determined from *in vitro* changes in dissolved O_2 after 24-h light and dark bottle incubations. Although *in situ* incubations would be preferred to deck incubations, they were prevented by cruise logistics (AMT cruises are cruises of opportunity where stations last ~ 2 h and are separated by ~ 350 km ($\sim 13,500$ km transect sampled over 42 days). Seawater was sampled daily, ~ 2 h before dawn, from three to six depths (median 5) using the rosette of 24×20 l Niskin bottles. The total number of suitable stations ranges from 8 (AMT17) to 37 (AMT22) (mean of 19). Water was collected into opaque polypropylene aspirators from 4 (AMT12), 5 (AMT11,14,15,16,17) or 6 (AMT13,18,21,22) light depths covering the PAR range from ~ 97 to 1% of surface irradiance. Actual light depths changed between stations and cruises, but typically included 97, 33, 14, 7 and 1% of surface irradiance. Irradiance levels were determined from light measurements made the previous day, and the assumption that the deep fluorescence maximum approximated the depth to which 1% of surface irradiance reached. The depth of the 1% light in the next noon PAR profile was always within 7% of our estimated depth. Twelve to 18 120-cm^3, gravimetrically calibrated, borosilicate glass bottles were carefully filled from each aspirator bottle using silicon tubing, overflowing by > 250 ml. From each depth, four to six replicate 'zero-time' bottles were fixed immediately with Winkler reagents (1 ml of 3 M MnCl or MnSO$_4$ and 1 ml of (8 M KOH + 4 M KI) solutions) added separately with an automatic multipipette. Two further sets of four to six replicates were incubated for 24 h in surface water-cooled deck incubators or in temperature controlled water baths at *in situ* temperatures. These numbers of replicates are enough to obtain average coefficients of variation of the O_2 concentration measurements in the zero, dark and light bottles <0.1%, while allowing the daily completion by one person of all the analyses corresponding to five to six depths. One set of replicates was incubated in the dark, the other set in the equivalent irradiance to that found at the *in situ* sampling depth using various combinations of neutral density and blue plastic filters. Owing to limitations of sample water volume and analysis time during AMT12 and 13, we assumed that plankton community structure was homogeneous within the surface 15 m and so incubated a set of replicates sampled from the 55% light depth at 97% of surface irradiance. Flow cytometry data from AMT13 confirmed that this was a reasonable assumption, since picoplankton cell abundance varied by less than 5% between the two light depths throughout the transect, and nanoplankton varied by 11% (ref. 24). Incubations always started at dawn, and during the hours of darkness the incubators were covered with opaque screens to prevent interference from the ship's deck lights. After the 24-h incubation period, the dark and light bottles were fixed with Winkler reagents. Dissolved oxygen concentration was determined with automated Winkler titration systems using potentiometric (Metrohm 716 DMS Titrino)[46] or photometric[47] end-point detection. For the potentiometric method, aliquots of fixed samples were delivered with a 50-ml overflow pipette. Fixing, storage and standardization procedures followed the recommendations by Grasshoff *et al.*[48] Production and respiration rates were calculated from the

difference between the means of the replicate light and dark incubated bottles and zero-time analyses: NCP = ΔO_2 in light bottles (mean of [O_2] in 24-h light—mean zero time [O_2]); R = ΔO_2 in dark bottles (mean zero time [O_2]—mean [O_2] in 24-h dark); P = NCP + R. The average s.e.m. of both the NCP and R rate measurements (which includes both experimental and biological variability) was 0.18 ($n = 875$ and 876, respectively) mmolO$_2$ m^{-3} per day. Integrated values were obtained by trapezoidal integration of the volumetric data down to the depth of 1% surface incident photosynthetically active irradiance. The s.d. of integrated P, R and NCP was calculated through propagation of the random error in the volumetric measurements. Samples were excluded (16 stations) when significant negative values of either P or R were measured, or when the variance of replicated measurements was anomalously high. The NAST and SATL P, R and NCP data are normality distributed (Lilliefors test, $\alpha = 0.05$).

The complete volumetric data set of P, R and NCP is available at http://www.uea.ac.uk/environmental-sciences/people/People/Faculty + and + Research + Fellow/robinsonc#research maintained by Carol Robinson. The complete data set of all the physical, chemical and biological variables from each cruise is available at BODC (http://www.bodc.ac.uk/projects/uk/amt). The conditions under which the data may be used are in line with the NERC Data Policy (http://www.nerc.ac.uk/research/sites/data/policy/), and will be explained following a request, before the delivery of data.

References

1. Falkowski, P. G. The role of phytoplankton photosynthesis in global biogeochemical cycles. *Photosynth. Res.* **39**, 235–258 (1994).
2. Falkowski, P. G., Fenchel, T. & Delong, E. F. The microbial engines that drive Earth's biogeochemical cycles. *Science* **320**, 1034–1039 (2008).
3. Ducklow, H. W. & Doney, S. C. What is the metabolic state of the oligotrophic ocean? A debate. *Ann. Rev. Mar. Sci.* **5**, 525–533 (2013).
4. Emerson, S. *et al.* Experimental determination of organic carbon flux from open-ocean surface waters. *Nature* **389**, 951–954 (1997).
5. Neuer, S. *et al.* Differences in the biological carbon pump at three subtropical ocean sites. *Geophys. Res. Lett.* **29**, 321–324 (2002).
6. Christian, J. R. Biogeochemical cycling in the oligotrophic ocean: Redfield and non-Redfield models. *Limnol. Oceanogr.* **50**, 646–657 (2005).
7. Polovina, J. J., Howell, E. A. & Abecassis, M. Ocean's least productive waters are expanding. *Geophys. Res. Lett.* **35**, L03618 (2008).
8. Bidigare, R. R. *et al.* Subtropical ocean ecosystem structure changes forced by North Pacific climate variations. *J. Plankton Res.* **31**, 1131–1139 (2009).
9. del Giorgio, P. A., Cole, J. J. & Cimbleris, A. Respiration rates in bacteria exceed phytoplankton production in unproductive aquatic systems. *Nature* **385**, 148–151 (1997).
10. Duarte, C. M., Regaudie-de-Gioux, A., Arrieta, J. M., Delgado-Huertas, A. & Agustí, S. The oligotrophic ocean is heterotrophic. *Ann. Rev. Mar. Sci.* **5**, 551–569 (2013).
11. Williams, P.J.le. B., Quay, P. D., Westberry, T. K. & Behrenfeld, M. J. The oligotrophic ocean is autotrophic. *Ann. Rev. Mar. Sci.* **5**, 535–549 (2013).
12. Emerson, S., Stump, C. & Nicholson, D. Net biological oxygen production in the ocean: remote *in situ* measurements of O_2 and N_2 in surface waters. *Global Biogeochem. Cy.* **22**, GB3023 (2008).
13. Lipschultz, F., Bates, N. R., Carlson, C. A. & Hansell, D. A. New production in the Sargasso Sea: History and current status. *Global Biogeochem. Cy.* **16**, 1–17 (2002).
14. Hansell, D. A., Ducklow, H. W., Macdonald, A. M. & Baringer, M. O. Metabolic poise in the North Atlantic Ocean diagnosed from organic matter transports. *Limnol. Oceanogr.* **49**, 1084–1094 (2004).
15. Kähler, P., Oschlies, A., Dietze, H. & Koeve, W. Oxygen, carbon, and nutrients in the oligotrophic eastern subtropical North Atlantic. *Biogeosciences* 7, 1143–1156 (2010).
16. Williams, P. J.le. B. The balance of plankton respiration and photosynthesis in the open oceans. *Nature* **394**, 55–57 (1998).
17. Duarte, C. M. & Agustí, S. The CO$_2$ balance of unproductive aquatic ecosystems. *Science* **281**, 234–236 (1998).
18. Regaudie-de-Gioux, A. & Duarte, C. M. Global patterns in oceanic planktonic metabolism. *Limnol. Oceanogr.* **58**, 977–986 (2013).
19. Westberry, T. B., Williams, P. J.le. B. & Behrenfeld, M. J. Global net community production and the putative net heterotrophy of the oligotrophic oceans. *Global Biogeochem. Cy.* **26**, GB4019 (2012).
20. Aristegui, J. & Harrison, W. G. Decoupling of primary production and community respiration in the ocean: implications for regional carbon studies. *Aquat. Microb. Ecol.* **29**, 199–209 (2002).
21. Karl, D. M., Laws, E. A., Morris, P., Williams, P.J.le.B. & Emerson, S. Metabolic balance of the open sea. *Nature* **426**, 32 (2003).
22. Williams, P. J.le. B., Morris, P. & Karl, D. M. Net community production and metabolic balance at the oligotrophic ocean site, station ALOHA. *Deep-Sea Res. I* **51**, 1563–1578 (2004).
23. Serret, P., Fernández, E. & Robinson, C. Biogeographic differences in the net ecosystem metabolism of the open ocean. *Ecology* **83**, 3225–3234 (2002).

24. Gist, N., Serret, P., Woodward, E. M. S., Chamberlain, K. & Robinson, C. Seasonal and spatial variability in plankton production and respiration in the Subtropical Gyres of the Atlantic Ocean. *Deep-Sea Res. II* **56**, 931–940 (2009).

25. Smith, S. V. & Hollibaugh, J. T. Annual cycle and interannual variability of ecosystem metabolism in a temperate climate embayment. *Ecol. Monogr.* **67**, 509–533 (1997).

26. Middelburg, J. J. Chemoautotrophy in the ocean. *Geophys. Res. Lett.* **38**, L24604 (2011).

27. Goericke, R. Bacteriochlorophyll a in the ocean: is anoxygenic bacterial photosynthesis important? *Limnol. Oceanogr.* **47**, 290–295 (2002).

28. Kirchman, D. L. & Hanson, T. E. Bioenergetics of photoheterotrophic bacteria in the oceans. *Environ. Microbiol. Rep.* **5**, 188–199 (2013).

29. Palovaara, J. *et al.* Stimulation of growth by proteorhodopsin phototrophy involves regulation of central metabolic pathways in marine planktonic bacteria. *Proc. Natl Acad. Sci. USA* **111**, E3650–E3658 (2014).

30. Koblížek, J. in *Microbial Carbon Pump in the Ocean.* (eds Jiao, N., Azam, F. & Sanders, S.) 49–51 (Science/AAAS, 2011).

31. Moore, C. M. *et al.* Processes and patterns of oceanic nutrient limitation. *Nat. Geosci.* **6**, 701–710 (2013).

32. Poulton, A. J. *et al.* Phytoplankton carbon fixation, chlorophyll-biomass and diagnostic pigments in the Atlantic Ocean. *Deep-Sea Res. II* **53**, 1593–1610 (2006).

33. Reynolds, S., Mahaffey, C., Roussenov, V. & Williams, R. G. Evidence for production and lateral transport of dissolved organic phosphorus in the eastern subtropical North Atlantic. *Global Biogeochem. Cy.* **28**, 805–824 (2014).

34. Björkman, K., Duhamel, S. & Karl, D. M. Microbial group specific uptake kinetics of inorganic phosphate and adenosine-5′-triphosphate (ATP) in the North Pacific Subtropical Gyre. *Front. Microbiol.* **3**, 1–17 (2012).

35. Karl, D. M., Bidigare, R. R. & Letelier, R. M. Long-term changes in plankton community structure and productivity in the North Pacific subtropical gyre: the domain shift hypothesis. *Deep-Sea Res. II* **48**, 1449–1470 (2001).

36. Saito, M. A. *et al.* Multiple nutrient stresses at intersecting Pacific Ocean biomes detected by protein biomarkers. *Science* **345**, 1173–1177 (2014).

37. Karl, D. M. Microbial oceanography: paradigms, processes and promise. *Nat. Rev. Microbiol.* **5**, 759–769 (2007).

38. Serret, P. *et al.* Predicting plankton net community production in the Atlantic Ocean. *Deep-Sea Res. II* **56**, 941–953 (2009).

39. Carr, M. E. *et al.* A comparison of global estimates of marine primary production from ocean color. *Deep-Sea Res. II* **53**, 741–770 (2006).

40. Chavez, F. P., Messié, M. & Pennington, J. T. Marine primary production in relation to climate variability and change. *Ann. Rev. Mar. Sci.* **3**, 227–260 (2011).

41. Longhurst, A. *Ecological Geography of the Sea* 2 (Academic Press, 2006).

42. Wolkovich, E. M., Cook, B. I., McLauchlan, K. K. & Davies, T. J. Temporal ecology in the Anthropocene. *Ecol. Lett.* **17**, 1365–1379 (2014).

43. Carpenter, S. R. & Turner, M. G. Hares and tortoises: interactions of fast and slow variables in ecosystems. *Ecosystems* **3**, 495–497 (2001).

44. Robinson, C. *et al.* The Atlantic Meridional Transect (AMT) Programme: a contextual view 1995-2005. *Deep Sea Res. II* **53**, 1485–1515 (2006).

45. Welschmeyer, N. A. Fluorometric analysis of chlorophyll-a in the presence of chlorophyll-b and phaeopigments. *Limnol. Oceanogr.* **39**, 1985–1992 (1994).

46. Oudot, C., Gerard, R., Morin, P. & Gningue, I. Precise shipboard determination of dissolved oxygen (Winkler procedure) for productivity studies with a commercial system. *Limnol. Oceanogr.* **33**, 146–150 (1988).

47. Williams, P. J.le. B. & Jenkinson, N. W. A transportable micro-processor controlled precise Winkler titration suitable for field station and shipboard use. *Limnol. Oceanogr.* **27**, 576–584 (1982).

48. (eds Grasshoff, K., Ehrhardt, M. & Kremling, K.) *Methods of Seawater Analysis* 2nd edn, 419 (Verlag Chemie, 1983).

Acknowledgements

We thank the principal scientists and all personnel on board during the AMT11 to AMT22 cruises. Special thanks to Sandy Thomalla, Mike Lucas, Alex Poulton, Patrick Holligan, Andy Rees and Claire Widdicombe for chlorophyll *a* measurements used to calibrate the CTD fluorescence sensor used to determine the depth of the DCM, and to Gavin Tilstone for support with the incubations. This study is a contribution to the international IMBER project and was supported by the UK Natural Environment Research Council National Capability funding to Plymouth Marine Laboratory and the National Oceanography Centre, Southampton. We acknowledge the UK NERC funding required to initiate and sustain the AMT programme: two Plankton Reactivity in the Marine Environment (PRIME) research grants and the Plymouth Marine Laboratory Core Strategic Research Programme sustaining AMT (1995–2000), the AMT consortium grant NER/O/S/2001/00680 awarded to C.R. (2001–2006) and the Oceans 2025 programme that funded AMT (2007–2012). We also acknowledge the Spanish MICINN grants CTM2009-0S069-E/MAR and CTM2011-29616 awarded to P.S., which funded the participation of M.A.-G., E.E.G.-M., J.L. and P.S. in AMT cruises. This is contribution number 255 of the AMT programme.

Author contributions

P.S. designed this work and wrote the manuscript, based on the ideas that were thoroughly discussed with C.R., who revised the manuscript. M.A.-G., E.E.G.-M. and V.K. provided additional insight to both the conception and review of the paper. P.S., C.R., M.A.-G., E.E.G.-M., N.G., V.K., J.L. and J.S. measured plankton metabolism on AMT cruises. C.H. and R.T. provided the AMT22 nutrients and chlorophyll *a* data, respectively.

Additional information

Competing financial interests: The authors declare no competing financial interests.

Glacial ice and atmospheric forcing on the Mertz Glacier Polynya over the past 250 years

P. Campagne[1,2,3], Xavier Crosta[1], M.N. Houssais[2], D. Swingedouw[1], S. Schmidt[1], A. Martin[2], E. Devred[3], S. Capo[1], V. Marieu[1], I. Closset[2] & G. Massé[2,3]

The Mertz Glacier Polynya off George V Land, East Antarctica, is a source of Adélie Land Bottom Water, which contributes up to ~25% of the Antarctic Bottom Water. This major polynya is closely linked to the presence of the Mertz Glacier Tongue that traps pack ice upstream. In 2010, the Mertz Glacier calved a massive iceberg, deeply impacting local sea ice conditions and dense shelf water formation. Here we provide the first detailed 250-year long reconstruction of local sea ice and bottom water conditions. Spectral analysis of the data sets reveals large and abrupt changes in sea surface and bottom water conditions with a ~70-year cyclicity, associated with the Mertz Glacier Tongue calving and regrowth dynamics. Geological data and atmospheric reanalysis, however, suggest that sea ice conditions in the polynya were also very sensitive to changes in surface winds in relation to the recent intensification of the Southern Annular Mode.

[1]EPOC, UMR CNRS 5805, Université de Bordeaux, Allée Geoffroy St Hilaire, Pessac 33615, France. [2]LOCEAN, UMR CNRS/UPMC/IRD/MNHN 7159, Université Pierre et Marie Curie, 4 Place Jussieu, Paris 75252, France. [3]TAKUVIK, UMI 3376 UL/CNRS, Université Laval, 1045 avenue de la Médecine, Quebec City, Quebec, Canada G1V 0A6. Correspondence and requests for materials should be addressed to X.C. (email: x.crosta@epoc.u-bordeaux1.fr) or to G.M. (email: guillaume.masse@takuvik.ulaval.ca).

Antarctic Bottom Water (AABW) is considered as the great ventilator of the world deep ocean and is a critical component of the global climate system[1,2]. AABW is sourced by dense saline shelf waters, locally produced in several polynyas, where low sea ice concentrations (SICs) and open water conditions are maintained during winter and favour local sea ice and brine production[3,4]. Coastal polynyas are usually associated with the presence of local topographic or glacial features and synoptic-scale gravity winds[4]. Although they only represent a small fraction of the area covered by sea ice, polynyas are considered of key importance for the Earth climate system[3,5]. Indeed, while the 13 major Antarctic coastal polynyas only represent $\sim 1\%$ of the maximum ice area, they are responsible for the production of 10% of the Southern Ocean sea ice[3]. Recently, it

has been shown that $\sim 25\%$ of the AABW is sourced from the export of Adélie Land Bottom Water (ALBW) produced in the George V Land polynyas[2,6]. ALBW mainly originates from the Adélie Depression off the coast of George V Land[3,6,7], where the Mertz Glacier (MG) Polynya (MGP) develops in winter along the western flank of the MG Tongue (MGT) and further extends to the West in the adjacent coastal bays (Commonwealth, Watt and Buchanan Bays; Fig. 1b; refs 8,9). The MGP constitutes the third most productive polynya in Antarctica, with a winter area of up to 6,000 km^2 and an annual sea ice production of 120 km^3 over the 1992–2001 period[3,4].

Early in 2010, the MG underwent massive calving and lost half its tongue after collision with the B09B iceberg. The generation of an 80-km-long iceberg and subsequent retreat of the MG front had a profound impact on both regional icescape and ocean conditions[9–11]. Satellite passive microwave data show that SIC increased by 50% (relative to the 1979–2009 period) in the MGP area after the calving (Fig. 1e). Advection of thick consolidated sea ice and presence of fast ice downstream the MG in the MGP (Fig. 1c,d) led to a large reduction in sea ice production in the MGP, which in turn affected dense water formation[9–11]. Within 2 years following the calving, bottom water salinity decreased by over 0.30 psu in Commonwealth Bay (CB)[9].

Located in the southern part of the polynya, CB is ideal for studying past dynamics of the MGP and dense water formation[9]. Indeed, mooring observations and satellite data showed that oceanic and sea ice conditions (Fig. 1e) over the CB and MGP areas have experienced similar and synchronous pre- and post-calving changes over the recent years[9]. In addition, recent observations highlighting the presence of the saltiest waters of the Adélie depression in CB suggest that large production rates of dense shelf water occur in the bay[9,12]. Based on past expeditions, it has been suggested that the MG previously underwent similar calving events in the early part of the twentieth century[13]. However, these sporadic observations are insufficient to reconstruct past MGT dynamics with a high level of confidence.

In the present study, we provide a detailed reconstruction of surface and bottom oceanic conditions in the MGP over the last 250 years. This reconstruction is based on high-resolution analyses of diatom assemblages, diatom-specific biomarkers and major element abundances along a well-dated interface sediment core retrieved in January 2010 at CB (Fig. 1a; 66°54,38′ S— 142°26,18′ E; 775 m depth). This study represents the first detailed assessment of the MG-MGP system dynamics over the last few centuries, and therefore provides essential information on how the system recovers after major perturbations such as calving events. We also evaluate the impact of the recent positive shift in the Southern Annular Mode (SAM) on MGP surface conditions through a century-long atmospheric reanalysis.

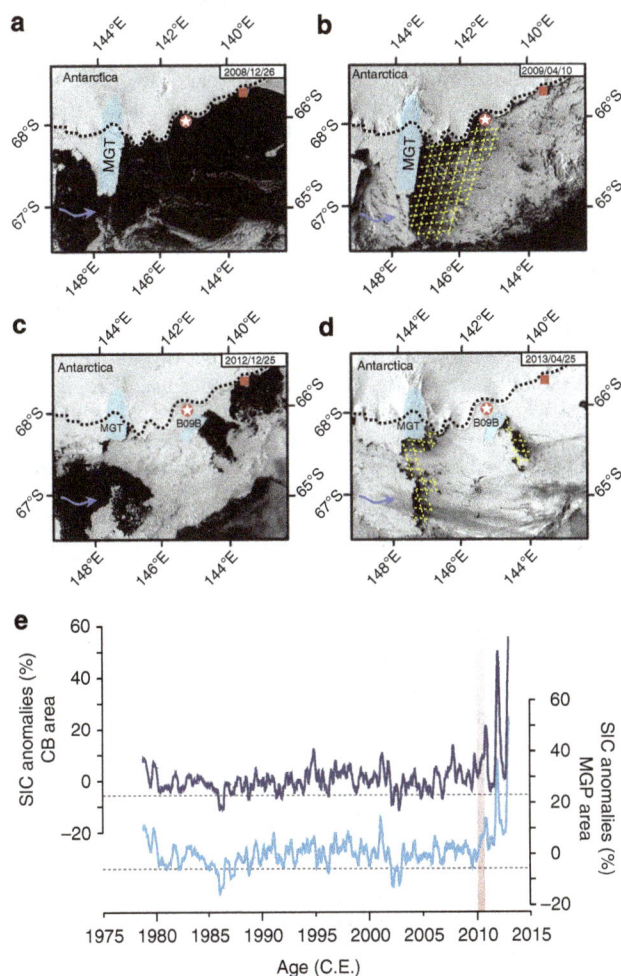

Figure 1 | Pre- and post-calving sea ice conditions in the George V Land. Aqua MODIS satellite images showing sea ice conditions in the study area during summer (**a**) and winter (**b**) before the 2010 calving event, and during summer (**c**) and winter (**d**) after the 2010 calving event. The white star in a red circle indicates the core location; the Dumont D'Urville station is indicated by a red square; blue shadings indicate glacial features, the MGT and the B09B iceberg; the MGP area in winter before and after the calving is designated by the yellow grid on **b,d**; and the blue arrow indicates the general direction of the East Wind Drift; coastline is marked by the black dotted line. (**e**) Three-month averaged SSM/I time series of daily SIC anomalies in CB (dark blue) and in the Mertz area (MGP—light blue), for the period 1978-2012 C.E. (Common Era) using 1978-2009 as the reference period (for the exact grid points locations, see Fig. 7). The red shading indicates the 2010 calving event.

Results

Multidecadal trends of sea surface conditions. Variations of *Fragilariopsis cylindrus* populations and di-unsaturated highly branched isoprenoid (HBI) lipid [HBI:2] abundances characterize spring sea ice occurrence[14,15], while summer open ocean conditions are inferred from high abundances of open water diatoms[16] (see Methods section). Titanium (Ti) content is used here to infer changes in terrigenous supply to the ocean[17], thus reflecting the variability of the terrigenous delivery by glacial melting, the dominant process in Antarctic coastal areas at such timescale[18,19]. High (low) relative abundances of open water diatoms are concomitant with high (low) contents of Ti and, conversely, with low (high) relative abundances of *F. cylindrus* and low (high) concentrations of [HBI:2] (Fig. 2). Minimum abundances of [HBI:2] and *F. cylindrus* associated with maxima

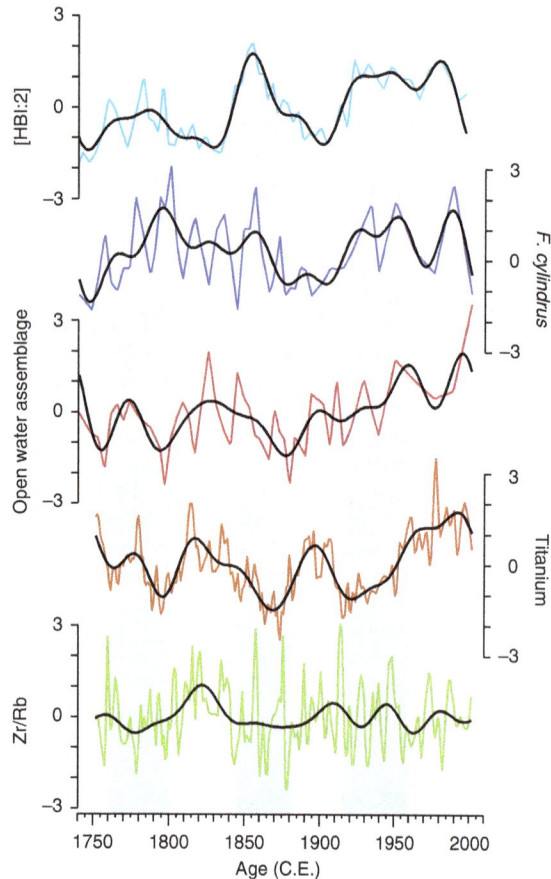

Figure 2 | Sea ice and ocean conditions in CB over the last 250 years. Sea ice proxies: standardized concentrations of HBI:2 (light blue) and relative abundances of *F. cylindrus* (dark blue). Open water proxies: standardized relative abundances of open water diatom group (red) and titanium content (orange). Bottom current variability: standardized Zr/Rb ratio (green). Black curves represent the low-pass-filtered data, with a cutoff frequency at 1/32 (in year). The blue shadings indicate decades following each calving event with marked increase in SIC in the CB area.

eighteenth century, surface ocean conditions and MGP activity alternated between periods of high sea ice presence and periods of prolonged open ocean conditions.

Cyclic pattern. These multidecadal trends of the MG-MGP system are further characterized by a strong asymmetric evolution with sharp increases in SIC at ~1760, ~1845 and ~1915, followed by a slow decrease over the subsequent decades (Fig. 2). A similar sharp increase observed after the 2010 calving event (Fig. 1e) suggests that, at least for the past 250 years, these asymmetric oscillations (~70 years periodicity) are related to a series of large calving events. Each of these events probably induced a rapid closing of the MGP as the loss of the ice tongue allowed for sea ice transported within the East Wind Drift to enter the area.

This scenario is further supported by the wavelet analysis of our data set, which shows that all variables exhibit a common high power in the 60–80 year band, throughout the entire record (Fig. 3). Although such periods (60–80 years) are close to the limit of detection (the bottom of the cone influence), they are characterized by a high statistical significance level of the period (95%, according to ref. 21). In addition, in the 60–80 year band, a positive correlation, characterized by rightward pointed arrows, is shown by cross-wavelet analyses (see Methods section) between sea ice proxies (Fig. 3a) and between open water proxies (Fig. 3b). In contrast, as indicated by leftward pointed arrows, negative correlations occur in the 60–80 year band when comparing sea ice versus open water proxies (Fig. 3c–f). Interestingly, continuous wavelet transforms indicate that open water and sea ice diatom signals (Supplementary Fig. 1) exhibit an additional ~20–25 year cyclicity, which is absent from the geochemical proxies. Although the causes remain unclear, differences in the spectral pattern of diatom and geochemical records may be attributed to the origin of the proxies themselves. The [HBI:2] is synthesized by diatoms living within or attached to sea ice. As such, we expect a direct response of [HBI:2] to SIC and the timing of ice waning[14]. Sea ice-related diatoms such as *F. cylindrus* thrive at the sea ice margin while open water diatoms thrive in the water column. Therefore, we believe that diatoms respond to both the timing of sea ice waning and oceanic biotic and abiotic factors such as stratification, upwelling and the injection of warmer waters[15–17], which are expressed at shorter timescales[22,23].

Barrier effect of the MG. Each calving event is characterized by relatively elevated abundances of the sea ice proxies for a couple of decades, followed by a slow decrease of the sea ice proxies and a concomitant increase of the open water proxies. These results suggest that each event was followed by a slow and constant re-advance of the glacier, but that a few decades are necessary for the tongue to reach a sufficient length to prevent sea ice advection in the area. As such, during few decades after 1800, 1880 and 1960 (Fig. 2), the tongue probably acted as a barrier and deflected the East Wind Drift pack ice northward, thus restoring favourable conditions for the establishment of the polynya. Although only limited observations exist to reconstruct the recent history of the MGP, it is clear that at least one major calving event occurred between 1912, when the Australasian Antarctic Expedition led by Mawson (1911–1914) measured a 150-km-long tongue from the grounding line, and 1958, when the Soviet Antarctic Expedition reported a 113-km-long tongue[13]. The sharp increase in sea ice proxies in 1915 strongly suggests that this event occurred only a few years after Mawson's expedition. Given the mean growth rate of the tongue of ~1 km per year[24], the glacier had presumably reached ~150–160 km at that time when it calved. This length is similar to the one reached by the glacier before the 2010

of open water diatoms and Ti during 1740–1760, 1800–1845 and 1880–1915 periods (Fig. 2), therefore indicate the presence of a well-developed polynya characterized by low SIC during the winter, early spring sea ice retreat and long summer season. The presence of a well-developed MGP promoted intense sea ice formation during winter and enhanced brine-induced convection. Variations in bottom current velocities were estimated from changes in sediment grain size inferred from Zirconium versus Rubidium (Zr/Rb) relative abundances[20]. Assuming that greater deep-reaching convection is associated with intensified bottom currents, the high Zr/Rb ratios (Fig. 2) recorded during periods of well-developed MGP suggest enhanced bottom currents favouring a coarse sedimentation at the core location. In contrast, the 1760–1800, 1845–1880 and 1915–1960 periods are characterized by maxima in [HBI:2] and *F. cylindrus* relative abundances and minima in open water diatoms, Ti and Zr/Rb records (Fig. 2). This reflects the presence of heavier sea ice conditions in spring, leading to shorter growing seasons and cooler summers, associated with reduced inputs from glacial runoffs and melting. During these intervals, a reduction of the MGP area and sea ice production coincided with a weaker bottom circulation, as revealed by the presence of finer (low Zr/Rb) sediments. Our records therefore suggest that since the mid-

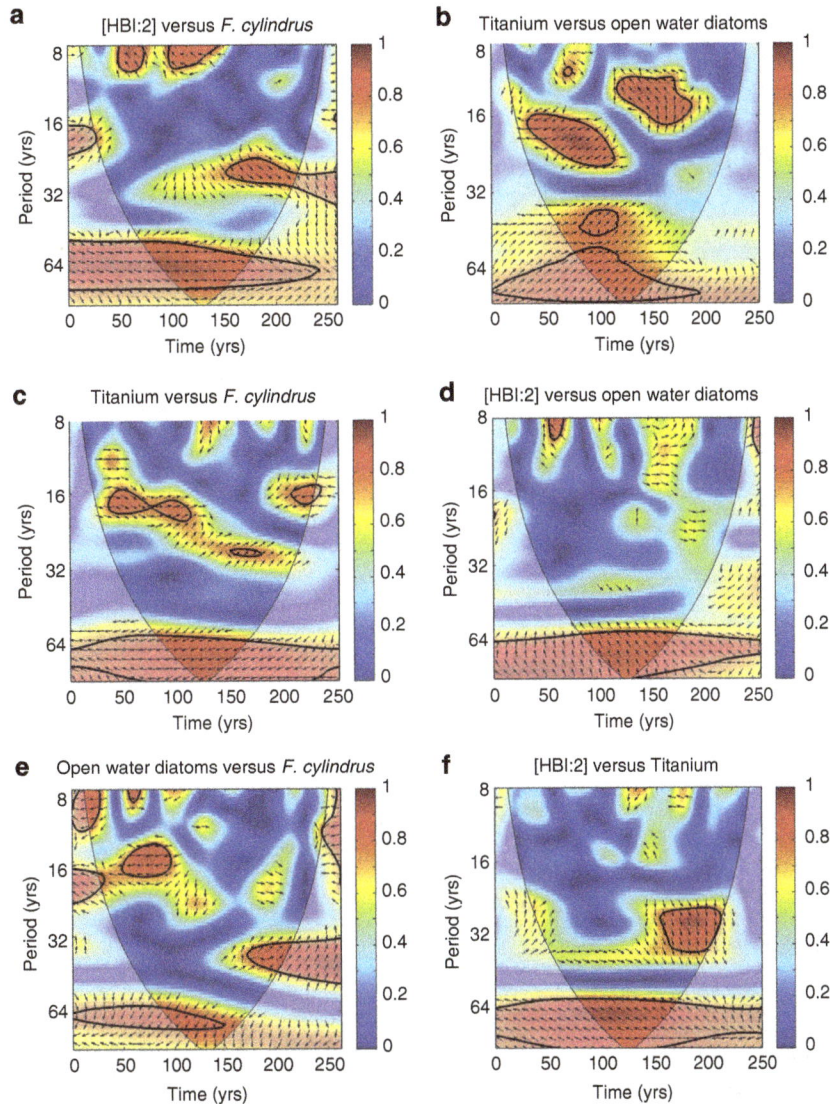

Figure 3 | Relationship between proxies and cyclicity of sea surface conditions in the MGP area. Cross-wavelet transform (XWT) on the sedimentary proxies (using Morlet wavelet and Monte Carlo methods[21]) between (**a**) [HBI:2] versus *F. cylindrus*; (**b**) Ti versus open water diatoms; (**c**) Ti versus *F. cylindrus*; (**d**) [HBI:2] versus open water diatoms; (**e**) open water diatoms versus *F. cylindrus*; and (**f**) [HBI:2] versus Ti. Statistically significant periods are identified by the black circled red zones. Rightward pointed arrows indicate positively correlated signals while leftward pointed arrows indicate negatively correlated signals. Yrs, years.

calving[25,26] and is in line with Mawson's observations in 1912. If the MGT lost half (80 km) of its length, similar to the 2010 calving event, it would have taken ~40 years for the tongue to grow back from ~80 km in 1915 to 113 km, based on the Soviet Antarctic Expedition in 1958. These results suggest that from the 1960 s, the ice tongue was then long enough to effectively act as a barrier for drifting ice. In the 1990s, signs of imminent calving were already detected with the formation of two major rifts near the glacier grounding line and the front of the glacier grounding on the Mertz Bank to the north[13,25]. The shallow bathymetry of the bank (Supplementary Fig. 2) enhances local tidal currents exerting lateral stress on the tip of the glacier, likely impacting the tongue along-flow velocity[26]. In addition, the presence of several icebergs released by upstream glaciers and grounded onto the Mertz bank[13,26] probably further increases the lateral distortion of the tongue. As such, it appears that beyond a threshold of ~150–160 km, lateral stresses exerted on the sides and at the lie of the tongue lead to its rupture.

Recent changes in seasonality. Interestingly, the analysis of the upper sediment section show that while the MGT was sufficiently long to promote open water conditions in the lie of the glacier during the last 50 years, both [HBI:2] and *F. cylindrus* remained relatively abundant in the sediments (Fig. 2). In addition, Zr/Rb values did not show a marked increase as expected during this phase (Fig. 2). This suggests that, in contrast to previous cycles, pre-calving conditions over recent decades were characterized by a more persistent sea ice cover during spring and were associated with weaker bottom water circulation than observed during previous cycles. We also note higher Ti contents (terrigenous inputs) along with greater abundances of open water diatom species (and in particular large centric diatoms; Fig. 4), suggesting strong glacial melting and more persistent open water conditions led by warmer surface waters during summer over the recent decades. Warmer open water conditions during the summer are further confirmed by the large peak of a specific HBI isomer [HBI:3] during the past 40 years (Supplementary Fig. 3). High

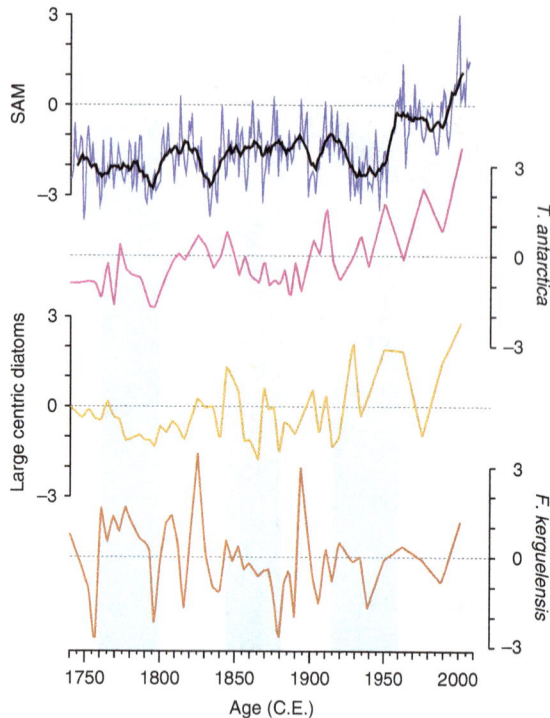

Figure 4 | Atmospheric forcing and impacts on the sea ice free season during the instrumental period in the MGP area. Evolution of the SAM according to the reconstructed Marshall index (blue, annual values and black, 11 years moving average) over the last 250 years[31]. Open water proxies: standardized relative abundances of *T. antarctica* sp (pink), large centric diatom group (yellow; see Methods section for species composition) and *F. kerguelensis* (orange). The blue shadings indicate decades following each calving event with marked increase in SIC in the CB area.

abundances of [HBI:3] in sediments reflect the contribution of phytoplanktonic-derived organic matter[14], and thus indicate a higher phytoplankton productivity during the summer (Supplementary Note 1). Overall, these results therefore provide strong evidence for changes occurring over the growing season in recent decades. From our records, this apparent shift in seasonality is characterized by cooler and icier springs, warmer and more open summers in line with recent atmospheric and oceanic observations from other areas in the Southern Hemisphere[22,27].

Discussion

Our results strongly suggest that surface oceanic conditions and dense shelf water production are closely related to the MGP presence and activity. Reasons for the ∼70 years cyclicity of the MGP are still not fully understood but, given the major constraints of the local topography, it is likely that these cycles are set by the rate of advance and along-flow velocity of the MG. However, according to several studies[13,25], icebergs released from upstream outlet glaciers (for example, Ninnis Glacier) could have impacted the stability of the MGT when they grounded and/or passed through the area. In addition, when the tongue is too short to constitute a barrier, these icebergs can impact the regional oceanography for several years, leading to a temporary increase in sea ice conditions, as observed since 2010 with the presence of the B09B iceberg in front of CB (Fig. 1c,d). The resolution of our records, however, does not allow us to capture such phenomena, if they ever occurred in or around CB in the past. Our data also indicate a close relationship between the MG history and MGP

dynamics between 1740 and 1960. However, since the 1960s, our records suggest unexpectedly cool and icy springs at a time where the MGT should have reached a sufficient length to promote the presence of a well-established MGP.

We propose that, superimposed on the large multidecadal oscillations generated by the MG dynamics, additional factors contributed to modulating sea surface conditions in the area. The SAM, principal mode of atmospheric variability over the Southern Ocean[28], has shown a steadily increasing index over the last 50 years (refs 27,29; and Figs 4 and 5a), with a more positive trend in summer[27,30]. Reconstructions indicate that this increase is unprecedented over the last few centuries[31] possibly due to ozone depletion and a rise in greenhouse gases[29,32], and recent investigations have argued that such trend had a direct influence on sea surface conditions in several regions around Antarctica through modulation of the wind pattern[22,23,29,33–35]. The recent positive trend in the SAM index is indeed associated with an intensification of the polar vortex[32] leading to a southward shift of enhanced circumpolar westerlies[28]. This, in turn, led to a more intense Antarctic Circumpolar Current (ACC)[35] and associated ocean eddy activity[36]. We postulate that such large-scale changes impacted the sea ice distribution in spring, modified the summer off-shelf ocean circulation due to changes in the large-scale wind stress pattern and are the dominant cause for the contrasted response of the Adélie Land continental shelf over the last 50 years.

Indeed, several studies have evidenced a close link between the planetary circulation in the southern polar atmosphere and the katabatic wind regime, the latter being part of the large-scale meridional tropospheric circulation over Antarctica[37–40]. Observations in East Antarctica revealed that an intensified tropospheric vortex was associated with weakened katabatic winds over the Antarctic margin[40], and analysis of the atmospheric wind fields during the last 140 years from the twentieth century reanalysis (20CR) reanalysis[41] confirms a significant westward shift of the wind pattern in the George V Land over recent decades (Fig. 5a, green curve; and Supplementary Note 2). A weakening of the meridional wind circulation and thus increasing zonal circulation over the last decades as suggested by refs 40,42 have also been observed in the region[34,43]. Indeed, at seasonal to inter-annual timescales, sea ice conditions in the MGP area have been shown to be sensitive to the latitudinal location of the Antarctic Circumpolar Trough and associated to more along shore wind transport[34,43]. In the present scenario, reduction of katabatic winds intensity over the MGP area would weaken the northward transport of sea ice, resulting in the retention of more ice within the area. The latter would delay the onset of sea ice melt in spring in agreement with higher sea ice proxies in our record since 1960. Although based on a restricted number of data points, a reduction of sea ice proxies in the uppermost sediment sections suggests a recent tendency towards a more reduced sea ice cover in spring, which is in line with the recent analysis of the sea ice seasonality in the area using the satellite records[44].

Our data also indicate warmer sea surface conditions during summer over recent decades. As recently observed in several Antarctic regions, it is argued here that intrusions of warm Circumpolar Deep Water (CDW) onto the continental shelf promoted open water conditions and higher sea surface temperatures[7,23,33,45]. Increasing CDW contribution onto the Antarctic Peninsula shelf was linked to the recent changes in the strength of the SAM[46]. Indeed, modelling studies have shown that advection of CDW onto the Antarctic continental shelf is linked to enhanced upwelling southward of the ACC, promoted by a southward shift and strengthening of the Southern Ocean Westerlies[35,47]. Surface wind stress curl calculated from the

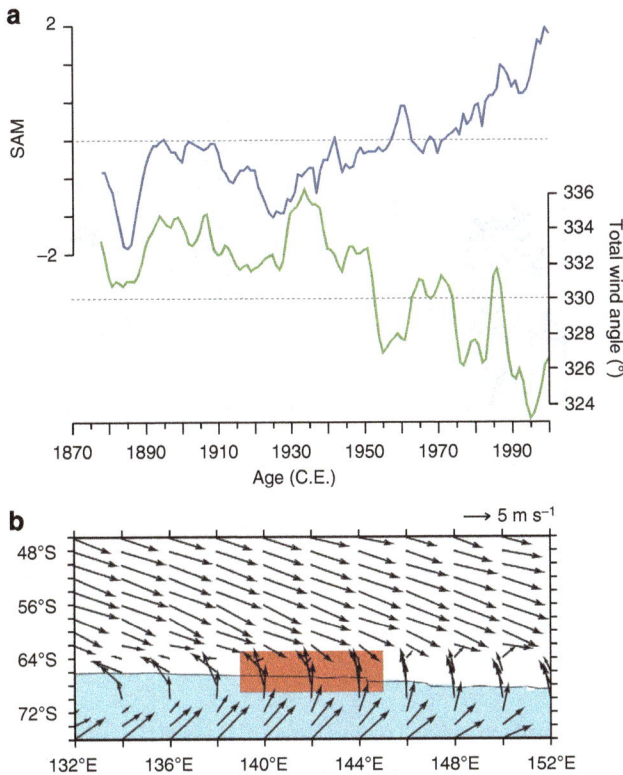

a

SAM axis values: 2, −2

Right axis (Total wind angle (°)): 336, 334, 332, 330, 328, 326, 324

X axis (Age (C.E.)): 1870, 1890, 1910, 1930, 1950, 1970, 1990

b

→ 5 m s⁻¹

Y axis: 48°S, 56°S, 64°S, 72°S

X axis: 132°E, 136°E, 140°E, 144°E, 148°E, 152°E

Figure 5 | Evolution of the SAM and of the wind pattern over the Adélie-George V Land since the nineteenth century. (**a**) Standardized values of SAM index (5 years running mean; blue), computed in the 20CR reanalysis following ref. 70 and the total wind direction angle at 2 m (green; °) computed in the red box of the lower panel using the same reanalysis. The North direction indicates the 0° modulo 360° and the angle is counted positively clockwise. (**b**) Annual mean wind speed from 20CR reanalysis averaged over the period 1871–2010. The red box indicates the study area where the wind have been computed in **a**.

20CR reanalysis indeed suggests long periods of strong positive trend in upwelling-favourable vertical velocities with positive values starting roughly after 1950, concomitant with the SAM trend (Supplementary Fig. 4 and Supplementary Note 2). In the Mertz region, in contrast with West Antarctica, flooding of the shelf by the CDW occurs as synoptic weather timescale intrusions rather than as a continuous flow[7]. However, the proximity of George V Land continental shelf to the southern ACC boundary facilitates transport of CDW into the Mertz region[48], as suggested in other Antarctic areas[23], providing that adequate dynamical forcing exists to drive this water mass on the shelf. According to ref. 49, a number of mechanisms involving interaction of the zonal flow with the topographic troughs disseminated along the Antarctic shelf break are likely to favour such transport. While some of these may be linked to the internal variability of the flow[50], some would be the result of enhanced wind-driven zonal flow. We note that, off the Adélie Land shelf, a more zonal wind pattern in the 20CR reanalysis (Fig. 5) over recent decades was concomitant to the southward shift (Supplementary Fig. 5) and the strengthening[35,47] of the Westerlies at mid latitudes. Increasing zonal wind circulation poleward may have contributed to accelerate the westward flowing Antarctic Slope Current, the southern branch of the Australian Antarctic Basin cyclonic circulation[51,52], thus favouring the inflow of CDW through nonlinear momentum advection onto the shelf trough. Recent observations suggest that a weaker and more zonal

circulation promote uplift and enhanced onshore intrusion of CDW[45], and studies attributed an increased abundance of terrigenous particles in the sediments to enhanced melting of both continental and sea ice during periods of increased advection of warm CDW onto the shelf[17–19,53]. The high terrigenous content recorded in CB2010 since 1960 (Fig. 2) can therefore be interpreted as a result of enhanced melting of regional glaciers in response to increased advection of warm CDW onto the shelf. This enhanced melting of glacial ice could also be contributing to further freshening the ALBW as reported by Lacarra *et al.*[9]

Our results demonstrate that, in response to glacial dynamics and local physiography, recurrent massive calving events of the MG occurred over the last 250 years. These events had profound impacts on ocean surface conditions and dense water production of the downstream polynya. Taking into consideration that many of these glacier-polynya systems are disseminated around Antarctica[4] and that dense shelf water is ultimately a precursor of AABW, whose impact on large-scale ocean circulation is well known[1,2], our study provides evidence that local processes may contribute to alter global ocean and climate systems. In contrast to previous cycles, our data indicate that during the last 50 years, the region was experiencing a more compact sea ice cover, a larger supply of glacial meltwater and a slowdown of bottom currents. These results suggest a reduction of dense water production and are consistent with the long-term freshening of the ALBW observed in the Australian Antarctic Basin since the late 1960s (refs 54,55). This contrasted response of the MGP region during the last Mertz cycle may just be a transitory phenomenon as the SAM increase over that period is thought to be partly due to ozone depletion in the polar vortex. Since studies are predicting that the ozone hole will replenish in the future decades[56], natural multidecadal glacier cycles such as those identified in our records are likely to take over.

Methods

Sediment material and chronology. A 30.5-cm-long interface core was retrieved aboard the R/V Astrolabe (66°54.38′ S; 142°26.18′ E; 775 m water depth) during the 2010 COCA cruise. Positive X-ray images performed on the SCOPIX image-processing tool[57] gave detailed information on sediment density and structure. SCOPIX images revealed that, in contrast to high accumulation sites from Dumont d'Urville Trough, sediments from CB were massive with no signs of laminations. The core was sampled continuously at 0.5 cm resolution and its chronological framework was determined based on ^{210}Pb excess (^{210}Pb$_{xs}$; $T_{1/2} = 22.3$ years), which is rapidly incorporated into the sediment from atmospheric fallout and water column scavenging. The activities of ^{210}Pb and ^{226}Ra were measured on dried sediments by non-destructive gamma spectrometry using a well-type, high-efficiency low-background detector equipped with a Cryo-cycle (CANBERRA). Activities are expressed in mBq g^{-1} and errors are based on 1 s.d. counting statistics (Fig. 6). ^{210}Pb$_{xs}$ was determined by subtracting the activity supported by its parent isotope, ^{226}Ra, from the total ^{210}Pb activity in the sediment. Mass accumulation rate (0.025 g cm^{-2} y^{-1}) was calculated from the sedimentary profiles of ^{210}Pb$_{xs}$, plotted against cumulative mass. The deposition time (in years) was obtained by dividing the cumulative dry mass per unit area by mass accumulation rate. The deposition year for each sediment layer was subsequently estimated based on the 2010 sampling date for the sediment–water interface.

Diatoms. Micropaleontological analyses were performed according to the methodology described in ref. 58. For each sample, 300–350 diatom valves were counted and data are presented as species relative abundances. Briefly, diatom identification was performed to the species or species group level at a ~5-year resolution. Sixty-eight diatom species were identified in down-core assemblages, from which only a dozen species presented abundances higher than 2% of the total diatom population. Only these species were considered relevant for reconstructing environmental changes. Diatom species or species groups we confront here experience similar ranges of variability in CB2010, ~2.5–13% for *F. cylindrus*, ~1–9% for *F. kerguelensis* and ~0–7% for the large centric group and *Thalassiosira antarctica* abundances, respectively (Supplementary Fig. 6).

F. cylindrus is one of the most common diatoms found along the Adélie Land margins[59] as it thrives within stratified sea ice-covered waters[15]. Large abundances of *F. cylindrus* in sediments indicate the presence of a sea ice cover persisting over 7.5 months per year[15]. In contrast, the open water diatom

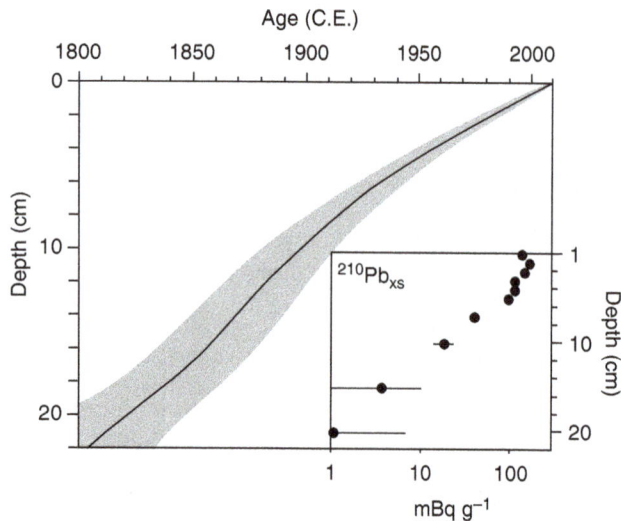

Figure 6 | CB2010 chronology. CB2010 age model (dark line) based on ^{210}Pb excess ([210]Pb$_{xs}$) and associated age-model errors (grey area). The inset corresponds to the down-core profile of [210]Pb$_{xs}$ (error bars correspond to 1 s.d.).

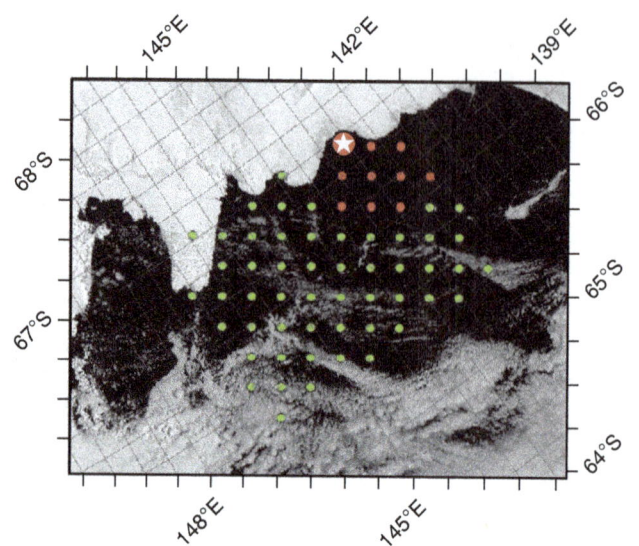

Figure 7 | Pixels locations for extraction of SICs. MODIS satellite image (2008/12/26) of the George V Land indicating the grid points used for the extraction of the daily SIC values: white star in a red circle indicates the core location; red spots correspond to the CB area and green spots represent the MGP area.

assemblage, composed by *F. kerguelensis* and large centric diatoms, characterizes open water conditions during summer[16].

F. kerguelensis dominates assemblages of the open ocean zone south of the Polar Front where sea ice is absent during summer[60]. Similarly, high abundances of large centric diatom species, such as *T. lentiginosa* or *T. oliverana*, commonly occur in the Southern Ocean south of the Polar Front in areas characterized by permanent open ocean conditions[16,61]. Relative abundances of *T. lentiginosa* show an inverse relationship with sea ice cover with high occurrences between 0 and 4 months of sea ice presence per year and a decline towards prolonged sea ice duration[16]. *T. oliverana* is clearly dominant in locations where open ocean conditions to sea ice edge during summer occur[16]. *T. antarctica* has been described as a dominant species within diatom assemblages in non-stratified or weakly stratified Antarctic surface waters[62]. *T. antarctica* blooms in open waters during summer–autumn, and produces resting spores at the end of the growing season when sea ice returns[63]. *T. antarctica* resting spores, the main form encountered in sediments, is most abundant in regions where sea ice is present for at least 6 months per year, and is believed to be induced under nutrient-stressed conditions or low light intensities[15].

Geochemical data. Few marine and freshwater diatoms belonging to *Haslea, Navicula, Pleurosigma* and *Rhizosolenia* genera were recently found to be synthesizing HBIs[14,64]. A di-unsaturated isomer [HBI:2] has been identified in Antarctic sea ice and isotopic analyses provide evidence for that this isomer to be synthesized by sea ice dwelling diatoms. In contrast, tri-unsaturated HBI isomers [HBI:3] have been identified in water column phytoplankton[14]. Recent studies have proposed the use of [HBI:2] and [HBI:3] to reconstruct variations of Holocene Antarctic sea ice duration as complementary sea ice proxy to diatom counts[65]. Biomarker analysis followed the technique described by ref. 14 and were performed every 0.5 cm through the core. Briefly, an internal standard was added to the freeze-dried sediments, lipids were extracted using a dichloromethane/methanol mixture to yield a total organic extract, which was then purified using open column chromatography (silica). Hydrocarbons were analysed using a gas chromatograph coupled to a mass spectrometry detector.

Ti and Zr are considered to be direct indicators of terrigenous inputs as these elements are not involved in biological cycles[17]. In the literature, variations in Ti content are largely used to infer past changes in terrigenous supply to the ocean[17]. Microfabric analysis of sediment in the Mertz-Ninnis and Adélie troughs show that both terrigenous content and Ti content are low in spring laminae and increase over the growing season (when open water diatoms dominate the siliceous assemblage) and during which the summer glacial melting is high[17,53,66]. Indeed, in Antarctic coastal areas, delivery of terrigenous particles is possible via several dominant modes as meltwater discharge, ice rafting, runoff from outlet glaciers and aeolian transport, although this latter source is considered negligible in coastal East Antarctic regions[17–19]. Variations in Zr to Rb content ratio track changes in sediment grain size, where Zr represents the coarsest sediment fraction and Rb represents the finest[20]. Ti, Zr and Rb contents were measured on slab sections at a 2 mm resolution along the entire core using an AAVATECH XRF core-scanner[67].

Satellite data. Daily SICs for the time period 1978–2012 were obtained from the National Snow and Ice Data Center data repository. The data set is based on passive microwave observations from the Nimbus-7 SSMR (1978–1987), DMSP

SSM/I (1987–2007) and SSMIS (2007–2012) radiometers processed with the NASA Team algorithm[68] at a spatial resolution of 25 × 25 km. Averaged concentrations were calculated over two specific domains, CB and the entire MGP domain (Fig. 7). Daily anomalies were calculated using the average of the pre-calving 1978–2009 period and then low-pass filtered using a 3-month moving average (Fig. 1e). SIC data were standardized. Anomalies represent differences between the daily value and the mean daily value calculated over the reference period.

Spectral analysis. Unlike many traditional mathematical methods (for example, Fourier analysis), the wavelet approach can be used to analyse time series that contain non-stationary spectral power at many different frequencies[21]. For geological time series, although visual comparison of plots is commonly used, cross-wavelet analysis permits detection, extraction and reconstruction of relationships between two non-stationary signals simultaneously in frequency (or scale) and time (or location)[69]. The continuous wavelet transform (CWT; Supplementary Fig. 1) of time series is its convolution with the local basis functions, or wavelets, which can be stretched and translated with flexible resolution in both frequency and time. The principle of cross-wavelet analysis and the complete method we used were described in ref. 21. We used the MATLAB package for cross-wavelet analysis written by Grinsted *et al.*[21] and applied the Morlet wavelet as the mother function on our data set. This method provides a good balance between time and frequency localization, and we used Monte Carlo simulations to provide frequency-specific probability distribution (global wavelet spectrum) that can be tested against wavelet coefficients. Statistical significance was estimated against a red noise model[21]. In this study, to test the relevance of our proxies, their statistical relationships and to examine periodicities in a frequency domain, we compared the two time series by their CWTs, which we hypothesized are linked in some way. The resulting cross wavelet transform (XWT, Fig. 3) exposed their common power and relative phase in time–frequency space of the two signals. Data were previously standardized, which did not introduce any change in the shape of the records but normalized the amplitude of the variations (Supplementary Fig. 6).

Atmospheric reanalysis. To analyse the Southern Hemisphere atmospheric circulation during the last 140 years, we used the recent 20CR Project version 2 (ref. 41), consisting of an ensemble of 56 realizations with 2° × 2° gridded 6-hourly weather data from 1871 to 2010. Each ensemble member was performed using the NCEP/GFS (National Center for Environmental Prediction/Global Forecast System) atmospheric model, prescribing the monthly sea surface temperature and sea ice changes from HadISST as boundary conditions, and assimilating sea level pressure data from the International Surface Pressure Databank version 2 (http://rda.ucar.edu/datasets/ds132.0). We used the ensemble mean to perform our analysis. An important caveat concerns the fact that few data were assimilated at the beginning of the reanalysis in the Southern Hemisphere, owing to the lack of available observations. Nevertheless, this product is one of the best data sets available for the evaluation of atmospheric circulation changes at a large scale in the Southern Hemisphere. Wind speed products were plotted over the Terre Adélie-George V Land (Fig. 5) and the offshore region ~ 55–60°S (Supplementary Fig. 5).

References

1. Johnson, G. C. Quantifying Antarctic Bottom Water and North Atlantic Deep Water volumes. *J. Geophys. Res.* **113,** C05027 (2008).
2. Jacobs, S. Bottom water production and its links with the thermohaline circulation. *Antarct. Sci.* **16,** 427–437 (2004).
3. Tamura, T., Ohshima, K. I. & Nihashi, S. Mapping of sea ice production for Antarctic coastal polynyas. *Geophys. Res. Lett.* **35,** L07606 (2008).
4. Arrigo, K. R. & van Dijken, G. L. Phytoplankton dynamics within 37 Antarctic coastal polynya systems. *J. Geophys. Res.* **108,** 3271 (2003).
5. Barber, D. G. & Massom, R. A. in *Polynyas: Windows to the World* Vol 74 (eds Smith, W. O. *et al.*) 1–54 (Elsevier Oceanography Series, 2007).
6. Rintoul, S. R. in *Ocean, Ice, and Atmosphere: Interactions at the Antarctic Continental Margin* Vol 75 (eds Jacobs, S. & Weiss, R.) 151–171 (Antarctic Research Series, 1998).
7. Williams, G. D., Bindoff, N. L., Marsland, S. J. & Rintoul, S. R. Formation and export of dense shelf water from the Adélie Depression, East Antarctica. *J. Geophys. Res.* **113,** C04039 (2008).
8. Massom, R. A., Harris, P. T., Michael, K. J. & Potter, M. J. The distribution of formative processes of latent heat polynyas in East Antarctica. *Ann. Glaciol.* **27,** 420–426 (1998).
9. Lacarra, M., Houssais, M. N., Herbaut, C., Sultan, E. & Beauverger, M. Dense shelf water production in the Adélie Depression 2004-2012: impact of the Mertz glacier calving. *J. Geoph. Res.* **119,** 5203–5220 (2014).
10. Tamura, T., Williams, G. D., Fraser, A. D. & Ohshima, K. I. Potential regime shift in decreased sea ice production after the Mertz Glacier calving. *Nat. Commun.* **3,** 826 (2012).
11. Kusahara, K., Hasumi, H. & Williams, G. Impact of Mertz Glacier Tongue calving on dense shelf water. *Nat. Commun.* **2,** 159 (2011a).
12. Williams, G. D. & Bindoff, N. L. Wintertime oceanography of the Adélie depression. *Deep-Sea Res. Pt. II* **50,** 1373–1392 (2003).
13. Frezzotti, M., Cimbelli, A. & Ferrigno, J. G. Ice-front change and iceberg behavior along Oates and George V coasts, Antarctica, 1912–96. *Ann. Glaciol.* **27,** 643–650 (1998).
14. Massé, G. *et al.* Highly branched isoprenoids as proxies for variable sea ice conditions in the Southern Ocean. *Antarct. Sci.* **23,** 487–498 (2011).
15. Armand, L. K., Crosta, X., Romero, O. & Pichon, J.-J. The biogeography of major diatom taxa in Southern Ocean sediments: 1. Sea ice related species. *Palaeogeog., Palaeoclim., Palaeoecol.* **223,** 93–126 (2005).
16. Crosta, X., Romero, O., Armand, L. K. & Pichon, J.-J. The biogeography of major diatom taxa in Southern Ocean sediments: 2. Open ocean related species. *Palaeogeog., Palaeoclim., Palaeoecol.* **223,** 66–92 (2005).
17. Denis, D. *et al.* Seasonal and sub-seasonal climate changes recorded in laminated diatom ooze sediments, Adélie Land, East Antarctica. *Holocene* **16,** 1137–1147 (2006).
18. Presti, M., De Santis, L., Busetti, M. & Harris, P. T. Late Pleistocene and Holocene sedimentation on the George V Continental Shelf, East Antarctica. *Deep-Sea Res. Pt. II* **50,** 1441–1461 (2003).
19. Escutia, C. *et al.* Sediment distribution and sedimentary processes across the Antarctic Wilkes Land margin during the Quaternary. *Deep-Sea Res. Pt. II* **50,** 3225–3226 (2003).
20. Dypvik, H. & Harris, N. B. Geochemical facies analysis of fine-grained siliciclastics using Th/U, Zr/Rb and (Zr + Rb)/Sr ratios. *Chem. Geol.* **181,** 131–146 (2001).
21. Grinsted, A., Moore, J. C. & Jevrejeva, S. Application of the cross wavelet transform and wavelet coherence to geophysical time series. *Nonlin. Proc. Geophys.* **11,** 561–566 (2004).
22. Stammerjohn, S. E., Martinson, D. G., Smith, R. C., Yuan, X. & Rind, D. Trends in Antarctic annual sea ice retreat and advance and their relation to El Niño–Southern Oscillation and Southern Annular Mode variability. *J. Geophys. Res.* **113,** C03S90 (2008).
23. Dinniman, M. S., Klinck, J. M. & Hofmann, E. E. Sensitivity of Circumpolar Deep Water Transport and Ice Shelf Basal Melt along the West Antarctic Peninsula to changes in the winds. *Amer. Meteor. Soc.* **25,** 4799–4816 (2012).
24. Wuite, J. *Spatial and Temporal Dynamics of Three East Antarctic Outlet Glaciers and Their Floating Ice Tongues* 190, PhD thesis, Ohio State Univ. (2006).
25. Lescarmontier, L. *et al.* Vibration of the Mertz Glacier ice tongue, East Antarctica. *J. Glaciol.* **58,** 665–676 (2012).
26. Legresy, B., Wendt, A., Tabacco, I., Remy, F. & Dietrich, R. Influence of tides and tidal current on Mertz Glacier, Antarctica. *J. Glaciol.* **50,** 427–435 (2004).
27. Marshall, G. J. Trends in the Southern Annular Mode from observations and reanalyzes. *J. Clim.* **16,** 4134–4143 (2003).
28. Thompson, D. W. J. & Wallace, J. M. Annular modes in the extratropical circulation. Part I: month-to-month variability. *J. Clim.* **13,** 1000–1016 (2000).
29. Thompson, D. W. J. & Solomon, S. Interpretation of recent Southern Hemisphere climate change. *Science* **296,** 895–899 (2002).
30. Fogt, R. L., Perlwitz, J., Monaghan, A. J., Bromwich, D. H., Jones, J. M. & Marshall, G. J. Historical SAM variability. part II: twentieth-century variability and trends from reconstructions, observations, and the IPCC AR4 models. *J. Clim.* **22,** 5346–5365 (2009).
31. Villalba, R. *et al.* Unusual Southern Hemisphere tree growth patterns induced by changes in the Southern Annular Mode. *Nat. Geosci.* **5,** 793–798 (2012).
32. Arblaster, J. M. & Meehl, G. A. Contributions of external forcings to Southern Annular Mode Trends. *J. Clim.* **19,** 2896–2905 (2006).
33. Jacobs, S. Observations of changes in the Southern Ocean. *Phil. Trans. R. Soc. A* **364,** 1657–1681 (2006).
34. Massom, R. A. *et al.* Fast ice distribution in Adélie Land, East Antarctica: Interannual variability and implications for Emperor penguins (*Aptenodytesforsteri*). *Mar. Ecol. Progr. Ser.* **374,** 243–257 (2009).
35. Hall, A. & Visbeck, M. Synchronous variability in the Southern Hemisphere atmosphere, sea ice, and ocean resulting from the Annular Mode. *J. Clim.* **15,** 3043–3057 (2002).
36. Meredith, M. P. & Hogg, A. M. Circumpolar response of Southern Ocean eddy activity to a change in the Southern Annular Mode. *Geophys. Res. Lett.* **33,** L16608 (2006).
37. Parish, N. J. P. & Bromwich, D. H. Continental-scale simulation of the Antarctic katabatic wind regime. *J. Clim.* **4,** 135–146 (1991).
38. Van der Broeke, M. R., van Lipzig, N. P. M. & van Meijgaard, E. Momentum budget of the East Antarctic atmospheric boundary layer: results of a Regional Climate Model. *J. Atmos. Sci.* **59,** 3117–3129 (2002).
39. Egger, J. Slope winds and the axisymmetric circulation over Antarctica. *J. Atmos. Sci.* **42,** 1859–1867 (1985).
40. Yasunari, T. & Kodama, S. Intraseasonal variability of katabatic wind over east Antarctica and planetary flow regime in the southern hemisphere. *J. Geophys. Res.* **98,** 13063–13070 (1993).
41. Compo, G. P. *et al.* The twentieth century reanalysis project. *Q.J.R. Meteor. Soc.* **137,** 1–28 (2011).
42. Van der Broeke, M. & van Lipzig, N. P. M. Changes in Antarctic temperature wind and precipitation in response to Antarctic oscillation. *Ann. Glaciol.* **39,** 119–126 (2004).
43. Massom, R. A. *et al.* An anomalous late-season change in the regional sea ice regime in the vicinity of the Mertz Glacier Polynya, East Antarctica. *J. Geophys. Res.* **108,** 3212 (2003).
44. Massom, R. A. *et al.* Change and variability in East Antarctic sea ice seasonality, 1979/80–2009/10. *PLoS ONE* **8,** e64756 (2013).
45. Schmidtko, S., Heywood, K. J., Thompson, H. & Shigeru, A. Multidecadal warming of antarctic waters. *Science* **346,** 1227–1231 (2014).
46. Martinson, D. G., Stammerjohn, S. E., Iannuzzi, R. A., Smith, R. C. & Vernet, M. Western Antarctic Peninsula physical oceanography and spatio-temporal variability. *Deep-Sea Res. Pt. II* **55,** 1964–1987 (2008).
47. Marini, C., Frankignoul, C. & Mignot, J. Links between the Southern Annular Mode and the Atlantic Meridional Overturning Circulation in a Climate Model. *J. Clim.* **24,** 624–640 (2011).
48. Orsi, A. H., Whitworth, III T. & Nowlin, Jr W. D. On the meridional extent and fronts of the Antarctic Circumpolar Current. *Deep-Sea Res. Pt. I* **42,** 641–673 (1995).
49. Klinck, J. M. & Dinniman, M. S. Exchange across the shelf break at high southern latitudes. *Ocean Sci.* **6,** 513–524 (2010).
50. St-Laurent, P., Klinck, J. M. & Dinniman, M. S. On the role of coastal troughs in the circulation of Warm Circumpolar Deep Water on Antarctic Shelves. *J. Phys. Ocean* **43,** 51–64 (2013).
51. Bindoff, N. L., Rosenberg, M. A. & Warner, M. J. On the circulation and water masses over the Antarctic continental slope and rise between 80 and 150°E. *Deep-Sea Res Pt. II* **47,** 2299–2326 (2000).
52. Aoki, S., Sasai, Y., Sasaki, H., Mitsudera, H. & Williams, G. D. The cyclonic circulation in the Australian Antarctic basin simulated by an eddy resolving general circulation model. *Ocean Dyn.* **60,** 743–757 (2010).
53. Maddison, E. J. *et al.* Post-glacial seasonal diatom record of the Mertz Glacier Polynya, East Antarctica. *Mar. Micropal.* **60,** 66–88 (2006).
54. Rintoul, S. R. Rapid freshening of Antarctic Bottom Water formed in the Indian and Pacific oceans. *Geophys. Res. Lett.* **34,** L06606 (2007).
55. Aoki, S., Rintoul, S., Ushio, S. & Watanabe, S. Freshening of the Adélie Land Bottom Water near 140°E. *Geophys. Res. Lett.* **32,** L23601 (2005).
56. Previdi, M. & Polvani, L. M. Climate system response to stratospheric ozone depletion and recovery. *Q.J.R. Meteorol. Soc.* **140,** 2401–2419 (2014).
57. Migeon, S., Weber, O., Faugères, J.-C. & Saint-Paul, J. SCOPIX: a new X-ray imaging system for core analysis. *Geo-Mar. Lett.* **18,** 251–255 (1999).
58. Crosta, X. & Koc, N. in *Methods in Late Cenozoic Paleoceanography* (eds Hilaire-Marcel, C. & de Vernal, A.) 327–369 (Elsevier, 2007).
59. Kang, S.-H. & Fryxell, G. A. *Fragilariopsis cylindrus* (Grunow) Krieger: the most abundant diatom in water column assemblages of Antarctic marginal ice edge zones. *Pol. Biol.* **12,** 609–627 (1992).

60. Froneman, J. W., McQuaid, G. D. & Perissinotto, R. Biogeographic structure of the microphytoplankton assemblages of the south Atlantic and Southern Ocean during austral summer. *J. Plankton Res.* **17,** 1791–1802 (1995b).

61. Johansen, J. R. & Fryxell, G. A. The genus *Thalassiosira* (Bacillariophyceae): Studies on species occurring south of the Antarctic Convergence Zone. *Phycologia* **24,** 155–179 (1985).

62. Gregory, T. *Holocene sea ice-ocean-climate variability from Adélie Land, East Antarctica* 237, PhD thesis, Cardiff University (2012).

63. Cunningham, W. L. & Leventer, A. Diatom assemblages in surface sediments of the Ross Sea: relationship to present oceanographic conditions. *Antarct. Sci.* **10,** 134–146 (1998).

64. Sinninghé Damsté, J. S. *et al.* The rise of the Rhizosolenoid diatoms. *Science* **304,** 584–587 (2004).

65. Collins, L. G. *et al.* Evaluating highly branched isoprenoid (HBI) biomarkers as a novel Antarctic sea-ice proxy in deep ocean glacial age sediments. *Quat. Sci. Rev.* **79,** 87–98 (2013).

66. Maddison, E. J., Pike, J. & Dunbar, R. Seasonally laminated diatom-rich sediments from Dumont d'Urville Trough, East Antarctic Margin: Late-Holocene Neoglacial sea-ice conditions. *Holocene* **22,** 857–875 (2012).

67. Jansen, J. H. F., Van der Gaast, S. J., Koster, B. & Vaars, A. J. CORTEX, a shipboard XRF-scanner for element analyses in split sediment cores. *Mar. Geol.* **151,** 143–153 (1998).

68. Cavalieri, D. J., St. Germain, K. & Swift, C. T. Reduction of weather effects in the calculation of sea ice concentration with the DMSP SSM/I. *J. Glaciol.* **41,** 455–464 (1995).

69. Prokoph, A. & El Bilali, H. Cross-Wavelet Analysis: a tool for detection of relationships between Paleoclimate Proxy Records. *Math. Geosci.* **40,** 575–586 (2008).

70. Gong, D. & Wang, S. Definition of Antarctic Oscillation Index. *Geophys. Res. Lett.* **26,** 459–462 (1999).

Acknowledgements

This research was funded by the ERC StG ICEPROXY project (203441), the ANR CLIMICE project and FP7 Past4Future project (243908). P.C. is supported by a CNRS studentship. The French Polar Institute provided logistical support for sediment and data collection (IPEV projects 452 and 1010). This is ESF PolarClimate HOLOCLIP contribution n°24 and Past4Future contribution n°83.

Author contributions

X.C., G.M. and M.N.H. designed the study; P.C. performed diatom and XRF analyses; I.C. performed the biomarker analysis; D.S. performed atmospheric reanalysis; M.N.H. and A.M. extracted daily sea ice concentrations and provided modern oceanographic insights; S.S. performed ^{210}Pb analyses and developed the age model of the core; S.C., E.D. and V.M. helped with satellite images and spectral analyses; and all the authors contributed to the redaction of the manuscript.

Additional information

Global pulses of organic carbon burial in deep-sea sediments during glacial maxima

Olivier Cartapanis[1,2], Daniele Bianchi[2,3,4], Samuel L. Jaccard[1] & Eric D. Galbraith[2,5,6]

The burial of organic carbon in marine sediments removes carbon dioxide from the ocean–atmosphere pool, provides energy to the deep biosphere, and on geological timescales drives the oxygenation of the atmosphere. Here we quantify natural variations in the burial of organic carbon in deep-sea sediments over the last glacial cycle. Using a new data compilation of hundreds of sediment cores, we show that the accumulation rate of organic carbon in the deep sea was consistently higher (50%) during glacial maxima than during interglacials. The spatial pattern and temporal progression of the changes suggest that enhanced nutrient supply to parts of the surface ocean contributed to the glacial burial pulses, with likely additional contributions from more efficient transfer of organic matter to the deep sea and better preservation of organic matter due to reduced oxygen exposure. These results demonstrate a pronounced climate sensitivity for this global carbon cycle sink.

[1] Institute of Geological Sciences and Oeschger Centre for Climate Change Research, University of Bern, 3012 Bern, Switzerland. [2] Department of Earth and Planetary Sciences, McGill University, Montreal, Canada H3A 2A7. [3] School of Oceanography, University of Washington, Seattle, Washington 98105, USA. [4] Department of Atmospheric and Oceanic Sciences, University of California Los Angeles, Los Angeles, California 90095-1565, USA. [5] Institució Catalana de Recerca i Estudis Avancats (ICREA), 08010 Barcelona, Spain. [6] Institut de Ciència i Tecnologia Ambientals and Department of Mathematics, Universitat Autonoma de Barcelona, 08193 Barcelona, Spain. Correspondence and requests for materials should be addressed to O.C. (email: olivier.cartapanis@geo.unibe.ch).

The climate history of the Earth offers a wide range of timescales on which the sensitivity of organic carbon burial to climate might be tested, including the relatively well-documented glacial–interglacial cycles triggered by insolation variations over the quaternary[1]. Much research has focused on changes in the partitioning of carbon between the ocean and atmosphere over these cycles, amplifying orbital forcing by transferring CO_2 from the atmosphere to the ocean during glacial times, and back to the atmosphere during interglacial periods[1-3]. However, interactions between the long-term storage of organic carbon in oceanic sediments and global climate variations on glacial–interglacial timescales have received little attention.

More than 60 years ago, the discovery of high concentrations of organic carbon in deep-sea sediments near the Galapagos Islands led Arrhenius to propose that exposure of the Galapagos Plateau during the low sea level of the last ice age had caused the resuspension and downslope transport of organic-rich coastal sediments[4]. Later, enhanced burial of organic carbon in glacial-age sediments from the nearby equatorial Pacific was interpreted as reflecting increased biological productivity accompanying intensified wind-driven upwelling[5], and more recently, reinterpreted as better preservation under reduced oxygenation[6]. Enhanced burial of organic carbon in glacial sediments was also observed in the equatorial Atlantic over several glacial cycles[7], and at low latitudes for the last glacial cycle[8], and interpreted as the response of primary productivity to either orbitally forced changes in ocean circulation or to changes in trade winds[7,8].

Here we make a quantitative analysis of several hundreds of high-quality sediment records to extend these pioneering observations to the global scale. We focus our analysis on organic carbon burial in deep-sea sediments, given that the deposition of sediments on continental shelves is complicated by pronounced spatial heterogeneity in biological production and sedimentation (see Methods). Our results suggest that organic carbon burial in deep-sea sediments increased during peak glacial conditions, demonstrating the sensitivity of this component of the global carbon cycle to climatic variation.

Results

Organic carbon burial rates during the Last Glacial Maximum.

We estimate global variations of total organic carbon (TOC) mass accumulation rate (MAR) over the past 150 kyr by combining the modern organic carbon burial distribution (see Methods and Supplementary Fig. 1) with time series of TOC MAR from a global compilation of 561 sediment cores extending from the present to the Marine Isotopic Stage 6 (MIS6; Methods). This analysis reveals the existence of robust geographical patterns (Fig. 1a). TOC burial was higher during the Last Glacial Maximum (LGM) in most provinces (Fig. 1a, see Methods and Supplementary Figs 2 and 3), including the tropical and subtropical Atlantic, the eastern and southern Pacific, and the Arabian Sea. The only provinces in which sedimentary TOC burial was lower during the LGM are the high latitudes of the Arctic and Antarctic oceans, the Bering and Okhotsk seas, the California Current, and the Caribbean region. Our reconstruction constrains the global deep-sea TOC MAR during the LGM to between 118 and 171% of the Holocene MAR (Table 1), with a mean estimate of $147 \pm 18\%$. Despite the uncertainties, none of the scenarios that we considered (Table 1, see Methods) allows lower burial during the LGM compared with the Holocene. On the basis of the downcore sediment MAR changes and modern burial map, significantly increased glacial burial occurred in the tropical regions of the Atlantic and eastern Pacific, and the Subantarctic (Fig. 1b). The elevated TOC burial during the LGM

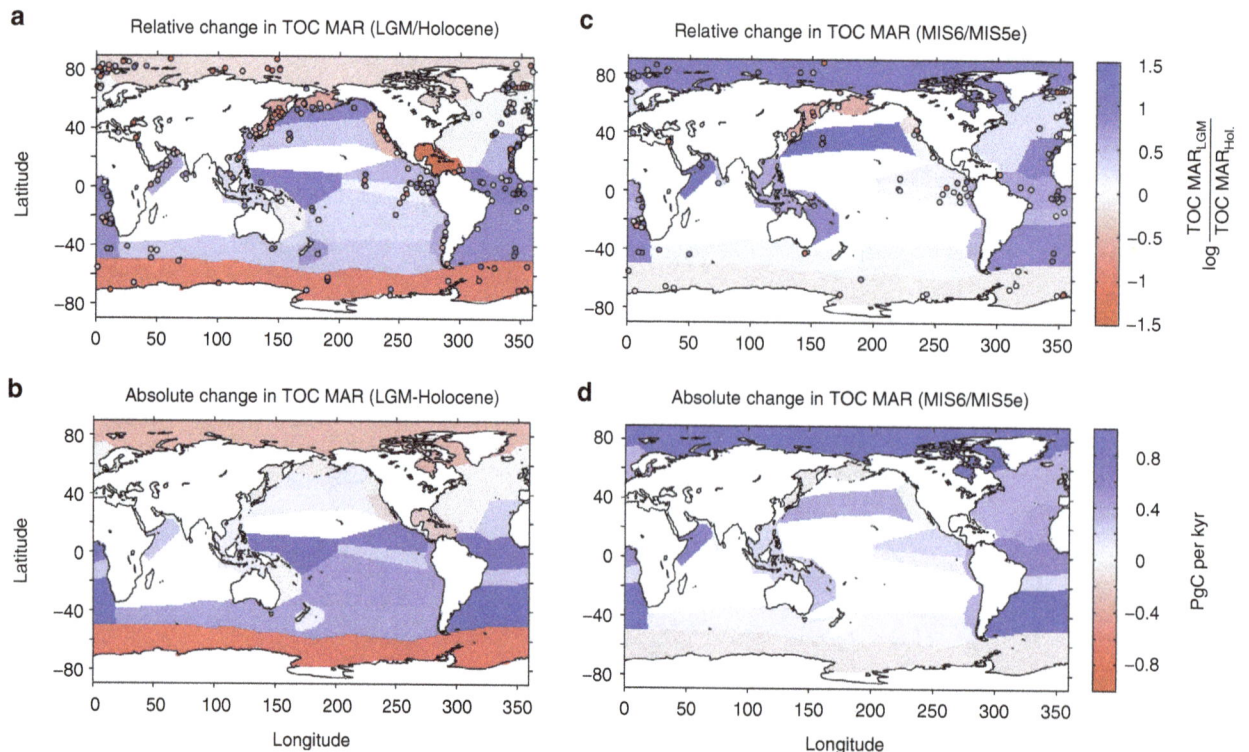

Figure 1 | Relative and absolute changes in deep-sea burial of organic carbon for the two last deglacial transition. Relative (**a,c**) and absolute (**b,d**; PgC per kyr) changes in deep-sea burial of organic carbon from MIS2 to Holocene transition (**a,b**) and MIS6 to MIS5e transition (**c,d**). Individual sedimentary ratios are shown as coloured circles in **a,c**. Shadings correspond to the mean ratio in each province (**a,c**) and to the absolute changes in **b,d**. Note that (**b,d**) show total burial changes in each province, and the absolute changes are not area-normalized.

Table 1 | Changes in organic carbon burial over the two last glacial–interglacial transitions.

Province map	Holocene MAR (PgC per kyr)	LGM/ Holocene (%)	LGM MAR (PgC per kyr)	MIS6/MIS5e (%)	MIS6 MAR (PgC per kyr)	Glacial excess burial (PgC)	Number of provinces
Ocean	17.1	117.7	20.1	152.0	26.0	196	7
Seas	17.1	134.7	23.0	153.2	26.2	242	101
Longhurst (L.)	16.8	159.3	26.8	148.2	24.9	338	56
Modified L. 1	17.1	141.0	24.1	155.6	26.6	262	30
Modified L. 1 + depth.	17.1	147.0	25.1	167.3	28.6	281	60
Modified L. 2	17.1	158.6	27.1	176.0	30.1	435	15
Modified L. 2 + depth.	17.1	170.7	29.2	183.0	31.3	501	30
Mean		147.00	25.06	162.19	27.67	322	
s.d.		17.75	2.99	13.38	2.37	110	

Holocene and LGM deep-ocean TOC burial (MAR) estimated for the different province maps (shown in Supplementary Fig. 2). A subdivision of the provinces following the 1,500 m isobath was added for province maps 5 and 7. The glacial excess burial was calculated between 80 ka, when global burial diverged from MIS5 values, and 10 ka, when global burial reached low Holocene values (Fig. 1).

is due, in similar proportions, to higher sedimentary organic matter concentrations (126 ± 8% of interglacial value), and to higher sedimentation rates (125 ± 15% of interglacial value, see Methods).

Organic carbon burial rates during MIS6. A range of factors could potentially bias our LGM/Holocene TOC MAR estimates, including sediment compaction in cores for which density measurements are not available (which would tend to overestimate Holocene MARs) and diagenetic remineralization of TOC (which would tend to underestimate LGM TOC). These biases should be minimized for the penultimate glacial termination, which is generally found at significantly greater depths below the seafloor, where vertical gradients in compaction and labile TOC are far more subdued. Thus, we performed the same analyses for 135 available records in our database that include the MIS6–MIS5e transition (between 135–143 ka and 119–127 ka). Depending on the province map used (Methods), MIS6 burial corresponded to between 148 and 183% of MIS5e TOC MAR, with a mean value of 162 ± 13% (Table 1). Thus, the MIS6/MIS5e ratio shows similar amplitude to the LGM/Holocene ratio. Considering that the climatic conditions during interglacials (MIS5e and the Holocene) were relatively similar, we assume that carbon burial during MIS5e was the same as that during the modern conditions and calculate the global TOC MAR during MIS6 using the MIS6/5e downcore MAR changes (Table 1, Fig. 1c,d). The distribution of changes is very similar to the LGM/Holocene transition, except for the Arctic (Fig. 1c,d.), where age models are arguably poorly constrained[9].

Organic carbon burial over the past 150 kyr. Finally, we reconstructed a time series of global organic carbon burial throughout the last glacial cycle by applying the same procedure each 1 kyr time step between 0 and 150 ka (see Methods). The most prominent feature of the reconstructed global organic carbon burial rate over the past 150 kyr (Fig. 2) is the increased burial during glacial maxima, regardless of the province selection strategy (Supplementary Fig. 2). It would therefore appear that the global organic carbon pulses reflect a very consistent response to glacial maxima, which could have resulted from any combination of more rapid export of organic matter from the surface mixed layer, more efficient transfer of organic matter from the upper ocean and continental margins to the seafloor, and better preservation of organic matter in the sediment. We estimate the 'excess' removal of C from the system that would have occurred during the last glacial period, above the baseline of interglacial C burial rates, by integrating the C burial between 80 and 10 ka, and

subtracting the baseline interglacial burial. Our estimate suggests that excess burial in the deep sea, that is, that which would have exceeded a constant interglacial burial flux, removed between 200 and 500 PgC from the ocean–atmosphere system (Table 1).

Discussion

In general, higher glacial organic carbon burial occurred in regions where a previous qualitative reconstruction, summarizing diverse proxies[10], inferred higher export production from the surface ocean. The potential importance of increased export is further suggested by similar geographic and temporal patterns between TOC and opal burial (see Methods and Supplementary Fig. 4), which also suggests some degree of increased silicic acid supply to the low latitudes[11], particularly in the Atlantic. Export production is limited by the supply of nitrogen to the mixed layer over most of the ocean, and by iron and/or other factors limiting growth in nitrate-rich regions[12]. The global nitrate inventory may have been larger during the LGM than at present, due to slower denitrification rates, and a potential fertilization by N_2 fixing cyanobacteria enhanced by the supply of iron from glacial dust[13–15]. However, isotopic constraints suggest that the fixed N inventory was not > 50% and likely not > 30% larger than present during the LGM[16,17], which makes it unlikely to have been the sole cause of the observed > 50% increase in TOC MAR. Despite the possibility of a larger nitrate inventory, reconstructions from the LGM show much less nitrate at the surface of the Southern Ocean and subarctic Pacific[18], with low TOC MAR in the coldest parts of these regions, while the high TOC MAR in the Subantarctic is consistent with accelerated export due to dust-borne iron inputs in this region[19,20] (see also Supplementary Fig. 5). Thus, increased export production due to higher dust supply could have contributed to the accelerated burial of organic carbon in presently iron-limited regions, drawing down the available nitrate, while expanded summer sea ice cover and reduced vertical nutrient resupply[21] likely throttled export production in the coldest oceanic realms. Meanwhile, an intensification of wind-driven upwelling could have provided an additional increase of export production by supplying more nutrients to the tropical Atlantic and Pacific[5]. Despite the potential of these multiple nutrient supply mechanisms to explain the glacial burial peak, they cannot obviously explain the lack of a global burial peak during MIS4, when dust flux was high[20] and many other features of the global climate were similar to MIS2 (Supplementary Fig. 5).

The transfer of organic matter from the sunlit surface and continental margins to the ocean floor could also have varied over glacial cycles. The transfer efficiency of organic detritus through

Global TOC MAR in deep sea sediment

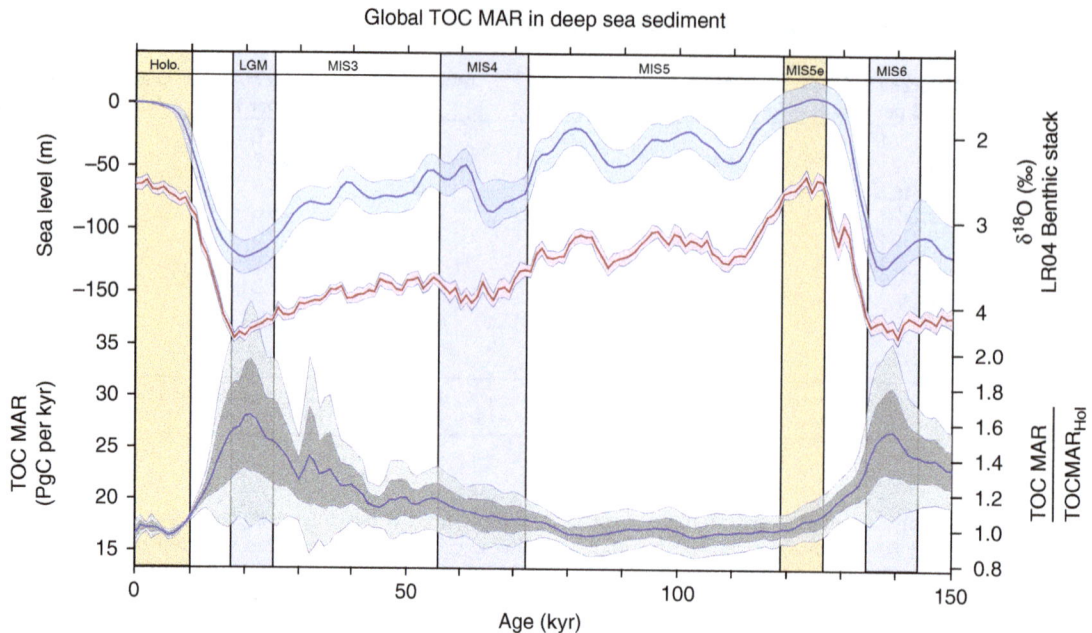

Figure 2 | Global TOC burial over the past 150 kyr. Sea-level reconstruction (and associated confidence interval[50]; LR04 benthic foraminifera δ^{18}O stack[51] (±2 s.e.), and reconstructed global organic carbon burial in deep-sea sediment over the past 150 kyr (PgC per kyr, ±1σ, ±2σ and mean scenario based on the different province scenarios, see Methods, Supplementary Fig. 2 and Supplementary Data 1). The right axis shows relative changes in TOC MAR as compared with Holocene values. The mean organic carbon burial is significantly correlated to both the sea-level reconstruction and the benthic stack (R = 0.85, P << 0.05 and R = 0.77, P << 0.05, respectively). Yellow vertical bands correspond to interglacial periods (Holocene and Marine Isotopic Stage (MIS) 5e), blue vertical bands correspond to glacial condition (Last Glacial Maximum (LGM), MIS4 and MIS6).

the water column depends on the types of organic particles produced by the phytoplankton community, the abundance and behaviour of the heterotrophic community that feeds on the organic matter, and the presence of ballast minerals that protect organic matter from bacterial respiration and/or increase sinking velocities of aggregates[22]. Changes in the transfer efficiency could be consistent with the similar spatial and temporal patterns observed in TOC and opal burial (Supplementary Fig. 4). Colder water temperatures may also have increased the transfer efficiency by reducing the metabolic rates of heterotrophs[11]. In addition, both marine and terrestrial organic matter can be transferred from the coastal zone to the deep sea through downslope transport and nepheloid layers. The exposure of the continental shelf during glacial times favoured direct transfer of coastal and riverine sediment to the deep ocean, bypassing temporary storage and partial remineralisation on shelves, and increasing burial efficiencies[23]. Indeed, terrestrial organic carbon deposited at the mouth of the Amazon River under present-day conditions was transferred almost entirely to the deep-sea fan during the LGM[24,25]. Increased downslope transport of fine-grained lithogenic material during the glacial sea-level lowstand could also help to explain the increased sedimentation rates evidenced in this study (Methods), and over most major deep-sea fan systems[26]. However, the glacial burial peaks do not appear to have been amplified closer to coasts, as might be expected were downslope transport of shelf sediments the main driver of the global burial pulses (see Methods, and Supplementary Figs 6, 7 and 8).

The third candidate explanation involves enhanced preservation of organic matter in glacial sediment. Increased transport of fine-grained material to the deep sea could have favoured organic matter preservation, which is enhanced with increasing surface area of lithogenic sediment[23]. In addition, organic matter preservation on the seafloor is related to oxygen exposure time[27], a function of sedimentation rates and bottom water oxygenation. Higher bulk accumulation rates during glacial

maxima, potentially due to enhanced dust flux, continental erosion by ice sheets and/or erosion of exposed shelves, would have reduced the oxygen exposure time. Furthermore, proxy reconstructions have shown that the global deep ocean was less oxygenated during the LGM[28,29]. A postulated decrease in the proportion of well-ventilated North Atlantic Deep water[30,31] could have contributed to reduced oxygenation of the deep Atlantic[28] (see Supplementary Fig. 5), and enhanced the burial of organic matter during glacial intervals[32]. However, changes in oxygen exposure time cannot explain covariations between opal and TOC burial (see Supplementary Fig. 4), and the non-linear relationship between oxygen exposure time and burial efficiency[27] suggests that this process may have less leverage in the deep sea, where sedimentation rates are low. It is possible that both enhanced TOC and opal burial reflect better preservation due to increased bulk sedimentation rates[33,34].

The glacial burial pulses documented here, and their correlation with sea level and global ice volume (Fig. 2), are evidence of a striking sensitivity of deep-sea organic carbon burial to climate. The excess burial of 200–500 PgC of organic carbon in deep-sea sediment during glacials is comparable to the excess carbon stored below ice sheets[35], or within soil, permafrost and peat deposits[36] during glacials. While our results constrain the deep-sea organic carbon sink during glacials, a closure of the long-term carbon budget in the ocean will require an estimate of deep-sea calcium carbonate burial, as well as burial on shelves. In addition, the fact that a single major deep-sea fan off of the Amazon basin accounts for a significant part of glacial organic matter burial points to the need for studies specifically targeting deep-sea fans and slopes.

Three general factors have been presented here that could have contributed to the burial pulses, acting individually or together. The apparent lack of an organic carbon burial peak during MIS4, an interval in many respects similar to MIS2 and MIS6, is an intriguing observation that deserves further attention. Disentangling the interwoven causes behind the burial pulses provides an important challenge for understanding the global carbon cycle, and its variations through Earth history.

Methods

Analytical strategy overview. Our estimate for carbon burial involves three consecutive steps. First, we use a global map of organic carbon burial in modern (core-top) sediments, based on ref. 37 (Supplementary Fig. 1). Second, in order to infer regional changes in burial patterns from individual sediment cores while accounting for the geographic and bathymetric variability in global organic carbon burial, we subdivide the world ocean into different sets of geographical provinces. Given that any such division will introduce inherent biases, we used a total of seven different subdivision strategies (Supplementary Fig. 2), and compare the results, as a test of the degree to which the results depend on the selection of provinces, and to evaluate uncertainties. Third, we use quality-controlled sediment records to reconstruct relative changes, as compared with the Holocene, of the burial rate in each province. These composite time series are then used, in combination with modern estimates of burial rates, to calculate absolute changes in the burial rates in each province.

Modern TOC MAR and associated uncertainties. The modern organic carbon MAR map (Supplementary Fig. 1) was generated by multiplying the TOC concentrations reported for surface sediments (core tops) by the corresponding bulk sediment accumulation rate. The TOC content map was generated by combining the map reported by ref. 38, augmented with unpublished data, while the MAR map was determined based on the geometric mean of two pre-existing maps (ref. 39 and other unpublished data, see details in ref. 37). It is important to note that both the MAR and the TOC maps have insufficient spatial resolution to adequately resolve continental margins. Even though a substantial portion of organic carbon burial may occur in coastal environments (up to 90% (ref. 40)), our maps with a 1×1 degree spatial resolution cannot resolve small-scale features such as shorelines, shelves and deep-sea fans. Uncertainties related to the modern burial estimates may have two impacts on our results:

(1) The absolute values of the modern burial conditions and the burial values inferred for the past. The modern map that we use is consistent with the most commonly cited value for deep-ocean organic carbon burial[41], but other studies have reported values that diverge by orders of magnitude (see details and references in ref. 42). For this reason, we reported relative changes in Table 1, based on the downcore records only. Similarly, the right y axis on the bottom panel of Fig. 2 shows relative changes as compared with the Holocene, rather than absolute values.

(2) A different spatial distribution of the modern burial would change the relative contribution of the provinces to the global budget, altering both the pattern and absolute value of the reconstructed global TOC burial. To test the impact of a modern map with TOC burial focused on continental slopes and coastal environments, we used the global map of modelled organic carbon flux to the seafloor from ref. 43. This map shows a similar pattern in the deep sea, but orders of magnitude higher flux in coastal regions compared with the map used in our study (note that the flux of TOC to the seafloor differs from burial, because of organic carbon remineralization on the seafloor; Supplementary Fig. 9, blue axis). Following exactly the same approach, but using this alternative map as modern reference, did not substantially modify the shape and amplitude of the relative global TOC burial changes (Supplementary Fig. 9, blue lines), as compared with the reference maps (Supplementary Fig. 9, red lines).

We further tested the influence of the modern burial map on our results, using an updated TOC content map, and only the Jahnke bulk MAR map[39] (Supplementary Fig. 9, green lines). Once again, the relative changes in global burial obtained were very similar to the one presented in this study, suggesting that our reconstruction is robust and relatively independent on our starting assumptions in terms of variability.

Because paleoceanographic studies favour using sedimentary archives that are continuous, and hence not perturbed by abrupt sedimentary processes such as mass flows or turbidity currents, or by hiatuses in sedimentation rates, the coring sites are generally expected to have been selected to avoid such processes. However, sedimentation over slopes and deep-sea fans probably occurs mostly as massive and sporadic, yet very localized events. Meanwhile, some portions of the ocean floor probably see no sediment deposition at all, or even erosion. These types of sedimentary environments are likely under sampled by the sediment cores used in this study.

Thus, our reconstruction applies mainly for deep-sea sediment, poorly resolving coastal sediment deposits. No MAR map is available for the Mediterranean Sea to our knowledge, and as a consequence this region was not considered further.

Province definition strategies. The first two strategies subdivide the global ocean into the major ocean basins (Supplementary Fig. 2a), as well as finer subdivisions of the oceans into seas (Supplementary Fig. 2b), based on ref. 44. Next, based on the assumption that the large-scale ocean features driving biogeochemical cycles in the modern ocean have remained relatively stationary over time, we used the annual climatology of the ocean biogeochemical provinces based on ref. 45, later updated by ref. 46. This map defines 56 coherent provinces from a biogeochemical

perspective (Supplementary Fig. 2c). To adapt the subdivision of the ocean to our sample distribution, the Longhurst map was redefined and simplified into two different maps with, respectively, 30 (Supplementary Fig. 2d) and 15 different provinces (Supplementary Fig. 2e). To investigate potential influences of the depth of the record, these two simplified maps were used to create two new sets of maps by dividing each province into a shallow and a deep component using the 1,500 m isobaths (Supplementary Fig. 2f,g).

Extracting burial rate changes from sediment core database. First, we created a database comprising surface and downcore sediment composition, retrieving available data from the NOAA (ftp://ftp.ncdc.noaa.gov/pub/data/paleo/paleocean/sediment_files/) and PANGAEA (http://www.pangaea.de) online repositories (Supplementary Data 2). Thus, any data set used in this study is available online in one of these repositories. All the data sets that contained TOC concentrations, TOC accumulation rates, age models, sediment density values, along other parameters were taken into consideration. Given that some records are reported with their original depths, while others can be reported with composite or corrected depth scales (from drilling disturbance, voids, or instantaneous deposits such as turbidites or tephras), we first verified the internal consistency of the depth scale, by comparing similar proxies from different records from the same core. When more than one combination between TOC content, age model and density was possible, the best one was selected. The objective criterion used to select the best combination was as follows: outliers detected visually from the cloud of different solutions were removed. More recent versions of age models were favoured over older versions. Age models with a high number of tie points and calibrated ^{14}C measurements were preferred, and a composite age scale was created for some of the records, using least square splines.

230-Thorium normalized sediment accumulation rates were used whenever available. If they were not available (the majority of records), sedimentation rates were inferred at the depth of each TOC measurement by calculating the linear sedimentation rate implied by the age model. We used measured dry bulk densities, when available, interpolated onto the TOC record sampling depth, to convert linear accumulation rate to MAR. When dry bulk densities were not directly available, we assumed constant values equal to $0.9\,g\,cm^{-3}$, corresponding to the mean dry bulk density for compacted marine sediments in our database.

Each sedimentary record of organic carbon accumulation was expressed as a ratio to the mean Holocene value of the same record. The mean Holocene to LGM ratio of sedimentary records was calculated for each province (Fig. 1a), and used to estimate LGM TOC burial by multiplying the province ratio with the corresponding province modern burial value (Fig. 1b, Table 1). The LGM global burial rate was then obtained by summing the inferred burial in each province (Table 1). Note that we used the geometric, rather than the arithmetic mean of LGM to Holocene ratios, given that the values are log-normally distributed.

Our database includes 561 total TOC MAR time series. Of these, 260 were of sufficient resolution and length for both the Holocene (defined as 0–10 ka; 454 records), and the LGM (18–25 ka; 303 records) to calculate average LGM to Holocene TOC MAR ratios (coloured circles, Fig. 1a). The provinces for which no downcore records are available account for a small fraction (0 to 23%, depending on the province strategy) of modern global burial. Significant variations of sedimentary LGM/Holocene ratios occur within a single province, indicating local variability in the factors affecting the production, transfer and sedimentation of organic matter (Supplementary Fig. 3).

The same procedure was used to calculate global TOC burial rate over the past 150 kyr with a 1 kyr time step. We used interpolated downcore MAR records with a 1 kyr time step, expressed as a fraction of Holocene MAR (0–10 kyr). Any time interval within a sediment record in which no measurement was present over a period > 10 kyr was excluded from the calculation of the regional stacks. When no record was available for a specific province, and/or for a specific period, the province flux was assumed to be constant and equal to modern values.

To evaluate the relative influence of organic carbon content and sedimentation rate variations on the results, we also calculated global burial assuming (1) constant sedimentation rates and (2) constant TOC content. This test indicates that the more elevated TOC burial during the LGM is due, in similar proportions, to higher sedimentary organic matter concentrations (126 ± 8% of interglacial value) and to higher sedimentation rates (125 ± 15% of interglacial value).

Given that the deglacial trend in organic matter burial in the Arctic is uncertain, and shows opposite behaviour from MIS6 to MIS5e, and LGM to Holocene transitions, we excluded the Arctic Ocean. As only a small proportion of global burial currently occurs in the Arctic Ocean (<8%), this exclusion has only a very minor impact on the results in terms of timing and amplitude.

Long-term trend in global organic matter burial. The mean of the different scenarios shows a pronounced increasing trend towards the present (Supplementary Fig. 10). There are different explanations to account for that trend, which are not related to actual changes. Given that density measurements were not always available, we performed our calculation assuming a constant density for the entire core. The consequence of this assumption is that the calculated MARs are slightly overestimated for the most recent sediments, for which the density is expected to be lower than for older sediments, because of sediment compaction. This can also partly explain the increasing trend during the late Holocene (Supplementary Figs 5 and 10).

In addition, the slow long-term diagenesis of refractory organic carbon contributes to the gradual decrease of apparent burial fluxes with age.

Another source of uncertainty arises from the use of raw ^{14}C ages to derive some sediment core-age models. Records for which only raw ^{14}C ages were available should slightly overestimate the most recent calculated TOC MAR, due to the difference between calendar and ^{14}C ages. Thus, these records can slightly reduce the LGM to Holocene amplitude, but cannot explain the decreasing trend. This is confirmed by similar patterns between MIS6 and MIS5, out of the range covered by ^{14}C-dating.

Finally, apparent sedimentation rates decrease as a power law function of the intervals between two age control[47], indicating that sedimentation rates could be overestimated during the Holocene and the deglaciation, when the age model resolution is generally higher, as compared with older periods.

To correct for these biases, we removed the long-term trend from each scenario (Supplementary Fig. 10) calculated using a least-squares spline modelling tool (MATLAB Shape Language Modeling toolbox, D' Errico, 2009, MATLAB Central File Exchange, retrieved online on February 2012). It is important to note that this correction implicitly assumes similar burial rates for the Holocene and MIS5.

Biogenic opal burial variations. To evaluate the changes in opal burial over the past 150 kyr, we applied the same method described in this paper for TOC burial to biogenic opal records available in our database. The number of sedimentary records for opal burial was lower than for TOC. However, we obtained remarkably similar results from a spatial and temporal point of view, in provinces with a sufficient number of records (Supplementary Fig. 4).

Impact of sea-level change on sediment redistribution. As sea level dropped during glacial periods, continental shelves were exposed and eroded, activating submarine canyons and rerouting coastal deposits directly to the deep sea[48]. Indeed, the present-day deposition of terrestrial organic carbon at the mouth of the Amazon River was transferred almost entirely to the deep-sea fan during the LGM, representing 3.7 PgC per kyr[24], or >13% of global LGM deep-sea burial rate (Table 1). Although this particular hot spot in marine accumulation rate is not resolved in our analysis, it is possible that downslope transport of organic matter does contribute to some continental slope records included in our database. Despite its importance, this mechanism is unlikely to explain more than a fraction of our reconstructed changes in TOC MAR, given that many of the records in our database are far from continental slopes. It is worth noting that erosion of coastal deposits during sea-level lowstands may have enhanced nutrient availability and productivity of the ocean[48,49], while increased organic matter reaching the seafloor may have contributed to reduced oxygenation of the deep sea[48].

To further test the potential influence of coastal deposit remobilization during sea-level lowstands, we performed additional analyses, following exactly the same procedure as outline above, using newly designed province maps for which we distinguished coastal and open ocean regions. Assuming that glacial–interglacial sediment remobilization/relocation was more important near shorelines, we used the ocean and seas maps, as well as the two simplified Longhurst maps (Supplementary Fig. 6) and split each province into a coastal and an open ocean province. Coastal regions were defined using distance thresholds of 500, 1,000 and 1,500 km from the closest point on the coastline (Supplementary Fig. 6). Changes of global TOC burial based on these nine new province maps were similar to the previous analyses, in terms of shape and absolute value (Supplementary Figs 6 and 7). Moreover, the burial in open ocean and in coastal regions displays similar patterns to the whole deep ocean regardless of the width of the coastal provinces, suggesting similar temporal patterns in both open ocean and coastal regions (Supplementary Fig. 7). Finally, we calculated the distance from the shelf, here defined as the 150 m isobath, for each individual sediment core (Supplementary Fig. 8). The absence of correlations between the MAR change between the Holocene and the LGM, and the distance from the shelf, further suggests that regional patterns of burial dominate over continental influences.

References

1. Jouzel, J. et al. Orbital and millennial Antarctic climate variability over the past 800,000 years. *Science* **317**, 793–796 (2007).
2. Broecker, W. S. Ocean chemistry during glacial time. *Geochim. Cosmochim. Acta* **46**, 1689–1705 (1982).
3. Petit, J. R. et al. Climate and atmospheric history of the past 420,000 years from the Vostok ice core, Antarctica. *Nature* **399**, 429–436 (1999).
4. Arrhenius, G., Nyberg, A., Blomqvist, N. & Svenska, D. *Sediment cores from the East Pacific* (Elanders boktryckeri aktiebolag, 1952).
5. Pedersen, T. F. Increased productivity in the eastern equatorial Pacific during the last glacial maximum (19,000 to 14,000 yr B.P). *Geology* **11**, 16–19 (1983).
6. Bradtmiller, L. I., Anderson, R. F., Sachs, J. P. & Fleisher, M. Q. A deeper respired carbon pool in the glacial equatorial Pacific Ocean. *Earth Planet Sci. Lett.* **299**, 417–425 (2010).
7. Lyle, M. Climatically forced organic carbon burial in equatorial Atlantic and Pacific Oceans. *Nature* **335**, 529–532 (1988).
8. Sarnthein, M., Winn, K., Duplessy, J.-C. & Fontugne, M. R. Global variations of surface ocean productivity in low and mid latitudes: Influence on CO_2 reservoirs of the deep ocean and atmosphere during the last 21,000 years. *Paleoceanography* **3**, 361–399 (1988).
9. Sundby, B., Lecroart, P., Anschutz, P., Katsev, S. & Mucci, A. When deep diagenesis in Arctic Ocean sediments compromises manganese-based geochronology. *Mar. Geol.* **366**, 62–68 (2015).
10. Kohfeld, K. E., Quéré, C. L., Harrison, S. P. & Anderson, R. F. Role of marine biology in glacial-interglacial CO_2 cycles. *Science* **308**, 74–78 (2005).
11. Matsumoto, K. Biology-mediated temperature control on atmospheric pCO_2 and ocean biogeochemistry. *Geophys. Res. Lett.* **34**, L20605 (2007).
12. Moore, C. M. et al. Processes and patterns of oceanic nutrient limitation. *Nat. Geosci.* **6**, 701–710 (2013).
13. Galbraith, E. D., Kienast, M., Albuquerque, A. L., Altabet, M. A. & Batista, F. et al. The acceleration of oceanic denitrification during deglacial warming. *Nat. Geosci.* **6**, 579–584 (2013).
14. Ganeshram, R. S., Pedersen, T. F., Calvert, S. E. & Murray, J. W. Large changes in oceanic nutrient inventories from glacial to interglacial periods. *Nature* **376**, 755–758 (1995).
15. Falkowski, P. G. Evolution of the nitrogen cycle and its influence on the biological sequestration of CO_2 in the ocean. *Nature* **387**, 272–275 (1997).
16. Deutsch, C., Sigman, D. M., Thunell, R. C., Meckler, A. N. & Haug, G. H. Isotopic constraints on glacial/interglacial changes in the oceanic nitrogen budget. *Global. Biogeochem. Cycles.* **18**, GB4012 (2004).
17. Eugster, O., Gruber, N., Deutsch, C., Jaccard, S. L. & Payne, M. R. The dynamics of the marine nitrogen cycle across the last deglaciation. *Paleoceanography* **28**, 116–129 (2013).
18. Galbraith, E. D. & Jaccard, S. L. Deglacial weakening of the oceanic soft tissue pump: global constraints from sedimentary nitrogen isotopes and oxygenation proxies. *Quat. Sci. Rev.* **109**, 38–48 (2015).
19. Martínez García, A. et al. Southern Ocean dust-climate coupling over the past four million years. *Nature* **476**, 312–315 (2011).
20. Winckler, G., Anderson, R. F., Fleisher, M. Q., McGee, D. & Mahowald, N. Covariant Glacial-Interglacial Dust Fluxes in the Equatorial Pacific and Antarctica. *Science* **320**, 93–96 (2008).
21. Sigman, D. M., Hain, M. P. & Haug, G. H. The polar ocean and glacial cycles in atmospheric CO_2 concentration. *Nature* **466**, 47–55 (2010).
22. Klaas, C. & Archer, D. E. Association of sinking organic matter with various types of mineral ballast in the deep sea: Implications for the rain ratio. *Global. Biogeochem. Cycles.* **16**, 1116 (2002).
23. Keil, R., Tsamakis, E., Wolf, N., Hedges, J. & Goñi, M. in *Proceedings of the Ocean Drilling Program. Scientific results* (1997).
24. Schlünz, B., Schneider, R. R., Müller, P. J., Showers, W. J. & Wefer, G. Terrestrial organic carbon accumulation on the Amazon deep sea fan during the last glacial sea level low stand. *Chem. Geol.* **159**, 263–281 (1999).
25. Goñi, M. A. in *Proceedings of the Ocean Drilling Program, Scientific Results* (1997).
26. Covault, J. A. & Graham, S. A. Submarine fans at all sea-level stands: Tectono-morphologic and climatic controls on terrigenous sediment delivery to the deep sea. *Geology* **38**, 939–942 (2010).
27. Hartnett, H. E., Keil, R. G., Hedges, J. I. & Devol, A. H. Influence of oxygen exposure time on organic carbon preservation in continental margin sediments. *Nature* **391**, 572–574 (1998).
28. Hoogakker, B. A. A., Elderfield, H., Schmiedl, G., McCave, I. N. & Rickaby, R. E. M. Glacial-interglacial changes in bottom-water oxygen content on the Portuguese margin. *Nat. Geosci.* **8**, 40–43 (2015).
29. Jaccard, S. L. & Galbraith, E. D. Large climate-driven changes of oceanic oxygen concentrations during the last deglaciation. *Nat. Geosci.* **5**, 151–156 (2012).
30. Böhm, E. et al. Strong and deep Atlantic meridional overturning circulation during the last glacial cycle. *Nature* **517**, 73–76 (2015).
31. Jonkers, L. et al. Deep circulation changes in the central South Atlantic during the past 145 kyrs reflected in a combined ^{231}Pa/^{230}Th, Neodymium isotope and benthic ^{13}C record. *Earth Planet Sci. Lett.* **419**, 14–21 (2015).
32. Koho, K. A. et al. Microbial bioavailability regulates organic matter preservation in marine sediments. *Biogeosciences* **10**, 1131–1141 (2013).
33. Ragueneau, O. et al. A review of the Si cycle in the modern ocean: recent progress and missing gaps in the application of biogenic opal as a paleoproductivity proxy. *Global Planet. Change* **26**, 317–365 (2000).
34. Sayles, F. L., Martin, W. R., Chase, Z. & Anderson, R. F. Benthic remineralization and burial of biogenic SiO_2, $CaCO_3$, organic carbon, and detrital material in the Southern Ocean along a transect at 170°West. *Deep-Sea Res. Part II* **48**, 4323–4383 (2001).
35. Zeng, N. Glacial-interglacial atmospheric CO_2 change —The glacial burial hypothesis. *Adv. Atmos. Sci.* **20**, 677–693 (2003).
36. Ciais, P. et al. Large inert carbon pool in the terrestrial biosphere during the Last Glacial Maximum. *Nat. Geosci.* **5**, 74–79 (2012).
37. Dunne, J. P., Sarmiento, J. L. & Gnanadesikan, A. A synthesis of global particle export from the surface ocean and cycling through the ocean interior and on the seafloor. *Global Biogeochem. Cycles* **21**, GB4006 (2007).

38. Seiter, K., Hensen, C., Schröter, J. & Zabel, M. Organic carbon content in surface sediments—defining regional provinces. *Deep-Sea Res. Part I* **51**, 2001–2026 (2004).

39. Jahnke, R. A. The global ocean flux of particulate organic carbon: Areal distribution and magnitude. *Global Biogeochem. Cycles* **10**, 71–88 (1996).

40. Sarmiento, J. L. & Gruber, N. *Ocean Biogeochemical Dynamics* (Princeton Univ. Press, 2006).

41. Burdige, D. J. Burial of terrestrial organic matter in marine sediments: A re-assessment. *Global Biogeochem. Cycles* **19**, 4, Gb4011 (2005).

42. Burdige, D. J. Preservation of organic matter in marine sediments: controls, mechanisms, and an imbalance in sediment organic carbon budgets? *Chem. Rev.* **107**, 467–485 (2007).

43. Dunne, J. P., Hales, B. & Toggweiler, J. R. Global calcite cycling constrained by sediment preservation controls. *Global Biogeochem. Cycles* **26**, GB3023 (2012).

44. International Hydrographic Organization. *Limits of Oceans and Seas*. Special Publication 23 (International Hydrographic Organization, 1953).

45. Longhurst, A. Seasonal cycles of pelagic production and consumption. *Prog. Oceanogr.* **36**, 77–167 (1995).

46. Reygondeau, G. *et al.* Dynamic biogeochemical provinces in the global ocean. *Global Biogeochem. Cycles* **27**, 1046–1058 (2013).

47. Schumer, R. & Jerolmack, D. J. Real and apparent changes in sediment deposition rates through time. *J Geophys Res.* **114**, F00A06 (2009).

48. Tsandev, I., Rabouille, C., Slomp, C. P. & Van Cappellen, P. Shelf erosion and submarine river canyons: implications for deep-sea oxygenation and ocean productivity during glaciation. *Biogeosciences* **7**, 1973–1982 (2010).

49. Menviel, L., Joos, F. & Ritz, S. P. Simulating atmospheric CO_2, ^{13}C and the marine carbon cycle during the Last Glacial–Interglacial cycle: possible role for a deepening of the mean remineralization depth and an increase in the oceanic nutrient inventory. *Quat. Sci. Rev.* **56**, 46–68 (2012).

50. Waelbroeck, C. *et al.* Sea-level and deep water temperature changes derived from benthic foraminifera isotopic records. *Quat. Sci. Rev.* **21**, 295–305 (2002).

51. Lisiecki, L. E. & Raymo, M. E. A Pliocene-Pleistocene stack of 57 globally distributed benthic $\delta^{18}O$ records. *Paleoceanography* **20**, PA1003 (2005).

Acknowledgements

We thank John Dunne for providing the modern burial flux maps, and three anonymous reviewers for insightful comments, which helped improve the quality of the manuscript. O.C. and S.L.J. were funded by the Swiss National Science Foundation (SNF grant PP00P2_144811). The Canadian Institute for Advanced Research (CIFAR), the Canadian Foundation for Innovation (CFI), and the Natural Sciences and Engineering Research Council (NSERC) supported O.C., D.B. and E.D.G. D.B. acknowledges funding from University of Washington.

Author contributions

O.C., D.B. and E.D.G. designed the proxy database and the data analysis. O.C., D.B., E.D.G. and S.L.J. wrote the manuscript. All authors contributed to the interpretation of the results.

Additional information

21

The absence of an Atlantic imprint on the multidecadal variability of wintertime European temperature

Ayako Yamamoto[1] & Jaime B. Palter[1,2]

Northern Hemisphere climate responds sensitively to multidecadal variability in North Atlantic sea surface temperature (SST). It is therefore surprising that an imprint of such variability is conspicuously absent in wintertime western European temperature, despite that Europe's climate is strongly influenced by its neighbouring ocean, where multidecadal variability in basin-average SST persists in all seasons. Here we trace the cause of this missing imprint to a dynamic anomaly of the atmospheric circulation that masks its thermodynamic response to SST anomalies. Specifically, differences in the pathways Lagrangian particles take to Europe during anomalous SST winters suppress the expected fluctuations in air–sea heat exchange accumulated along those trajectories. Because decadal variability in North Atlantic-average SST may be driven partly by the Atlantic Meridional Overturning Circulation (AMOC), the atmosphere's dynamical adjustment to this mode of variability may have important implications for the European wintertime temperature response to a projected twenty-first century AMOC decline.

[1] Department of Atmospheric and Oceanic Sciences, McGill University, 805 Sherbrooke Street West, Montreal, Quebec, Canada H3A 2K6. [2] Graduate School of Oceanography, University of Rhode Island, Narragansett Bay Campus, Narragansett, Rhode Island 02882, USA. Correspondence and requests for materials should be addressed to A.Y. (email: ayako.yamamoto@mail.mcgill.ca).

Large-scale, multidecadal variability in North Atlantic sea surface temperature (SST), frequently referred to as the Atlantic Multidecadal Oscillation (AMO), is a prominent feature of Northern Hemisphere climate[1,2]: Sahel drought[3], Atlantic hurricanes[4], large-scale atmospheric circulation[2,5–7] and summertime European temperature and precipitation[8,9] all respond sensitively to this low-frequency variability in North Atlantic SST. A number of studies suggest that the cause of this SST oscillation is internal variation in ocean heat transport, possibly related to the Atlantic Meridional Overturning Circulation (AMOC) variability[10–12], with the role of external and/or atmospheric stochastic forcing provoking recent controversy[13–16]. Evidence in support of the AMO variability being driven internally comes in the form of proxy evidence of a persistent oscillation throughout the past 8,000 years[17] and the reconstruction of this mode of variability in a number of modelling studies, even in the absence of external forcing[18–20].

It is well known that the North Atlantic strongly influences western European climate, with the most obvious manifestation being the anomalous wintertime warmth of the region relative to the zonal mean at equivalent latitudes[21]. Moreover, a recent study showed that temporal variability in western European wintertime temperature is set largely by the size of the air–sea turbulent fluxes along the trajectories of Lagrangian air parcels en route to Europe[22]. Coupled with evidence that variability in air–sea heat fluxes over the Atlantic is controlled by the ocean on decadal and longer time scales[23], it is natural to expect that decadal, basin-scale SST fluctuations should translate to variability in European temperature. Indeed, in all seasons besides winter, the imprint of the AMO is evident in European temperature[8,9]. The SST anomaly associated with the AMO persists throughout the year[5,10], making the absence of a wintertime AMO signal in western Europe all the more puzzling (Fig. 1a).

We propose here that the AMO is closely associated with variability in the position and strength of the storm track, which suppresses the influence of the anomalous SST on the heat fluxes seen by Lagrangian parcels transiting to Europe. To evaluate this possibility, we examine the National Centers for Environmental Prediction/National Center for Atmospheric Research 20th-Century Reanalysis (20CR)[24] from complementary Eulerian and Lagrangian perspectives. The 20CR is one of the longest reanalysis products currently available, and is faithful with independent observations of northeast Atlantic storminess from 1940 onward[25]. Thus, we analyse storm track and Lagrangian pathway variability in the period from 1940 to 2011, which encompasses one full AMO cycle.

Results

Eulerian perspective. The AMO index, computed by taking an area-weighted mean of the linearly detrended SST field over the North Atlantic[9,26], is generally positive from 1940 to 1963 and 1996 to 2011 and negative from 1966 to 1994 (Fig. 1a). There are various approaches to defining an AMO index[27,28], yet the main features remain almost identical to those identified here regardless of the method chosen[18,20,27]. In particular, both modelling[6,10,19] and observational studies[2,6,9,11], utilizing different methodologies to isolate only the internal mode of variability, show spatial patterns of SST anomalies similar to that shown in Fig. 2a, and the approximate timing of transitions between AMO phases is not sensitive to its definition[28].

During each multidecadal period characterized by a given phase of the AMO index, short-term variability gives rise to months in which the basin-averaged SST anomaly is near zero or of the opposite sign relative to the decade in which it is embedded (Fig. 1a). To expose the association of the atmospheric anomaly

pattern with SST anomalies, we made composite periods using only the January months with the most extreme AMO index. The extreme AMO months are chosen such that they meet the criteria, $|AMO index| > 0.15$, which is nearly one standard deviation beyond zero (Fig. 1a). All of these extreme months fall within a longer period where the 10-year running mean AMO index has the appropriate sign. In this manner, 17 positive and 18 negative AMO January months are selected (Fig. 1a). We repeated the analysis using all years within the corresponding AMO phase and present the results in the Supplementary Materials, where it is apparent that the key results and interpretation are essentially unchanged, although their statistical power is slightly weaker.

The large-scale atmospheric flow varies with the AMO index (Fig. 2b). The difference in the 500-hPa geopotential height (Z500) field, which is analogous to streamlines, shows that the direction of winds arriving in western Europe changes between the two AMO phases: winds are more northerly during the anomalous AMO-positive years, whereas they are more zonal during the AMO-negative years (Fig. 2b). The more tightly spaced isohypses during the AMO-negative years indicate a swifter flow relative to the AMO-positive years. Accordingly, the AMO-negative years see an elongated and more zonal January storm track (Supplementary Fig. 1), which is consistent with results from a free-running climate model[7]. Composite Z500 maps constructed with more complete sampling of the longer decadal periods associated with the AMO show similar, albeit weaker, anomaly patterns (Supplementary Fig. 2a).

Lagrangian perspective. The impact of the modulation of the large-scale atmospheric flow on temperature in western Europe is best evaluated in a Lagrangian framework, where the dynamic variability of the atmosphere and the variability in the air–sea turbulent exchange can be assessed simultaneously for the atmospheric particles that influence Europe. Therefore, we launch virtual Lagrangian particles from the surface of forty-one uniformly distributed points over land in western Europe (Fig. 1b) and track them backward in time for 10 days using the atmospheric dispersion model, FLEXPART[29] (see Methods for details). Ten atmospheric particles are released twice a day in January from 1940 to 2011 from each of the forty-one release points. A climatological two-dimensional histogram of the positions of the resulting Lagrangian particles is shown in Fig. 3a, in which their trajectories are seen spreading out over the North Atlantic, many stretching back to the Labrador Sea and northern Canada. The statistical significance of these results is strengthened when separating the particle launch locations into northern and southern sub-regions of western Europe (see Supplementary Fig. 3 for details).

Our previous work showed that air–sea turbulent fluxes at the base of the atmospheric planetary boundary layer (PBL) govern variability in the potential temperature change along these particle trajectories in January almost entirely: turbulent fluxes alone explain more than 80% of the variability in the potential temperature change along 10-day back trajectories from western Europe[22]. Although the fluxes through the top of the planetary boundary layer are important for closing the heat budget of the layer, they are not crucial for understanding the low-frequency variability of the temperature tendency along Lagrangian trajectories. Therefore, we track the ocean-atmosphere turbulent fluxes along each particle's trajectory by interpolating these fluxes from 20CR to each particle's hourly position when the particle is within PBL. The turbulent fluxes are a function of the temperature and moisture gradients at the air–sea interface and the surface wind speed (see Methods for details).

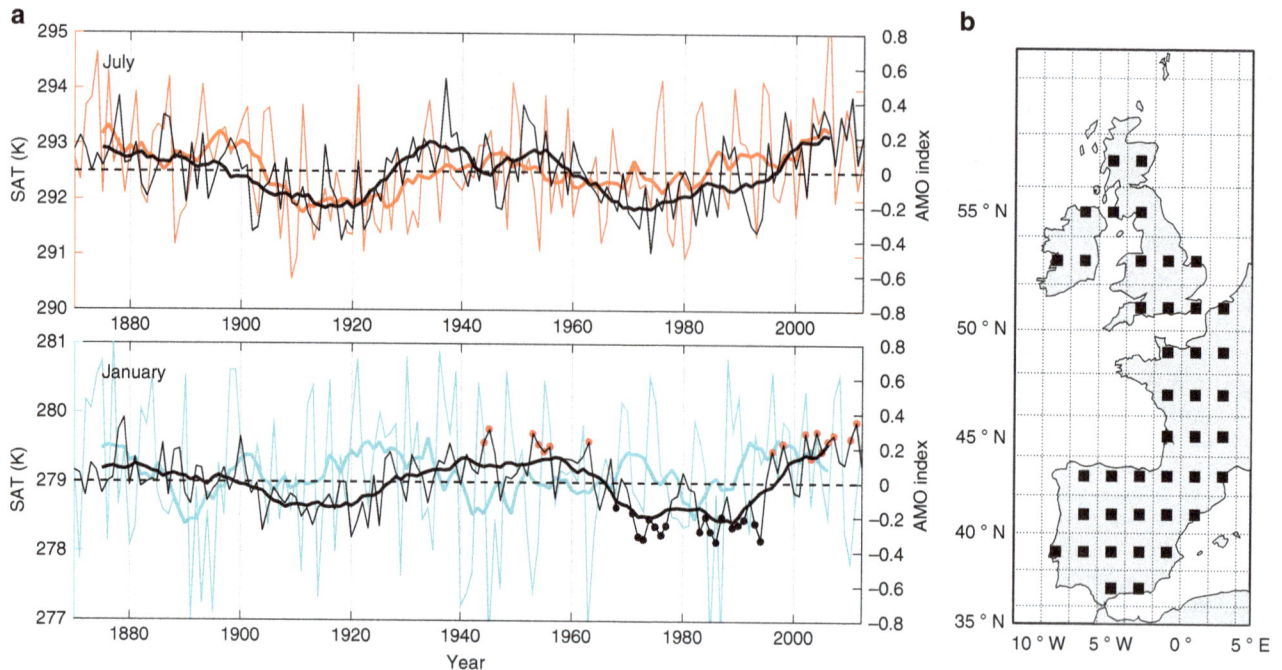

Figure 1 | Decadal variability in North Atlantic SST and western European SAT. (a) Time series of the linearly detrended North Atlantic SST (black lines, referred to as the AMO index) and SAT averaged over western Europe ([36N 60N] × [10W 3E]; shown in coloured lines) in July (top panel) and January (bottom panel). Bold lines show 10-year running means. The correlation coefficient between the 10-year running mean of the detrended SAT and AMO index is 0.61 in July (statistically significant at 10% confidence level even after accounting for the reduced effective degrees of freedom due to autocorrelation of the time series) and − 0.02 in January; these correlations are insensitive to the averaging region chosen for western Europe. The red circles on January plot indicate the AMO-positive years chosen for the composite analysis, whereas the blue circles indicate the AMO-negative years chosen. **(b)** Study region encompassing western Europe ([36N 60N] × [10W 3E]) and locations for the backtracked Lagrangian particle release (black squares).

The turbulent fluxes are calculated in 20CR with the product's wind speeds and temperature and moisture gradients, and the size of the fluxes is influenced by correlations between the wind speed and temperatures[22].

In winter in the North Atlantic, SST is almost always warmer than the surface air temperature (SAT), so the ocean loses heat rapidly to the atmosphere over the entirety of the basin (that is, positive fluxes in our convention; Fig. 3b and Supplementary Fig. 3b). The fluxes over the warm Gulf Stream and its North Atlantic Current extension are generally a factor of five higher than found elsewhere. However, a view of the fluxes weighted by the fraction of time the particles spend in each location on their journey to western Europe (Fig. 3c and Supplementary Fig. 3c) suggests a reduced role of these strong flux regions in establishing western European wintertime temperature.

The difference in the number density of the particle positions between the composite AMO periods (Fig. 3d) shows a significant distinction in the preferred pathways, with the statistical significance increasing when results are separated by particles launched from northern and southern sub-regions of western Europe (Supplementary Fig. 3d). In the AMO-positive years, particles spend more of their 10-day trajectory recirculating locally to the southwest of Iceland. During the AMO-negative years, the pathways are anomalously long, and a greater number of trajectories originate from North America and the Arctic, before transiting over the Labrador Sea and mid-latitude North Atlantic. These differences in the atmospheric trajectories are explained mechanistically by the difference in the Z500 anomalies associated with the AMO, which shows swifter, more zonal winds during AMO-negative years (Fig. 2b). This largely barotropic anomaly pattern has been noted in a number of modelling and observational studies[5,10,30,31], and is somewhat similar

to the atmospheric circulation patterns associated with the North Atlantic Oscillation (NAO)[5]. To explore whether this atmospheric anomaly pattern is linked with anomalous SST conditions of the AMO regardless of the NAO phase, we performed an additional analysis excluding the strong NAO years. This exclusion only amplifies the signal of intensified zonal flow during negative AMO years relative to positive years (c.f. Fig. 3d and Supplementary Fig. 4a). The cause of the linkage between AMO and NAO has been the subject of debate, with several papers arguing that North Atlantic SST anomalies force an atmosperic NAO response[32–34], and others arguing the reverse[12,35]. Regardless of what drives the relationship, the association between the atmospheric circulation and the AMO index is clear in the Lagrangian trajectory composites (Fig. 3d).

The difference map of turbulent fluxes along these Lagrangian trajectories points towards a leading cause of the missing AMO imprint on European wintertime temperatures. During AMO-positive years, the shorter trajectories arriving from the north and southwest (Fig. 3d and Supplementary Fig. 3d) are accompanied by high fluxes (Fig. 3e and Supplementary Fig. 3e). However, the long and zonal trajectories associated with the AMO-negative years are accompanied by even stronger turbulent fluxes over much of the mid-latitude North Atlantic. Therefore, there are partially compensating regions of elevated and depressed flux during both phases of the AMO. The same pattern is found when the difference maps are constructed from composites of the full decadal periods associated with the AMO, but with slightly weaker statistical strength (Supplementary Fig. 2b,c).

Time series constructed by averaging along Lagrangian back trajectories (Fig. 4) further reveal the net effect of the combined

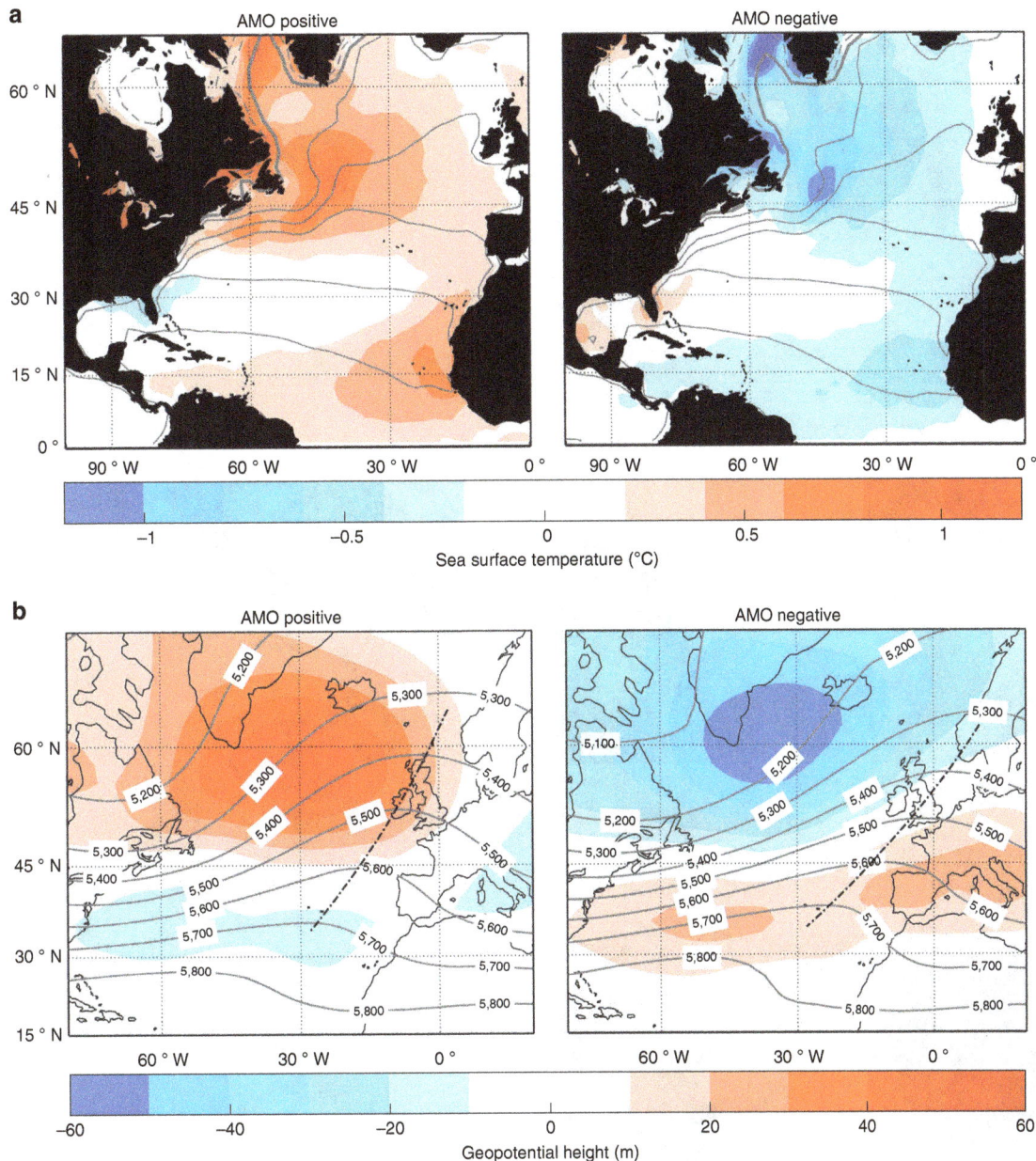

Figure 2 | The spatial pattern of the AMO index and its relationship with the atmospheric flow in January. Composite maps of (**a**) sea surface temperature (SST) field and (**b**) 500 hPa geopotential height field (Z500) for AMO anomalously positive years (left panel) and negative years (right panel). The January mean field is shown in contours, and its departure from the 72-year climatology is represented by colour shading. The thick grey contour line in **a** denotes 0 °C, whereas thin (dashed) lines denote positive (negative) SST every 5 °C. The black dashed lines in **b** are drawn through the local maxima of the geopotential height field at each latitude, which is the point where the wind changes direction from south-westerly to north-westerly.

changes to Lagrangian pathways and the properties along them. Notably, the SST sampled along the Lagrangian trajectories lacks a clear AMO signal (Fig. 4a), because decadal variability in atmospheric trajectories, which travel over a spatially variable SST field, swamps the temporal variability of the North Atlantic average SST. The SAT is highly correlated with the SST (not shown), as turbulent fluxes work to bring the surface boundary layers of the atmosphere and ocean towards equilibrium. Yet, during the negative AMO years from the mid-1970s to 1990, the air advected along Lagrangian trajectories is more anomalously cold than the SST, producing strengthened air–sea temperature gradients (Fig. 4b). Further heightened by stronger winds (Fig. 4c), the largest turbulent fluxes are achieved during these

AMO-negative years (Fig. 4d). We acknowledge that at the onset of the AMO-negative period around 1968, both SST and SAT averaged along the trajectories were elevated and fluxes were approximately average. Nevertheless, the overall effect is that the 10-year running mean turbulent fluxes sampled along Lagrangian trajectories are weakly anticorrelated with the AMO index ($r = -0.39$, not significant given the few effective degrees of freedom in the smoothed time series; Fig. 4d). We conclude that, in winter, the dynamic modulation of Lagrangian pathways and the atmospheric properties transported with them oppose the influence of basin-scale SST fluctuations on turbulent air–sea fluxes, thereby concealing the temperature expression of the AMO in atmospheric particles arriving in Europe.

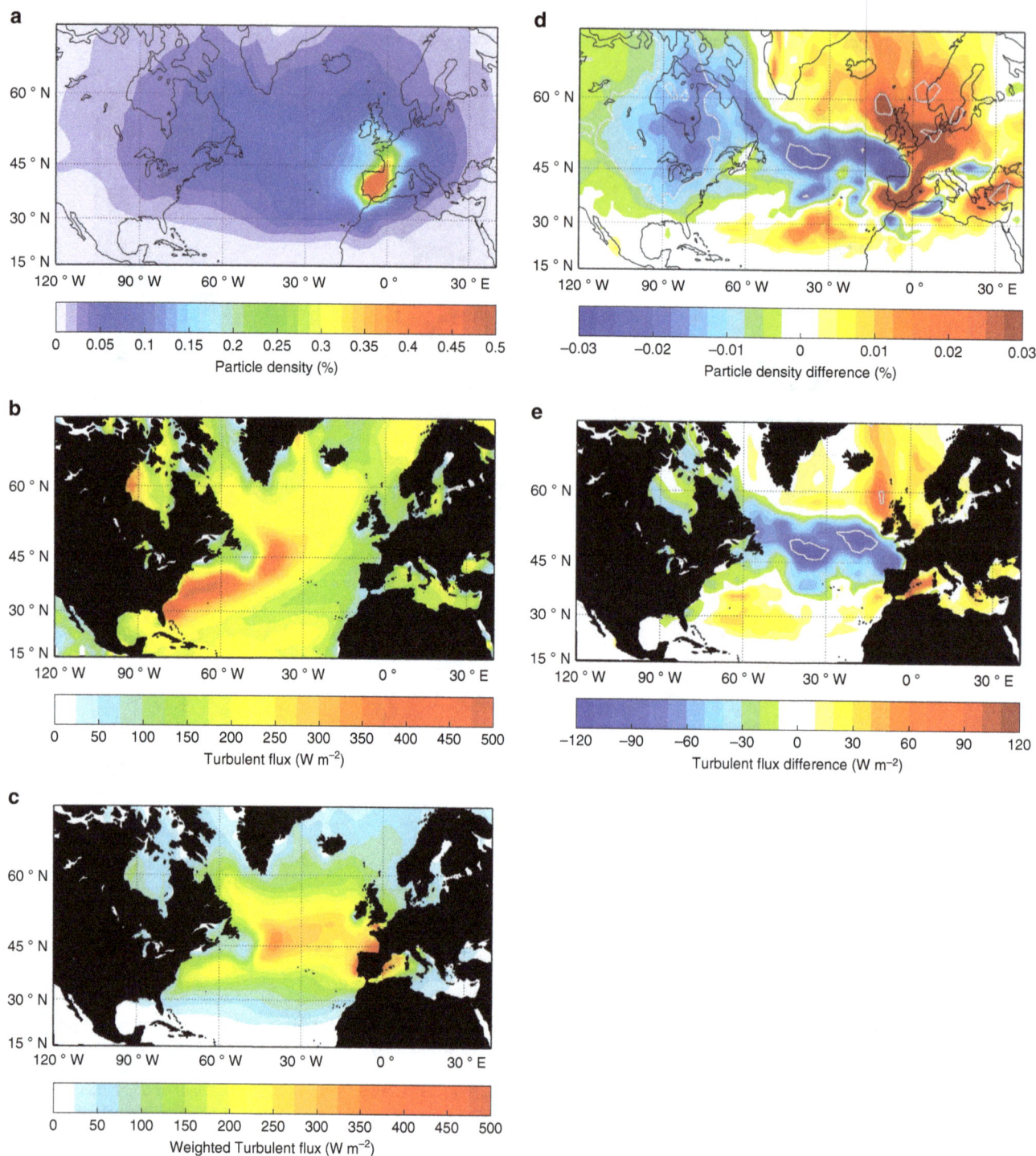

Figure 3 | Climatological mean and AMO influence on backward Lagrangian trajectories and the properties along them. (**a**) Lagrangian particle climatological number density, given as the percentage of all hourly positions that were spent in any 2° × 2° grid cell. (**b**) Climatological turbulent fluxes (sensible + latent; W m^{-2}) calculated by averaging the fluxes along the Lagrangian trajectories. (**c**) Turbulent fluxes as in **b** but weighted by the fraction of hourly particle positions spent in each grid cell, and normalized to have an equal spatial mean as the unweighted fluxes (see Methods for details; W m^{-2}). (**d**) The difference in number density for AMO-positive state minus AMO-negative state (% particle's hourly positions). (**e**) The difference in turbulent fluxes (W m^{-2}) for AMO-positive state minus AMO-negative state, weighted and normalized as in **c**. In **d**, statistical significance at 10 and 15% is shown, whereas in **e**, the 15% significance level is shown in grey contours. All the significance levels were obtained using a bootstrapping method (see Methods for details).

To further assess the degree to which the dynamic modulation of the trajectories is responsible for suppressing the AMO imprint on the fluxes, we re-ran our Lagrangian simulations with randomly selected, unvarying trajectories (see Methods for details). The time series of the 10-year running mean conditions along these random trajectories is plotted in blue in Fig. 4. The SST (Fig. 4a) averaged along these randomly selected trajectories vary in phase with the AMO index, and the anticorrelation of the

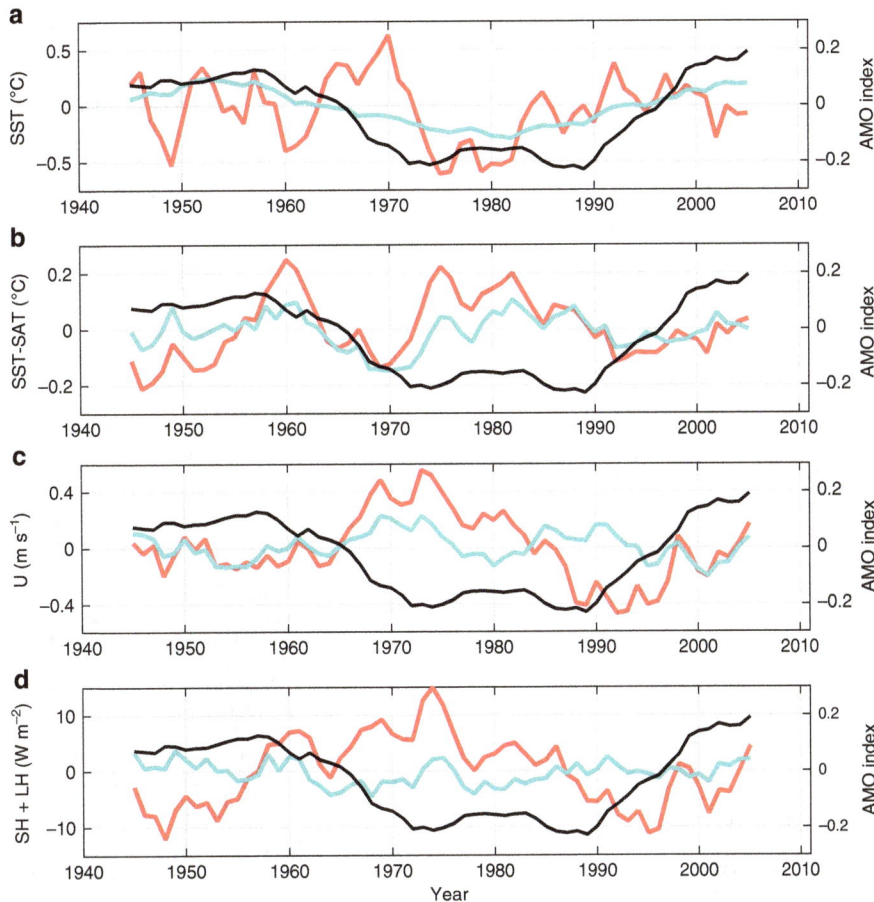

Figure 4 | Time series of properties and heat fluxes averaged along Lagrangian trajectories. 10-Year running mean of each linearly detrended variable along true trajectories (red), and along 10 sets of 10 random trajectories (blue), averaged across all 41 release locations. The January 10-year running mean AMO index is overlaid in black in each panel. (**a**) SST (°C), (**b**) SST − SAT (°C), (**c**) wind speed (m s^{-1}) and (**d**) turbulent fluxes (W m^{-2}).

fluxes with the AMO is eliminated (Fig. 4d). Finally, we confirm that, in summer, when the European temperature reflects AMO variability, the turbulent fluxes vary in phase with the AMO. In July, there is a minimal difference between the preferred pathways by AMO phase (Supplementary Fig. 5). Hence, the dominant signature of the AMO-positive phase appears to be due to higher basin-scale SST, which allows for a broad region of enhanced fluxes, consistent with an extended analysis of air–sea fluxes over the past century[23].

Discussion

The strengthening and lengthening of the storm track in sync with anomalously cooler North Atlantic SSTs has important implications for future climate. Given that decadal variability in North Atlantic SSTs may be driven partly by fluctuations in the strength of the AMOC[10-12], our result suggests the possibility of a stabilizing feedback for ocean circulation: Cooler SSTs associated with a sluggish AMOC is linked with an atmospheric adjustment that enhances turbulent heat fluxes over oceanic convective regions in winter. These larger fluxes could possibly reinvigorate convection, deep water formation and the AMOC. Moreover, the observed link of the atmospheric circulation with the cool SST anomalies of the late 1970s to early 1990s is much like the predicted change of the storm track in response to a decline of the AMOC under global warming[36]. A weakened AMOC has long been thought to cause anomalous cooling in western Europe via a decline in oceanic heat transport and associated atmospheric feedbacks[21]. However, the changes we describe here in

atmospheric Lagrangian trajectories and the heat fluxes along them could provide a mechanism that reduces the magnitude of European wintertime cooling on decadal time scales, even as they might stabilize the oceanic circulation.

Methods

FLEXPART. We adapted the Lagrangian atmospheric dispersion model FLEXPART version 9.02 (ref. 29) for use with the Twentieth Century Atmospheric Reanalysis product (20CR)[24] in order to simulate the atmospheric particles released from 41 equally spaced western European locations (Fig. 1b). These release points are chosen from an evenly spaced 2° × 2° grid over the study region [36N 60N] × [10W 3E], when these points fall on land. Every January from 1940 to 2011, ten particles are released from the surface of each of these points twice daily at 0 coordinated universal time (UTC) and 12 UTC and advected backward in time for the duration of 10 days, following the three-dimensional wind field. There are three components to this wind field: (i) resolved wind, (ii) turbulent wind fluctuations and (iii) mesoscale wind fluctuations. FLEXPART accounts for the turbulent wind fluctuations by adding a perturbation to the velocity field for atmospheric particles in the PBL, where these random motions are calculated by solving Langevin equations for Gaussian turbulence. Mesoscale velocity, whose spectral interval falls between the resolved flow and the turbulent flow, is included by solving an independent Langevin equation. The PBL height is diagnosed at each particle's hourly position. The duration of 10 days for the back trajectories was chosen based on the fact that the Lagrangian decorrelation time scale is ∼3 days[22]; thus, the choice of 10 days is long enough for the memory of each particle's initial temperature to be erased under the effect of diabatic processes along the trajectory.

20CR is one of the longest reanalysis products currently available. It has 6-h temporal resolution and 2° × 2° spatial resolution. The product assimilates only observations of surface pressure, monthly SST and sea-ice distributions, and we only use the ensemble mean fields. To assess the reliability of our FLEXPART results with use of 20CR, the trajectories computed using 20CR was compared with those with a default input for FLEXPART, Climate Forecast System Reanalysis

(CFSR) forecast and reanalysis data sets from National Centers for Environmental Prediction[37], which has hourly temporal and $0.5° \times 0.5°$ spatial resolution, for the period of 1981–2009 under the same set up as Yamamoto et al.[22] We found that the trajectory paths computed with 20CR are generally very close to those computed with CFSR especially over the ocean, with particles in 20CR taking slightly more northern paths relative to CFSR (Supplementary Fig. 6). We note that 20CR assimilates monthly mean SST data, whereas CFSR assimilates SST every 6 h. The agreement of the amplitude and variability of the turbulent fluxes along Lagrangian trajectories constructed from the two reanalysis products (Supplementary Fig. 7) suggests that the missing sub-monthly SST variability in 20CR has a minimal impact on these fluxes on interannual and longer time scales.

Bootstrapping. Bootstrapping was used in order to gauge the statistical significance of the difference in the spatial patterns of Lagrangian trajectory pathways and the fluxes along the trajectories for the two AMO phases (Fig. 3d,e). We sample the Lagrangian particle trajectories or the fluxes along them from randomly selected January months ('pseudo-periods') for the same number of years as each AMO phase (17 years for AMO positive and 18 years for AMO negative, respectively). We then take a difference between the composite pseudo-periods. This operation was repeated 500 times. We consider differences of the Lagrangian particle density and the fluxes along them from the true AMO composites significant at the 10% (15%) level when this true difference exceeds the 90th (85th) percentile of the pseudo-period differences. We repeated the same procedure to make the composites with entire AMO periods (40 AMO positive years and 29 AMO negative years) and show these results in the Supplementary Materials.

Surface fluxes along the trajectories. Along the particle trajectories simulated using FLEXPART, the surface turbulent fluxes are interpolated from 20CR sensible heat (SH) and latent heat flux (LH) fields, whenever the particle's hourly position falls within PBL, under the assumption that the turbulent fluxes influence the entire air mass within the PBL. The turbulent fluxes in 20CR are computed using bulk formulae with a typical formulation[38]:

$$SH = \rho_a c_p C_h U(T_s - T_a) \quad (1)$$

$$LH = \rho_a L_e C_e U(q_s - q_a) \quad (2)$$

where ρ_a is the atmospheric density, c_p is the atmospheric heat capacity, C_h and C_e are transfer coefficients, U is the mean value of wind speed relative to the surface ocean current, T_s is sea surface temperature, T_a is the atmospheric potential temperature at a reference height, L_e is latent heat of evaporation, q_s is interfacial value of water vapour mixing ratio, and q_a is the atmospheric water vapour mixing ratio at a reference height.

Time series of the mean accumulated surface heat fluxes along the trajectories using 20CR are highly correlated with those using CFSR (Supplementary Fig. 7), with the average correlation coefficient being $r = 0.92$.

Weighting of composite fluxes. In the composite figures of weighted surface fluxes along the trajectories (Fig. 3c,e), weights are proportional to the fraction of all particle positions that pass over a each $2° \times 2°$ grid cell (that is, the number density, given as a percentage in Fig. 3a). These weights are scaled so that the mean value of the climatological map (Fig. 3c) is equal to the mean of the unweighted climatology (Fig. 3b). The spatial mean of both mapped fields used in this scaling includes only those grid cells visited by at least 0.01% of the climatological particle positions; these collectively contain 90.1% of all hourly particle positions.

Randomly selected trajectories. The unvarying particle trajectory pathways used to produce Fig. 4 (blue lines) were chosen by randomly selecting ten particles from the set of all possible pathways generated from the full 72-year Lagrangian simulation. The truly varying surface fluxes are interpolated along these random trajectories. We then repeat this process ten times, each time by picking a different random set of ten trajectories from each release point, thereby creating a spread of particle positions. In total, 100 particles (10 particles × 10 realizations) are selected for each release location.

References
1. Schlesinger, M. E. & Ramankutty, N. An oscillation in the global climate system of period 65–70 years. Nature 367, 723–726 (1994).
2. Kushnir, Y. Interdecadal variations in North Atlantic sea surface temperature and associated atmospheric conditions. J. Clim. 7, 141–157 (1994).
3. Rowell, B. D. P., Folland, C. K., Maskell, K. & Ward, M. N. Variability of summer rainfall over tropical north Africa (1906-92): Observations and modelling. Quat. J. R. Meteorol. Soc. 121, 669–704 (1995).
4. Goldenberg, S. B., Landsea, C. W., Mestas-Nunez, A. M. & Gray, W. M. The recent increase in Atlantic hurricane activity: causes and implications. Science (New York, N.Y.) 293, 474–479 (2001).
5. Sutton, R. T. & Hodson, D. L. R. Influence of the ocean on North Atlantic climate variability 1871-1999. J. Clim. 16, 3296–3313 (2003).
6. Gastineau, G., D'Andrea, F. & Frankignoul, C. Atmospheric response to the North Atlantic Ocean variability on seasonal to decadal time scales. Clim. Dyn. 40, 2311–2330 (2012).
7. Zhang, R. & Delworth, T. L. Impact of the Atlantic Multidecadal Oscillation on North Pacific climate variability. Geophys. Res. Lett. 34, 2–7 (2007).
8. Arguez, A., O'Brien, J. J. & Smith, S. R. Air temperature impacts over Eastern North America and Europe associated with low-frequency North Atlantic SST variability. Int. J. Climatol. 10, 1–10 (2009).
9. Sutton, R. T. & Dong, B. Atlantic Ocean influence on a shift in European climate in the 1990s. Nat. Geosci. 5, 788–792 (2012).
10. Delworth, T. L. & Mann, M. E. Observed and simulated multidecadal variability in the Northern Hemisphere. Clim. Dyn. 16, 661–676 (2000).
11. Latif, M., Roeckner, E., Botzet, M. & Esch, M. Reconstructing, monitoring, and predicting multidecadal-scale changes in the North Atlantic thermohaline circulation with sea surface temperature. J. Clim. 17, 1605–1614 (2004).
12. McCarthy, G. D., Haigh, I. D., Hirschi, J. J.-M., Grist, J. P. & Smeed, D. A. Ocean impact on decadal Atlantic climate variability revealed by sea-level observations. Nature 521, 508–510 (2015).
13. Otterå, O. H., Bentsen, M., Drange, H. & Suo, L. External forcing as a metronome for Atlantic multidecadal variability. Nat. Geosci. 3, 688–694 (2010).
14. Booth, B. B. B., Dunstone, N. J., Halloran, P. R., Andrews, T. & Bellouin, N. Aerosols implicated as a prime driver of twentieth-century North Atlantic climate variability. Nature 484, 228–232 (2012).
15. Zhang, R. et al. Have Aerosols Caused the Observed Atlantic Multidecadal Variability? J. Atmos. Sci. 70, 1135–1144 (2013).
16. Clement, A. et al. The Atlantic Multidecadal Oscillation without a role for ocean circulation. Science 350, 320–324 (2015).
17. Knudsen, M. F., Seidenkrantz, M.-S., Jacobsen, B. H. & Kuijpers, A. Tracking the Atlantic Multidecadal Oscillation through the last 8,000 years. Nat. Commun. 2, 178 (2011).
18. Ting, M., Kushnir, Y., Seager, R. & Li, C. Forced and Internal Twentieth-Century SST Trends in the North Atlantic. J. Clim. 22, 1469–1481 (2009).
19. Ting, M., Kushnir, Y., Seager, R. & Li, C. Robust features of Atlantic multidecadal variability and its climate impacts. Geophys. Res. Lett. 38, L17705 (2011).
20. DelSole, T., Tippett, M. K. & Shukla, J. A Significant Component of Unforced Multidecadal Variability in the Recent Acceleration of Global Warming. J. Clim. 24, 909–926 (2011).
21. Palter, J. B. The Role of the Gulf Stream in European Climate. Annu. Rev. Marine Sci. 7, 113–137 (2015).
22. Yamamoto, A., Palter, J. B., Lozier, M. S., Bourqui, M. S. & Leadbetter, S. J. Ocean versus atmosphere control on western European wintertime temperature variability. Clim. Dyn. 45, 3593–3607 (2015).
23. Gulev, S. K., Latif, M., Keenlyside, N., Park, W. & Koltermann, K. P. North Atlantic Ocean control on surface heat flux on multidecadal timescales. Nature 499, 464–467 (2013).
24. Compo, G. P. et al. The Twentieth Century Reanalysis Project. Quat. J. R. Meteorol. Soc. 137, 1–28 (2011).
25. Krueger, O., Schenk, F., Feser, F. & Weisse, R. Inconsistencies between Long-Term Trends in Storminess Derived from the 20CR Reanalysis and Observations. J. Clim. 26, 868–874 (2013).
26. Enfield, D. B., Mestas-Nunez, A. M. & Trimble, P. J. The Atlantic multidecadal oscillation and its relation to rainfall and river flows in the continental U.S. Geophys. Res. Lett. 28, 2077–2080 (2001).
27. Trenberth, K. E. & Shea, D. J. Atlantic hurricanes and natural variability in 2005. Geophys. Res. Lett. 33, L12704 (2006).
28. Nigam, S., Guan, B. & Ruiz-Barradas, A. Key role of the Atlantic Multidecadal Oscillation in 20th century drought and wet periods over the Great Plains. Geophys. Res. Lett. 38, L16713 (2011).
29. Stohl, A., Forster, C., Frank, A., Seibert, P. & Wotawa, G. Technical note: The Lagrangian particle dispersion model FLEXPART version 6.2. Atmos. Chem. Phys. 5, 2461–2474 (2005).
30. Ting, M., Kushnir, Y. & Li, C. North Atlantic Multidecadal SST Oscillation: External forcing versus internal variability. J. Marine Sys. 133, 27–38 (2014).
31. Gastineau, G. & Frankignoul, C. Influence of the North Atlantic SST on the atmospheric circulation during the twentieth century. J. Clim. 28, 1396–1416 (2015).
32. Frankignoul, C., Chouaib, N. & Liu, Z. Estimating the observed atmospheric response to SST anomalies: maximum covariance analysis, generalized equilibrium feedback assessment, and maximum response estimation. J. Clim. 24, 2523–2539 (2011).
33. Peings, Y. & Magnusdottir, G. Forcing of the wintertime atmospheric circulation by the multidecadal fluctuations of the North Atlantic ocean. Environ. Res. Lett. 9, 034018 (2014).
34. Omrani, N.-E., Keenlyside, N. S., Bader, J. & Manzini, E. Stratosphere key for wintertime atmospheric response to warm Atlantic decadal conditions. Clim. Dyn. 42, 649–663 (2014).

35. Mecking, J. V., Keenlyside, N. S. & Greatbatch, R. J. Stochastically-forced multidecadal variability in the North Atlantic: a model study. *Clim. Dyn.* **43,** 271–288 (2014).

36. Woollings, T., Gregory, J. M., Pinto, J. G., Reyers, M. & Brayshaw, D. J. Response of the North Atlantic storm track to climate change shaped by ocean—atmosphere coupling. *Nat. Geosci.* **5,** 313–317 (2012).

37. Saha, S. *et al.* The NCEP climate forecast system reanalysis. *Bull. Amer. Meteor. Soc.* **91,** 1015–1057 (2010).

38. Fairall, C. W., Bradley, E. F., Rogers, D. P., Edson, J. B. & Young, G. S. Bulk parameterization of air-sea fluxes for Tropical Ocean-Global Atmosphere Coupled-Ocean Atmosphere Response Experiment difference relative analysis. *J. Geophys. Res.* **101,** 3747–3764 (1996).

Acknowledgements

This work was supported by a funding from the McGill University, the NSERC Discovery Program, FQRNT's Programme Etablissement de Nouveaux Chercheurs Universitaires and Quebec-Ocean. We thank three anonymous reviewers and E. Galbraith and M. Gervais for their constructive comments that greatly improved earlier versions of this manuscript.

Author contributions

A.Y. and J.B.P. designed the experiments, analysed the results and wrote the manuscript together. A.Y. adapted FLEXPART for 20CR input and performed the experiments.

Additional information

22

Evidence for an ice shelf covering the central Arctic Ocean during the penultimate glaciation

Martin Jakobsson[1,2,3], Johan Nilsson[2,4], Leif Anderson[5], Jan Backman[1,2], Göran Björk[5], Thomas M. Cronin[6], Nina Kirchner[7], Andrey Koshurnikov[8,9], Larry Mayer[10], Riko Noormets[3], Matthew O'Regan[1,2], Christian Stranne[1,2,10], Roman Ananiev[8,9], Natalia Barrientos Macho[1,2], Denis Cherniykh[8,11], Helen Coxall[1,2], Björn Eriksson[1,2], Tom Flodén[1], Laura Gemery[6], Örjan Gustafsson[2,12], Kevin Jerram[10], Carina Johansson[1,2], Alexey Khortov[8], Rezwan Mohammad[1,2] & Igor Semiletov[8,11]

The hypothesis of a km-thick ice shelf covering the entire Arctic Ocean during peak glacial conditions was proposed nearly half a century ago. Floating ice shelves preserve few direct traces after their disappearance, making reconstructions difficult. Seafloor imprints of ice shelves should, however, exist where ice grounded along their flow paths. Here we present new evidence of ice-shelf groundings on bathymetric highs in the central Arctic Ocean, resurrecting the concept of an ice shelf extending over the entire central Arctic Ocean during at least one previous ice age. New and previously mapped glacial landforms together reveal flow of a spatially coherent, in some regions >1-km thick, central Arctic Ocean ice shelf dated to marine isotope stage 6 (~140 ka). Bathymetric highs were likely critical in the ice-shelf development by acting as pinning points where stabilizing ice rises formed, thereby providing sufficient back stress to allow ice shelf thickening.

[1] Department of Geological Sciences, Stockholm University, Stockholm 106 91, Sweden. [2] Bolin Centre for Climate Research, Stockholm University, Stockholm 106 91, Sweden. [3] UNIS - The University Centre in Svalbard, Longyearbyen N-9171, Svalbard. [4] Department of Meteorology, Stockholm University, Stockholm 106 91, Sweden. [5] Department of Marine Sciences, University of Gothenburg, Gothenburg 405 30, Sweden. [6] US Geological Survey Reston, 12201 Sunrise Valley Drive, Reston, Virginia 20192, USA. [7] Department of Physical Geography, Stockholm University, Stockholm 106 91, Sweden. [8] National Research Tomsk Polytechnic University, Tomsk 634050, Russia. [9] Department of Geocryology, Moscow State University, Moscow 119991, Russia. [10] Center for Coastal and Ocean Mapping, University of New Hampshire, 24 Colovos Road, Durham, New Hampshire 03824, USA. [11] Russian Academy of Sciences, Pacific Oceanological Institute, 43 Baltiiskaya Street, Vladivostok 690041, Russia. [12] Department of Environmental Science and Analytical Chemistry, Stockholm University, Stockholm 106 91, Sweden. Correspondence and requests for materials should be addressed to M.J. (email: martin.jakobsson@geo.su.se).

ce conditions in the Arctic Ocean during glacial maxima have been much debated, with hypotheses formulated long before direct observational data existed. In 1888, Sir William Thomson speculated about extensive and thick floating ice, and elaborated on possible effects of isolating Arctic Ocean water masses from the remaining World Ocean[1]. Nearly a century later, speculations ranged from a sea-ice-free Arctic Ocean during glacial maxima[2] to one where an extensive and thick ice shelf persisted[3–5]. In the mid-1960s, it was proposed that large portions of the Barents Sea had been covered by a marine ice sheet during the last glacial maximum (LGM)[6]. In 1970, Mercer pointed out striking similarities between the glacial-age Arctic Ocean and today's West Antarctic ice sheet, where extensive ice shelves exist, and he stressed that the idea of an Arctic Ocean filled by thick ice was glaciologically sound and should be taken seriously[3]. Additional support for Arctic Ocean ice caps with huge floating parts in the form of ice shelves was shortly thereafter provided by Broecker[4] and Hughes et al.[5], who proposed a thick, floating and dynamic ice shelf during the LGM on the basis that such an ice shelf may have been necessary to stabilize the inherently unstable marine ice sheets located on the Arctic's continental margins (Fig. 1a). The concept of a dynamic ice shelf was developed further in some Arctic Ocean ice sheet reconstructions for the LGM[7,8].

When the ice-shelf theory was developed in the 1970s and 1980s, the climatic implications of a huge floating Arctic Ocean ice shelf and extensive sea ice were addressed[8] along with its potential effect on the global ocean $\delta^{18}O$ record measured in benthic foraminifera[4,9,10]. It was during this period specifically noted that the amplitude of $\delta^{18}O$ variations in benthic foraminifera predicts more ice volume than available sea-level records indicate[11], a discrepancy that may be explained if ^{16}O is stored in a huge floating ice shelf that, once melted, has only a minor effect on sea level[4,9,10]. Lack of direct evidence, however, destined the notion of an Arctic Ocean-wide ice shelf eventually to relative obscurity, and the discrepancy between ice volume inferred from $\delta^{18}O$ and geological sea-level records was in

addition suggested to be caused mainly by massive Antarctic ice shelves[12].

Seafloor mapping of the Yermak Plateau off northern Svalbard provided the first evidence of thick glacial ice grounding in the Arctic Ocean[13] (Fig. 1b). This was followed by the discovery of seafloor ice erosion at depths approaching 1,000 m on the Lomonosov Ridge (LR) and the Chukchi Borderland[14,15]. These results, together with dating of sediment cores from the eroded areas, pointed to an ice shelf constrained to the Amerasian Basin during marine isotope stage (MIS) 6 ($\sim$140–160 ka)[16] (Fig. 1b). There were two main reasons for limiting this ice shelf to the Amerasian Basin: (1) previous mapping of <1,000 m deep sectors of the LR between 84°30′ N and the Siberian margin had not revealed ice grounding[17], and (2) the ice erosion mapped on the central LR was assumed to have been caused by armadas of icebergs rather than a coherent ice shelf[16,18].

Here we present new multibeam bathymetry and sub-bottom profiles documenting ice scours and other glacial landforms extending across the central Arctic Ocean that compel us to assess the concept of a coherent $\sim$1-km thick ice shelf extending over the entire Arctic Ocean. Our new observations from the LR off the Siberian margin, the Arlis Plateau and the continental slope north off Herald Canyon (Fig. 1b), merged with published observations, require an Arctic ice shelf close to the 'maximum ice' scenario, in terms of thickness, area and flow pattern, as hypothesized by Hughes et al.[5]. The new seafloor mapping data were collected during the SWERUS-C3 (Swedish–Russian–US Arctic Ocean Investigation of climate–cryosphere–carbon interactions) expedition in 2014.

Results

Geophysical mapping. Two sets of highly parallel streamlined submarine landforms cross the southern LR crest at about 81° N 143° E (Fig. 2). These features consist of approximately 10–15 high ridges that are spaced between 400 and 800 m apart. Morphologically, the mapped landforms closely match mega-scale lineations that are widely found in formerly glaciated continental

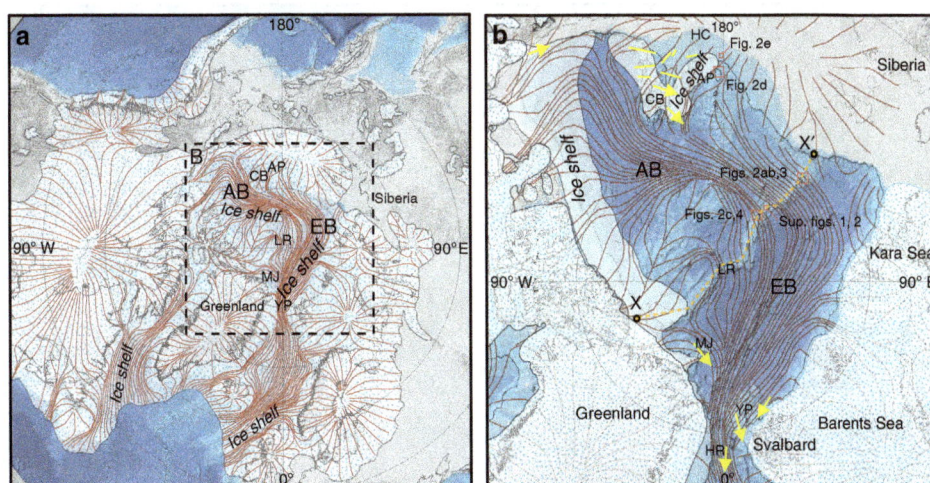

Figure 1 | Ice-sheet reconstructions during glacial conditions involving ice shelves in the Arctic Ocean. (**a**) LGM ice-sheet reconstruction by Hughes et al.[5] with an ice shelf that covers the entire Arctic Ocean and extends into the North Atlantic. Brown lines represent inferred ice-sheet flow. The modern coastline is used as a reference. (**b**) The limited ice shelf proposed by Jakobsson et al.[16] is shown as white semi-transparent area. The extent of the MIS 6 (Late Saalian) Barents–Kara Sea ice sheet[49] is shown as white semi-transparent blue dotted area. The North American ice sheet (late Wisconsinan[50], also blue dotted area, is assumed to have been similar to the Illinoian ice sheet (MIS 6). Yellow arrows represent previously published evidence of ice-shelf grounding and interpreted flow direction[16,22,34,51]. Flow lines from Hughes et al.[5] are shown also in **b** for comparison with ice-sheet flow inferred from mapped landforms. The orange dashed line X to X′ marks the bathymetric profile in Fig. 7a. The black contour line in **b** represent the present day 1,000 m isobaths (Sup. Figs 1 and 2 refer to Supplementary Figs 1 and 2). The modern coastline is used as a reference. AB, Amerasian Basin; AP, Arlis Plateau; CB, Chukchi Borderland; EB, Eurasian Basin; HC, Herald Canyon; HR, Hovgaard Ridge; LR, Lomonosov Ridge; MJ, Morris Jesup Rise; YP, Yermak Plateau.

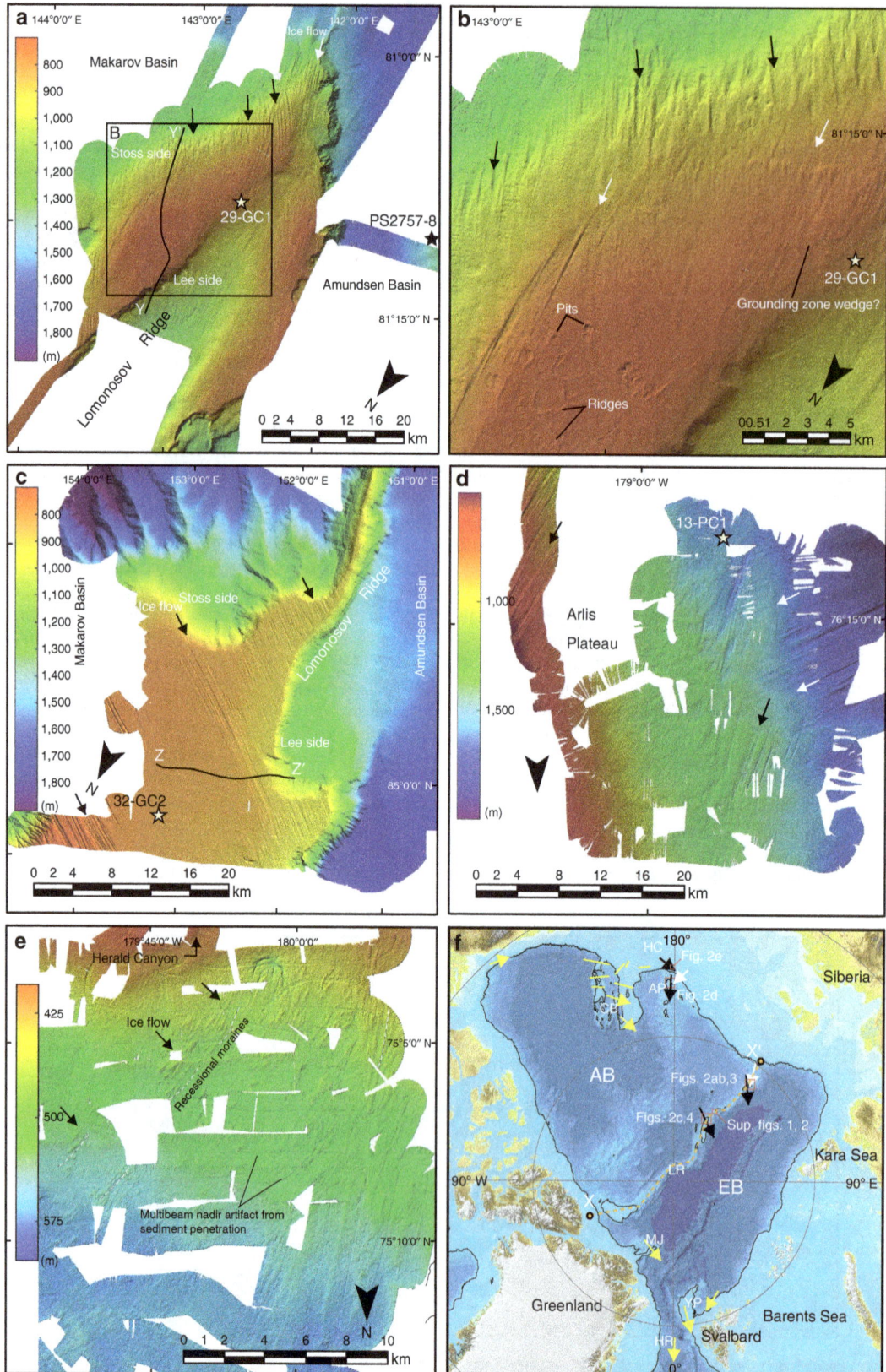

Figure 2 | Multibeam bathymetry of submarine glacial landforms mapped during the SWERUS-C3 expedition. The SWERUS (Swedish–Russian–US Arctic Ocean Investigation of Climate–Cryosphere–Carbon Interactions) exhibition, 2014 data from bathymetric highs are interpreted to signify ice-shelf grounding. (**a–c**) Lomonosov Ridge (**b** is a detail of **a**), (**d**) Arlis Plateau and (**e**) the slope north of Herald Canyon. The locations of all inset maps are shown in Fig. 1b as well as in **f**, where the present day bathymetry from the International Bathymetric Chart of the Arctic Ocean (IBCAO) is shown[52]. The 1,000 m isobaths is shown as bathymetric reference in black in **f**. Yellow arrows in **f** represent previously published evidence of ice-shelf grounding and interpreted flow direction[16,22,34,51]. Chirp sonar profiles between Y–Y' and Z–Z' are shown in Figs 3 and 4, respectively. The location of SWERUS-C3 cores used to date the ice-shelf grounding are marked with yellow stars and the stratigraphically correlated core PS2757-8 (ref. 30) is shown with a black star. See caption of Fig. 1 for used abbreviations undersea features.

margins where they are interpreted to signify fast-flowing ice streams[19]. The LR crest is in the area of 81° N 143° E and is shaped by ice grounding with a gently sloping stoss side towards the Makarov Basin and a steep lee side facing the Amundsen Basin (Fig. 2a,b). The general ice-flow direction is diagonally across the LR towards northwest, from Makarov Basin to Amundsen Basin. Lineations extend as deep as 1,280 m below present sea level on the stoss side. The flattened ridge crest contains small arcuate ridges with a relief of about 6 m and their pointed edges facing southward, towards the youngest ice-flow direction across the ridge. There are faint indications of what may be grounding zone wedges on the flat ridge crest (Fig. 2b). The flat-topped nature of the ridge crest is caused at least in part by ice grounding, as evident by an unconformity visible in sub-bottom profiles (Fig. 3). Emergence and subsidence could generate a similar flat-topped appearance of the ridge crest, but only on much longer time scales[20]. Further north at about 85° N and 153° E, the LR shows a more accentuated flat-topped ridge crest formed by ice grounding between about 1,000 and 700 m water depth (Figs 2c and 4). Also in this area, the ridge slope facing the Makarov Basin is gentler than towards the Amundsen Basin, suggesting a stoss and lee side with respect to the ice-shelf flow. Remarkably consistent parallel lineations extend diagonally across the ridge towards west–northwest. At about 84°15′ N there is a section of the LR crest at around 890 m present water depth where the ice shelf apparently did not ground and where giant pockmarks dominate the seafloor morphology (Supplementary Figs 1 and 2).

The Arlis Plateau (Fig. 2d) was also mapped during SWERUS-C3 to complement previously mapped glacial lineations extending to ~1,200 m depth, interpreted to represent grounding of an ice shelf extending from the East Siberian margin[21]. We mapped the intersection between two distinct sets of lineations on the Arlis Plateau crest (Fig. 2d). Superposition demonstrates that lineations having directions towards east–northeast, rather than northeast, are older (white arrows in Fig. 2d).

In addition, the seabed of the Chukchi Borderland is heavily affected by ice grounding[14,22]. Mapped patterns of glacial landforms have been interpreted to show a large ice rise belonging to the MIS 6 Amerasian ice shelf and marine outlet glaciers emanating from hypothesized East Siberian ice sheets[21–23]. Mapping of the slope west of the Chukchi Borderland north of Herald Canyon during the SWERUS-C3 expedition shows distinct sets of ridges diagonal to the dip of the slope in water depths from about 390 to >600 m (Fig. 2e). Morphologically, these ridges resemble recessional moraines perpendicular to the past ice flow[24]. Their direction suggests ice flow from the Chukchi Sea margin where the Herald Canyon

ends. Similar ridges have been previously mapped further downslope at about 700 m water depth[21].

Dating of ice grounding. Sediments deposited atop the ice-grounded surfaces on bathymetric highs in the central Arctic Ocean have been dated using several methods: radiocarbon dating of MIS 3-1 (refs 25,26), astronomical tuning of sediment physical and chemical properties[27,28], calcareous nannofossil biostratigraphy of MIS 5 (ref. 29), inter-core correlation based on dinoflagellate cysts[30], and benthic and planktic foraminifera[31,32].

New data for SWERUS-C3 sediment cores from the ice-grounded areas are shown in Fig. 5 and their locations are shown in Fig. 2 and Supplementary Fig. 3. Acoustically laminated sediments overlie the eroded surfaces. On the central LR, core SWERUS-L2-32-GC2 (32-GC2 on map, Fig. 2c, and on sub-bottom profile in Supplementary Fig. 3) can be accurately correlated to core 96/12-1PC, which has a well-constrained age model back to MIS 6 (ref. 28; Fig. 5). The correlations are based on physical property variations, that is, magnetics susceptibility and bulk density, captured in high-resolution multi-sensor core-logging measurements. The developed correlation indicates that the 2.5 m gravity core recovered an undisturbed sedimentary section back to MIS 5.5, constraining the ice scouring event at this site to MIS 6 or older. The correlation, which places MIS 5 between 1.5 and 2.35 m.b.s.f. (metre below seafloor), is supported by rare occurrences of the calcareous nannofossil *Emiliania huxleyi* in three samples at 1.69–1.72 m.b.s.f.

On the Southern LR, the 4.66-m long core SWERUS-L2-29-GC1 (29-GC1 on map, Fig. 2a,b, and on sub-bottom profile in Supplementary Fig. 3) sampled acoustically stratified sediments deposited on top of the ice-scoured surface. This core was collected in 824 m water depth. Physical properties from this core are correlated to neighbouring core collected by the *Polarstern* in 1995 (PS2757-8) (Fig. 5). This core was recovered from a water depth of 1,241 m, and is from just below the maximum ice-grounding surface on this portion of the LR (Fig. 2a). The entire recovered sedimentary sequence in SWERUS-L2-29-GC1 is mirrored in the deeper lying core, indicating that the erosional surface lies at the base of this core. Calcareous nannofossils indicate a Holocene age in the uppermost 5 cm of this core, with rare occurrences of *E. huxleyi*, *C. leptoporus* and *G. muellerae*. No age diagnostic microfossils were observed below 0.05 m.b.s.f., although one potential observation of *E. huxleyi* was made in the sample at 3.81 m.b.s.f. The base of PS-2757 has previously been assigned an MIS 6 age through correlation of organic geochemical parameters and magnetic susceptibility measurements to better-dated records on the Laptev and Barents Sea slope[30]. Although this age assignment remains speculative, radiocarbon dating of

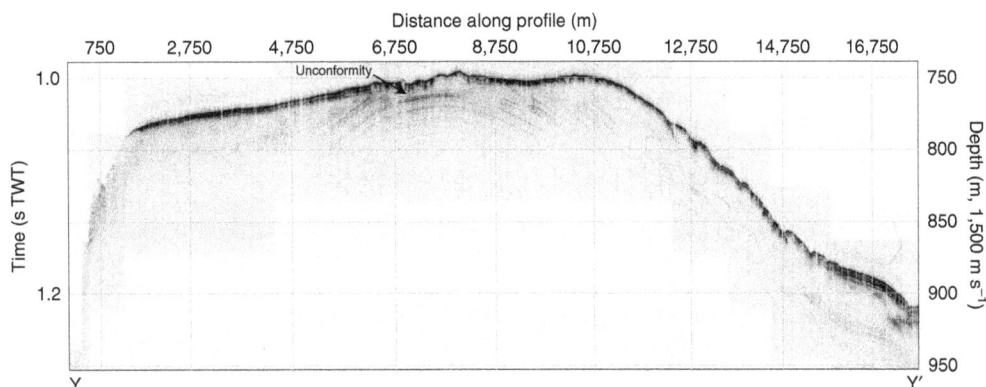

Figure 3 | Chirp sonar sub-bottom profile from the southern Lomonosov Ridge. The location of the profile between Y and Y′ is shown in Fig. 2a. The erosional unconformity, formed by ice grounding, is indicated in the profile.

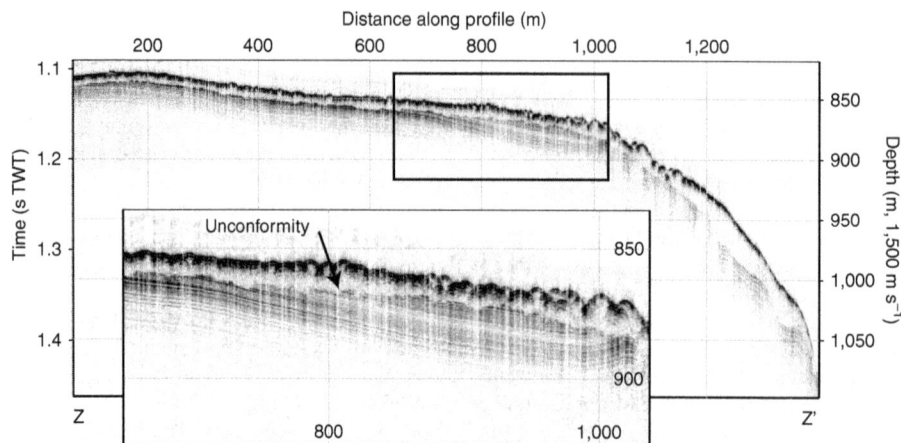

Figure 4 | Chirp sonar sub-bottom profile from the central Lomonosov Ridge. The location of the profile between Z and Z' is shown in Fig. 2c. The erosional unconformity, formed by ice grounding, is indicated in the profile.

the upper 10 cm of PS2757-8, and dinocyst abundance and assemblage data, firmly place the base of the Holocene at 0.6 m.b.s.f., suggesting sedimentation rates between 5 and 7 cm per ka (ref. 30). Extrapolating these sedimentation rates downcore suggests that the base of PS2757-8 is younger than 200 ka. Conclusively, available data show that the ice-scoured surface is older than the LGM, and likely occurred during MIS 6.

On the Arlis Plateau, core SWERUS-L2-13-PC1 (13-PC1 on map, Fig. 2d, and on sub-bottom profile in Supplementary Fig. 3) recovered 6.14 m of sediment at a water depth of 1,119 m where the seafloor has been subjected to ice grounding. A dark brown layer between 2.64 and 2.96 m.b.s.f. had rare nannofossils in three samples. The 2.86 m.b.s.f. sample yielded rare *E. huxleyi* and *Gephyrocapsa* spp., suggesting a MIS 5 age.

Discussion

Taken together, the new results suggest that an ice shelf existed during MIS 6 that was thicker and covered substantially more of the Arctic Ocean than previously suggested (Fig. 1b)[16]. A minimum scenario suggests an ice shelf during MIS 6 that covered most of, if not the entire, Amerasian Basin. Bathymetric highs generally shallower than ~1,000 m present water depth (at southern LR as deep as 1,280 m) acted as stabilizing pinning points through the formation of ice rises/rumples. Not all <1,000 m parts of the LR acted as pinning points, since a few sections are untouched by the ice shelf, but these may be explained by an uneven ice thickness (Fig. 6). It may at first seem reasonable to limit this ice shelf to the Amerasian Basin and the LR, however, previously mapped lineations on the Yermak Plateau[16,33] fit well with the hypothesized flow pattern suggested by Hughes *et al.*[5]. Suggested causes for the lineations on the Yermak Plateau include grounding of a larger ice-shelf fragment originating from the Amerasian Basin, an armada of large icebergs, or an ice-sheet component extending northward into the Arctic Ocean from the Barents Sea ice sheet[16,33]. The morphological similarity between the lineations previously mapped on the Yermak Plateau and those mapped on the LR during SWERUS-C3, together with their flow direction, suggest that they all originate from grounding of an Arctic Ocean-wide ice shelf, similar to the suggestion of Hughes *et al.*[5]. If several smaller ice shelves at different times during the MIS 6 glaciation instead were responsible for the mapped glacial landforms, the spatially coherent pattern over the central Arctic Ocean is difficult to explain. The recently discovered deep scours reaching water depths >1,200 m on the Hovgaard Ridge, located south of the Fram Strait, are interpreted to indicate a massive outflow of large

deep-drafting icebergs from the Arctic Ocean[34]. Additional detailed mapping of this ridge and other bathymetric highs south of the Fram Strait is required to rule out the possibility that the Arctic Ocean ice shelf did not extend into the Norwegian–Greenland seas as suggested by Hughes *et al.*[5] With no data at hand, we assume that the MIS 6 ice shelf was limited to the central Arctic Ocean.

The age(s) of deep ice grounding in the central Arctic Ocean have been discussed since the first evidence was mapped on the central LR and sediment cores from this area were dated[14,35]. The assigned MIS 6 age stems from the fact that there is a rather systematic drape of sediment beginning from MIS 5.5 atop the mapped glacial landforms and glacially eroded surfaces. However, it should be noted that Chukchi Borderland generally has a more complicated sediment stratigraphy with indications of additional glacial erosional events younger than MIS 6; the most recent is suggested to have occurred during MIS 4 (ref. 36). We cannot at present exclude the possibility of occurrences of thinner ice shelves younger than MIS 6 over large areas of the central Arctic Ocean that did not reach bathymetric highs as deep as ~1,000 m. Neither can we rule out that large ice shelves existed during older glacials than MIS 6 (for example, MIS 8 or 12) since evidence in the form of glacial landforms on bathymetric highs may have been erased by the most recent event during MIS 6.

While the morphological evidence suggests an Arctic Ocean-wide ice shelf at MIS 6, is such a feature oceanographically possible? The answer to this question lies in the details of the oceanographic conditions at the time. The present influx of >3 Sv $(1 Sv = 10^6 m^3 s^{-1})$ of warm (>0 °C) Atlantic water between about 200 and 600 m, would strongly impact negatively on ice-shelf development[16]. It has been suggested that Atlantic water was forced deeper in the central Arctic Ocean during glacials[16,37], thus limiting Atlantic flow across the LR into the Amerasian Basin. Oceanographic conditions in the Amerasian Basin during MIS 6 may be characterized as an extreme version of conditions in present-day cavities below Antarctic ice shelves[38]. However, as our mapping results from the LR suggest that the MIS 6 ice shelf extended into the Amundsen Basin, it likely was in contact with warm Atlantic water. This would result in a circulation where water warmer than the freezing point of ice melts the underside of the ice shelf to produce cool fresher water that strives to rise towards the surface. Could this kind of circulation have reached all the way across the LR and into the Amerasian basin due to that warmer Atlantic water flowed over the ridge? Wide bathymetric passages in the ridge below the grounded ice shelf (Fig. 6) may have formed conduits for warm

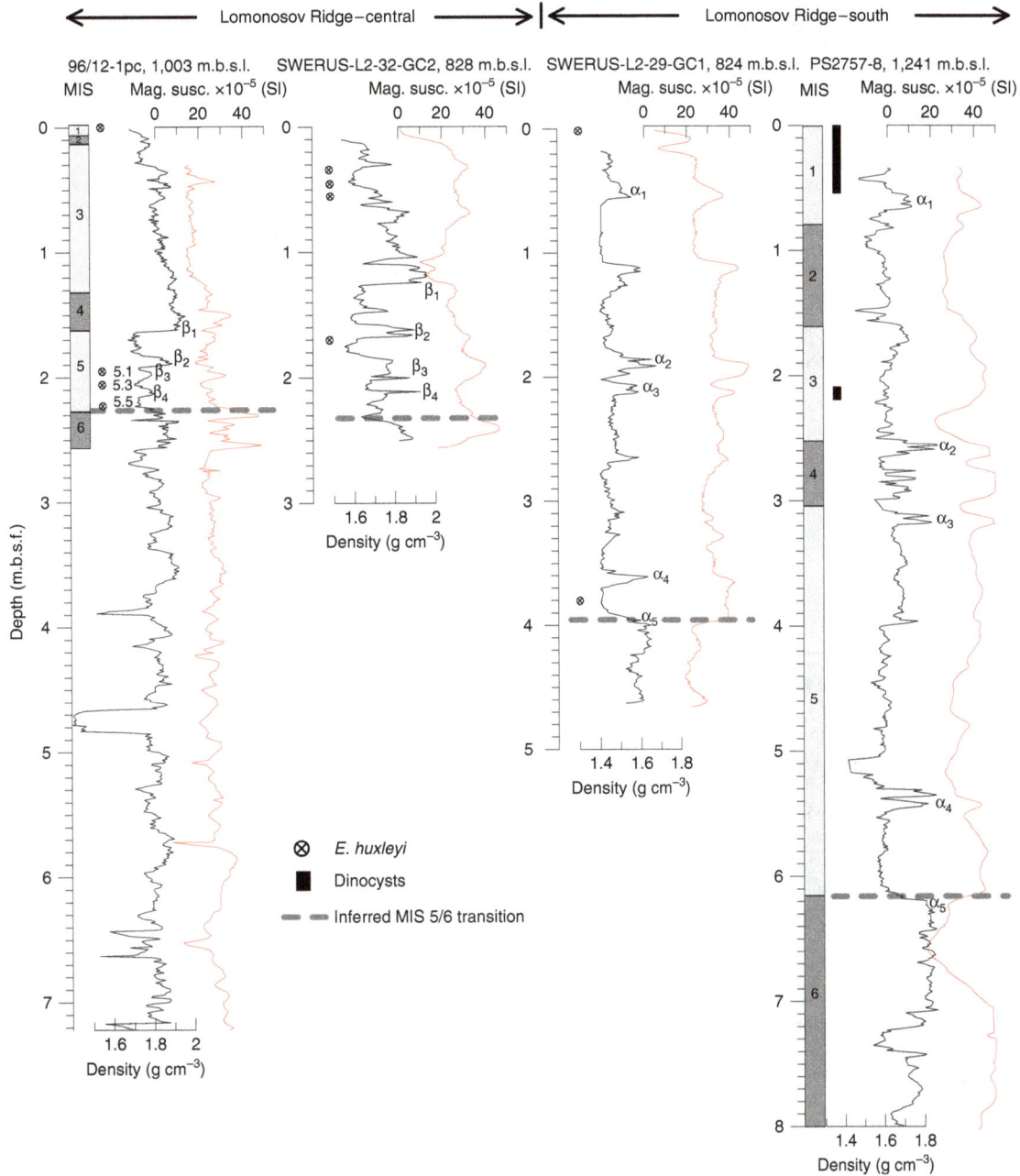

Figure 5 | Stratigraphic correlations between cores from SWERUS-C3 and previously collected and dated sediments cores from the Lomonosov Ridge. Bulk density and magnetic susceptibility of core 96/12-1pc from the central Lomonosov Ridge is shown with previously inferred bulk density stratigraphic tie points β_1–β_4 (ref. 35) linking it to nearby SWERUS-L2-32-GC2 (short name 32-GC2 on map in Fig. 2c). Bulk density and magnetic susceptibility records from SWERUS-L2-29-GC1 (short name 29-GC1 shown on map in Fig. 2a) correlated to core PS2757-8, both from the southern Lomonosov Ridge off the Siberian continental margin. The age models for 96/12-1pc (ref. 28) and PS2757-8 (ref. 30) are shown by the bar to the left with inferred MIS 1-6.

water causing melting also in the Amerasian Basin. This type of circulation is known from Antarctica to have large horizontal variability, and uneven melting might have formed an irregular underside of the ice shelf explaining the uneven grounding depth indicated by seafloor mapping data (Fig. 6).

Based on previous work[16], we formulate a conceptual two-layer ocean model on the flow underneath the ice shelf, that is, the ice cavity (Fig. 6). The aim is to obtain a simple model of the melt rate in a large ice cavity with a highly restricted exchange flow. For the sake of simplicity, we assume a steady state and ignore spatial variations below the ice shelf, having an essentially constant thickness over the Amerasian Basin. The water in the

cavity that is in contact with the ice shelf is assumed to be at freezing point, which decreases with increasing depth (pressure) of the base of the ice shelf. Continuity of volume is given by

$$M = M_A + F, \qquad (1)$$

where M is volume outflow from the ice cavity, M_A the oceanic inflow, and F the net freshwater supply due to ice melt or growth in the cavity. The remaining components of this model, based on conservation of salt and heat, are described in the Methods. For a reasonable choice of parameters, the resulting volume flow M is on the order of 1 Sv and the oceanic heat flux to the base of the shelf is on the order of 1 W m^{-2}, corresponding to an ice melt of

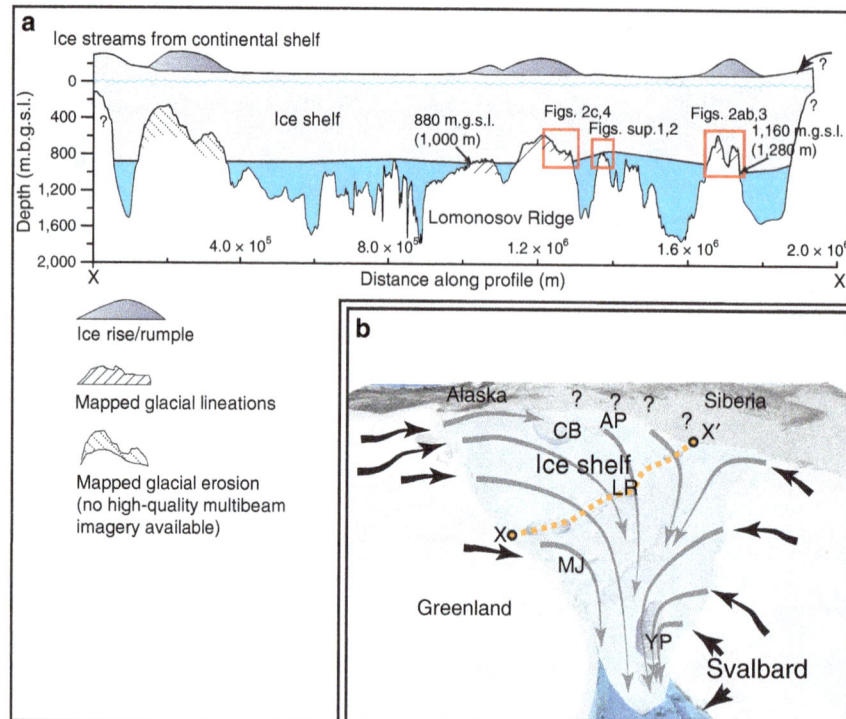

Figure 6 | Conceptual sketches of ice shelf covering the entire central Arctic Ocean. (**a**) Bathymetric profile along the LR, from the Greenland continental margin (X) to the margin off the New Siberian Islands (X'). See profile location in **b** and Fig. 1b. (**b**) Sketch of an ice shelf covering the entire central Arctic Ocean with flow lines generalized from mapped glacial landforms. The locations of ice rises/rumples on bathymetric highs, inferred from mapped landforms, are shown with the darker grey shading (see legend in **a**). m.g.s.l. refers to depth in metres below glacial sea level, here defined as 120 m below present sea level. The inferred general flow lines of the floating ice shelf (grey) are tapered to illustrate that the shelf thickness would thin with the distance from feeding ice streams (black arrows) and grounding lines, except for where the shelf grounded on bathymetric highs to form local ice rises/rumples. A positive mass balance of the ice shelf would be sustained if accumulation rates through precipitation in the central Arctic Ocean are higher than basal melting and mass loss at the calving front (see model of ocean–ice-cavity interaction). The locations of undersea features where ice grounding and additional glacial landforms have been mapped are marked with abbreviations: AP, Arlis Plateau; CB, Chukchi Borderland; LR, Lomonosov Ridge; MJ, Morris Jesup Rise; YP, Yermak Plateau.

about 0.1 m per year. Thus, this highly simplified model suggest that if the turbulent mixing intensity in the cavity is weak, as is reasonable to assume, then the basal melting of the shelf would likely be less than snow accumulation on the top of the ice shelf. In turn, this indicates that the ocean-induced basal melting was weak enough to allow for a 1-km thick ice shelf over the Arctic Ocean as indicated by geophysical mapping data from bathymetric highs.

The geometry of the nearly landlocked Arctic Ocean likely played a major role for the formation of the MIS 6 ice shelf[39]. The Ross Ice Shelf, West Antarctica, fills its embayment even in today's 'warm' climate. Ice flux over its grounding line and calving flux are approximately in balance ($\sim$150 Gt per year), while surface mass balance outweighs underside melt by $\sim$4 Gt per year, implying an overall positive mass balance[40]. Major ice streams in the Arctic drained the Laurentide Ice and Barents–Kara ice sheets and fed the MIS 6 ice shelf (Fig. 1). Stabilized by perennial sea ice during inception, and by grounding on bathymetric highs, the ice shelf would have successively filled the entire central Arctic Ocean embayment. MIS 6 climate modelling studies indicate surface mass accumulation rates on the order 0.15–0.2 m per year for an ice shelf covering the entire Arctic Ocean[41], further supporting that a positive mass balance could be sustained if basal melt is less than 0.1 m per year. Furthermore, simulations of the MIS 6 glacial maximum, $\sim$140 ka, using a coupled Atmosphere–Ocean–Sea-Ice–Land model yield air surface temperatures over the Arctic Ocean that were 12–16 °C colder than pre-industrial temperatures, leading to perennial

sea-ice cover reaching >10 m thickness over the entire Arctic Ocean[42]. These simulations, although made without an ice shelf, can be used in the following physical consideration: over the long timescales considered here, the ice-shelf temperature distribution may be approximated to be constant at a certain time and certain ice-shelf thickness. As a consequence, the vertical heat flux is constant through the ice shelf, which from the heat conduction law implies a linear temperature profile[43]. At the base of the ice shelf, the temperature equals the local freezing point of the sea water, which is about -2.5 °C for a 1,000-m thick ice shelf. Taking the simulated surface temperatures[42] and the physically constrained basal temperature, the heat conduction law gives a vertical heat flux of about 0.1 W m^{-2} through the MIS 6 ice shelf. The simple oceanographic model presented here suggests that the heat flux from the ocean to the ice is an order of magnitude greater. Thus, melting, rather than accretion, is expected at the base of the MIS 6 ice shelf.

We propose that the area of the MIS 6 ice shelf roughly coincided with the central Arctic Ocean basin area, from the shelf break to the North Pole (Fig. 5b). This area is here calculated to about 4.25×10^6 km^2. The glacial landforms signifying ice-shelf grounding on the LR suggest that the ice shelf reached as deep as 1,280 m below present sea level where the ridge approaches the Siberian continental margin and 1,200 m on the Arlis Plateau. However, there are areas in the central Arctic Ocean where the ice shelf was thinner (Fig. 5b), and it was likely thicker near grounding lines. Assuming a sea level of $\sim$121 m lower than today for MIS 6 (ref. 44), we may approximate the average ice

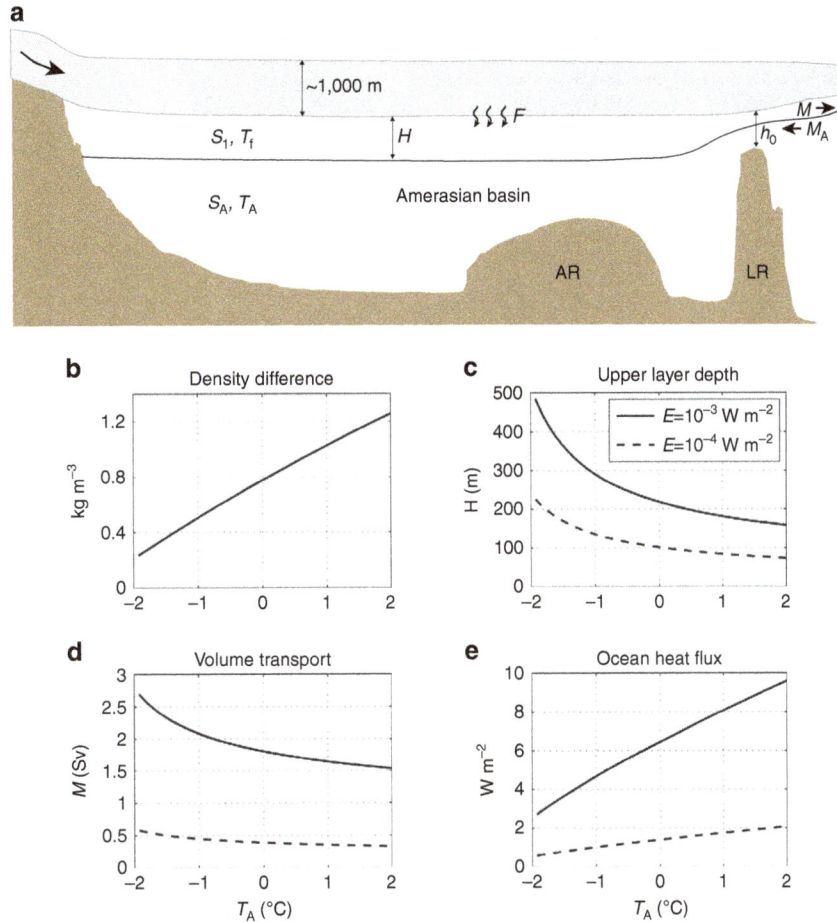

Figure 7 | Basic components of the conceptual oceanographic model developed for the 'ice cavity' in the Amerasian Basin underneath an ice shelf and calculated features of the cavity flow. (**a**) The ice shelf oceanographic model is assuming that the Lomonosov Ridge (LR) act as a barrier with a few open passages (h_0) for water exchange to the Eurasian Basin, in turn connected to the World Ocean through the 2,500 m deep Fram Strait. (**b-e**) Calculated features of the cavity flow beneath an ~1,000 m thick ice shelf as a function of the ambient ocean temperature T_A; the minimum value of T_A is set to the surface freezing temperature for $S_A = 35$, which is about $-2\,°C$. The oceanic heat flux is computed as $c\Delta TM/A$, where $A = 3 \times 10^{12}\,m^2$ is the approximate area of the Amerasian Basin. A heat flux per unit area of $1\,Wm^{-2}$ corresponds to a basal ice-shelf melt of about 0.1 m per year.

shelf draft to 1,000 m in the central Arctic Ocean basin. This implies an average thickness of 1,121 m using an ocean density of 1,028 kg m^{-3} and ice density of 917 kg m^{-3}. Adopting the Archimedes principle, the net sea-level-rise effect from melting a floating ice mass is equivalent to the difference between density of sea water and that of the melted ice[45]. The Arctic Ocean ice shelf with an average thickness of 1,121 m and an aerial extent as outlined in Fig. 5b has a volume of ~4.67 × 10^6 km^3, which if melted would approximately provide a sea-level rise of ~0.34 m. This rough estimate is derived using a world ocean area compensated for a 121-m lower glacial sea level than at present and assuming an equal spread of the meltwater, which we recognize is a simplified view of the sea-level change. We note that the estimated volume of the MIS 6 Arctic Ocean ice shelf amounts to approximately seven times of the volume of all ice shelves on Earth today.

Shackleton and Opdyke[46] showed in the early 1970s that benthic foraminifera from the equatorial Pacific recorded glacial/interglacial δ^{18}O differences on the order of 1.7‰. In light of these results and subsequent δ^{18}O paleoceanographic records derived from deep-sea sediment cores, the question of how much a thick Arctic Ocean ice shelf would affect the global ocean δ^{18}O, while only having a minor effect on sea level, was addressed[4,9,10]. Mix[9] used a δ^{18}O value for shelf ice of -40 ± 10‰ (SMOW) to

estimate that a 700 ± 300-m thick ice shelf covering the entire Arctic Ocean would increase the δ^{18}O ocean value by 0.12 ± 0.09‰ (SMOW). It follows that using a δ^{18}O value for shelf ice of -40 ± 10‰ (SMOW), and the estimated MIS 6 ice-shelf volume of 4.67 × 10^6 km^3, the MIS 6 δ^{18}O ocean can be calculated to have increased by 0.14 ± 0.03‰ (SMOW). This difference in calculated δ^{18}O ocean compared with Mix[9] is due to the larger volume of the MIS 6 ice shelf. Since the MIS 6 ice shelf only affects the eustatic sea level by ~0.34 m, this should be taken into account when interpreting the global ocean δ^{18}O value and its relation to global eustatic sea level of that time.

Finally, we emphasize that our results do not exclude the possibility that large ice shelves developed in the central Arctic Ocean during other glaciations than MIS 6. On the contrary, it seems likely that ice shelves were reoccurring components in most of the Quaternary glaciations considering the land-locked nature of the Arctic Ocean. Furthermore, the difficulty in precisely dating Arctic Ocean sediments leaves a high degree of uncertainty regarding the timing of several glacial landforms in general. However, the development of the large MIS 6 ice shelf may, in part, have been catalysed by the long duration of this glacial stage, where the eccentricity of Earth's orbit during the glacial maximum ~140 ka caused particularly cold springs and summers[47].

Methods

Geophysical mapping. The new seafloor mapping data presented in this work were collected during a two-leg 90-day long expedition in 2014 with Swedish *IB Oden* expedition within the Swedish–Russian–US Investigation of Climate, Cryosphere and Carbon interaction (SWERUS-C3) program. The expedition started/ended in Tromsö, Norway, July 5/October 3. Rotation between the two legs was carried out in Barrow, Alaska. The data presented in this work were collected during Leg 2 and outside the Russian Exclusive Economic Zone.

Bathymetric mapping was carried out using the Kongsberg EM 122 (12 kHz, $1° \times 1°$) multibeam echosounder hull mounted in *IB Oden*. This system has a Seatex Seapath 330 unit for integration of GPS navigation, heading and attitude. Sound velocity control was achieved through regular CTD (conductivity, temperature and depth) stations supplemented with XBT (expendable bathythermograph). All data were acquired using Kongsberg Seafloor Information System (SIS) and processed using a combination of the software Caris and Fledermaus-QPS. The processed data were gridded to a horizontal resolution ranging between 15×15 and 30×30 m. Seafloor morphology was interpreted in the three-dimensional environment of Fledermaus and maps were subsequently produced in the GIS software ArcMap. Sub-bottom profiles were collected using the Kongsberg SBP 120 $3° \times 3°$ chirp sonar integrated with the multibeam in *IB Oden*. The chirp sonar was operated continuously using a 2.5-7-kHz pulse. Any use of trade, firm or product names is for descriptive purposes only and does not imply endorsement by the US Government.

A simple model of ocean-ice-cavity interaction. Following equation (1), the remaining components of our two-layer ocean model are as follows: conservation of salt can be written as

$$\Delta SM = S_A F, \tag{2}$$

where $\Delta S \equiv S_A - S$ here is the salinity difference between inflowing water with salinity S_A and the outflowing ice-shelf water with salinity S. The thermodynamic balance of the ice-shelf water layer, which governs the basal melt F, is given by

$$FL = c\Delta T M_A - QA, \tag{3}$$

where c is the heat capacity of sea water, $\Delta T \equiv T_A - T_f$ the temperature difference, T_f the freezing temperature at the hydrostatic pressure of the shelf ice base, Q the upward heat flow per unit area through the ice, A the shelf area and L the latent heat of freezing. By combining the thermodynamic relation with equation (1) it is possible to obtain

$$F = \frac{c\Delta TM - QA}{L + c\Delta T}. \tag{4}$$

Combining this with the salinity equation yields

$$(\chi S_A - \Delta S)M = S_A \frac{QA}{L + c\Delta T}, \tag{5}$$

where we have introduced

$$\chi \equiv \frac{c\Delta T/L}{1 + c\Delta T/L} \approx c\Delta T/L \tag{6}$$

Here, χS_A represents a characteristic salinity difference; $\Delta T \sim 2\,°C$, gives $\chi \sim 0.03$. We will here consider the limiting case where the conductive heat flux through the ice is small compared with the ocean heat flux, that is, assuming that $c\Delta TM \gg QA$. This is reasonable for thick shelf ice and with this limit equations (5) and (6) yield

$$\Delta S \equiv S_A \frac{c\Delta T/L}{1 + c\Delta T/L}, \tag{7}$$

implying that the density difference is given by

$$\Delta\rho = \rho(S, T_A, D) - \rho(S_A - \Delta S, T_f, D), \tag{8}$$

where D is the pressure at the ice-ice base. Thus the salinity and density difference depends only on the salinity and temperature outside the cavity and the pressure at the shelf base. As $\Delta\rho$ is taken to be known, we can use the results of ref. 16 to compute the volume flow M and the upper-layer depth H in the cavity. We provide here a short summary of the physical explanation: The turbulence intensity in the ice cavity should be very weak, as there is no wind forcing and presumably weak tidal currents. Thus, it is reasonable to assume that the buoyant layer of the freezing-point water in contact with ice shelf is fairly shallow, having a depth that is small compared with the height of the openings in the LR, typically about 800 m. The widths of the gaps in the ridge are also generally wider than the internal Rossby radius, which should be less than 10 km. In this regime, the outflow in the ridge gaps will be subjected to rotational hydraulic control and a rough upper bound on the volume transport is given by ref. 48

$$M = \frac{g\Delta\rho H^2}{2\rho_0 f}, \tag{9}$$

where g is the acceleration of gravity, f the Coriolis parameter and $\rho_0 = 1,000\,\text{kg m}^{-3}$ a constant reference density. The diapycnal upwelling into the upper layer in the ice cavity is equal to M_A, the inflow of water from the outside. We use a simple but physically well-founded representation of M_A

$$M_A = \frac{\rho_0 EA}{g\Delta\rho H}, \tag{10}$$

where E is the supply of mixing energy per unit area. Further, when Q is negligible, Equation (4) implies that $F/M \approx c\Delta T/L \ll 1$. Thus, to a good approximation $M = M_A$, which yields the following steady-state relations for the upper-layer depth and volume flow

$$H = \left(\frac{AE2f\rho_0}{g^2\Delta\rho^2}\right)^{1/3}, \quad M = \left(\frac{A^2E^2}{g^2\Delta\rho 2f\rho_0}\right)^{1/3}. \tag{11}$$

Here, we assume that these relations provide rough upper bounds on the cavity upper layer depth and the exchange flow for given values of the ambient water temperature T_A and the mixing energy supply E. Figure 7 shows how the density difference, upper-layer depth, volume flow and oceanic heat transports vary with the temperature of the water outside the ice cavity. We have no information on the diapycnal mixing intensity. Therefore, the flow features are calculated for two values of the mixing energy input: $E = 10^{-3}\,\text{W m}^{-2}$, a value representative for the present-day Arctic Ocean, and $E = 10^{-4}\,\text{W m}^{-2}$. Given the absence of wind-generated turbulence in the cavity, the former value is certainly too large. Focusing on the more feasible case with $E = 10^{-4}\,\text{W m}^{-2}$, one finds that the results are not strongly sensitive to the ambient water temperature.

References

1. Thomson, W. Polar ice-caps and their influence on changing sea levels. *Trans. Geol. Soc. Glasgow* **8**, 322–340 (1888).
2. Donn, W. L. & Ewing, M. A theory of Ice Ages III. *Science* **152**, 1706–1712 (1966).
3. Mercer, J. H. A former ice sheet in the Arctic Ocean? *Palaeogeogr. Palaeoclimatol. Palaeoecol.* **8**, 19–27 (1970).
4. Broecker, W. S. Floating glacial ice caps in Arctic Ocean. *Science* **188**, 1116–1118 (1975).
5. Hughes, T. J., Denton, G. H. & Grosswald, M. G. Was there a late-Würm Arctic ice sheet? *Nature* **266**, 596–602 (1977).
6. Schytt, V., Hoppe, G., Blake, Jr W. & Grosswald, M. G. The extent of the Würm glaciation in the European Arctic. *Meddelanden från Naturgeografiska Institutionen vid Stockholms universitet* **79**, 207–216 (1966).
7. Grosswald, M. G. Late Weichselian ice sheets of northern Eurasia. *Quat. Res.* **13**, 1–32 (1980).
8. Denton, G. H. & Hughes, T. J. in *The Last Great Ice Sheets.* (eds Denton, G. H. & Hughes, T. J.) Ch. 8, 437–467 (Wiley Interscience, 1981).
9. Mix, A. in *The Geology of North America* Vol. K-3 (eds Ruddiman, W. F. & Wright, Jr H. E.) (The Geological Society of America, 1987).
10. Williams, D. F., Moore, W. S. & Fillon, R. H. Role of glacial Artic Ocean ice sheets in Pleistocene oxygen isotope and sea level records. *Earth Planet. Sci. Lett.* **56**, 157–166 (1981).
11. Johnson, R. G. & Andrews, J. T. Rapid ice-sheet growth and initiation of the last glaciation. *Quat. Res.* **12**, 119–134 (1979).
12. Johnson, R. G. & Andrews, J. T. Glacial terminations in the oxygen isotope record of deep sea cores: hypothesis of massive Antarctic ice-shelf destruction. *Palaeogeogr. Palaeoclimatol. Palaeoecol.* **53**, 107–138 (1986).
13. Vogt, P. R., Crane, K. & Sundvor, E. Deep Pleistocene iceberg plowmarks on the Yermak Plateau: sidescan and 3.5 kHz evidence for thick calving ice fronts and a possible marine ice sheet in the Arctic Ocean. *Geology* **22**, 403–406 (1994).
14. Polyak, L., Edwards, M. H., Coakley, B. J. & Jakobsson, M. Ice shelves in the Pleistocene Arctic Ocean inferred from glaciogenic deep-sea bedforms. *Nature* **410**, 453–459 (2001).
15. Jakobsson, M. First high-resolution chirp sonar profiles from the central Arctic Ocean reveal erosion of Lomonsov Ridge sediments. *Mar. Geol.* **158**, 111–123 (1999).
16. Jakobsson, M. *et al.* An Arctic Ocean ice shelf during MIS 6 constrained by new geophysical and geological data. *Quat. Sci. Rev.* **29**, 3505–3517 (2010).
17. Jokat, W. The sedimentary structure of the Lomonosov Ridge between 88° N and 80° N. *Geophys. J. Int.* **163**, 698–726 (2005).
18. Kristoffersen, Y. *et al.* Seabed erosion on the Lomonosov Ridge, central Arctic Ocean: a tale of deep draft icebergs in the Eurasia Basin and the influence of Atlantic water inflow on iceberg motion? *Paleoceanography* **19**, PA3006 (2004).
19. Clark, C. D. Mega-scale glacial lineations and cross-cutting ice-flow landforms. *Earth Surf. Processes Landforms* **18**, 1–29 (1993).
20. Jokat, W., Weigelt, E., Kristoffersen, Y., Rasmussen, T. & Schöne, T. New insigths into the evolution of the Lomonosov Ridge and the Eurasaian Basin. *Geophys. J. Int.* **122**, 378–392 (1995).
21. Niessen, F. *et al.* Repeated Pleistocene glaciation of the East Siberian continental margin. *Nat. Geosci.* **6**, 842–846 (2013).

22. Dove, D., Polyak, L. & Coakley, B. Widespread, multi-source glacial erosion on the Chukchi margin, Arctic Ocean. *Quat. Sci. Rev.* **92,** 112–122 (2014).

23. Jakobsson, M. *et al.* Arctic Ocean glacial history. *Quat. Sci. Rev.* **92,** 40–67 (2014).

24. Ottesen, D. & Dowdeswell, J. A. An inter-ice-stream glaciated margin: Submarine landforms and a geomorphic model based on marine-geophysical data from Svalbard. *Geol. Soc. Am. Bull.* **121,** 1647–1665 (2009).

25. Hanslik, D. *et al.* Quaternary Arctic Ocean sea ice variations and radiocarbon reservoir age corrections. *Quat. Sci. Rev.* **29,** 3430–3441 (2010).

26. Poirier, R. K., Cronin, T. M., Briggs, Jr W. M. & Lockwood, R. Central Arctic paleoceanography for the last 50 kyr based on ostracode faunal assemblages. *Mar. Micropaleontol.* **88–89,** 65–76 (2012).

27. O'Regan, M. *et al.* Constraints on the Pleistocene chronology of sediments from the Lomonosov Ridge. *Paleoceanography* **23,** PA1S19 (2008).

28. Jakobsson, M. *et al.* Manganese and color cycles in Arctic Ocean sediments constrain Pleistocene chronology. *Geology* **28,** 23–26 (2000).

29. Backman, J., Fornaciari, E. & Rio, D. Biochronology and paleoceanography of late Pleistocene and Holocene calcareous nannofossils across the Arctic Basin. *Mar. Micropaleontol.* **72,** 86–98 (2009).

30. Matthiessen, J., Knies, J., Nowaczyk, N. R. & Stein, R. Late Quaternary dinoflagellate cyst stratigraphy at the Eurasian continental margin, Arctic Ocean: indications for Atlantic water inflow in the past 150,000 years. *Global Planet. Change* **31,** 65–68 (2001).

31. Polyak, L., Curry, W. B., Darby, D. A., Bischof, J. & Cronin, T. M. Contrasting glacial/interglacial regimes in the western Arctic Ocean as exemplified by a sedimentary record from the Mendeleev Ridge. *Palaeogeogr. Palaeoclimatol. Palaeoecol.* **203,** 73–93 (2004).

32. Cronin, T. M. *et al.* Quaternary ostracode and foraminiferal biostratigraphy and paleoceanography in the western Arctic Ocean. *Mar. Micropaleontol.* **111,** 118–133 (2014).

33. Dowdeswell, J. A. *et al.* High-resolution geophysical observations from the Yermak Plateau and northern Svalbard margin: implications for ice-sheet grounding and deep-keeled icebergs. *Quat. Sci. Rev.* **29,** 3518–3531 (2010).

34. Arndt, J. E., Niessen, F., Jokat, W. & Dorschel, B. Deep water paleo-iceberg scouring on top of Hovgaard Ridge–Arctic Ocean. *Geophys. Res. Lett.* **41,** 2014GL060267 (2014).

35. Jakobsson, M. *et al.* Pleistocene stratigraphy and paleoenvironmental variation from Lomonosov Ridge sediments, central Arctic Ocean. *Global Planet. Change* **31,** 1–22 (2001).

36. Polyak, L., Darby, D., Bischof, J. & Jakobsson, M. Stratigraphic constraints on late Pleistocene glacial erosion and deglaciation of the Chukchi margin, Arctic Ocean. *Quat. Res.* **67,** 234–245 (2007).

37. Cronin, T. M. *et al.* Deep Arctic Ocean warming during the last glacial cycle. *Nat. Geosci.* **5,** 631–634 (2012).

38. Nøst, O. A. *et al.* Eddy overturning of the Antarctic slope front controls glacial melting in the Eastern Weddell Sea. *J. Geophys. Res. C: Oceans* **116,** C11014 (2011).

39. Kirchner, N., Furrer, R., Jakobsson, M., Zwally, H. J. & Robbins, J. W. Statistical modeling of a former Arctic Ocean ice shelf complex using Antarctic analogies. *J. Geophys. Res. Earth Surf.* **118,** 1–13 (2013).

40. Depoorter, M. A. *et al.* Calving fluxes and basal melt rates of Antarctic ice shelves. *Nature* **502,** 89–92 (2013).

41. Colleoni, F., Krinner, G., Jakobsson, M., Peyaud, V. & Ritz, C. Influence of regional parameters on the surface mass balance of the Eurasian ice sheet during the peak Saalian (140 kya). *Global Planet. Change* **68,** 132–148 (2009).

42. Colleoni, F., Werkele, C. & Masina, S. *Long-Term Safety of a Planned Geological Repository for Spent Nucleaer Fuel in Forsmark - Estimate of Maximum Ice Sheet Thicknesses* 1–96 (Swedish Nuclear Fuel and Waste Management Co, 2014).

43. Pierrehumbert, R. T., Abbot, D. S., Voigt, A. & Koll, D. Climate of the Neoproterozoic. *Ann. Rev. Earth Planet. Sci.* **39,** 417–460 (2011).

44. Siddall, M. *et al.* Sea-level fluctuations during the last glacial cycle. *Nature* **423,** 853–858 (2003).

45. Noerdlinger, K. W. & Brower, K. R. The melting of floating ice raises the ocean level. *Geophys. J. Int.* **170,** 145–150 (2007).

46. Shackleton, N. J. & Opdyke, N. D. Oxygen isotope and Palaeomagnetic stratigraphy of equatorial pacific core V28-238: oxygen isotope temperatures and ice volumes on a 10^5 year and 10^6 year scale. *Quat. Res.* **3,** 39–55 (1973).

47. Colleoni, F. *On the Late Saalian Glaciation (160 - 140 ka) – A Climate Modeling Study PhD thesis* (Stockholm University/Université Joseph Fourier, 2009).

48. Gill, A. E. The hydraulics of rotating-channel flow. *J. Fluid Mech.* **80,** 641–671 (1977).

49. Svendsen, J. I. *et al.* Late quaternary ice sheet history of northern Eurasia. *Quat. Sci. Rev.* **23,** 1229–1271 (2004).

50. Dyke, A. S. *et al.* The Laurentide and Innuitian ice sheets during the last glacial maximum. *Quat. Sci. Rev.* **21,** 9–31 (2002).

51. Engels, J. L., Edwards, M. H., Polyak, L. & Johnson, P. D. Seafloor evidence for ice shelf flow across the Alaska–Beaufort margin of the Arctic Ocean. *Earth Surf. Processes Landforms* **32,** 1–17 (2008).

52. Jakobsson, M. *et al.* The International Bathymetric Chart of the Arctic Ocean (IBCAO) Version 3.0. *Geophys. Res. Lett.* **39,** L12609 (2012).

Acknowledgements

The SWERUS-C3 expedition was financed by Knut and Alice Wallenberg Foundation, Swedish Polar Research Secretariat and Stockholm University. Research grants to individual scientist were provided by the Swedish Research Council (VR), National Science Foundation, US Geological Survey Climate and Land Use R&D Program, Government of the Russian Federation (grant no. 2013-220-04157/no. 14, Z50.31.0012/03.19.2014), the Russian Foundation for Basic Research (Nos. 13-05-12028, 13-05-12041, 12-05-12029 and 13-05-12015) and The Russian Science Foundation (grant No. 15-17-20032). The data reported in this paper are archived at the Bolin Centre Database: http://bolin.su.se/data/.

Author contributions

M.J. and J.N. developed the concept of this paper and led the writing. J.B., M.O. and T.C. led the sediment core stratigraphic work. All authors contributed to the development of the paper either by directly taking part in the writing process or in interpretation of results.

Additional information

Recent increases in Arctic freshwater flux affects Labrador Sea convection and Atlantic overturning circulation

Qian Yang[1], Timothy H. Dixon[1], Paul G. Myers[2], Jennifer Bonin[3], Don Chambers[3] & M.R. van den Broeke[4]

The Atlantic Meridional Overturning Circulation (AMOC) is an important component of ocean thermohaline circulation. Melting of Greenland's ice sheet is freshening the North Atlantic; however, whether the augmented freshwater flux is disrupting the AMOC is unclear. Dense Labrador Sea Water (LSW), formed by winter cooling of saline North Atlantic water and subsequent convection, is a key component of the deep southward return flow of the AMOC. Although LSW formation recently decreased, it also reached historically high values in the mid-1990s, making the connection to the freshwater flux unclear. Here we derive a new estimate of the recent freshwater flux from Greenland using updated GRACE satellite data, present new flux estimates for heat and salt from the North Atlantic into the Labrador Sea and explain recent variations in LSW formation. We suggest that changes in LSW can be directly linked to recent freshening, and suggest a possible link to AMOC weakening.

[1] School of Geosciences, University of South Florida, 4202 E Fowler Avenue, Tampa, Florida 33620, USA. [2] Department of Earth and Atmospheric Sciences, University of Alberta, 1-26 ESB, Edmonton, Alta, Canada T6G 2E3. [3] College of Marine Science, University of South Florida, St. Petersburg, Florida 33701, USA. [4] Institute for Marine and Atmospheric Research Utrecht, Utrecht University, P.O. Box 80.005, 3508 TA, Utrecht, The Netherlands. Correspondence and requests for materials should be addressed to Q.Y. (email: qianyang@mail.usf.edu).

It has long been accepted that the Atlantic Meridional Overturning Circulation (AMOC) has two stable modes[1-3], and that anthropogenic warming could weaken or shut down the AMOC[4,5]. Recent accelerated melting of the Greenland ice sheet is freshening the North Atlantic[6-10]. So-called 'hosing experiments', where freshwater may be distributed over broad or narrow regions of the North Atlantic in numerical models, have been used to study the sensitivity of the AMOC to freshwater flux[11-17]. Some of these studies suggest that AMOC strength is sensitive to Greenland melting[11,17], while others do not[12,14,16]. A few studies suggest that freshwater additions of 0.1 Sv (100 mSv)[18-20] or possibly less[11,17] could affect the AMOC.

Changes in the AMOC are difficult to measure directly: currents that comprise the deeper, southward flowing portions can be diffuse and/or spatially and temporally variable, and instrumental drift can mask subtle, long-term changes in oceanic properties. It is also challenging to separate changes forced by anthropogenic warming from natural variability. The AMOC is difficult to model numerically: model grids may be too coarse to reflect realistic oceanic processes and geographic constraints, and feedbacks among atmosphere, ocean and cryosphere (land and sea ice) are poorly known.

The Labrador Sea is a key location for the formation of one of the dense, deep-water components of the AMOC via winter convection; however, the process is sensitive to surface conditions[21]. Wood et al.[5] suggest the possibility of a shutdown in Labrador Sea convection in response to global warming. Kuhlbrodt et al.[22] provide a theoretical stability analysis, and suggest that winter convection in the Labrador Sea can be turned off by increased freshwater input. Unfortunately, winter convection here is difficult to observe directly because of extreme conditions and its small spatial scale.

Here we consider recent Labrador Sea changes associated with an increased freshwater flux. We derive a new estimate for recent increased freshwater flux into the sub-polar North Atlantic, and suggest that because of the clockwise nature of ocean circulation around Greenland[23], most of this increase is being focused towards the Labrador Sea (Fig. 1), magnifying its impact and increasing the likelihood of significant effects on the AMOC.

Results

Recent accelerated melting of the Greenland ice sheet. Numerous studies have described recent acceleration of Greenland's ice mass loss[6-10]. We use GRACE data updated to October 2014 to derive a new acceleration estimate and its onset time (Methods). GRACE data and uncertainty estimates follow Bonin and Chambers[24]. We fit a constant acceleration model to the data, and extrapolate the best-fit model back to the time of zero mass loss rate, obtaining 20-Gt per square year acceleration with a start time of 1996 ± 1.4 years (Fig. 2). Several lines of evidence suggest that the ice sheet was relatively stable from 1980 to the early 1990s (refs 25,26), and we use that assumption in our modelling of GRACE data and freshwater flux calculations (below and Methods section).

Irminger Water heat and salt fluxes. Warming of sub-polar mode waters including Irminger Water in the mid- to late-1990s (refs 27,28) is thought to influence coastal mass loss in Greenland by increasing submarine melting of outlet glaciers and related dynamic effects[29-31]. Here we examine the variability of heat and salt fluxes of Irminger Water along three sections (Fig. 1) offshore southwest coastal Greenland for the period 1949–2013 (Methods). Currents associated with the sub-polar gyre here are quite compact as they round the southern tip of Greenland, limiting spatial variability and facilitating accurate flux

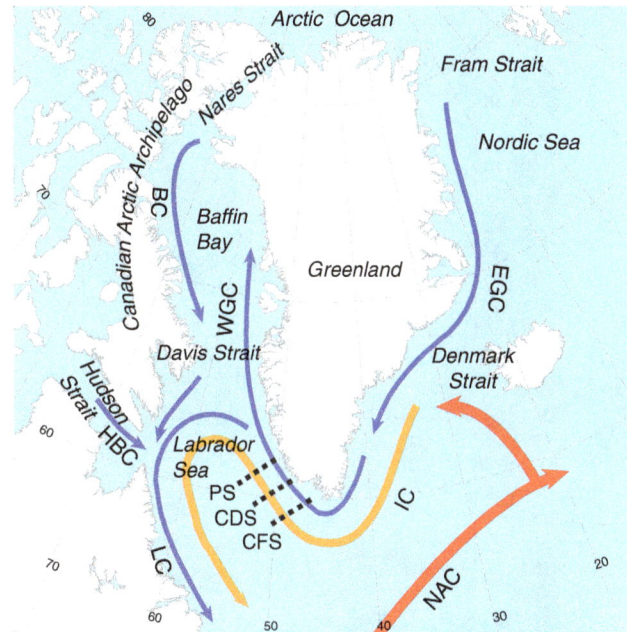

Figure 1 | Study region showing oceanographic sections and major currents around Greenland. Red and orange arrows indicate Atlantic-origin water and blue arrows indicate Arctic-origin water. BC, Baffin Current; CDS, Cape Desolation Section; CFS, Cape Farewell Section; EGC, East Greenland Current; HBC is Hudson's Bay Current; IC, Irminger Current; LC, Labrador Current; NAC, North Atlantic Current; PS, Paamiut Section; WGC is West Greenland Current. Three-dimensional structure of major water masses in the Labrador Sea is shown in Supplementary Fig. 1.

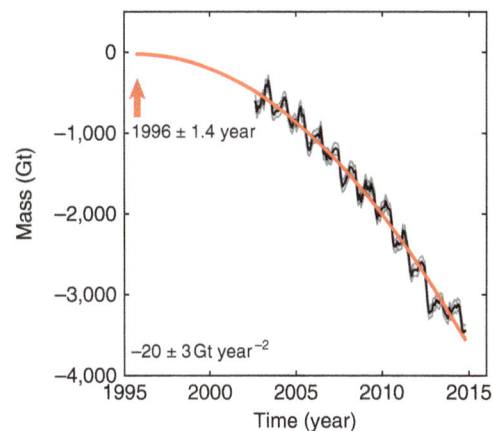

Figure 2 | Mass change of Greenland estimated from GRACE for the period 2002-2014. Black curve shows data, grey shading indicates monthly uncertainty and red curve shows the best fitting constant acceleration model. Onset time of acceleration defined when the rate of mass change is zero, in 1996 (red arrow), with mass arbitrarily set to zero.

measurements because the cross-section area of current is well defined. Note that, while the flux (sensu stricto) is flow rate per unit area and transport (or total flux) represents the flux integrated over the larger area of interest, the terms 'flux' and 'transport' are often used interchangeably in the oceanographic literature. We follow the broader (sensu lato) usage here.

We carry out our analysis on the upper 700 m, the greatest depth common to all years, binned on a 2-m vertical grid. Time series of heat and salt fluxes at the three sections are shown in Fig. 3. At the southernmost Cape Farewell section, both heat and salt fluxes experienced a large multi-year anomaly around 1995,

a

b

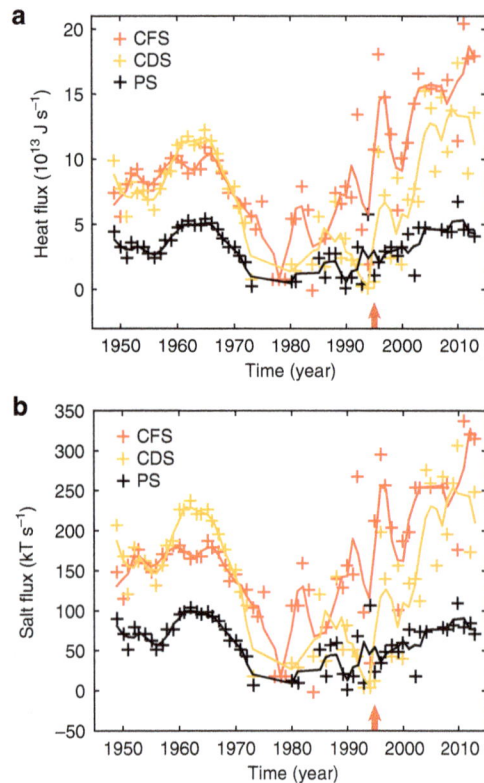

Figure 3 | Heat and salt fluxes of Irminger Water for the period 1949–2013. (**a**) Heat and (**b**) salt fluxes of Irminger Water are measured at three sections in southwest Greenland. Locations of three sections are shown in Fig. 1. CDS, Cape Desolation Section; CFS, Cape Farewell Section; PS, Paamiut Section. Solid line represents a 3-year running average, yearly data shown by plus signs. Red arrow marks the onset time of accelerated mass loss for Greenland estimated from GRACE (Fig. 2).

followed by another in the late 1990s. The heat flux was 80% higher than a previous multi-year anomaly in the 1960s. Similar variability is seen at the more northerly Cape Desolation section, although salinities and heat are generally lower, and only exceed previous levels after 2000. No significant anomalies were observed at the northernmost Paamiut section during these times; however, the heat and salt fluxes are still roughly 50% higher after 2000 than they were in the 1980s, and approach levels that are not seen since the 1960s. Thus, we conclude that Irminger Water became warmer and saltier in the mid-late 1990s, which agrees well with the onset time of recent accelerated Greenland mass loss (Fig. 2). This is consistent with the idea that accelerating ice mass loss in the mid-late 1990s reflects, at least in part, the appearance of warmer Irminger Water on the peripheral continental shelf at that time[29]. The anomalous heat flux we observe off southern Greenland in the mid-1990s can be directly tied to warming of the North Atlantic (Supplementary Fig. 2; see also ref. 31).

Northward reduction in heat and salt transport between the Cape Desolation and Paamiut sections likely reflects strong offshore eddy transport[32], advecting Irminger Water into the interior of the Labrador Sea. However, since the sections are only occupied once a year in summer, some seasonal aliasing is possible. The eddies also transport fresh shelf water into the Labrador Sea[33].

Estimates of the freshwater flux into the Labrador Sea. Major sources of freshwater entering the Labrador Sea include precipitation, oceanic transport and melt from the Greenland ice

sheet, glaciers in the Canadian Arctic Archipelago (CAA) and Arctic sea ice. Precipitation in the Labrador Sea region is about 20–30 mSv (ref. 34), and there has been a general increase over the North Atlantic region in the last few decades as the hydrologic cycle accelerates[35]. Oceanic transport from the Arctic Ocean is the largest source of Labrador Sea freshwater and is derived from several sources, including the difference between precipitation and evaporation, river discharge and fractionation associated with annual sea ice formation. Peterson et al.[36] show that the average annual river discharge from six rivers in Eurasia into the Arctic Ocean has increased by 7% (~4 mSv) from 1936 to 1999. The Arctic Ocean exports low-salinity water to the North Atlantic through two main pathways: Fram Strait east of Greenland and the CAA west of Greenland. The CAA pathway has three main routes: Barrow Strait, Nares Strait and Cardigan Strait-Hell Gate. Roughly, 100 mSv of freshwater is exported through each of the east and west pathways, relative to a reference salinity of 34.80 (ref. 37). Within broad error bars, oceanic transport from the Arctic Ocean is relatively stable on the decadal timescale, although there has been some reduction through the CAA and then Davis Strait, and shorter-term fluctuations are common[37–39].

Here we focus on three Arctic freshwater sources that are undergoing rapid increases, which likely contribute freshwater to the Labrador Sea, and which can be estimated from remote observations: the Greenland ice sheet, CAA glaciers and Arctic sea ice. We also consider snowmelt runoff from tundra in Greenland and the CAA as they follow directly from the same models used to quantify Greenland ice sheet and CAA glacier melt[40,41]. As we are not considering the large Arctic oceanic transport term and several other sources, our estimate is a minimum estimate.

The freshwater flux from Greenland is composed of ice and tundra runoff plus ice discharge; this quantity is equal to accumulation minus mass balance (Methods). We derive mass balance for Greenland from GRACE, while accumulation is obtained from the RACMO2.3 model[42,43]. Our GRACE data suggest that mass loss of the Greenland ice sheet accelerates from 1996 onwards (Fig. 2; Methods). Our mass balance estimate agrees with the estimate of Box and Colgan[26], with the Greenland ice sheet in near balance from 1980 to about 1996, after which it starts to lose mass (Supplementary Fig. 3). Therefore, we assume that between 1980 and 1996 the freshwater flux from Greenland is approximately equal to accumulation; after 1996, the freshwater flux from Greenland equals the sum of mass loss and accumulation (Supplementary Fig. 3). Since the accumulation is highly variable from year to year, we smooth it with a 5-year running mean. Figure 4 shows the resulting freshwater flux estimates from Greenland. This approach yields freshwater flux estimates that agree with those of Bamber et al.[40] during the period of data overlap, once a correction for solid ice discharge is applied[8] (Supplementary Fig. 4). Freshwater from the CAA is approximated by ice and tundra runoff predicted by RACMO2.3 since ice discharge (0.16 mSv) is negligible[44].

Large amounts of Arctic sea ice and freshwater are exported from the Arctic Ocean to the North Atlantic through several pathways. Of these, Fram Strait and the CAA are the major ones; nearly all (~98%) Arctic Ocean exports drain through them[37]. However, there are large uncertainties in these fluxes[37]. We focus on changes in the freshwater flux as inferred from recent accelerated melting of Arctic sea ice, assuming that the change is partitioned the same way as the total export, that is, 98% of the change is advected through Fram Strait and the CAA. Changes in the annual minimum of Arctic sea ice volume are a relevant indicator (see Methods and Supplementary Methods). We first use the annual minimum volume predicted by the Pan-Arctic Ice

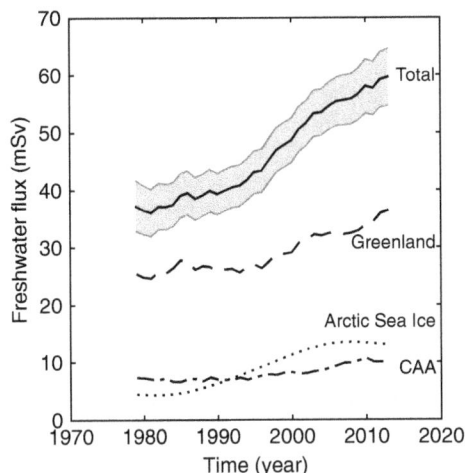

Figure 4 | Freshwater flux from Greenland and CAA and Arctic sea ice for the period 1979–2013. For Arctic sea ice, we plot only changes in flux (see text). The sum of these sources (Total) is also plotted. Grey shading indicates propagated uncertainty (see Supplementary Note 1).

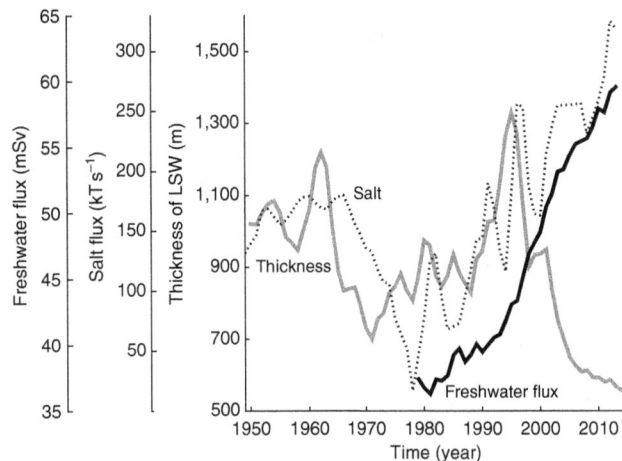

Figure 5 | Thickness of LSW and total freshwater flux and salt flux of Irminger Water. Grey solid line indicates the thickness of LSW, black solid line indicates total freshwater flux and dotted line indicates salt flux of Irminger Water. Thickness and salt flux are smoothed with a 3-year running mean. Thickness is obtained from the objective analysis of EN4.0.2 data set from the UK Met Office Hadley Center[52]. Thickness is averaged over 50° N–65° N and 38° W–65° W. Expression of salt flux in terms of freshwater flux is shown in Supplementary Fig. 7.

Ocean Modeling and Assimilation System (PIOMAS) model[45]. We also apply the same method to the Arctic sea ice extent and sea ice area data sets[46], where 'extent' defines a region as either 'ice-covered' or 'not ice-covered' using a threshold of 15%; 'area' is a more conservative estimate, defined as the percentage of actual sea ice within a given data cell. We assume a standard ice thickness of 2 m (ref. 47) to convert ice extent and ice area to volume, obtaining results that are somewhat smaller than the PIOMAS volume model. Figure 4 shows results from the PIOMAS volume model. Results from the other two approaches are shown in Supplementary Figs 5 and 6.

Figure 4 also shows the summed result from these various freshwater sources (recall that this summed value does not include several major sources and is therefore a minimum estimate), which is our estimate of the freshwater flux into the sub-polar North Atlantic. The freshwater flux from Greenland is relatively stable from 1979 to the mid-late 1990s and then increases. The freshwater flux from the CAA is relatively stable until the early 2000s and then increases. Freshwater flux from Arctic sea ice increases mainly during the period 1990–2000. The total freshwater flux for the sub-polar North Atlantic from these sources is about 60 mSv by 2013, with an increase of 20 mSv during the last two decades. Of this, ~12 mSv comes from the Greenland ice sheet and CAA glaciers, whereas ~8 mSv represents excess melting of Arctic sea ice.

Focused freshwater flux into the Labrador Sea has the potential to disrupt the AMOC by increasing the buoyancy of surface waters and reducing the formation of dense, deep water[13]. How much of the enhanced freshwater flux that we calculate actually winds up in the Labrador Sea?

Myers et al.[33,48] showed that a significant fraction of freshwater originating in and around Greenland is transported to the Labrador Sea: melt water from eastern Greenland is entrained in the East Greenland Current, where it moves south and merges with the Irminger Current as it rounds Cape Farewell; melt water from southwestern Greenland joins the West Greenland Current, similarly merging with the Irminger Current (Fig. 1). Melt water from the CAA enters the Labrador Sea through Davis and Hudson straits, either directly or indirectly[49]. The pattern of boundary currents and eddy activity around Greenland and Labrador insures that at least 75 per cent of the freshwater flux from the Greenland ice sheet and CAA eventually winds up in the Labrador Sea (Supplementary Methods). Freshwater and sea ice

drained from the Arctic Ocean moves south through Fram Strait and the CAA[37], also contributing to freshening of the Labrador Sea both remotely and locally[50,51]. We estimate that at least 65 per cent of freshwater and sea ice exported from the Arctic Ocean through Fram Strait and the CAA ultimately makes it to the Labrador Sea (Supplementary Methods). Assuming that these estimates are correct, of the 20-mSv freshwater flux increase that we estimate, at least 14 mSv (70%) winds up in the Labrador Sea (Supplementary Methods). Given typical coastal current velocities, most of this freshwater is transported to the Labrador Sea within 3–12 months. Some freshwater from the CAA may take 2–3 years to reach the Labrador Sea due to recirculation and storage in Baffin Bay and/or recirculation in the sub-polar gyre.

Impact of increased the freshwater flux on deep water formation. To investigate effects of increased freshwater flux on deep water formation in the Labrador Sea, we can either look at the mean density of Labrador Sea Water (LSW) within a given depth range or look at the thickness of LSW as defined by a given density range. We used both approaches, obtaining similar results. Figure 5 shows results from the second approach, where we calculate the thickness of LSW, defined by $\sigma_\theta = 27.74 - 27.80 \, \text{kg m}^{-3}$, from 1950 to 2013, using the objective analyses of the EN4.0.2 data set from the UK Met Office Hadley Center[52]. The data set includes monthly temperature and salinity, with a spatial resolution of 1° × 1° and 42 depth intervals (5–5,350 m) from 1900 to present. Results for density over a fixed depth range (1,000–2,500 m) are shown in Supplementary Fig. 8.

Figure 5 shows the time series of LSW thickness, compared with our estimate of freshwater flux and with the Irminger salt flux time series. From 1950 to the mid-1990s, Irminger salt flux and LSW thickness are weakly correlated ($R = 0.3$, $P = 0.03$), and both show multidecadal oscillations, with highs in the 1960s, lows in the 1980s and highs in the 1990s. In particular, LSW thickness increased significantly (by 65%) between 1990 and 1995 when the salt flux increased, consistent with the idea that dense deep water in the Labrador Sea originates from warm, saline North Atlantic water that subsequently experiences winter cooling. However, this relationship begins to break down in the mid- to late-1990s, when the freshwater flux from Greenland and other sources increased rapidly. Since then, LSW thickness decreased continuously, reaching lows not observed since the early 1970s, despite continued high salt flux. One interpretation of this is that the increased freshwater flux has now overwhelmed increased salt flux from the Atlantic, and has begun to influence LSW formation. Recall that the increased salt flux from the Atlantic is accompanied by an increased heat flux (Fig. 3), which promotes melting of marine-terminating outlet glaciers in southern Greenland[29,53] and an increased freshwater flux.

Our data are consistent with recent studies, showing a decline in the thickness of the dense mode of LSW since 1994/95, with a switch to a less dense and presumably fresher and warmer upper mode[54,55]. Yashayaev et al.[56] show declining upper salinity since the mid-2000s and suggest that salinity is the controlling factor for ocean stratification in this region. Declining upper layer salinity would weaken or even prevent Labrador Sea convection. However, cold winter air also plays a role in LSW formation. Severe winter conditions and strong air–sea heat exchange for the period 1990–1995 may have contributed to the increased LSW thickness[57], while milder winter conditions and weaker cooling since 1995 may have contributed to LSW decline[58]. The Labrador Sea is also sensitive to multidecadal climate variations. Hydrographic properties in the Labrador Sea exhibit multidecadal variability that resemble the Atlantic Multidecadal Oscillation and the North Atlantic Oscillation[56], and these variations are obvious in the flux (Fig. 3) and LSW thickness (Fig. 5) time series. Bidecadal variability in the Labrador Sea forced by volcanic activity has also been proposed[59]. Despite these complications, our data clearly show a steep, recent increase in the freshwater flux into the Labrador Sea and a steep decline in LSW thickness (and density) at the same time (Fig. 5), which is inconsistent with the estimated salt flux into the region. This suggests a potential impact on the formation of North Atlantic Deep Water.

Discussion

Our reconstructed annual freshwater flux for the sub-polar North Atlantic reaches 60 mSv in 2013, with an increase of 20 mSv in the last two decades (Fig. 4). At least 70 per cent (14 mSv) of this increased freshwater is focused towards the Labrador Sea (Supplementary Methods). This is a minimum estimate since we do not consider other major sources. LSW formation may reflect a delicate balance between this cold freshwater and warm, salty North Atlantic water from the Irminger Current. The flux of freshwater from Greenland may in turn be influenced by warm Atlantic water and its influence on the regional ocean and atmosphere, a potentially important feedback in the system.

Since LSW is an important component of the dense southward return flow of the AMOC, factors influencing LSW formation may in turn have an impact the AMOC. Hosing experiments show different sensitivities of the AMOC to freshwater fluxes at high latitudes[11–17]. Hu et al.[14] suggest that freshwater inputs much larger than we observe are required. On the other hand, Fichefet et al.[11] suggest that freshwater flux anomalies as small as 15 mSv will affect the AMOC. Brunnabend et al.[17] suggest that freshwater flux anomalies as small as 7 mSv applied over 30 years could have an impact on the AMOC. Different model outcomes partly reflect their spatial resolution, the degree to which freshwater fluxes are focused towards the Labrador Sea, and the timescale over which the anomalous flux is applied. Swingedouw et al.[15] compared different model responses to freshwater release around Greenland, assuming freshwater focusing into the Labrador Sea. They show significant AMOC weakening after several decades with a flux anomaly of 100 mSv.

If our inference that the sub-polar gyre's coastal currents focus melt water from Greenland, CAA glaciers and Arctic sea ice into the Labrador Sea is correct, then present rates of increased freshwater flux may be sufficient to influence convection in the Labrador Sea and, by implication, the AMOC. Northward decreases in heat and salt fluxes across our three sections in southwest Greenland indicate a strong mixing of coastal water and westward advection into the Labrador Sea. Eddy kinetic energy reaches a local maximum offshore Cape Desolation and Paamiut, where a front develops between Irminger Water and fresh shelf water, promoting baroclinic instability and eddy formation; these eddies propagate westwards into the Labrador Sea. Local bathymetric structures, especially the sill at Davis Strait, also promote westward propagation of coastal water from southwestern Greenland. Recent high-resolution eddy-permitting or eddy-resolving numerical models support this type of spatial focusing, and indicate decline or even shutdown of Labrador Sea convection with an enhanced freshwater flux from Greenland[60] or from the Arctic Ocean through the CAA[61]. Since freshwater lenses can retain their integrity for some time, 'temporal focusing' may also be important. Summer (June, July and August) freshwater fluxes from Greenland and CAA's ice and snow runoff greatly exceed the annual mean. Summer freshwater flux from Greenland and the CAA increased by ~50 mSv from mid-late 1990s to 2013, reaching a height of 150 mSv in 2012 (Supplementary Fig. 9), potentially limiting convection during the subsequent winter.

We suggest that recent freshening in the vicinity of Greenland is reducing the formation of dense LSW, potentially weakening the AMOC. Recent observations are beginning to document declines in the AMOC[62–64], consistent with our hypothesis. Longer time series will be required to confirm this link, but our preliminary results suggest that detailed studies of Labrador Sea hydrography and proximal sources of freshwater, including Greenland, have the potential to improve our understanding of AMOC variability and sensitivity to anthropogenic warming.

Methods

GRACE data. The GRACE time series were created via the least squares inversion method described in ref. 24. Release-05 GRACE data from the Center for Space Research were used, with the standard post-processing applied as described in that paper: C_{20} is replaced by Satellite Laser Ranging estimates, a geocentre model is added, GIA is corrected for and the monthly averages of the Atmosphere and Ocean Dealiasing product are restored.

The inversion technique is designed to localize the mass change signal, such that coastal mass loss from Greenland does not leak into the ocean or into interior Greenland because of GRACE's inherently low spatial resolution. Briefly, the method involves breaking Greenland and the surrounding area into pre-defined regions (Greenland drainage basins; Supplementary Fig. 10). Each region is assumed to have a uniform mass distribution when gridded as $1° \times 1°$-binned kernels. The transformation to degree/order 60 spherical harmonics is then made on each individual regional kernel, resulting in a smoothed version of each region that mimics what GRACE would see from its limited resolution, if a uniform mass of 1 was placed over the kernel, with zeroes elsewhere.

The goal is to find a set of multipliers for each region that most closely describes mass distribution over Greenland, given the assumption of uniform weights across each pre-defined shape. A least squares method is used to fit an optimal multiplier to each basin simultaneously, such that the combination of the multiplier times the smoothed basin kernels best fits the actual (smoothed) GRACE data for that month. An optimal amount of process noise is added to stabilize the solution[24].

The GRACE mass balance in this paper is the sum of the individual signals from the 16 Greenland regions (Supplementary Fig. 10).

Irminger Water heat and salt flux analysis. Details of the data collection and analysis are discussed in Myers et al.[27] and summarized here. Before 1984, the estimates are based on a climatological analysis of the Labrador Sea. The 1984–2013 observations are collected on a set of standard sections by the Danish Meteorological Institute. Each section (Fig. 1) involves the same five stations; however, in some years only three or four stations could be occupied. The sections are occupied annually in most years, in late June or early July. Direct sampling using bottles was performed in 1984–1987, while conductivity–temperature–depth data were collected in later years. We carry out our analysis on the upper 700 m, the deepest depth common to all years, binned on a 2-m vertical grid. For current motions, we determine the geostrophic velocity, relative to 700 dbar ($\sim$700 m depth) or the bottom in shallower water, for each pair of stations at each depth, and add an estimate of the barotropic velocity[33]. If data are missing, we do not include that point in the calculation. We calculate heat flux (Q_t) and salt flux (Q_s) at each depth and then sum those whose temperature and salinity are consistent with Irminger Water to obtain the total transport:

$$Q_t = \rho \cdot C_p \cdot \int_{s=1}^{s=5} \int_{z=-700}^{z=0} v(s,z) \cdot (T(s,z) - T_{ref}) dz ds \tag{1}$$

$$Q_s = \int_{s=1}^{s=5} \int_{z=-700}^{z=0} v(s,z) \cdot (S(s,z) - S_{ref}) dz ds \tag{2}$$

where ρ and C_p are ocean water density and heat capacity, respectively; $v(s,z)$, $T(s,z)$ and $S(s,z)$ are velocity, temperature and salinity in station s at depth z, respectively; T_{ref} is the reference temperature (0 °C) and S_{ref} is the reference salinity (34.80). Here we choose a broad definition including both pure and modified Irminger Water, with temperatures warmer than 3.5 °C and salinity higher than 34.88 (ref. 27).

Freshwater flux. To estimate the freshwater flux from Greenland, we first use a simple constant acceleration model to fit the monthly GRACE mass balance data:

$$M(t_i) = a + bt_i + \frac{1}{2}ct_i^2 \tag{3}$$

where $M(t_i)$ ($i = 1,2,3\ldots..n$) are GRACE monthly solutions, a is the initial mass of Greenland, b is the initial mass balance and c is the acceleration term. Given the estimated parameters, the mass balance (MB) of Greenland can be represented by:

$$MB(t_i) = b + ct_i \tag{4}$$

The start time of recent accelerated mass loss is the time t_i when $MB(t_i)$ is zero. The mass balance of Greenland is:

$$MB = SMB - D \tag{5}$$

where SMB is surface mass balance and D is discharge, related to freshwater flux (FWF) by:

$$SMB = A - R \tag{6}$$

$$FWF = R + D \tag{7}$$

where A is the accumulation and R is runoff.

We then calculate freshwater flux from Greenland using the above relations, rewriting them as:

$$FWF = A - MB \tag{8}$$

where accumulation (A) is predicted by RACMO2.3 and MB is estimated from the GRACE data. Note that accumulation is defined over ice and tundra, and mass balance is the total mass balance of Greenland, including ice and tundra. Therefore, freshwater flux from Greenland is composed of ice mass loss and tundra runoff (Supplementary Fig. 11). Mass balance is considered equal to zero before the recent acceleration phase, beginning in 1996. Since mass balance is the long-term average, accumulation is smoothed with 5-year running average.

For the CAA, we assume FWF = R when estimating freshwater flux since ice discharge from the CAA is negligible compared with runoff[44]. Thus, freshwater flux from the CAA is derived from runoff predicted by RACMO2.3. Note that both ice runoff and tundra runoff are considered in the freshwater flux calculation (Supplementary Fig. 12).

For Arctic sea ice, we focus just on recent accelerated melting of multi-year ice, which results in the loss of ice area and extent, rather than the much larger contribution from the annual freeze–thaw cycle, which forms significant freshwater through fractionation (Supplementary Methods), but is more difficult to quantify with remote methods. We use three data sets (area, extent and volume; see Supplementary Methods and Supplementary Fig. 5) to estimate freshwater flux from accelerated melting of Arctic sea ice. All three approaches give similar results (Supplementary Fig. 6). The one based on volume is shown in Fig. 4. To convert area and extent to mass, we assume that sea ice thickness is 2 m (ref. 47) and sea ice density is 900 kg m^{-3}. Annual melting of Arctic sea ice is estimated by fitting annual minimum Arctic sea ice mass estimates with a linear state space model (Supplementary Methods).

References

1. Stommel, H. Thermohaline convection with two stable regimes of flow. *Tellus* **13**, 224–230 (1961).
2. Rooth, C. Hydrology and ocean circulation. *Prog. Oceanogr.* **11**, 131–149 (1982).
3. Broecker, W. S., Peteet, D. M. & Rind, D. Does the ocean-atmosphere system have more than one stable mode of operation. *Nature* **315**, 21–26 (1985).
4. Broecker, W. S. Unpleasant surprises in the greenhouse? *Nature* **328**, 123–126 (1987).
5. Wood, R. A., Keen, A. B., Mitchell, J. F. B. & Gregory, J. M. Changing spatial structure of the thermohaline circulation in response to atmospheric CO2 forcing in a climate model. *Nature* **401**, 508–508 (1999).
6. Jiang, Y., Dixon, T. H. & Wdowinski, S. Accelerating uplift in the North Atlantic region as an indicator of ice loss. *Nat. Geosci.* **3**, 404–407 (2010).
7. Rignot, E., Velicogna, I., van den Broeke, M. R., Monaghan, A. & Lenaerts, J. T. M. Acceleration of the contribution of the Greenland and Antarctic ice sheets to sea level rise. *Geophys. Res. Lett.* **38**, L05503 (2011).
8. Enderlin, E. M. *et al.* An improved mass budget for the Greenland ice sheet. *Geophys. Res. Lett.* **41**, 866–872 (2014).
9. Velicogna, I., Sutterley, T. C. & van den Broeke, M. R. Regional acceleration in ice mass loss from Greenland and Antarctica using GRACE time-variable gravity data. *Geophys. Res. Lett.* **41**, 8130–8137 (2014).
10. Yang, Q., Dixon, T. H. & Wdowinski, S. Annual variation of coastal uplift in Greenland as an indicator of variable and accelerating ice mass loss. *Geochem. Geophys. Geosyst.* **14**, 1569–1589 (2013).
11. Fichefet, T. *et al.* Implications of changes in freshwater flux from the Greenland ice sheet for the climate of the 21st century. *Geophys. Res. Lett.* **30**, 1911 (2003).
12. Jungclaus, J. H., Haak, H., Esch, M., Roeckner, E. & Marotzke, J. Will Greenland melting halt the thermohaline circulation? *Geophys. Res. Lett.* **33**, L17708 (2006).
13. Stouffer, R. J. *et al.* Investigating the causes of the response of the thermohaline circulation to past and future climate changes. *J. Clim.* **19**, 1365–1387 (2006).
14. Hu, A. X., Meehl, G. A., Han, W. Q. & Yin, J. J. Effect of the potential melting of the Greenland Ice Sheet on the Meridional Overturning Circulation and global climate in the future. *Deep Sea Res. II* **58**, 1914–1926 (2011).
15. Swingedouw, D. *et al.* Decadal fingerprints of freshwater discharge around Greenland in a multi-model ensemble. *Clim. Dyn.* **41**, 695–720 (2013).
16. Ridley, J. K., Huybrechts, P., Gregory, J. M. & Lowe, J. A. Elimination of the Greenland ice sheet in a high CO2 climate. *J. Clim.* **18**, 3409–3427 (2005).
17. Brunnabend, S. E., Schröter, J., Rietbroek, R. & Kusche, J. Regional sea level change in response to ice mass loss in Greenland, the West Antarctic and Alaska. *J. Geophys. Res. Oceans* **120**, 7316–7328 (2015).
18. Rahmstorf, S. Bifurcations of the Atlantic thermohaline circulation in response to changes in the hydrological cycle. *Nature* **378**, 145–149 (1995).
19. Rahmstorf, S. *et al.* Thermohaline circulation hysteresis: a model intercomparison. *Geophys. Res. Lett.* **32**, L23605 (2005).
20. Hawkins, E. *et al.* Bistability of the Atlantic overturning circulation in a global climate model and links to ocean freshwater transport. *Geophys. Res. Lett.* **38**, L10605 (2011).
21. Yashayaev, I. & Loder, J. W. Enhanced production of Labrador Sea Water in 2008. *Geophys. Res. Lett.* **36**, L01606 (2009).
22. Kuhlbrodt, T., Titz, S., Feudel, U. & Rahmstorf, S. A simple model of seasonal open ocean convection. *Ocean Dyn.* **52**, 36–49 (2001).
23. Joyce, T. M. & Proshutinsky, A. Greenland's Island Rule and the Arctic Ocean circulation. *J. Mar. Res.* **65**, 639–653 (2007).
24. Bonin, J. & Chambers, D. Uncertainty estimates of a GRACE inversion modelling technique over Greenland using a simulation. *Geophys. J. Int.* **194**, 212–229 (2013).
25. Howat, I. M. & Eddy, A. Multi-decadal retreat of Greenland's marine-terminating glaciers. *J. Glaciol* **57**, 389–396 (2011).
26. Box, J. E. & Colgan, W. Greenland ice sheet mass balance reconstruction. part III: marine ice loss and total mass balance (1840–2010). *J. Clim.* **26**, 6990–7002 (2013).
27. Myers, P. G., Kulan, N. & Ribergaard, M. H. Irminger water variability in the west greenland current. *Geophys. Res. Lett.* **34**, L17601 (2007).
28. Thierry, V., de Boisséson, E. & Mercier, H. Interannual variability of the subpolar mode water properties over the Reykjanes Ridge during 1990–2006. *J. Geophys. Res.* **113**, C04016 (2008).
29. Holland, D. M., Thomas, R. H., de Young, B., Ribergaard, M. H. & Lyberth, B. Acceleration of Jakobshavn Isbrae triggered by warm subsurface ocean waters. *Nat. Geosci.* **1**, 659–664 (2008).
30. Joughin, I., Alley, R. B. & Holland, D. M. Ice-sheet response to oceanic forcing. *Science* **338**, 1172–1176 (2012).
31. Straneo, F. & Heimbach, P. North Atlantic warming and the retreat of Greenland's outlet glaciers. *Nature* **504**, 36–43 (2013).
32. Jakobsen, P. K., Ribergaard, M. H., Quadfasel, D., Schmith, T. & Hughes, C. W. Near-surface circulation in the northern North Atlantic as inferred from Lagrangian drifters: variability from the mesoscale to interannual. *J. Geophys. Res. Oceans* **108**, 3251 (2003).
33. Myers, P. G., Donnelly, C. & Ribergaard, M. H. Structure and variability of the West Greenland Current in Summer derived from 6 repeat standard sections. *Prog. Oceanogr.* **80**, 93–112 (2009).

34. Myers, P. G., Josey, S. A., Wheler, B. & Kulan, N. Interdecadal variability in Labrador Sea precipitation minus evaporation and salinity. *Prog. Oceanogr.* **73**, 341–357 (2007).

35. Josey, S. A. & Marsh, R. Surface freshwater flux variability and recent freshening of the North Atlantic in the eastern subpolar gyre. *J. Geophys. Res. Oceans* **110**, C05008 (2005).

36. Peterson, B. J. *et al.* Increasing river discharge to the Arctic Ocean. *Science* **298**, 2171–2173 (2002).

37. Haine, T. W. N. *et al.* Arctic freshwater export: status, mechanisms, and prospects. *Global Planet. Change* **125**, 13–35 (2015).

38. Castro de la Guardia, L., Hu, X. M. & Myers, P. G. Potential positive feedback between Greenland Ice Sheet melt and Baffin Bay heat content on the west Greenland shelf. *Geophys. Res. Lett.* **42**, 4922–4930 (2015).

39. Curry, B., Lee, C. M., Petrie, B., Moritz, R. E. & Kwok, R. Multiyear volume, liquid freshwater, and sea ice transports through Davis Strait, 2004-10*. *J. Phys. Oceanogr.* **44**, 1244–1266 (2014).

40. Bamber, J., van den Broeke, M., Ettema, J., Lenaerts, J. & Rignot, E. Recent large increases in freshwater fluxes from Greenland into the North Atlantic. *Geophys. Res. Lett.* **39**, L19501 (2012).

41. Lenaerts, J. T. M. *et al.* Irreversible mass loss of Canadian Arctic Archipelago glaciers. *Geophys. Res. Lett.* **40**, 870–874 (2013).

42. Ettema, J. *et al.* Higher surface mass balance of the Greenland ice sheet revealed by high-resolution climate modeling. *Geophys. Res. Lett.* **36**, L12501 (2009).

43. Noël, B. *et al.* Summer snowfall on the Greenland Ice Sheet: a study with the updated regional climate model RACMO2.3. *Cryosphere Disc.* **9**, 1177–1208 (2015).

44. Gardner, A. S. *et al.* Sharply increased mass loss from glaciers and ice caps in the Canadian Arctic Archipelago. *Nature* **473**, 357–360 (2011).

45. Zhang, J. L. & Rothrock, D. A. Modeling global sea ice with a thickness and enthalpy distribution model in generalized curvilinear coordinates. *Mon. Weather Rev.* **131**, 845–861 (2003).

46. Fetterer, F., Knowles, K., Meier, W. & Savoie, M. *Sea Ice Index* (National Snow and Ice Data Center, 2002).

47. Laxon, S., Peacock, N. & Smith, D. High interannual variability of sea ice thickness in the Arctic region. *Nature* **425**, 947–950 (2003).

48. Myers, P. G. Impact of freshwater from the Canadian Arctic Archipelago on Labrador Sea Water formation. *Geophys. Res. Lett.* **32**, L06605 (2005).

49. McGeehan, T. & Maslowski, W. Evaluation and control mechanisms of volume and freshwater export through the Canadian Arctic Archipelago in a high-resolution pan-Arctic ice-ocean model. *J. Geophys. Res.* **117**, C00D14 (2012).

50. Koenigk, T., Mikolajewicz, U., Haak, H. & Jungclaus, J. Variability of Fram Strait sea ice export: causes, impacts and feedbacks in a coupled climate model. *Clim. Dyn.* **26**, 17–34 (2006).

51. Peterson, B. J. *et al.* Trajectory shifts in the Arctic and subarctic freshwater cycle. *Science* **313**, 1061–1066 (2006).

52. Good, S. A., Martin, M. J. & Rayner, N. A. EN4: Quality controlled ocean temperature and salinity profiles and monthly objective analyses with uncertainty estimates. *J. Geophys. Res. Oceans* **118**, 6704–6716 (2013).

53. Rignot, E. & Kanagaratnam, P. Changes in the velocity structure of the Greenland ice sheet. *Science* **311**, 986–990 (2006).

54. Rhein, M. *et al.* Deep water formation, the subpolar gyre, and the meridional overturning circulation in the subpolar North Atlantic. *Deep Sea Res. II* **58**, 1819–1832 (2011).

55. Kieke, D. & Yashayaev, I. Studies of Labrador Sea Water formation and variability in the subpolar North Atlantic in the light of

56. international partnership and collaboration. *Prog. Oceanogr.* **132**, 220–232 (2015).

56. Yashayaev, I., Seidov, D. & Demirov, E. A new collective view of oceanography of the Arctic and North Atlantic basins. *Prog. Oceanogr.* **132**, 1–21 (2015).

57. Lazier, J., Hendry, R., Clarke, A., Yashayaev, I. & Rhines, P. Convection and restratification in the Labrador Sea, 1990-2000. *Deep Sea Res. I* **49**, 1819–1835 (2002).

58. Vage, K. *et al.* Surprising return of deep convection to the subpolar North Atlantic Ocean in winter 2007-2008. *Nat. Geosci.* **2**, 67–72 (2009).

59. Swingedouw, D. *et al.* Bidecadal North Atlantic ocean circulation variability controlled by timing of volcanic eruptions. *Nat. Commun.* **6**, 6545 (2015).

60. Weijer, W., Maltrud, M. E., Hecht, M. W., Dijkstra, H. A. & Kliphuis, M. A. Response of the Atlantic Ocean circulation to Greenland Ice Sheet melting in a strongly-eddying ocean model. *Geophys. Res. Lett.* **39**, L09606 (2012).

61. McGeehan, T. & Maslowski, W. Impact of shelf basin freshwater transport on deep convection in the western Labrador sea. *J. Phys. Oceanogr.* **41**, 2187–2210 (2011).

62. Robson, J., Hodson, D., Hawkins, E. & Sutton, R. Atlantic overturning in decline? *Nat. Geosci.* **7**, 2–3 (2013).

63. Smeed, D. A. *et al.* Observed decline of the Atlantic meridional overturning circulation 2004-2012. *Ocean Sci.* **10**, 29–38 (2014).

64. Rahmstorf, S. *et al.* Exceptional twentieth-century slowdown in Atlantic Ocean overturning circulation. *Nature Clim. Change* **5**, 475–480 (2015).

Acknowledgements

West Greenland Current data were provided by the Danish Meteorological Institute (DMI) as well as the Greenland Institute of Natural Resource (GINR) for more recent years. We thank three anonymous reviewers for their insightful comments. This research was funded by NASA grants to T.H.D. and D.C.

Author contributions

Q.Y. conducted freshwater flux calculation and wrote the manuscript. T.H.D. designed the study and edited the manuscript. P.G.M. calculated the heat and salt fluxes of Irminger Water. J.B. and D.C. processed the GRACE data. M.R.v.d.B. provided the RACMO2.3 data. All authors discussed and commented on the manuscript.

Additional information

24

Evidence for link between modelled trends in Antarctic sea ice and underestimated westerly wind changes

Ariaan Purich[1,2,3], Wenju Cai[1], Matthew H. England[2,3] & Tim Cowan[1,4]

Despite global warming, total Antarctic sea ice coverage increased over 1979–2013. However, the majority of Coupled Model Intercomparison Project phase 5 models simulate a decline. Mechanisms causing this discrepancy have so far remained elusive. Here we show that weaker trends in the intensification of the Southern Hemisphere westerly wind jet simulated by the models may contribute to this disparity. During austral summer, a strengthened jet leads to increased upwelling of cooler subsurface water and strengthened equatorward transport, conducive to increased sea ice. As the majority of models underestimate summer jet trends, this cooling process is underestimated compared with observations and is insufficient to offset warming in the models. Through the sea ice-albedo feedback, models produce a high-latitude surface ocean warming and sea ice decline, contrasting the observed net cooling and sea ice increase. A realistic simulation of observed wind changes may be crucial for reproducing the recent observed sea ice increase.

[1] CSIRO Oceans and Atmosphere, Aspendale, Victoria 3195, Australia. [2] Climate Change Research Centre, University of New South Wales, Sydney, New South Wales 2052, Australia. [3] ARC Centre of Excellence for Climate System Science, University of New South Wales, Sydney, New South Wales 2052, Australia. [4] School of GeoSciences, University of Edinburgh, Edinburgh EH9 3FE, UK. Correspondence and requests for materials should be addressed to A.P. (email: ariaan.purich@csiro.au).

espite regional melting, total Antarctic sea ice has been expanding over the past 35 years[1–3]. Such changes have an impact on surface albedo and deep water formation, and thus are important to global climate. Spatial analysis of sea ice concentration (SIC) trends reveals opposing regional changes since satellite observations began in 1979; decreasing sea ice in the Amundsen and Bellingshausen Seas is outweighed by increasing sea ice in the Ross Sea and around eastern Antarctica, leading to an overall increase[4–6] (Fig. 1a). Although the circumpolar ice increase is statistically significant, it may still be within the range of natural variability[6–11]. However, as the most recent years of the sea ice record are included, the strength and statistical significance of the trend has increased[12].

The sea ice increase has been attributed to regional-scale wind trends causing both dynamic and thermodynamic changes[5,10,13–16]. On the other hand, models have linked hemispheric-scale wind changes associated with the positive trend in the Southern Annular Mode[17] (SAM), attributed to increasing greenhouse gases and stratospheric ozone depletion[18–20], to Southern Ocean warming and a sea ice decline[21–23]. This contrasts interannual variations, in which a positive SAM intensifies the westerly jet and shifts it polewards, resulting in cool sea surface temperature (SST) and increased sea ice extent (SIE) at most longitudes due to enhanced Ekman drift[24–26]. An exception is along the Antarctic Peninsula, where a positive SAM is associated with reduced sea ice, due to circulation changes associated with the Amundsen Sea low[4,27].

Most Coupled Model Intercomparison Project phase 5 (CMIP5) models fail to simulate the observed SIE increase in their historical experiments[7–9,11,28,29]. The vast majority of models produce a decrease in SIE and simulate considerable bias in mean-state SIE and its seasonal cycle[28]. The observed increase is suggested to lie within the range of modelled natural variability[7,8,29], although modelled Antarctic sea ice variability tends to be overestimated[9,28]; in addition, when the spatial pattern of sea ice trends are considered, the observed changes are distinguishable from the modelled pattern during austral summer and autumn[30]. To date, few studies have proposed physical mechanisms that may be responsible for the difference between observed and simulated Antarctic sea ice trends[29,31,32].

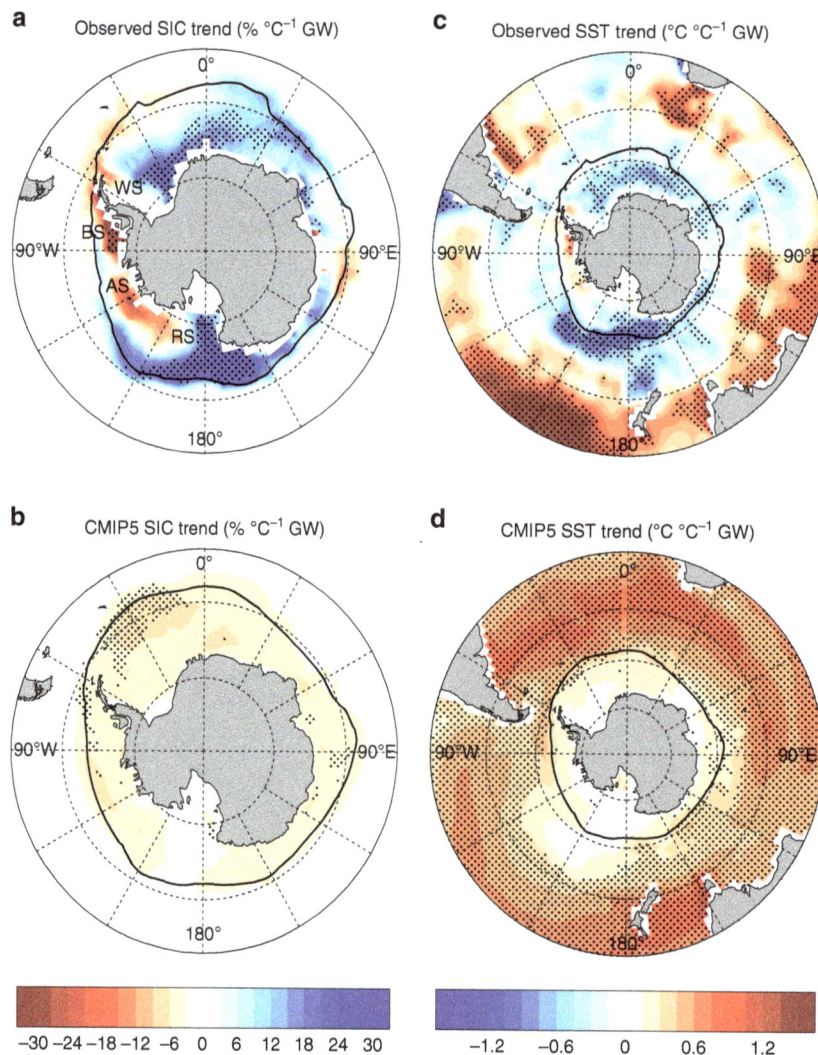

Figure 1 | Annual SIC and SST trends over 1979–2013. (**a**) Observed SIC from the National Snow and Ice Data Center (NSIDC) Bootstrap algorithm, (**b**) CMIP5 multi-model mean SIC, (**c**) observed SST from Hadley Centre Sea Ice and Sea Surface Temperature (HadISST) and (**d**) CMIP5 multi-model mean SST. Trends are expressed as a change per degree of global warming (°C^{-1} GW). Multi-model means are calculated using the first available ensemble member for each model. Stippling indicates significance: (**a,c**) above the 95% level as determined by a two-sided Student's *t*-test and (**b,d**) where 80% of models agree on the sign of the mean trend[33], which corresponds to 33 out of 41 models. The mean-state 15% SIC contour is shown in black. In (**a**) AS, Amundsen Sea; BS, Bellingshausen Sea; RS, Ross Sea; WS, Weddell Sea.

This study compares physical mechanisms affecting Antarctic sea ice in the CMIP5 models and observations, with the aim of explaining the difference between the observed increase and the modelled decline. We analyse monthly mean observations and output from 41 CMIP5 models with 87 realizations, over 1979–2013, the period for which regular satellite observations are available. Observed and modelled trends are assessed, and inter-model relationships used to gain insight into why models overall generate too great a sea ice loss and what the important processes behind this are. As the CMIP5 models underestimate recent changes in the SAM and the westerly wind jet intensification[7,33–35], we investigate the influence of jet trends on Antarctic sea ice and Southern Ocean SST. We find that underestimated changes in wind-induced ocean circulation in the models may contribute, in part, to their large Antarctic sea ice decline.

Results

Sea ice and SST trends. In contrast to the observed (Fig. 1a), multi-model mean SIC trends (Fig. 1b) show a decrease in all sea ice regions. The majority of models show an overall decrease in sea ice, despite inter-model variations in simulated spatial patterns, with many models showing small regions of increasing SIC (Supplementary Fig. 1). The multi-model mean regional decrease lacks broad coherence, except in the Bellingshausen and northern Weddell Seas, and in isolated pockets of eastern Antarctica and the Ross Sea. Coincident with the observed increase in

Antarctic sea ice, high-latitude SST has also decreased over 1979–2013 (Fig. 1c)[10,31,36], with cooling strongest in the Ross Sea. In contrast, the CMIP5 models show Southern Ocean surface warming over most regions (Fig. 1d), although there is no inter-model consensus in terms of warming at high latitudes.

Comparing SIC and SST trend patterns in individual models (Supplementary Figs 1 and 2) reveals that models with stronger SST warming show a larger SIC decrease, as expected[7]. A strong inter-model relationship exists between trends in area-averaged high-latitude (south of 55°S) SST and in circumpolar SIE: models that simulate greater warming produce a greater reduction in ice (Fig. 2). This relationship is highly statistically significant ($P<0.001$) and is evident in all seasons (shown for summer (December–February (DJF)) and winter (June–August (JJA); Fig. 2a,b, respectively; all seasons in Supplementary Table 1). The observed trends fit the tail-end of the spread in model trends. When trends are scaled by global mean temperature trends to take into account differences in climate sensitivity between observations and models and between models (Fig. 2), the observed SIE trend lies outside the 95% confidence interval of model trends. This is despite that in absolute terms the observed Antarctic sea ice trend is not statistically distinguishable from the modelled trends at the 95% confidence level. As such, natural internal variability remains a possible contributing factor for the observed trend. However, it is still of interest to investigate mechanisms that lead to the range in observed and modelled SIE trends. The SIE–SST relationship (Fig. 2) suggests that ocean

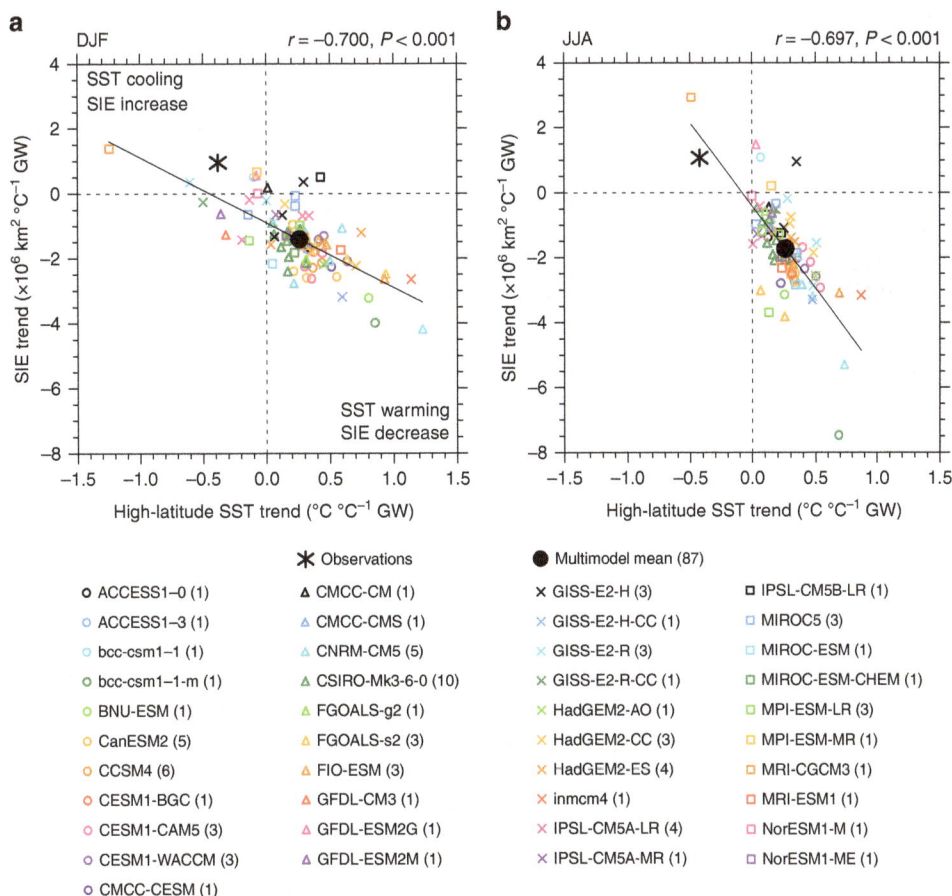

Figure 2 | Trends in SIE versus trends in high-latitude SST over 1979–2013. (**a**) DJF and (**b**) JJA. Trends are expressed as a change per degree of global warming. All available model ensemble members are shown (87 realizations). Observed SST from HadISST and SIE from NSIDC. Each model is shown by a marker with the number of runs per model indicated in the legend, the multi-model mean is shown by a black dot and observations are shown by a black asterisk. The inter-model correlation coefficient and *P*-value are shown above each panel. For *P* < 0.05, the inter-model regression is shown by a black line.

changes may influence sea ice trends. As such, to explain why the majority of CMIP5 models simulate a decrease in Antarctic sea ice in contrast to the observed increase, we must understand why modelled high-latitude SST warms too fast.

Sea ice and SST relationships with the westerly jet. Given the links between the westerly wind jet, sea ice and SST[21-26], we next investigate the influence of jet intensification trends. There is a significant inter-model relationship between jet strength trends and SIE trends during summer and autumn ($P < 0.01$; Supplementary Table 1). During these seasons, there is also a strong and significant relationship between high-latitude SST and jet strength ($P < 0.001$; Fig. 3a and Supplementary Table 1). In contrast, during winter there is no inter-model relationship between jet strength trends and SIE trends, although the relationship between jet strength trends and SST trends persists ($P < 0.001$; Fig. 3b). This relationship shows that models with a more intensified jet cool, or warm less, whereas models with a weaker intensification, or weakened jet, warm more. In summer, significant relationships are also found between trends in jet position and high-latitude SST, with a stronger poleward shift in the jet associated with high-latitude SST cooling or weaker warming (Supplementary Table 1).

Processes embedded in the inter-model trend relationships appear to also operate in the inter-model relationship in the mean state: models with stronger mean-state zonal winds south of 55°S tend to have a larger ice area, in particular during autumn[7]. On interannual timescales, an intensified summer and autumn jet is associated with above-average SIE in the observations ($P < 0.05$; Supplementary Table 2 and Supplementary Fig. 3). However, the majority of CMIP5 models do not capture this interannual relationship (median $P > 0.2$), indicating a failure to simulate wind–ice interactions adequately. This may explain the somewhat weak inter-model relationship between trends in jet strength and SIE (Supplementary Table 1).

Within individual CMIP5 models, interannual jet strength is more strongly correlated with high-latitude SST (Supplementary Table 2 and Supplementary Fig. 3). An intensified jet is associated with cooler high-latitude SST[24,25]. The observed relationship is significant in summer ($P < 0.05$), whereas for the majority of CMIP5 models it is statistically significant in both spring and summer (median $P < 0.05$ and $P < 0.01$, respectively). Thus, relative to observations, variations in the modelled jet have a

weaker influence on variations in sea ice, yet more influence on SST. As such, we focus on jet–SST dynamics, noting the strong relationship between SST and SIE (Fig. 2) that links wind changes back to sea ice.

Mechanism for wind-induced effects on SST. We hypothesize that the jet–SST trend relationship in the CMIP5 models is conducted through a high-latitude Ekman response to changing winds. There, the wind stress forces equatorward Ekman transport and the wind stress curl forces upward Ekman pumping[36,37] (see Methods). As such, an intensified jet results in strengthened equatorward Ekman transport and usually increased Ekman upwelling at high latitudes.

Ekman upwelling has a strong cooling effect on SST during summer when warm water resides at the surface forming a cap over cool Winter Water at depths ~ 20–150 m (Fig. 4a). The warm surface water results from short-wave radiation being received by the summer ice-free surface waters as sea ice melts. Beneath this, the permanent pycnocline with cold, fresh water overlying warm, salty water is apparent. By contrast, during winter surface waters are colder than water below (Fig. 4b), consistent with the typical temperature profile described for the high-latitude Southern Ocean[37], caused by seasonal sea ice melt/freeze and advection processes that freshen the surface layer. Because of the seasonal stratification, during summer enhanced Ekman upwelling brings cooler waters to the surface and this surface cooling spreads further north due to enhanced equatorward transport.

Consistently, summer Ekman pumping trends are significantly correlated with SST trends at high latitudes ($P < 0.001$; Fig. 4c): models with a strong increase in Ekman upwelling show SST cooling or weak warming, whereas models with weak trends in Ekman pumping show strong SST warming. Ekman transport trends are also significantly correlated with high-latitude SST trends in summer ($P < 0.001$; Supplementary Fig. 4): models with a strong increase in equatorward Ekman transport show cooling or weak warming. Scale analysis[38] has suggested that horizontal Ekman transport should initially dominate over vertical Ekman upwelling, although here we find that south of ~ 60°S, both Ekman transport and Ekman pumping are important (see Methods and Supplementary Fig. 5).

The CMIP5 models underestimate the summer intensification (Fig. 3a) and poleward shift in the jet[7,34], and therefore also

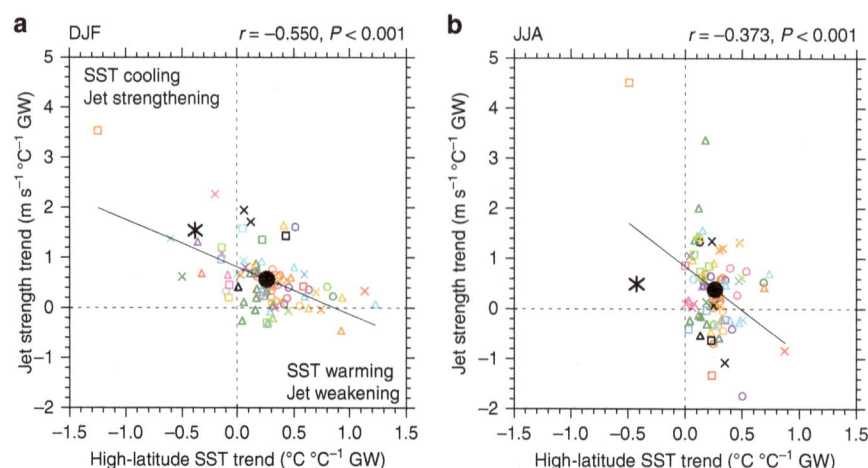

Figure 3 | Trends in jet strength versus trends in high-latitude SST over 1979–2013. (**a**) DJF and (**b**) JJA. Trends are expressed as a change per degree of global warming. All available model ensemble members are shown. Observed jet strength from European Centre for Medium-Range Weather Forecasts Interim Reanalysis (ERA-Interim). Figure details as per Fig. 2.

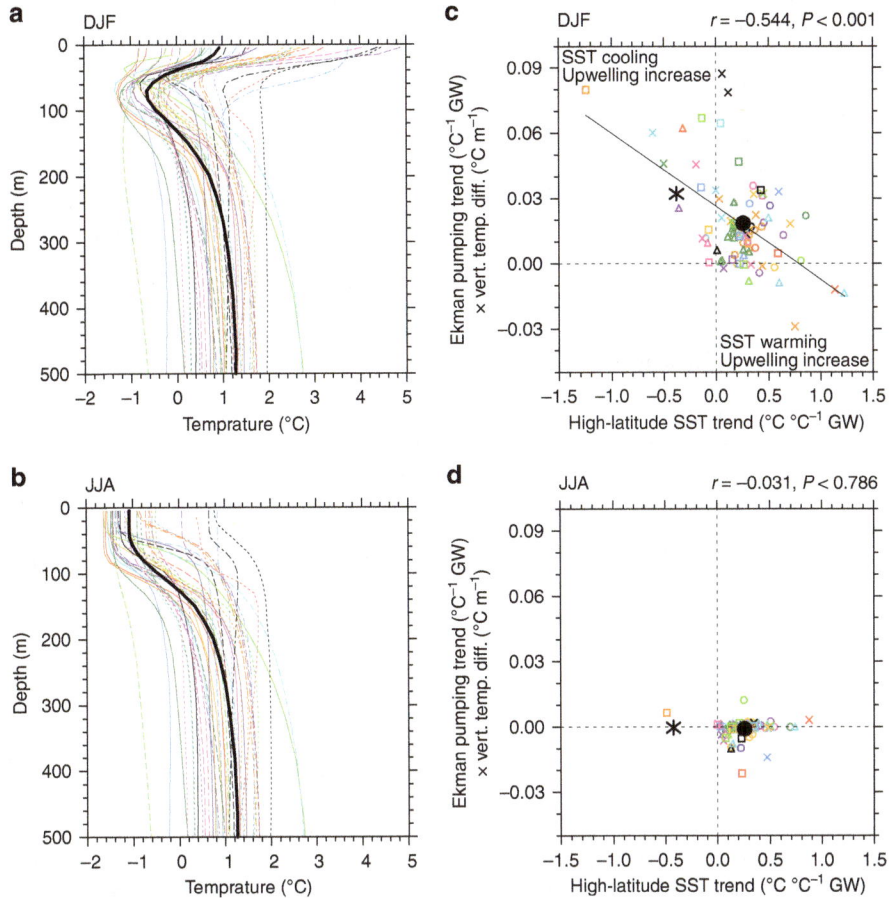

Figure 4 | Potential temperature profiles and Ekman pumping trends over 1979–2013. Seasonal zonal-mean potential temperature profiles averaged over 60-70°S for (**a**) DJF and (**b**) JJA. The first available ensemble member for each model is shown. The observed profile (black) is an average of Simple Ocean Data Assimilation (SODA) and Ishii reanalyses over 1979–2011. Trends in Ekman pumping versus trends in high-latitude SST for (**c**) DJF and (**d**) JJA. Trends are expressed as a change per degree of global warming. Ekman pumping trends are calculated as the trend in the Ekman pumping velocity principal component (PC) multiplied by the mean-state vertical temperature gradient near the surface. All available model ensemble members are shown. Details in (**c,d**) as per Fig. 2.

underestimate the increased upward Ekman pumping (Fig. 4c) and equatorward Ekman transport (Supplementary Fig. 4c) compared with observed trends. Many models underestimate the vertical temperature advection despite overestimating the surface stratification during summer (Fig. 4a). This contributes to their high-latitude SST warming trends in contrast to observed cooling.

There is considerable uncertainty in the observed jet trend, owing to sparse observations over the high-latitude Southern Hemisphere[19,35]. Although only ERA-Interim jet trends are presented here, stronger jet intensification is also seen in three other reanalyses (see Methods[39]). Increased wind speed is also evident in station-based wind observations[40,41]. However, satellite-based wind observations available over the shorter 1988–2011 period may cast some doubt over reanalysis trends[39], although the satellite products themselves contain uncertainty[42]. Overall, the analyses presented here depend on an accurate wind trend estimate over the Southern Ocean, however the mechanisms described remain robust. Further, although the observed (ERA-Interim) jet intensification is stronger than the multi-model mean, it does lie within the model spread (Fig. 3a). Nevertheless, as discussed above, the strong inter-model relationship shown here suggests that the strength of jet intensification is an important process influencing high-latitude SST in observations and coupled models, and the

majority of models produce a weaker than observed intensification.

No significant relationship between trends in Ekman pumping and SST exists during winter (Fig. 4d and Supplementary Table 1). This is to be expected, given the seasonal variation in vertical temperature stratification: during winter, increased upwelling would cause warming, offsetting cooling from increased equatorward transport (Supplementary Table 1).

Timescales of Ekman response. On interannual timescales, a positive SAM is associated with cool high-latitude SST[24–26], whereas over longer periods a positive trend in the SAM has been linked to high-latitude SST warming[21–23]. The apparent contradiction between the SAM–SST relationship over inter-annual versus multidecadal timescales has been explained by a two-timescale SST response to high-latitude wind changes[36,38]. Initially, a positive SAM trend is associated with short-term cooling, by increasing the northward Ekman transport of cold surface waters in the prevailing westerly wind regions[25,36,38], consistent with the inter-model trend relationship here (Supplementary Fig. 4). Over time, however, the cooling is replaced with a warming, accounted for by prolonged enhanced upwelling in a region where a temperature inversion occurs[21–23]. Our results are consistent with these previous studies in that most

CMIP5 models simulate a positive SAM trend (that is, jet intensification; Fig. 3a), increased upwelling and high-latitude SST warming (upper right quadrant of Fig. 4c). Our results suggest, however, that over the 35-year period assessed, Ekman upwelling is not responsible for the surface warming during summer, because due to the seasonal stratification profile models that simulate a stronger upwelling trend show a weaker rate of surface warming. Instead, the inter-model relationship suggests that cooling due to Ekman upwelling offsets other warming factors. Importantly, most models underestimate the increase in Ekman upwelling, resulting in a weaker cooling effect that is insufficient to offset warming from other processes, most notably surface heat fluxes.

The seasonal variation in the subsurface temperature profile is not discussed in previous studies and is key to interpreting our results: in contrast to previous studies that link initial cooling to equatorward Ekman transport only[36,38], our results suggest that Ekman pumping during summer is also important. It is plausible that the cooling associated with summer upwelling may eventually be replaced by warming, as water from below the mixed layer is entrained[36,38]; however, over the timescale assessed here (1979–2013), this does not appear to be the case. In previous model experiments, the time required for this temperature-trend transition varies from an order of years to a couple of decades[38]. If the timescale to a complete transition from initial cooling to later warming was at the longer end of this estimate, then the surface cooling seen in the observations and in some models could be consistent with this mechanism. The seasonal variation in the temperature profile may also contribute to a longer transitional timescale.

Ekman contribution to observed modelled disparity. The mechanism identified above is present during summer. We estimate that >25% of the difference in the CMIP5 SIE trends can be attributed to their underestimated jet intensification (see Methods). As such, the underestimation of westerly wind trends in CMIP5 models probably contributes to the sea ice decrease simulated by the majority of models.

Owing to the thermal inertia of the ocean, SST anomalies in summer are likely to persist and exert an influence beyond this season. Observed spring ice tendencies have been found to persist until the following winter[43,44], and here we find that in both observations and models summer ice tendencies persist significantly during autumn and winter. This confirms that the summer Ekman–SST mechanism can influence trends throughout much of the year[23,38].

Positive feedbacks associated with the wind-induced circulation changes could also contribute to the difference between observed and modelled trends. In observations, the magnitude of Ekman pumping and transport increases such that the associated cooling more than offsets the high-latitude heat flux increase. Cooler SST leads to increased sea ice, in particular in the Ross Sea sector. Through the sea ice-albedo feedback, solar radiation decreases, conducive to further cooling and increased sea ice[43]. The consequence is decreased zonal-mean downward heat flux in the sea-ice zone (Supplementary Fig. 6c). In the majority of the models, although the magnitude of Ekman pumping and transport also increases, it is to a smaller extent such that the associated cooling is not sufficient to offset the heat flux increase. The above sea ice-albedo feedback process operates in reverse, leading to an increased heat flux into the ocean in the sea-ice zone (Supplementary Fig. 6d). Based on the occurrence of opposite reinforcing feedback mechanisms occurring in observations and the majority of CMIP5 models, the difference between observed and modelled jet strength trends can lead to very different sea ice changes.

Considering other mechanisms. Further support for the importance of Ekman upwelling and transport comes from considering other potential mechanisms. Here we explore other possible processes and find that none of these contradict or offer an alternative explanation for the results described above.

It could be hypothesized that models with stronger jet trends show reduced high-latitude SST warming as a result of changed cloud cover or evaporative cooling, and that the correlations presented above between Ekman pumping and SST (Fig. 4c) are coincidental. No inter-model relationship is found between trends in SST and trends in overlying cloud cover during summer ($P > 0.15$; Supplementary Table 3), possibly due to differing cloud–jet relationships present in the models[45]. A significant inter-model relationship is found between high-latitude SST trends and evaporation trends ($P < 0.001$; Supplementary Table 3), in which models with increasing SST show an increase in evaporation and vice-versa. This suggests that SST anomalies are driving evaporation variations, as warmer waters evaporate more readily, whereas evaporative cooling would have the opposite effect on SST. Thus, evaporative fluxes are not the cause of the excessive warming in most CMIP5 models, instead they are a response to this warming. These results support our hypothesis above; namely, that the relationship between the westerly wind jet and SST trends occurs due to changes in ocean circulation.

Outside of the summer sea-ice zone, spatial trends in total downward heat flux oppose those in SST (Supplementary Fig. 6), that is, regions of SST cooling are associated with increased heat flux into the ocean, suggesting that changes in SST are driving changes in heat flux and not the other way around. During summer, the inter-model relationship between trends in SST and surface heat fluxes over these predominantly ice-free areas (55–65°S) is insignificant, although the sense of the relationship suggests that models with SST cooling show increased heat flux into the ocean (Supplementary Fig. 7), again suggesting that changes in SST are driving changes in heat flux. Thus, net heat flux trends are of the wrong sign to account for model SST trends. It is also noted that area-averaged trends in observed heat fluxes, although uncertain, are comparable to those in CMIP5 models over this region.

Southern Ocean freshening due to accelerated Antarctic ice shelf and/or ice sheet melting, not simulated by CMIP5 models, was proposed as a possible mechanism contributing to the Antarctic sea ice increase[31]. However, further model experiments found the influence of ice sheet melt on sea-ice trends to be minimal[29]. As such, this deficiency alone in CMIP5 models cannot account for the disparity between observed and simulated Antarctic sea ice trends. Nevertheless, reduced Southern Ocean convection in CMIP5 models has been linked with surface freshening[46], suggesting that overall changes in freshwater fluxes may be important for surface temperature trends[47].

Inter-model trends in SIE are strongly related with trends in sea surface salinity ($P < 0.001$; Supplementary Table 3 and Supplementary Fig. 8a): models with increasing salinity show a strong decrease in sea ice, whereas models with surface freshening show an increase, or weaker decrease, in sea ice. The sense of this relationship suggests that ocean surface salinity is influencing sea ice, not the other way around, as sea-ice-driven surface salinity changes would see freshening correspond to higher rates of sea-ice melt. Instead, greater sea-ice coverage is linked to fresher surface conditions and increased surface stratification, which suppresses convective overturning and vice versa for reduced sea-ice coverage.

Sea-surface salinity trends cannot be explained by trends in high-latitude precipitation minus evaporation $(P - E)$, as surprisingly increasing $P - E$ is associated with increasing salinity ($P = 0.05$ during summer; Supplementary Table 3 and Supplementary Fig. 8b), opposite to what would be expected, because both $P - E$ and surface salinity trends co-vary with the jet. Again, changes in wind-induced ocean circulation provide the probable explanation. Namely, surface salinity trends are weakly related with Ekman pumping trends ($P < 0.1$ during summer; Supplementary Table 3 and Supplementary Fig. 8c): models with a stronger increase in Ekman upwelling show an increase in surface salinity, whereas models with a weaker increase (or decrease) in Ekman upwelling show a decrease in surface salinity, due to the upwelling (or lack thereof) of saltier water from depth (Supplementary Fig. 8d). The temperature response associated with the Ekman upwelling of salty subsurface water (warming due to decreased stability and increased convective overturning) dampens the direct Ekman–SST response (upwelling of cool Winter Water) and this may be the reason why no significant Ekman–SIE relationship is found directly in the models over the 35-year period assessed.

Discussion

Both the observed and CMIP5 SIE trends are linked to the westerly wind jet intensification through the influence of SST. The models underestimate the observed jet intensification during summer, although we caution that the observed jet trend is uncertain. This causes a weaker strengthening of high-latitude Ekman pumping and transport than observed. Although increased Ekman upwelling of cool Winter Water and the associated equatorward Ekman transport contribute to the observed SST cooling, because their trends are underestimated in the models, these terms are insufficient to offset warming from increased surface heat fluxes. This leads to faster surface warming and a decreasing sea-ice trend in most of the models (summarized in Fig. 5). Once these trends are initiated, the sea ice-albedo positive feedback ensures the trend is sustained. These findings demonstrate the importance of accurately simulating changes in the wind[7,48]. By contrast, in the observations, the cooling effect from the wind changes appears to be sufficient to offset the warming tendency resulting in an initial cooling. The same sea ice-albedo positive feedback operates in reverse, leading to further cooling and an increasing sea ice trend. Although

analyses are largely conducted over circumpolar regions, when repeated for the Ross Sea sector, where observed SIE has increased most substantially, results remain robust.

Our finding that underestimated wind trends contribute to the discrepancy between the observed and model sea-ice changes occurs despite the fact these models do not resolve eddies, although almost all include a suitable eddy-induced advection scheme to approximate their effects. In the real world eddy compensation would partially counteract wind-induced changes[38,49], although Ekman changes still dominate in the surface layer[50,51]. In the presence of an eddy compensation effect, the underestimation of the impact in the models could be even larger. This influence and others such as the role of deep ocean overturning or convection will provide fertile ground for further research into the recent Southern Ocean circulation and sea ice changes.

Methods

Data. CMIP5 data from the historical and Representative Concentration Pathway 8.5 (RCP8.5; high-emission scenario) experiments are concatenated to match the observational period. The choice of RCP scenario over 2006–2013 has minimal influence on results, as all forcing scenarios are very similar over this time frame. We analyse all CMIP5 models that have SIC data available for both the historical and RCP8.5 experiments. This includes 41 CMIP5 models, with a total of 87 realizations (between one and ten runs are available per model), listed in the legend of Fig. 2. We also make use of SST, potential temperature, sea surface salinity, subsurface salinity (historical experiment only), surface air temperature, zonal wind, surface wind stresses, evaporation, precipitation, total cloud cover, mean sea level pressure and surface heat fluxes from the CMIP5 archive. At the time of analysis, potential temperature was not available for Flexible Global Ocean-Atmosphere-Land System model spectral version 2 and First Institute of Oceanography Earth System Model, and salinity was not available for Hadley Centre Global Environment Model version 2 Atmosphere-Ocean. Various surface heat flux terms were not available for Centro Euro-Mediterraneo sui Cambiamenti Climatici Climate Model with a resolved Stratosphere, First Institute of Oceanography Earth System Model, Goddard Institute for Space Studies ModelE/Russell (r1i1p2 only), Hadley Centre Global Environment Model version 2 Atmosphere-Ocean, Max Planck Institute Earth System Model Low Resolution (r2i1p1 and r3i1p1 only) and Meteorological Research Institute Earth System Model version 1.

For comparison with observations, we use passive microwave SIC processed using the NSIDC Bootstrap algorithm[2]. The possibility for spurious trends in this SIC data set has been identified[52]; thus, the results are compared with those obtained with SIC processed using the National Aeronautics and Space Administration Team algorithm and are found to be very similar. For area-averaged SIE, we make use of the pre-calculated NSIDC SIE index[53], as this index is commonly used in other studies[11,12]. Results calculated using the NSIDC Bootstrap

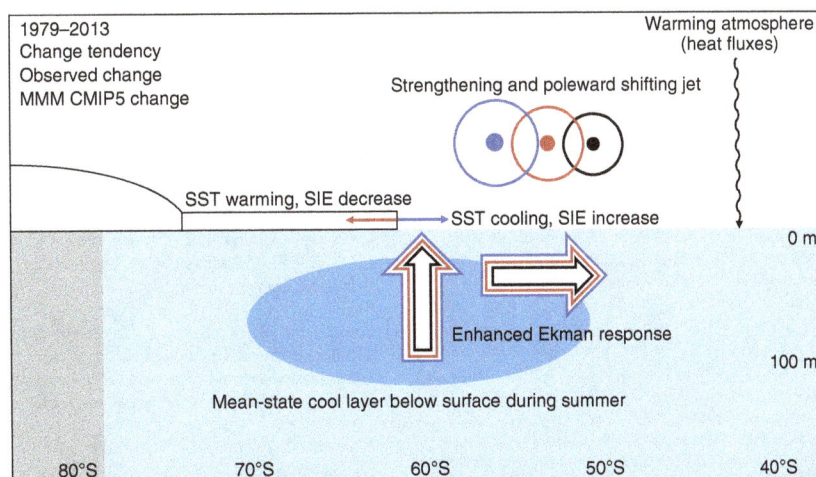

Figure 5 | Schematic of decadal scale wind induced surface layer changes during summer. Over recent decades, the observed westerly wind jet has strengthened and shifted poleward during austral summer (circles with dots). This has increased the upward Ekman pumping and equatorward Ekman transport (large arrows). During summer, increased upwelling at high latitudes brings cooler Winter Water to the surface. Combined with equatorward transport, this leads to SST cooling in observations. Multi-model mean (MMM) CMIP5 changes (red) are weaker than observed changes (blue). Under global warming, these weaker Ekman changes are insufficient to offset warming from other factors (curvy arrow). As such, multi-model mean CMIP5 high-latitude SST has warmed rather than cooled and Antarctic sea ice has declined rather than expanded (small arrows).

SIC are very similar. We use SST data from the HadISST data set[54]. For ocean temperature and salinity, we take an average of the SODA v2.2.4 (ref. 55) and Ishii[56] reanalyses. For atmospheric variables, we use the ERA-Interim reanalysis[57], regarded as the most reliable reanalysis over the Amundsen and Bellingshausen Seas[34], and over Antarctica[58,59]. Uncertainty exists in ERA-Interim wind trends[39]; however, considering the trends evident in the National Centers for Environmental Prediction (NCEP)/National Center for Atmospheric Research reanalysis, NCEP/Department of Energy reanalysis and Twentieth Century reanalysis v2, ERA-Interim winds may modestly underestimate the jet intensification, as this product yields the weakest trend among these four reanalysis products[39]. Weaker jet intensification is seen in National Aeronautics and Space Administration Modern Era-Retrospective analysis for Research and Applications and NCEP Climate Forecast System Reanalysis, although these products have previously been excluded when examining Southern Ocean wind strength trends, due to possible issues with reanalysis data assimilation[35]. The overall balance of evidence suggests that ERA-Interim winds provide one of the best estimates of wind trends over the Southern Ocean for the full study period of 1979–2013.

All data are bilinearly interpolated to a standard $2° \times 2°$ grid. This resolution is chosen to avoid overextrapolating low-resolution data from some models to higher resolutions. Potential temperature is converted from σ to z-levels where required and vertically interpolated to 40-depth levels (matching the SODA reanalysis). Data are stratified into seasonal and annual mean fields. The year of an austral summer corresponds to the year of the January–February.

Metrics. A number of metrics are calculated for observations and each model. Time series of metrics are used to investigate and compare linear trends and interannual variability between models and observations. For inter-model relationships, each ensemble member is included in the analysis and weighted evenly. Linear trends are calculated using the least squares regression method and are scaled by linear trends in global-mean temperature to take into account the different climate sensitivity of the models. Statistical significance is determined using the two-sided Student's t-test. When assessing the significance of interannual correlation coefficients, the lag-1 autocorrelation is accounted for by estimating the effective sample size, N_{eff}, as:

$$N_{eff} = N\left(\frac{1 - r_1 r_2}{1 + r_1 r_2}\right) \quad (1)$$

where N is the sample size, and r_1 and r_2 are the lag-1 autocorrelations of the time series of interest[60].

SIE in the models is defined as the circumpolar area where SIC exceeds 15% (ref. 28). We focus on SIE as an area-averaged metric, as it is commonly assessed[7–9,28,29], although we also present SIC trends to display regional trend characteristics (Fig. 1). High-latitude metrics (for example, SST, sea surface salinity and $P - E$) are defined as the area-averaged field south of 55°S (except where noted otherwise). The choice of latitude is assessed and results are found to be robust over a range of high latitudes. Only ocean grid points are considered in area averages. In HadISST, SST in grid cells partially covered by sea ice is determined based on a statistical relationship between SST and SIC[54]. In the CMIP5 models, SST is defined as the temperature of the uppermost model layer. Jet strength is defined as the maximum 925-hPa westerly wind between 35–70°S, where a cubic spline approximation is applied to the zonal–mean zonal wind[34].

Meridional Ekman transport, V_E, is calculated from the surface zonal wind stress ($V_E = -\tau_x/\rho f$) and Ekman pumping, w_E, from the curl of surface wind stresses ($w_E = \nabla \times (\tau/f)/\rho$), where τ is the wind stress, ρ is the density of seawater and f is the Coriolis parameter. Trends calculated from area-averaged Ekman transport and pumping time series are sensitive to the choice of latitude band, owing to variations in the wind fields among models and observations. To allow for spatial variations among models, we calculate the first empirical orthogonal function (EOF) of both Ekman transport and pumping over 55–70°S and use the standardized PCs to represent the Ekman transport and pumping time series, respectively. During summer, the first EOFs for both Ekman transport and pumping are well separated[61] from subsequent patterns in all models and observations. The first EOFs are related to the SAM, the leading mode of atmospheric variability in the extratropical Southern Hemisphere[62], and have a more coherent influence on SST (Supplementary Fig. 9). To account for the effect that equatorward Ekman transport has on SST, we estimate horizontal temperature advection by multiplying the Ekman transport PC by the mean-state horizontal temperature difference between 55–60°S, and 65–70°S, calculated for the zonal–mean surface layer (0–25 m). Likewise, to account for the effect that Ekman pumping has on SST, we estimate the vertical temperature advection by multiplying the Ekman pumping PC by the mean-state vertical temperature difference between the surface layer (0–25 m) and the layer just below the summer thermocline (70–80 m), calculated for the zonal–mean over 55–70°S.

Scale analysis. Horizontal temperature advection due to Ekman transport is compared with vertical temperature advection due to Ekman upwelling (Supplementary Fig. 5). We use area-averaged terms for calculations so that various latitude bands can be assessed; as we are interested in mean-state orders of magnitude rather than linear trends, comparing area averages rather than PCs is reasonable.

Horizontal and vertical advection terms are compared as follows:

$$\alpha = \frac{V_E \overline{T}_y/h}{w_E \overline{T}_z} \quad (2)$$

where $\overline{T}_y$ is the horizontal temperature gradient over the latitude bands assessed, h is the depth of the surface layer (25 m) and $\overline{T}_z$ is the vertical temperature gradient at the base of the summer mixed layer. The inclusion of h in the numerator is necessary to directly compare terms as Ekman transport is a depth-integrated flow in the upper h metres, whereas Ekman pumping is a velocity.

Over all latitude bands assessed $V_E >> w_E$; however, $\overline{T}_y/h << \overline{T}_z$. As a result, we find the numerator and denominator in equation (2) to be of similar orders of magnitude. Further equatorward, where τ_x is relatively large and $\nabla \times \tau$ small, northward Ekman transport is larger and $\alpha \sim 5$–10, but between 60–75°S, where the zonal wind transitions from westerly to easterly, $\alpha \sim 0.5$–2. As such, we conclude that over these latitude bands, both Ekman transport and Ekman pumping terms are important in driving variations in SST, and hence sea ice.

Estimating the Ekman contribution to modelled disparity. We use the difference between the observed (ERA-Interim) and multi-model mean jet strength trends and the sensitivity of SST trends to jet trends (that is, the inverse of Fig. 3a), to estimate the Ekman contribution to SST trends. From this, we use the sensitivity of SIE to SST (Fig. 2a), to further estimate the Ekman contribution to SIE trends.

References

1. Cavalieri, D. & Parkinson, C. Antarctic sea ice variability and trends, 1979-2006. *J. Geophys. Res.* **113**, C07004 (2008).
2. Comiso, J. C. & Nishio, F. Trends in the sea ice cover using enhanced and compatible AMSR-E, SSM/I, and SMMR data. *J. Geophys. Res. Oceans* **113**, C02S07 (2008).
3. Parkinson, C. & Cavalieri, D. Antarctic sea ice variability and trends, 1979-2010. *Cryosphere* **6**, 871–880 (2012).
4. Stammerjohn, S., Martinson, D. G., Smith, R., Yuan, X. & Rind, D. Trends in Antarctic annual sea ice retreat and advance and their relation to El Niño-Southern Oscillation and Southern Annular Mode variability. *J. Geophys. Res.* **113**, C03S90 (2008).
5. Holland, P. R. & Kwok, R. Wind-driven trends in Antarctic sea-ice drift. *Nat. Geosci.* **5**, 872–875 (2012).
6. Simpkins, G. R., Ciasto, L. M. & England, M. H. Observed variations in multidecadal Antarctic sea ice trends during 1979-2012. *Geophys. Res. Lett.* **40**, 3643–3648 (2013).
7. Mahlstein, I., Gent, P. R. & Solomon, S. Historical Antarctic mean sea ice area, sea ice trends, and winds in CMIP5 simulations. *J. Geophys. Res. Atmos.* **118**, 5105–5110 (2013).
8. Polvani, L. M. & Smith, K. L. Can natural variability explain observed Antarctic sea ice trends? New modeling evidence from CMIP5. *Geophys. Res. Lett.* **40**, 3195–3199 (2013).
9. Zunz, V., Goosse, H. & Massonnet, F. How does internal variability influence the ability of CMIP5 models to reproduce the recent trend in Southern Ocean sea ice extent? *Cryosphere* **7**, 451–468 (2013).
10. Fan, T., Deser, C. & Schneider, D. P. Recent Antarctic sea ice trends in the context of Southern Ocean surface climate variations since 1950. *Geophys. Res. Lett.* **41**, 2419–2426 (2014).
11. Gagné, M.-E., Gillett, N. P. & Fyfe, J. C. Observed and simulated changes in Antarctic sea ice extent over the past 50 years. *Geophys. Res. Lett.* **42**, 90–95 (2015).
12. Simmonds, I. Comparing and contrasting the behaviour of Arctic and Antarctic sea ice over the 35 year period 1979-2013. *Ann. Glaciol.* **56**, 18–28 (2015).
13. Turner, J. et al. Non-annular atmospheric circulation change induced by stratospheric ozone depletion and its role in the recent increase of Antarctic sea ice extent. *Geophys. Res. Lett.* **36**, L08502 (2009).
14. Li, X., Holland, D. M., Gerber, E. P. & Yoo, C. Impacts of the north and tropical Atlantic Ocean on the Antarctic Peninsula and sea ice. *Nature* **505**, 538–542 (2014).
15. Simpkins, G. R., McGregor, S., Taschetto, A. S., Ciasto, L. M. & England, M. H. Tropical connections to climatic change in the extratropical Southern Hemisphere: the role of Atlantic SST trends. *J. Clim.* **27**, 4923–4936 (2014).
16. Turner, J. et al. Recent changes in Antarctic Sea ice. *Phil. Trans. R. Soc. A* **373**, 1–13 (2015).
17. Marshall, G. J. Trends in the Southern Annular mode from observations and reanalyses. *J. Clim.* **16**, 4134–4143 (2003).
18. Arblaster, J. M. & Meehl, G. A. Contributions of external forcings to southern annular mode trends. *J. Clim.* **19**, 2896–2905 (2006).
19. Son, S.-W. et al. Impact of stratospheric ozone on Southern Hemisphere circulation change: a multimodel assessment. *J. Geophys. Res. Atmos.* **115**, D00M07 (2010).
20. Polvani, L. M., Waugh, D. W., Correa, G. J. & Son, S.-W. Stratospheric ozone depletion: the main driver of twentieth-century atmospheric circulation changes in the Southern Hemisphere. *J. Clim.* **24**, 795–812 (2011).

21. Sigmond, M. & Fyfe, J. C. Has the ozone hole contributed to increased Antarctic sea ice extent? *Geophys. Res. Lett.* **37,** L18502 (2010).
22. Bitz, C. M. & Polvani, L. M. Antarctic climate response to stratospheric ozone depletion in a fine resolution ocean climate model. *Geophys. Res. Lett.* **39,** L20705 (2012).
23. Smith, K. L., Polvani, L. M. & Marsh, D. R. Mitigation of 21st century Antarctic sea ice loss by stratospheric ozone recovery. *Geophys. Res. Lett.* **39,** L20701 (2012).
24. Hall, A. & Visbeck, M. Synchronous variability in the Southern Hemisphere atmosphere, sea ice, and ocean resulting from the annular mode. *J. Clim.* **15,** 3043–3057 (2002).
25. Sen Gupta, A. & England, M. H. Coupled ocean-atmosphere-ice response to variations in the Southern Annular Mode. *J. Clim.* **19,** 4457–4486 (2006).
26. Lefebvre, W. & Goosse, H. An analysis of the atmospheric processes driving the large-scale winter sea ice variability in the Southern Ocean. *J. Geophys. Res. Oceans* **113,** C02004 (2008).
27. Turner, J., Phillips, T., Hosking, J. S., Marshall, G. J. & Orr, A. The Amundsen Sea low. *Int. J. Climatol.* **33,** 1818–1829 (2013).
28. Turner, J., Bracegirdle, T. J., Phillips, T., Marshall, G. J. & Hosking, J. S. An initial assessment of Antarctic Sea ice extent in the CMIP5 models. *J. Clim.* **26,** 1473–1484 (2013).
29. Swart, N. & Fyfe, J. C. The influence of recent Antarctic ice sheet retreat on simulated sea ice area trends. *Geophys. Res. Lett.* **40,** 4328–4332 (2013).
30. Hobbs, W. R., Bindoff, N. L. & Raphael, M. N. New perspectives on observed and simulated Antarctic Sea ice extent trends using optimal fingerprinting techniques. *J. Clim.* **28,** 1543–1560 (2015).
31. Bintanja, R., van Oldenborgh, G., Drijfhout, S., Wouters, B. & Katsman, C. Important role for ocean warming and increased ice-shelf melt in Antarctic sea-ice expansion. *Nat. Geosci.* **6,** 376–379 (2013).
32. Haumann, F. A., Notz, D. & Schmidt, H. Anthropogenic influence on recent circulation-driven Antarctic sea ice changes. *Geophys. Res. Lett.* **41,** 8429–8437 (2014).
33. Purich, A., Cowan, T., Min, S.-K. & Cai, W. Autumn precipitation trends over Southern Hemisphere midlatitudes as simulated by CMIP5 models. *J. Clim.* **26,** 8341–8356 (2013).
34. Bracegirdle, T. J. *et al.* Assessment of surface winds over the Atlantic, Indian, and Pacific ocean sectors of the Southern Ocean in CMIP5 models: historical bias, forcing response, and state dependence. *J. Geophys. Res. Atmos.* **118,** 547–562 (2013).
35. Swart, N. & Fyfe, J. Observed and simulated changes in the Southern Hemisphere surface westerly wind-stress. *Geophys. Res. Lett.* **39,** L16711 (2012).
36. Marshall, J. *et al.* The ocean's role in polar climate change: asymmetric Arctic and Antarctic responses to greenhouse gas and ozone forcing. *Phil. Trans. R. Soc. A* **372,** 20130040 (2014).
37. Marshall, J. & Speer, K. Closure of the meridional overturning circulation through Southern Ocean upwelling. *Nat. Geosci.* **5,** 171–180 (2012).
38. Ferreira, D., Marshall, J., Bitz, C. M., Solomon, S. & Plumb, A. Antarctic Ocean and sea ice response to ozone depletion: a two-time-scale problem. *J. Clim.* **28,** 1206–1226 (2015).
39. Swart, N. C., Fyfe, J. C., Gillett, N. & Marshall, G. J. Comparing trends in the Southern Annular Mode and surface westerly jet. *J. Clim.* **28,** 8840–8859 (2015).
40. Yang, X.-Y., Huang, R. X. & Wang, D. X. Decadal changes of wind stress over the Southern Ocean associated with antarctic ozone depletion. *J. Clim.* **20,** 3395–3410 (2007).
41. Handle, L. B., Siems, S. T. & Manton, M. J. Observed trends in wind speed over the Southern Ocean. *Geophys. Res. Lett.* **39,** L11802 (2012).
42. Foreman, M. G. G., Pal, B. & Merryfield, W. J. Trends in upwelling and downwelling winds along the British Columbia shelf. *J. Geophys. Res.* **116,** C10023 (2011).
43. Stammerjohn, S., Massom, R. A., Rind, D. & Martinson, D. G. Regions of rapid sea ice change: an inter-hemispheric seasonal comparison. *Geophys. Res. Lett.* **39,** 1–8 (2012).
44. Holland, P. R. The seasonality of Antarctic sea ice trends. *Geophys. Res. Lett.* **41,** 1–8 (2014).
45. Grise, K. M. & Polvani, L. M. Southern Hemisphere cloud-dynamics biases in CMIP5 models and their implications for climate projections. *J. Clim.* **27,** 6074–6092 (2014).
46. deLavergne, C., Palter, J. B., Galbraith, E. D., Bernardello, R. & Marinov, I. Cessation of deep convection in the open Southern Ocean under anthropogenic climate change. *Nat. Climate Change* **4,** 278–282 (2014).
47. Morrison, A. K., England, M. H. & Hogg, A. M. Response of Southern Ocean convection and abyssal overturning to surface buoyancy perturbations. *J. Clim.* **28,** 4263–4278 (2015).
48. Sen Gupta, A. *et al.* Projected changes to the Southern Hemisphere ocean and sea ice in the IPCC AR4 climate models. *J. Clim.* **22,** 3047–3078 (2009).
49. Downes, S. M. & Hogg, A. M. Southern Ocean circulation and eddy compensation in CMIP5 models. *J. Clim.* **26,** 7198–7220 (2013).
50. Meredith, M. P., Naveira Garabato, A. C., Hogg, A. M. & Farneti, R. Sensitivity of the overturning circulation in the Southern Ocean to decadal changes in wind forcing. *J. Clim.* **25,** 99–110 (2012).
51. Morrison, A. K. & Hogg, A. M. On the relationship between Southern Ocean overturning and ACC transport. *J. Phys. Oceanogr.* **43,** 140–148 (2013).
52. Eisenman, I., Meier, W. & Norris, J. A spurious jump in the satellite record: has Antarctic sea ice expansion been overestimated? *Cryosphere* **8,** 1289–1296 (2014).
53. Fetterer, F., Knowles, K., Meier, W. & Savoie, M. *Sea Ice Index* (National Snow and Ice Data Center, Digital media: Boulder, Colorado USA, 2002). Available at http://dx.doi.org/10.7265/N5QJ7F7W.
54. Rayner, N. *et al.* Global analyses of sea surface temperature, sea ice, and night marine air temperature since the late nineteenth century. *J. Geophys. Res.* **108,** 4407 (2003).
55. Carton, J. A. & Giese, B. S. A reanalysis of ocean climate using Simple Ocean Data Assimilation (SODA). *Mon. Weather Rev.* **136,** 2999–3017 (2008).
56. Ishii, M. & Kimoto, M. Reevaluation of historical ocean heat content variations with time-varying XBT and MBT depth bias corrections. *J. Oceanogr.* **65,** 287–299 (2009).
57. Dee, D. P. *et al.* The ERA-Interim reanalysis: configuration and performance of the data assimilation system. *Q. J. R. Meteorol. Soc.* **137,** 553–597 (2011).
58. Bromwich, D. H., Nicolas, J. P. & Monaghan, A. J. An assessment of precipitation changes over Antarctica and the Southern Ocean since 1989 in contemporary global reanalyses. *J. Clim.* **24,** 4189–4209 (2011).
59. Bracegirdle, T. J. & Marshall, G. J. The reliability of Antarctic tropospheric pressure and temperature in the latest global reanalyses. *J. Clim.* **25,** 7138–7146 (2012).
60. Ciasto, L. M. & Thompson, D. W. Observations of large-scale ocean-atmosphere interaction in the Southern Hemisphere. *J. Clim.* **21,** 1244–1259 (2008).
61. North, G. R., Bell, T. L. & Cahalan, R. F. Sampling errors in the estimation of empirical orthogonal functions. *Mon. Weather Rev.* **110,** 699–706 (1982).
62. Thompson, D. W. *et al.* Signatures of the Antarctic ozone hole in Southern Hemisphere surface climate change. *Nat. Geosci.* **4,** 741–749 (2011).

Acknowledgements

This work was supported by CSIRO Oceans and Atmosphere, and the Australian Research Council (ARC) including the ARC Centre of Excellence in Climate System Science. A.P. was supported by an Australian Postgraduate Award and a CSIRO Office of the Chief Executive Science Leader scholarship. W.C. was supported by the Australian Climate Change Science Programme and a CSIRO Office of the Chief Executive Science Leader award. M.H.E. was supported by an ARC Laureate Fellowship. We thank Steve Rintoul for the thumbnail image. We acknowledge the World Climate Research Programme's Working Group on Coupled Modelling, which is responsible for CMIP, and thank the climate modelling groups (listed in Fig. 2 of this paper) for producing and making available their model output.

Author contributions

A.P., W.C. and M.H.E. conceived the study and undertook the initial analyses. A.P. assembled and analysed the observational and CMIP5 data, and wrote the first draft of the manuscript. T.C. assisted with data analysis. All authors contributed to the development of ideas, writing and revising the manuscript.

Additional information

25

Oxygen depletion recorded in upper waters of the glacial Southern Ocean

Zunli Lu[1], Babette A.A. Hoogakker[2], Claus-Dieter Hillenbrand[3], Xiaoli Zhou[1], Ellen Thomas[4], Kristina M. Gutchess[1], Wanyi Lu[1], Luke Jones[2] & Rosalind E.M. Rickaby[2]

Oxygen depletion in the upper ocean is commonly associated with poor ventilation and storage of respired carbon, potentially linked to atmospheric CO_2 levels. Iodine to calcium ratios (I/Ca) in recent planktonic foraminifera suggest that values less than $\sim 2.5\,\mu mol\,mol^{-1}$ indicate the presence of O_2-depleted water. Here we apply this proxy to estimate past dissolved oxygen concentrations in the near surface waters of the currently well-oxygenated Southern Ocean, which played a critical role in carbon sequestration during glacial times. A down-core planktonic I/Ca record from south of the Antarctic Polar Front (APF) suggests that minimum O_2 concentrations in the upper ocean fell below $70\,\mu mol\,kg^{-1}$ during the last two glacial periods, indicating persistent glacial O_2 depletion at the heart of the carbon engine of the Earth's climate system. These new estimates of past ocean oxygenation variability may assist in resolving mechanisms responsible for the much-debated ice-age atmospheric CO_2 decline.

[1] Department of Earth Sciences, Syracuse University, Syracuse, New York 13244, USA. [2] Department of Earth Sciences, University of Oxford, Oxford OX1 3AN, UK. [3] British Antarctic Survey, Cambridge CB3 0ET, UK. [4] Department of Geology and Geophysics, Yale University, New Haven, Connecticut, USA. Correspondence and requests for materials should be addressed to Z.L. (email: zunlilu@syr.edu) or to R.E.M.R. (email: rosr@earth.ox.ac.uk).

The Southern Ocean is widely considered to be critical to global nutrient and carbon cycling, including over glacial-interglacial time scales[1]. As an area of incomplete nutrient utilization, it is a major source of CO_2 to the atmosphere today[2]. At present, old (CO_2- and nutrient-rich and relatively O_2- depleted) deep waters upwell along most of the Antarctic continental margin[3,4] (Fig. 1), release CO_2 into, and recharge O_2 from surface waters before they down-well in distinct areas, such as the Weddell and Ross seas, to form Antarctic Bottom Water (AABW). In the glacial Southern Ocean, strengthening of the biological pump due to enhanced iron supply[5,6], increased stratification[7], and expanded sea–ice cover[8], were among the dominant players in reducing atmospheric CO_2 by ~ 90 p.p.m.V. Each of these mechanisms could counterbalance the increased O_2 solubility due to lower glacial temperatures, leading to a reduction in the O_2 concentration of the seawater. Since the Southern Ocean is thought to have reduced its CO_2 leakage during glacial periods[1], it provides an ideal location to search for evidence of deoxygenation linked to CO_2 sequestration in the upper ocean.

During the last glacial period, deep waters surrounding Antarctica were less ventilated, and older than today (relative to the atmosphere)[9]. A recent quantitative O_2 proxy study based on benthic foraminiferal $\delta^{13}C$ indicates that decreased ventilation linked to a reorganization of glacial ocean circulation and a strengthened global biological pump significantly enhanced the ocean storage of respired carbon in the deep North Atlantic[10]. Early box-models hypothesized very low-oxygen levels in the high latitude Southern Ocean[11,12]. Proxies did not paint a clear picture for bottom-water O_2 concentrations in the glacial Southern Ocean[13]. Only a few studies on marine sediment cores south of the APF have found evidence for substantially lowered bottom water O_2 concentrations. There, authigenic uranium concentrations were elevated in sediments deposited during glacial Marine Isotope Stages (MIS) 2 and 6 (refs 14,15). By contrast, another study highlighted a transient stagnation event during the early stage of the last interglacial (MIS 5e)[16].

Bottom water or porewater redox proxies cannot capture upper ocean O_2 levels far from the continental shelf, so there is scant constraint on upper ocean oxygenation conditions in vast tracts of the open ocean[13]. A novel proxy, the I/Ca composition of marine carbonates, especially planktonic and benthic foraminiferal tests, has demonstrated its potential to reconstruct paleo-oxygenation levels in both the upper ocean[17-20] and bottom waters[21], respectively. The thermodynamically stable forms of iodine in seawater are iodate (IO_3^-) and iodide (I^-)[22].

The total concentrations of IO_3^- and I^- are relatively uniform in the world ocean at around $0.45\,\mu mol\,l^{-1}$ due to the residence time of ~ 300 kyr (ref. 23), supported by a more recent compilation of iodine concentrations in global rivers[24]. Therefore, the total iodine concentration in the global ocean likely remained invariant over the duration of a glacial termination (~ 6 kyr).

Iodate is taken up by marine organisms as a micronutrient in surface waters[25], but its concentration does not increase during the aging of deep waters[26,27], in contrast to those of the major nutrients nitrate and phosphate, probably due to the low I/C_{org} ratio of plankton[25]. Iodine speciation is strongly redox sensitive. IO_3^- is completely converted to I^- when oxygen is depleted[28]. Because IO_3^- is the only chemical form of iodine that is incorporated into the structure of carbonate[17], calcareous tests precipitated closer to an oxygen minimum zone (OMZ) will record lower I/Ca and vice versa. An OMZ is defined by $O_2 < 20\,\mu mol\,kg^{-1}$ in the Pacific Ocean and $O_2 < 50\,\mu mol\,kg^{-1}$ in the Atlantic Ocean[29].

In this paper, we use recent planktonic foraminifera and modern water column data to establish typical I/Ca values for the presence of an OMZ or O_2-depleted water. On the basis of this proxy development, the down-core record of planktonic foraminifera I/Ca obtained at site TC493/PS2547 indicates the persistent presence of oxygen-depletion in the upper waters of high latitude Southern Ocean during the last two glacial periods.

Results

Site selection. We measured I/Ca values on eleven planktonic foraminiferal species in modern to Holocene samples, and in one sample from a previous interglacial (Supplementary Table 1 and Supplementary Figs 1 and 2). We chose sites from well-oxygenated areas (for example, the North and sub-Antarctic South Atlantic), and sites located beneath OMZs, including Ocean Drilling Program (ODP) Sites 658, 709, 720 (Site 720: last interglacial samples), 849 and 1242. First, we use these data to further establish the foundations of the I/Ca proxy. Subsequently, we focus on an I/Ca down-core record on *Neogloboquadrina pachyderma* sinistral deposited during the last two glacial cycles in two sediment cores (PS2547 and TC493) recovered from the same location (71°09′ S, 119°55′ W, water depth 2,096 m) on a seamount in the Amundsen Sea (Fig. 1)[30]. The excellent carbonate preservation at this site[30] provides a unique window to reconstruct past upper ocean conditions south of the APF. Site

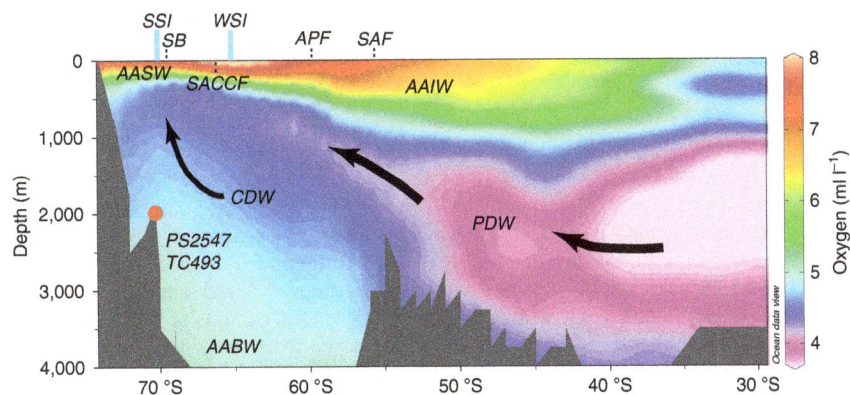

Figure 1 | Hydrographic section of Southern Ocean in the Pacific sector. Dissolved oxygen concentrations showing major water masses[50] and boundaries, average modern summer (SSI) and winter (WSI) sea-ice extent[62], and core site PS2547/TC493. The locations of the Antarctic Circumpolar Current (ACC) fronts are marked as SB, Southern Boundary of the ACC; SACCF, Southern ACC Front; APF, Antarctic Polar Front; SAF, Sub-Antarctic Front. AABW, Antarctic Bottom Water; AAIW, Antarctic Intermediate Water; AASW, Antarctic Surface Water; CDW, Circumpolar Deep Water; PDW, Pacific Deep Water. This graph is generated in Ocean Data View, using the Southern Ocean Atlas data set[63].

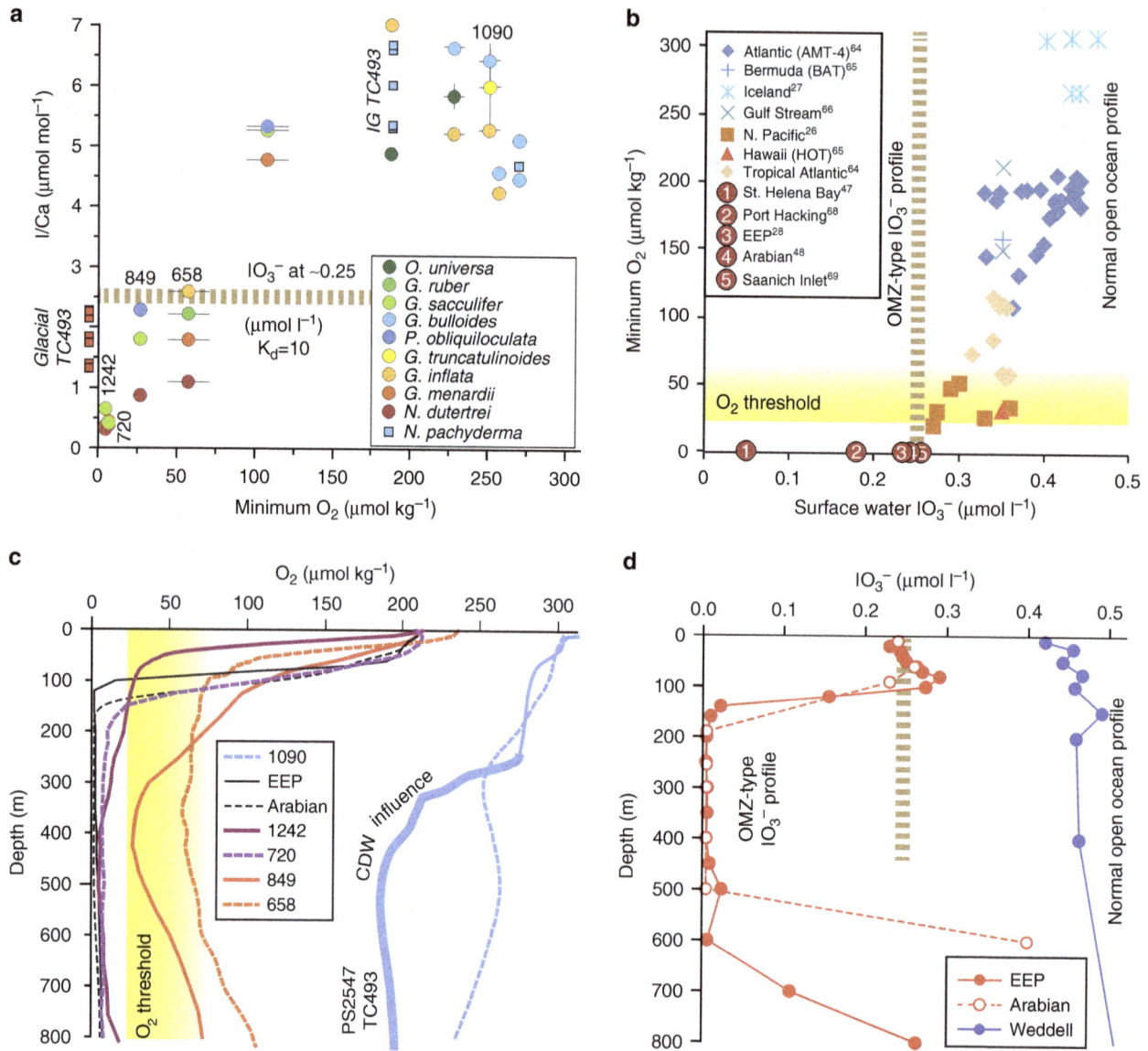

Figure 2 | I/Ca and modern OMZs. (a) Modern and late Holocene I/Ca in planktonic foraminiferal tests versus minimum O_2 concentrations in the upper 500 m of the water column (Note: I/Ca at Site 720 is from a MIS 5 sample). Error bars for y axis indicate the s.d. (1 s.d.) of triplicate measurements. Blue squares show down-core interglacial (IG) I/Ca data on *N. pachyderma* (s) from site TC493/PS2547 plotted against minimum O_2 concentrations in the modern water column, indicating well-oxygenated conditions. I/Ca for glacial *N. pachyderma* (s) tests are marked as red squares, indicating O_2 depletion. (b) Compilation of modern ocean surface water IO_3^- concentrations compared with minimum O_2 concentrations[26-28,47-48,64-69]. Brown dashed line indicates the surface water IO_3^- concentration of ~0.25 μmol I^{-1} as a threshold value for differentiating OMZ-type and normal open ocean type of IO_3^- depth profiles. (c) O_2 depth profiles. Yellow shading marks 20-70 μmol kg^{-1} O_2 concentration as the threshold for complete iodate reduction. (d) IO_3^- depth profiles at OMZ sites from the Eastern Equatorial Pacific (EEP)[28] and the Arabian Sea (station N8)[48] and at a well-oxygenated high-latitude site near the Weddell Sea (station PS71/179-1)[54].

TC493/PS2547 is currently bathed by Circumpolar Deep Water (CDW), which is overlain by a layer of Antarctic Surface Water (AASW), or Winter Water[31-33], and is located on the edge of the average modern summer sea–ice limit[34] (Fig. 1). During the Last Glacial Maximum (LGM), the sea–ice boundaries within the Southern Ocean shifted significantly northwards[35,36]. Thus, it is highly likely that site TC493/PS2547 was located within the permanent sea–ice zone during past glacial periods[34].

Age model and glacial polynyas. The sediments of core TC493/PS2547 consist mainly of foraminiferal ooze and sandy mud, with *N. pachyderma* (s) tests forming the primary carbonate component[30]. The age model of the record is based on

magnetostratigraphy combined with benthic foraminiferal (*Cibicides* cf. *wuellerstorfi*) oxygen isotope ($\delta^{18}O$) stratigraphy[30], tuned to the global benthic $\delta^{18}O$ stack[37]. Continuous deposition of foraminifera[30] indicates at least episodic opening of polynyas during glacial periods[34], because of its seamount location[38,39]. This scenario is consistent with the occurrence of the benthic foraminifera species *Epistominella exigua*, which is adapted to highly episodic phytodetritus supply[40].

I/Ca in foraminifera. I/Ca values in the modern and late Holocene samples are lower than ~2.5 μmol mol^{-1} at sites with O_2 minima <70 μmol kg^{-1} in the upper ocean (0–500 m) (Fig. 2a).

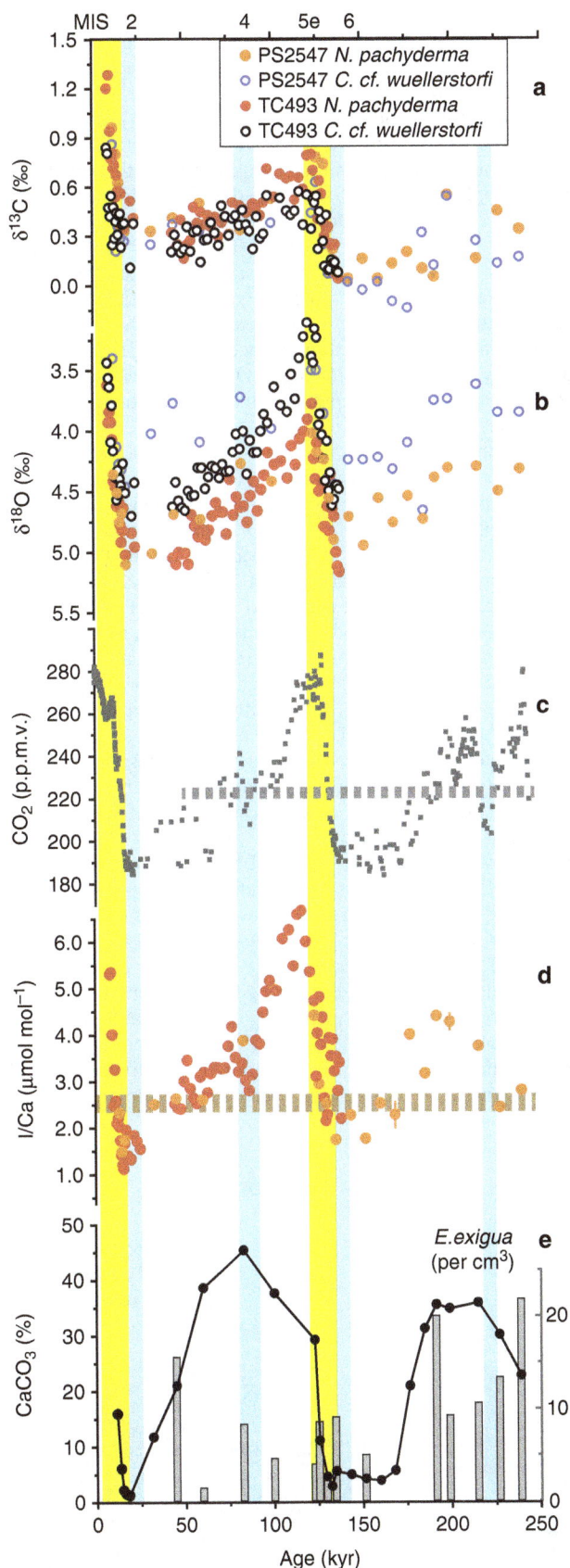

In contrast, recent planktonic foraminifera at sites with O_2 minima $> 100 \, \mu mol \, kg^{-1}$ have I/Ca $> 4 \, \mu mol \, mol^{-1}$, regardless of species (Fig. 2a). At site TC493/PS2547, the *N. pachyderma* (s) I/Ca ratio is high (5–7 $\mu mol \, mol^{-1}$) during the Holocene and MIS 5 relative to the lowest values ($< 2 \, \mu mol \, mol^{-1}$) during glacial MIS 2 and 6 (Fig. 3 and Supplementary Table 2).

Discussion

A tremendous amount of work has been devoted to developing foraminiferal proxies for temperature and pH, using global calibrations derived from core-top samples (for example, the Mg/Ca seawater temperature proxy[41]). Low I/Ca ratios of planktonic foraminifera unambiguously reveal the presence of low-oxygen waters, but a global calibration approach cannot establish planktonic foraminifera I/Ca as a linearly quantitative proxy for the continuum of dissolved O_2 concentration. Due to the stepwise nature of redox reactions[42], quantitative IO_3^- reduction does not occur before the dissolved oxygen is depleted to a certain threshold value, triggering nitrate reduction[43]. IO_3^- concentrations at water depths matching planktonic foraminiferal habitats are often not in equilibrium with the *in situ* O_2 concentrations, and O_2 contents which are sufficiently low to initiate major IO_3^- reduction may be detrimental to many species[44]. Instead, the I/Ca (recording the *in situ* IO_3^- concentration) is determined by the depth habitat of the foraminifera and the upper ocean IO_3^- mixing gradient. This mixing gradient is largely controlled by the surface water IO_3^- concentration and the depth of the IO_3^- reduction zone[28]. Nonetheless, a planktonic foraminifera proxy that semi-quantitatively approximates dissolved O_2 concentrations, indicative of the presence of an OMZ, can still be highly valuable for the paleoceanography community.

Before interpreting the down-core record from site TC493/PS2547, we identify the characteristic I/Ca signals for modern OMZs. IO_3^- depth profiles in the open ocean generally fall into two types (Fig. 2d): (1) the OMZ-type, with low surface water values and near-zero subsurface values in the OMZ; and (2) the normal open ocean type (for example, in a well-oxygenated water column), with relatively high surface water values and even higher subsurface values. A threshold O_2 concentration will cause complete IO_3^- reduction in the subsurface, and there may be a surface water IO_3^- threshold concentration below which complete IO_3^- reduction is likely to happen in the water column. Combined with modern water column IO_3^- and O_2 data, the I/Ca values measured on modern and late Holocene planktonic foraminifera consistently indicate that I/Ca $< 2.5 \, \mu mol \, mol^{-1}$ is equivalent to a surface water IO_3^- concentration of $< 0.25 \, \mu mol \, l^{-1}$, thus providing a marker for the presence of oxygen-depleted water with a subsurface O_2 concentration $< 20-70 \, \mu mol \, kg^{-1}$ (Fig. 2a–c).

Figure 3 | Down-core records of studied sites. (a,b) Stable carbon and oxygen isotopes measured on benthic (*C.* cf. *wuellerstorfi*) and planktonic (*N. pachyderma*) foraminiferal tests[30]. It is well documented that $\delta^{13}C$ of *N. pachyderma* (s) is offset by $-1.0‰$ south of the APF[49] and the values plotted here are uncorrected. **(c)** Atmospheric pCO$_2$ record at EPICA Dome C is plotted for comparison, with dashed line indicating the long-term mean value of CO_2 (50–270 ka) following Luethi *et al.*[56]. **(d)** I/Ca measured on *N. pachyderma* (s) tests from cores PS2547 and TC493. **(e)** Bulk sediment $CaCO_3$ content from PS2547. Grey columns show the abundance of the benthic foraminifera *E. exigua* as number of tests per cm^3 in core PS2547. Yellow shading highlights peak interglacial periods (including deglaciations), and blue shading marks glacial maxima and cooling intervals.

Modern surface water IO_3^- concentrations are influenced by productivity and the presence of a subsurface OMZ[25,28]. To visualize this relationship, we compiled surface water IO_3^- concentrations from the literature and plotted them against the minimum O_2 concentrations in the subsurface water (Fig. 2b). The IO_3^- concentration broadly increases with the minimum O_2 concentration when the surface water IO_3^- concentration is $> 0.25 \,\mu mol \, l^{-1}$ (Fig. 2b). This correlation is likely a reflection of surface productivity versus subsurface respiration, because lower productivity leads to lower iodine uptake in surface water and less oxygen consumption by subsurface organic matter decomposition. In areas with a strong OMZ and near-zero O_2 values, the surface water IO_3^- concentrations are below $0.25 \,\mu mol \, l^{-1}$ (Fig. 2b). A partition coefficient K_d ($K_d = [I/Ca]/[IO_3^-]$ with units of $[\mu mol \, mol^{-1}]/[\mu mol \, l^{-1}]$) of ~ 10 was reported from abiological calcite synthesis experiments[17,20]. Using this K_d value, an IO_3^- concentration $< \sim 0.25 \,\mu mol \, l^{-1}$ results in I/Ca values $< \sim 2.5 \,\mu mol \, mol^{-1}$ in calcite. This estimate is consistent with modern I/Ca at OMZ Sites 658, 849 and 1242, as well as the last interglacial I/Ca value at Site 720 (Fig. 2a). Therefore, a surface water I/Ca value $< 2.5 \,\mu mol \, mol^{-1}$ indicates that a pronounced subsurface O_2 minimum exerted the dominant control on the upper ocean IO_3^- profile. This I/Ca threshold value does not seem to depend on foraminiferal species (Fig. 2a).

The O_2 threshold for maintaining an OMZ-type IO_3^- profile is useful for the paleoceanographic application of the planktonic I/Ca proxy. At O_2 concentrations $< 20 \,\mu mol \, kg^{-1}$, microbial processes become dominant[29], and IO_3^- likely would be completely reduced to I^- since the reaction is biologically mediated[45] (for example, ODP Sites 1242, 720 and 849 in Fig. 2a,c). ODP Site 658 is located at the northern edge of a shallow pocket of distinctively low-oxygen water with mean O_2 concentrations of $\sim 70 \,\mu mol \, kg^{-1}$ in the upper 200 m (ref. 46), which may be sufficiently low to generate an OMZ-type iodate profile. Three species of planktonic foraminifera analysed at ODP Site 1242 show exceptionally low I/Ca ratios around $0.5 \,\mu mol \, mol^{-1}$, corresponding to an IO_3^- concentration of $\sim 0.05 \,\mu mol \, l^{-1}$. Such a low IO_3^- concentration is comparable to that reported for a location where an extreme hypoxic event occurred[47]. Moreover, this low IO_3^- concentration implies that IO_3^- reduction should occur shallower than at Site 849 and at two sites with classic OMZ-type IO_3^- profiles (Eastern Equatorial

Pacific[28] and Arabian Sea[48]; Fig. 2c). A comparison of the O_2 profiles of these sites reveals that the O_2 threshold needs to be $> 50 \,\mu mol \, kg^{-1}$ to achieve a shallower IO_3^- reduction at Site 1242. Therefore, we suggest that I/Ca values lower than $\sim 2.5 \,\mu mol \, mol^{-1}$ indicate O_2 minima $< 20–70 \,\mu mol \, l^{-1}$. This O_2 range cannot be further narrowed down with the available information, and we refer to this range as the O_2 threshold for an OMZ-type IO_3^- profile. However, the threshold behaviour of IO_3^- reduction (relative to O_2) in subsurface waters does not necessarily lead to step changes in down-core records of planktonic I/Ca. This is because planktonic foraminifera typically record the IO_3^- mixing gradient in the top part of water column, above the O_2-depleted zone where rapid step changes in IO_3^- concentrations occur. Low planktonic I/Ca values may be driven by shoaling of O_2-depleted water, and/or by increasing productivity, both of which could change gradually over time.

The available data from modern and late Holocene planktonic foraminifera suggest that the I/Ca ratio acts as a robust (paleo-) proxy for determining the signature of O_2-depletion in the upper ocean (Fig. 2). At site TC493/PS2547, I/Ca was high ($5–7 \,\mu mol \, mol^{-1}$) during the Holocene and interglacial MIS 5 when compared with the lowest values ($< 2 \,\mu mol \, mol^{-1}$) characterizing peak glacial periods MIS 2 and 6 (Fig. 3). Changes in salinity, temperature and foraminiferal habitat, most likely, are not the main drivers for this record (Supplementary Discussion). The glacial I/Ca values of *N. pachyderma* (s) are best explained by the presence of a water mass with a dissolved O_2 content $< 70 \,\mu mol \, kg^{-1}$ close to, i.e., above or near, this site (Figs 2 and 3). We reiterate that the low I/Ca does not necessarily imply O_2-depleted seawater within the foraminiferal habitat.

At present, CDW wells up to a water depth of approximately 250–300 m in the Amundsen Sea[31] and has O_2 concentrations notably lower than the top 200 m of the water column (Fig. 2c). Although the interpretation of absolute values of planktonic $\delta^{13}C$ is far from straightforward in the seasonal ice zone (for example, disequilibrium from seawater[49]), it is reasonable to assume that CDW had a strong influence on the local water column during glacial periods, as its upwelling along the continental margin was probably responsible for the opening of the glacial polynyas. The CDW upwelling at site TC493/PS2547 today partly originates from Pacific Deep Water (PDW) moving southward from the equator, with a low-oxygen and high nutrient signature

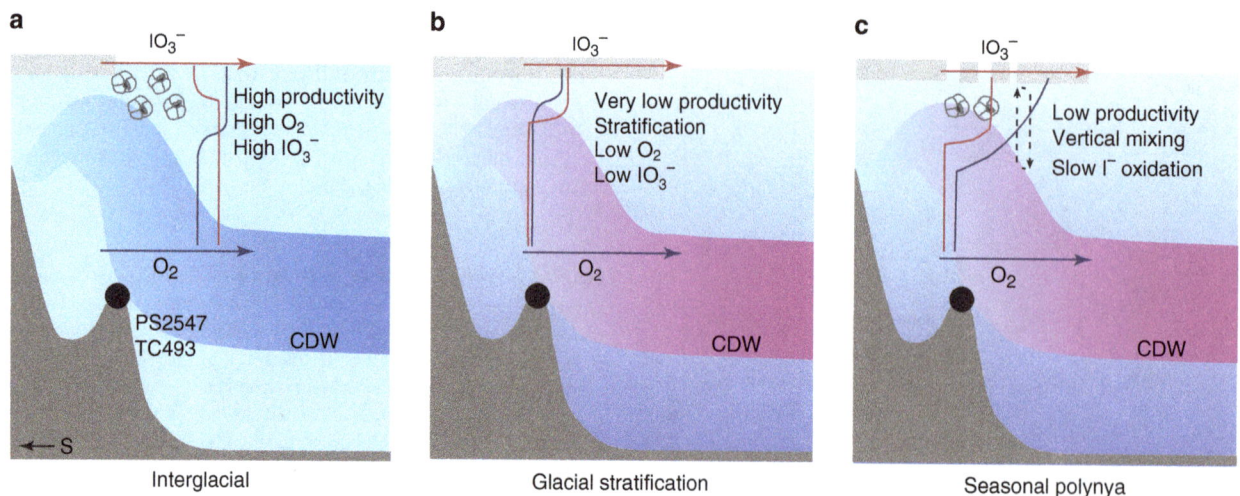

High productivity
High O_2
High IO_3^-

PS2547
TC493

CDW

Interglacial

Very low productivity
Stratification
Low O_2
Low IO_3^-

CDW

Glacial stratification

Low productivity
Vertical mixing
Slow I^- oxidation

CDW

Seasonal polynya

Figure 4 | Conceptual illustration of paleo-environmental changes. Upper ocean IO_3^- and O_2 profiles were influenced by circulation, productivity and polynyas over glacial cycles. (**a**) Well-oxygenated interglacial condition; (**b**) Relatively oxygen-depleted glacial conditions with expanded sea-ice cover; (**c**) Episodic polynya opening during glacials.

(Fig. 1)[50,51]. δ^{30}Si data from fossil diatoms and sponges indicate higher silicic acid concentrations in the Pacific sector of the Southern Ocean during the LGM, which further imply that either the southward transport of PDW was more efficient or PDW was less ventilated than today[52]. So glacial CDW was likely more O_2 depleted than during interglacials, and upwelling of this water contributed to the glacial I/Ca signal at site TC493/PS2547.

The oxidation of I^- to IO_3^- is thought to take from a few months up to 40 years[53]. Long-distance transport of well-oxygenated deep water with low IO_3^- concentrations ($< 0.25 \mu mol\, l^{-1}$) has not been documented in the modern ocean, but this scenario should be tested with further work on I^- oxidation kinetics. Today our site is bathed in CDW transported from a Pacific OMZ and the interglacial I/Ca values at site TC493/PS2547 do not show any remnant signal of the OMZ from the Pacific Ocean. On the basis of the knowledge about iodine speciation change in modern ocean, we interpret the observed glacial I/Ca values as a local signal, in principle, indicating the presence of a water mass with low O_2 and low IO_3^- vertically or horizontally close to the planktonic foraminiferal habitat.

In the setting of site TC493/PS2547 a coherent conceptual model for N. pachyderma (s) recording the presence/absence of O_2-depletion needs to integrate changes in productivity, sea-ice extent and the opening/closing of polynyas on time scales of glacial to seasonal cycles (Fig. 4). Although the polynyas complicate the interpretation of the proxy data, their presence arguably provides the only window for sufficient accumulation of planktonic microfossils to record upper ocean conditions during glacial periods at such high latitudes.

The modern O_2 profile at site TC493/PS2547 is defined by equilibration with the atmosphere at 0–250 m, and CDW influence below 250 m, as shown by the distinctively low O_2 concentrations (Fig. 2c). With O_2 above the threshold for complete IO_3^- reduction in the entire water column, the IO_3^- profile at site TC493/PS2547 should be similar to those at other high latitude locations, for example, site PS71/179-1 at 69°31′ S and 0°3′ W in the Weddell Sea[54] (Fig. 2d). An interglacial scenario of relatively high seasonal productivity, high O_2 and surface water IO_3^- ($> 0.3 \mu mol\, l^{-1}$) concentrations (Fig. 4a), is well described for the modern Atlantic sector of Southern Ocean[54].

Relative to the interglacial periods, the Southern Ocean experienced expanded sea-ice cover during glacial periods, and was less ventilated[9,36]. A more dynamic seasonal sea-ice cycle during ice ages would have increased water column stratification. Increased winter sea–ice formation (spatially and volumetrically) may have generated waters dense enough to sink ultimately to the bottom of the ocean[55]. On the other hand, melting of thicker sea ice during glacial-time summers in the seasonal sea–ice zone would have strengthened the halocline (not considering the influence of polynyas). So, the glacial seasonal stratification was likely stronger than today. These factors overall should have lowered the glacial O_2 concentrations in the Southern Ocean (Fig. 4b). At site TC493/PS2547, glacial I/Ca demonstrate that the IO_3^- profile was OMZ-like with complete IO_3^- reduction near the foraminiferal habitat (Fig. 2d). However, the dynamics of polynyas must be considered when interpreting the location of the low O_2 water mass, and the means by which the signal was recorded by N. pachyderma (s).

Without a polynya above site TC493/PS2547, glacial phytoplankton productivity under perennial sea–ice cover would have been relatively low due to the scarcity of light[34], and planktonic foraminifera depending on algae could not flourish. The water column would have been relatively poorly ventilated and strongly stratified during these times, creating the ideal environment for developing low O_2 conditions and an OMZ-type IO_3^- profile

(Fig. 4b). The episodic opening of a polynya re-established primary production (mainly by diatoms) and thus a planktonic foraminiferal habitat, vertical mixing and oxygenation in, at least, the uppermost part of the water column (Fig. 4c). While overall glacial-time production was reduced[30,34], the planktonic foraminifera preserved in the glacial sediments probably recorded transient I/Ca changes in the water column associated with polynya-induced peaks in glacial productivity. Modern open ocean productivity pulses do not lower IO_3^- concentrations to $< 0.25 \mu mol\, l^{-1}$ in oxygenated water (Supplementary Discussion)[54], thus the glacial I/Ca signal is most likely driven by changes in O_2 and not productivity.

The likely short-lived nature of glacial polynyas makes it difficult to envisage that very brief plankton blooms alone could produce a utilization-driven O_2 depletion in a cold, well-oxygenated Southern Ocean. For the same reason, it is difficult to imagine that the vertical mixing cells restricted by the size of the polynya could rapidly oxygenate voluminous nearby waters outside of the polynya, if most of the sea–ice covered areas were O_2-depleted. The more likely scenario is that the O_2 concentrations in the deep and abyssal Southern Ocean were generally lower during glacial periods than during interglacial periods. Upwelling of a more O_2-depleted CDW in the generally stratified upper ocean was mainly responsible for the IO_3^- reduction at site TC493/PS2547 (Fig. 4b), while the episodic opening of polynyas created habitable conditions for planktonic foraminifera to record the deoxygenation in the upper ocean (Fig. 4c). We suggest that the I/Ca proxy should be used as a local proxy, in principle. However, it is probably a reasonable speculation that this record (Fig. 3) shows oxygenation changes integrated over a regional volume of water (e.g. CDW).

The timing of glacial deoxygenation and deglacial reoxygenation at site PS2547 shows potential linkages to global climate changes (Fig. 3). The appearance of OMZ-type I/Ca values ($< \sim 2.5 \mu mol\, mol^{-1}$) during past glacial periods coincided with the lowering in atmospheric pCO_2 level below the long-term mean value[56]. Identical timing was reported for a strongly stratified Antarctic Zone coincident with pCO_2 decrease under the same threshold value (225 p.p.m.) in the Atlantic sector of the Southern Ocean[57]. Stronger stratification may be the common driving force for the productivity change (ODP Site 1094) and oxygenation change (PS2547/TC493) in the Antarctic zone. Furthermore, during the last interglacial period, the recovery of N. pachyderma (s) I/Ca values is offset from the $\delta^{18}O$ trend, with peak I/Ca occurring about 10 kyr after the peak $\delta^{18}O$ (Fig. 3), an observation worthy of future investigation.

Our I/Ca results build on other evidence[52,58,59] to make a stronger case for lower oxygen concentrations in CDW (and very likely PDW) during glacial periods. Altogether with the reconstructed O_2 content of deep waters in the glacial North Atlantic[10], these observations seem to allude to large scale deoxygenation in the glacial global ocean interior[60]. Future work providing quantitative reconstructions of bottom water O_2 concentrations in the Southern Ocean, especially south of the APF, and in other major ocean basins will shed new light on the mechanisms of sequestering atmospheric CO_2 during ice ages.

Methods

Foraminifera cleaning. Sediments were sampled from the split core sections and wet sieved. Approximately 40 tests of N. pachyderma sinistral were picked from the 200–250 μm size fraction of each sample. The cleaning procedure followed the Mg/Ca protocol of Barker et al.[61]. Cleaned glass slides were used to gently crack open all chambers. Clay particles were removed in an ultrasonic water bath. After adding NaOH-buffered 1% H_2O_2 solutions the samples were heated in boiling water for 10–20 min to remove organic matter. Calcareous microfossils were then thoroughly rinsed with de-ionized water. Reductive cleaning was not applied because contribution of iodine from Mn-oxides is deemed negligible[19].

ICP-MS measurements. The cleaned samples were dissolved in 3% nitric acid, and diluted to solutions with 50 p.p.m. Ca for analyses. Iodine calibration standards were freshly prepared also with 50 p.p.m. Ca. 0.5% tertiary amine solution (Spectrasol, CFA-C) was added to stabilize iodine within a few minutes after the sample dissolution. The measurements were performed immediately after that to minimize potential iodine loss. The sensitivity of iodine was tuned to above 80 kcps for a 1 p.p.b. standard. The precision for ^{127}I is typically better than 1%. The long-term accuracy is guaranteed by frequently repeated analyses of the reference material JCp-1 (ref. 17). The detection limit of I/Ca is on the order of 0.1 μmol mol^{-1}. The I/Ca measurements were performed using a quadrupole ICP-MS (Bruker M90) at Syracuse University.

References

1. Sigman, D. M., Hain, M. P. & Haug, G. H. The polar ocean and glacial cycles in atmospheric CO_2 concentration. *Nature* **466**, 47–55 (2010).
2. Marinov, I., Gnanadesikan, A., Toggweiler, J. R. & Sarmiento, J. L. The Southern Ocean biogeochemical divide. *Nature* **441**, 964–967 (2006).
3. Orsi, A. H., Whitworth, T. & Nowlin, W. D. On the meridional extent and fronts of the Antarctic Circumpolar Current. *Deep Sea Res. I* **42**, 641–673 (1995).
4. Schmidtko, S., Heywood, K. J., Thompson, A. F. & Aoki, S. Multidecadal warming of Antarctic waters. *Science* **346**, 1227–1231 (2014).
5. Martin, J. H. GLACIAL-interglacial CO_2 change: the iron hypothesis. *Paleoceanography* **5**, 1–13 (1990).
6. Martinez-Garcia, A. *et al.* Iron fertilization of the subantarctic ocean during the last ice age. *Science* **343**, 1347–1350 (2014).
7. Toggweiler, J. R. Variation of atmospheric CO2 by ventilation of the ocean's deepest water. *Paleoceanography* **14**, 571–588 (1999).
8. Stephens, B. B. & Keeling, R. F. The influence of Antarctic sea ice on glacial-interglacial CO_2 variations. *Nature* **404**, 171–174 (2000).
9. Skinner, L. C., Fallon, S., Waelbroeck, C., Michel, E. & Barker, S. Ventilation of the deep southern ocean and deglacial CO_2 rise. *Science* **328**, 1147–1151 (2010).
10. Hoogakker, B. A. A., Elderfield, H., Schmiedl, G., McCave, I. N. & Rickaby, R. E. M. Glacial-interglacial changes in bottom-water oxygen content on the Portuguese margin. *Nat. Geosci.* **8**, 40–43 (2015).
11. Knox, F. & McElroy, M. B. Changes in atmospheric CO_2 - influence of the marine biota at high-latitude. *J. Geophys. Res. Atmos.* **89**, 4629–4637 (1984).
12. Archer, D. E. *et al.* Model sensitivity in the effect of Antarctic sea ice and stratification on atmospheric pCO(2). *Paleoceanography* **18** (2003).
13. Jaccard, S. L. & Galbraith, E. D. Large climate-driven changes of oceanic oxygen concentrations during the last deglaciation. *Nat. Geosci.* **5**, 151–156 (2012).
14. Ceccaroni, L. *et al.* Late Quaternary fluctuations of biogenic component fluxes on the continental slope of the Ross Sea, Antarctica. *J. Mar. Syst.* **17**, 515–525 (1998).
15. Frank, M. *et al.* Similar glacial and interglacial export bioproductivity in the Atlantic sector of the Southern Ocean: Multiproxy evidence and implications for glacial atmospheric CO_2. *Paleoceanography* **15**, 642–658 (2000).
16. Hayes, C. T. *et al.* A stagnation event in the deep South Atlantic during the last interglacial period. *Science* **346**, 1514–1517 (2014).
17. Lu, Z., Jenkyns, H. C. & Rickaby, R. E. M. Iodine to calcium ratios in marine carbonate as a paleo-redox proxy during oceanic anoxic events. *Geology* **38**, 1107–1110 (2010).
18. Hardisty, D. S. *et al.* An iodine record of Paleoproterozoic surface ocean oxygenation. *Geology* **42**, 619–622 (2014).
19. Zhou, X. L., Thomas, E., Rickaby, R. E. M., Winguth, A. M. E. & Lu, Z. L. I/Ca evidence for upper ocean deoxygenation during the PETM. *Paleoceanography* **29**, 964–975 (2014).
20. Zhou, X. *et al.* Upper ocean oxygenation dynamics from I/Ca ratios during the Cenomanian-Turonian OAE 2. *Paleoceanography* **30** (2015).
21. Glock, N., Liebetrau, V. & Eisenhauer, A. I/Ca ratios in benthic foraminifera from the Peruvian oxygen minimum zone: analytical methodology and evaluation as a proxy for redox conditions. *Biogeosciences* **11**, 7077–7095 (2014).
22. Wong, G. T. F. The marine geochemistry of iodine. *Rev. Aquat. Sci.* **4**, 45–73 (1991).
23. Broecker, W. S. & Peng, H. T. *Tracers in the Sea* 690 (Lamont-Doherty Geological Observatory, 1982).
24. Moran, J. E., Oktay, S. D. & Santschi, P. H. Sources of iodine and iodine 129 in rivers. *Water Resources Research* **38** (2002).
25. Elderfield, H. & Truesdale, V. W. On the biophilic nature of iodine in seawater. *Earth Planet. Sci. Lett.* **50**, 105–114 (1980).
26. Nakayama, E., Kimoto, T., Isshiki, K., Sohrin, Y. & Okazaki, S. Determination and distribution of iodide-iodine and total-Iodine in the North Pacific Ocean - by using a new automated electrochemical method *Mar. Chem.* **27**, 105–116 (1989).
27. Waite, T. J., Truesdale, V. W. & Olafsson, J. The distribution of dissolved inorganic iodine in the seas around Iceland. *Mar. Chem.* **101**, 54–67 (2006).
28. Rue, E. L., Smith, G. J., Cutter, G. A. & Bruland, K. W. The response of trace element redox couples to suboxic conditions in the water column. *Deep Sea Res. I* **44**, 113–134 (1997).
29. Gilly, W. F., Beman, J. M., Litvin, S. Y. & Robison, B. H. in *Annual Review of Marine Science, Annual Review of Marine Science* Vol. 5 (eds Carlson, C. A. & Giovannoni, S. J.) 393–420 (Annual Reviews, 2013).
30. Hillenbrand, C. D., Futterer, D. K., Grobe, H. & Frederichs, T. No evidence for a Pleistocene collapse of the West Antarctic Ice Sheet from continental margin sediments recovered in the Amundsen Sea. *Geo-Mar. Lett.* **22**, 51–59 (2002).
31. Wahlin, A. K. *et al.* Some Implications of ekman layer dynamics for cross-shelf exchange in the amundsen sea. *J. Phys. Oceanogr.* **42**, 1461–1474 (2012).
32. Nakayama, Y., Schroder, M. & Hellmer, H. H. From circumpolar deep water to the glacial meltwater plume on the eastern Amundsen Shelf. *Deep Sea Res. I* **77**, 50–62 (2013).
33. Jacobs, S. *et al.* Getz Ice Shelf melting response to changes in ocean forcing. *J Geophys. Res. Oceans* **118**, 4152–4168 (2013).
34. Thatje, S., Hillenbrand, C. D., Mackensen, A. & Larter, R. Life hung by a thread: Endurance of antarctic fauna in glacial periods. *Ecology* **89**, 682–692 (2008).
35. Collins, L. G., Pike, J., Allen, C. S. & Hodgson, D. A. High-resolution reconstruction of southwest Atlantic sea-ice and its role in the carbon cycle during marine isotope stages 3 and 2. *Paleoceanography* **27** (2012).
36. Gersonde, R., Crosta, X., Abelmann, A. & Armand, L. Sea-surface temperature and sea ice distribution of the Southern Ocean at the EPILOG Last Glacial Maximum—a circum-Antarctic view based on siliceous microfossil records. *Quart. Sci. Rev.* **24**, 869–896 (2005).
37. Lisiecki, L. E. & Raymo, M. E. A Pliocene-Pleistocene stack of 57 globally distributed benthic delta O-18 records. *Paleoceanography* **20** (2005).
38. Bersch, M., Becker, G. A., Frey, H. & Koltermann, K. P. Topographic effects of the maud rise on the stratification and circulation of the weddell gyre. *Deep Sea Res. A* **39**, 303–331 (1992).
39. Martin, S. in *Encyclopedia of ocean sciences.* (eds Steele, H., Turekian, K. K. & Thorpe, S. A.) 2241–2247 (Academic Press, 2001).
40. Smith, J. A., Hillenbrand, C. D., Pudsey, C. J., Allen, C. S. & Graham, A. G. C. The presence of polynyas in the Weddell Sea during the Last Glacial Period with implications for the reconstruction of sea-ice limits and ice sheet history. *Earth Planet. Sci. Lett.* **296**, 287–298 (2010).
41. Elderfield, H. & Ganssen, G. Past temperature and delta O-18 of surface ocean waters inferred from foraminiferal Mg/Ca ratios. *Nature* **405**, 442–445 (2000).
42. Stumm, W. & Morgan, J. J. *Aquatic Chemistry,* 2nd edn 477 (Wiley, 1981).
43. Luther, G. W. *et al.* Simultaneous measurement of O-2, Mn, Fe, I-, and S(-II) in marine pore waters with a solid-state voltammetric microelectrode. *Limnol. Oceonogr.* **43**, 325–333 (1998).
44. Hull, P. M., Osborn, K. J., Norris, R. D. & Robison, B. H. Seasonality and depth distribution of a mesopelagic foraminifer, Hastigerinella digitata, in Monterey Bay, California. *Limnol. Oceonogr.* **56**, 562–576 (2011).
45. Kuepper, F. C. *et al.* Commemorating two centuries of iodine research: an interdisciplinary overview of current research. *Angew. Chem. Int. Ed.* **50**, 11598–11620 (2011).
46. Stramma, L., Huttl, S. & Schafstall, J. Water masses and currents in the upper tropical northeast Atlantic off northwest Africa. *J Geophys. Res. Oceans* **110** (2005).
47. Truesdale, V. W. & Bailey, G. W. Dissolved iodate and total iodine during an extreme hypoxic event in the Southern Benguela system. *Estuar. Coast. Shelf Sci.* **50**, 751–760 (2000).
48. Farrenkopf, A. M. & Luther, G. W. Iodine chemistry reflects productivity and denitrification in the Arabian Sea: evidence for flux of dissolved species from sediments of western India into the OMZ. *Deep Sea Res. II* **49**, 2303–2318 (2002).
49. Kohfeld, K. E., Anderson, R. F. & Lynch-Stieglitz, J. Carbon isotopic disequilibrium in polar planktonic foraminifera and its impact on modern and Last Glacial Maximum reconstructions. *Paleoceanography* **15**, 53–64 (2000).
50. Talley, L. D. Closure of the global overturning circulation through the indian, pacific, and southern oceans: schematics and transports. *Oceanography* **26**, 80–97 (2013).
51. McCave, I. N., Carter, L. & Hall, I. R. Glacial-interglacial changes in water mass structure and flow in the SW Pacific Ocean. *Quart. Sci. Rev.* **27**, 1886–1908 (2008).

52. Ellwood, M. J., Wille, M. & Maher, W. Glacial Silicic Acid Concentrations in the Southern Ocean. *Science* **330**, 1088–1091 (2010).
53. Chance, R., Baker, A. R., Carpenter, L. & Jickells, T. D. The distribution of iodide at the sea surface. *Environ. Sci. Process. Impacts* **16**, 1841–1859 (2014).
54. Bluhm, K., Croot, P. L., Huhn, O., Rohardt, G. & Lochte, K. Distribution of iodide and iodate in the Atlantic sector of the southern ocean during austral summer. *Deep Sea Res. II* **58**, 2733–2748 (2011).
55. Mackensen, A. Strong thermodynamic imprint on Recent bottom-water and epibenthic delta C-13 in the Weddell Sea revealed: Implications for glacial Southern Ocean ventilation. *Earth Planet. Sci. Lett.* **317**, 20–26 (2012).
56. Luthi, D. *et al.* High-resolution carbon dioxide concentration record 650,000-800,000 years before present. *Nature* **453**, 379–382 (2008).
57. Jaccard, S. L. *et al.* Two modes of change in southern ocean productivity over the past million years. *Science* **339**, 1419–1423 (2013).
58. Moy, A. D., Howard, W. R. & Gagan, M. K. Late Quaternary palaeoceanography of the Circumpolar Deep Water from the South Tasman Rise. *J. Quat. Sci.* **21**, 763–777 (2006).
59. Galbraith, E. D. & Jaccard, S. L. Deglacial weakening of the oceanic soft tissue pump: global constraints from sedimentary nitrogen isotopes and oxygenation proxies. *Quart. Sci. Rev.* **109**, 38–48 (2015).
60. Jaccard, S. L., Galbraith, E. D., Martínez-García, A. & Anderson, R. F. Covariation of deep Southern Ocean oxygenation and atmospheric CO_2 through the last ice age. *Nature* **530**, 207–210 (2016).
61. Barker, S., Greaves, M. & Elderfield, H. A study of cleaning procedures used for foraminiferal Mg/Ca paleothermometry. *Geochem. Geophys. Geosyst.* **4** (2003).
62. Comiso, J. C. *Large-Scale Characteristics and Variability of the Global Sea Ice Cover* 112–142 (Blackwell, 2003).
63. Olbers, D., Gouretski, V., Seiss, G. & Schroeter, J. *Hydrographic Atlas of the Southern Ocean* (Alfred-Wegener Institute (AWI), 1992).
64. Truesdale, V. W., Bale, A. J. & Woodward, E. M. S. The meridional distribution of dissolved iodine in near-surface waters of the Atlantic Ocean. *Progress Oceonagr.* **45**, 387–400 (2000).
65. Campos, M., Farrenkopf, A. M., Jickells, T. D. & Luther, G. W. A comparison of dissolved iodine cycling at the Bermuda Atlantic Time-Series station and Hawaii Ocean Time-Series Station. *Deep Sea Res. II* **43**, 455–466 (1996).
66. Wong, G. T. F. Dissolved iodine across the Gulf Stream Front and in the South Atlantic Bight. *Deep Sea Res. I* **42**, 2005–2023 (1995).
67. Farrenkopf, A. M., Dollhopf, M. E., NiChadhain, S., Luther, G. W. & Nealson, K. H. Reduction of iodate in seawater during Arabian Sea shipboard incubations and in laboratory cultures of the marine bacterium Shewanella putrefaciens strain MR-4. *Mar. Chem.* **57**, 347–354 (1997).
68. Smith, J. D., Butler, E. C. V., Airey, D. & Sandars, G. Chemical-properties of a low-oxygen water column in port-hacking (Australia) - arsenic, iodine and nutrients. *Mar. Chem.* **28**, 353–364 (1990).
69. Emerson, S., Cranston, R. E. & Liss, P. S. Redox species in a reducing fjord - equilibrium and kinetic considerations. *Deep Sea Res. A* **26**, 859–878 (1979).

Acknowledgements

Z.L. thanks NSF OCE 1232620. B.A.A.H. is supported by Natural Environment Research Council (NERC) grant NE/I020563/1. C.-D.H. is supported by NERC. Z.L. and R.E.M.R. were supported by NERC NE/E018432/1 during the initial development of the I/Ca proxy. We thank the captains, officers, crews and shipboard scientists of expeditions JR179 and ANT-XI/3 for recovering cores TC493 and PS2547, Professor David Hodell and Professor Andreas Mackensen for analysing stable isotopes on these cores, and Dr Hannes Grobe for providing samples from core PS2547. We acknowledge Professor Nick McCave for providing some of the core top materials and Mr Thomas Williams provided assistance for Fig. 1. This research used samples provided by the Ocean Drilling Program (ODP). ODP is sponsored by the US National Science Foundation and participating countries under management of the Joint Oceanographic Institutions (JOI). This manuscript greatly benefited from comments by Dr Nicolaas Glock and two anonymous reviewers.

Author contributions

Z.L., X.Z., K.M.G. and W.L. carried out the I/Ca analysis. C.-D.H. provided all samples from core TC493 and $CaCO_3$ and stable isotope data for cores TC493 and PS2547. B.A.A.H. and L.J. contributed the core top samples. E.T. identified *E. exigua*. All authors participated in data interpretation and manuscript preparation.

Additional information

Permissions

All chapters in this book were first published in NC, by Nature Publishing Group; hereby published with permission under the Creative Commons Attribution License or equivalent. Every chapter published in this book has been scrutinized by our experts. Their significance has been extensively debated. The topics covered herein carry significant findings which will fuel the growth of the discipline. They may even be implemented as practical applications or may be referred to as a beginning point for another development.

The contributors of this book come from diverse backgrounds, making this book a truly international effort. This book will bring forth new frontiers with its revolutionizing research information and detailed analysis of the nascent developments around the world.

We would like to thank all the contributing authors for lending their expertise to make the book truly unique. They have played a crucial role in the development of this book. Without their invaluable contributions this book wouldn't have been possible. They have made vital efforts to compile up to date information on the varied aspects of this subject to make this book a valuable addition to the collection of many professionals and students.

This book was conceptualized with the vision of imparting up-to-date information and advanced data in this field. To ensure the same, a matchless editorial board was set up. Every individual on the board went through rigorous rounds of assessment to prove their worth. After which they invested a large part of their time researching and compiling the most relevant data for our readers.

The editorial board has been involved in producing this book since its inception. They have spent rigorous hours researching and exploring the diverse topics which have resulted in the successful publishing of this book. They have passed on their knowledge of decades through this book. To expedite this challenging task, the publisher supported the team at every step. A small team of assistant editors was also appointed to further simplify the editing procedure and attain best results for the readers.

Apart from the editorial board, the designing team has also invested a significant amount of their time in understanding the subject and creating the most relevant covers. They scrutinized every image to scout for the most suitable representation of the subject and create an appropriate cover for the book.

The publishing team has been an ardent support to the editorial, designing and production team. Their endless efforts to recruit the best for this project, has resulted in the accomplishment of this book. They are a veteran in the field of academics and their pool of knowledge is as vast as their experience in printing. Their expertise and guidance has proved useful at every step. Their uncompromising quality standards have made this book an exceptional effort. Their encouragement from time to time has been an inspiration for everyone.

The publisher and the editorial board hope that this book will prove to be a valuable piece of knowledge for researchers, students, practitioners and scholars across the globe.

List of Contributors

Teresa S. Catalá and Isabel Reche
Departamento de Ecología and Instituto del Agua, Universidad de Granada, 18071 Granada, Spain

Antonio Fuentes-Lema
epartamento de Ecoloxía e Bioloxía Animal, Universidade de Vigo, 36208 Vigo, Spain

X. Antón Álvarez-Salgado and Mar Nieto-Cid
CSIC Instituto de Investigacións Mariñas, 36208 Vigo, Spain

Cristina Romera-Castillo
CSIC Instituto de Investigacións Mariñas, 36208 Vigo, Spain
CSIC Institut de Ciencies del Mar, 08003 Barcelona, Spain

Eva Ortega-Retuerta, Eva Calvo and Cèlia Marrasé
CSIC Institut de Ciencies del Mar, 08003 Barcelona, Spain

Marta Álvarez
IEO Centro Oceanográfico de A Coruña, 15006 A Coruña, Spain

Colin A. Stedmon
National Institute of Aquatic Resources, Technical University of Denmark, 2920 Charlottenlund, Denmark

Bror F. Jönsson
Department of Geosciences, Princeton University, Princeton, New Jersey 08544, USA

James R. Watson
College of Earth, Ocean and Atmospheric Sciences, Oregon State University, Corvallis, Oregon 97331-5503, USA
The Stockholm Resilience Centre, Stockholm University, 118 14 Stockholm, Sweden

Daniel Pauly and Dirk Zeller
Sea Around Us, Global Fisheries Cluster, University of British Columbia, 2202 Main Mall, Vancouver, British Columbia, Canada V6T 1Z4

Karen Grace V. Bondoc and Georg Pohnert
Institute for Inorganic and Analytical Chemistry, Bioorganic Analytics, Department of Chemistry and Earth Sciences, Friedrich-Schiller-Universität Jena, Lessingstrasse 8, D-07743 Jena, Germany
Max Planck Institute for Chemical Ecology, Max Planck Fellow, Hans-Knöll-Str. 8, D-07745 Jena, Germany

Jan Heuschele
Centre for Ocean Life, National Institute of Aquatic Resources, Technical University of Denmark, Charlottenlund Slot, Jægersborg Allé, DK-2920

Charlottenlund, Denmark
Department of Biology, Aquatic Ecology Unit, Lund University, SE-22362 Lund, Sweden

Wim Vyverman
Laboratory of Protistology and Aquatic Ecology, Department of Biology, University Gent, Krijgslaan 281, S8, 9000 Gent, Belgium

Jeroen Gillard
Laboratory of Protistology and Aquatic Ecology, Department of Biology, University Gent, Krijgslaan 281, S8, 9000 Gent, Belgium
California State University, Department of Biology, 9001 Stockdale Hwy, Bakersfield, California 93311, USA

Jamison M. Gove and Jeffrey J. Polovina
Ecosystems and Oceanography Program, Pacific Islands Fisheries Science Center, 1845 Wasp Blvd Building 176, Honolulu, 96818 Hawaii, USA

Margaret A. McManus, Anna B. Neuheimer, Jeffrey C. Drazen and Craig R. Smith
Department of Oceanography, University of Hawaii at Ma%noa, 1000 Pope Road, Marine Sciences Building, Honolulu, 96822 Hawaii, USA

Mark A. Merrifield
Department of Oceanography, University of Hawaii at Ma%noa, 1000 Pope Road, Marine Sciences Building, Honolulu, 96822 Hawaii, USA
Joint Institute for Marine and Atmospheric Research, University of Hawaii at Ma%noa, 1000 Pope Road, Marine Sciences Building, Honolulu, 96822 Hawaii, USA

Julia S. Ehses, Charles W. Young and Amanda K. Dillon
Joint Institute for Marine and Atmospheric Research, University of Hawaii at Ma%noa, 1000 Pope Road, Marine Sciences Building, Honolulu, 96822 Hawaii, USA
Coral Reef Ecosystem Program, Pacific Islands Fisheries Science Center, 1845 Wasp Blvd Building 176, Honolulu, 96818 Hawaii, USA

Alan M. Friedlander
Fisheries Ecology Research Laboratory, Department of Biology, University of Hawaii at Ma%noa, 2538 McCarthy Mall, Honolulu, 96822 Hawaii, USA
Pristine Seas, National Geographic Society, 1145 17th St NW, Washington, DC 20036, USA
Coral Reef Ecosystem Program, Pacific Islands Fisheries Science Center, 1845 Wasp Blvd Building 176, Honolulu, 96818 Hawaii, USA

Gareth J. Williams
Center for Marine Biodiversity and Conservation, Scripps Institution of Oceanography, UC San Diego, 9500 Gilman Drive, La Jolla, 92093 California, USA
School of Ocean Sciences, Bangor University, Menai Bridge, LL59 5AB Anglesey, UK

Paul B. Goddard and Jianjun Yin
Department of Geosciences, University of Arizona, Tucson, Arizona 85721, USA

Stephen M. Griffies and Shaoqing Zhang
Geophysical Fluid Dynamics Laboratory, NOAA, Princeton, New Jersey 08540, USA

Julia Gottschalk and Luke C. Skinner
Godwin Laboratory for Palaeoclimate Research, Earth Sciences Department, University of Cambridge, Downing Street, Cambridge CB2 3EQ, UK

Jörg Lippold, Hendrik Vogel and Samuel L. Jaccard
Institute of Geological Sciences and Oeschger Center for Climate Change Research, University of Bern, Baltzerstr. 1-3, Bern 3012, Switzerland

Norbert Frank
Institute of Environmental Physics, University of Heidelberg, Im Neuenheimer Feld 229, Heidelberg 69120, Germany

Claire Waelbroeck
Laboratoire des Sciences du Climat et de l'Environnement, LSCE/IPSL, CNRS-CEA-UVSQ, Université de Paris-Saclay, Domaine du CNRS, bât. 12, Gif-sur-Yvette 91198, France

Grażyna M. Durak
Marine Biological Association, The Laboratory, Citadel Hill, Plymouth, Devon PL1 2PB, UK
University of Konstanz, Department of Chemistry, Physical Chemistry, Universitätsstr. 10, Box 714, D-78457 Konstanz, Germany

Charlotte E. Walker, Declan Schroeder and Glen L. Wheeler
Marine Biological Association, The Laboratory, Citadel Hill, Plymouth, Devon PL1 2PB, UK

Alison R. Taylor
Department of Biology and Marine Biology, University of North Carolina Wilmington, 601 South College Road, Wilmington, North Carolina, 28403-5915, USA

Ian Probert, Colomban de Vargas and Stephane Audic
Station Biologique de Roscoff, Place Georges Teissier, 29680 Roscoff, France

Colin Brownlee
Marine Biological Association, The Laboratory, Citadel Hill, Plymouth, Devon PL1 2PB, UK
School of Ocean and Earth Sciences, University of Southampton, National Oceanography Centre, Southampton SO14 3ZH, UK

Armin Köhl and Detlef Stammer
Institute of Oceanography, Center for Earth System Research and Sustainability (CEN), University of Hamburg (UHH), Hamburg 20146, Germany

Chuanyu Liu
Institute of Oceanography, Center for Earth System Research and Sustainability (CEN), University of Hamburg (UHH), Hamburg 20146, Germany
Key Lab of Ocean Circulation and Waves (KLOCAW), Institute of Oceanology, Chinese Academy of Sciences (IOCAS), Nanhai Road 7, Qingdao 266071, China
Function Laboratory for Ocean and Climate Dynamics, Qingdao National Laboratory for Marine Science and Technology (QNLM), Qingdao 266237, China

Fan Wang
Key Lab of Ocean Circulation and Waves (KLOCAW), Institute of Oceanology, Chinese Academy of Sciences (IOCAS), Nanhai Road 7, Qingdao 266071, China
Function Laboratory for Ocean and Climate Dynamics, Qingdao National Laboratory for Marine Science and Technology (QNLM), Qingdao 266237, China

Zhiyu Liu
State Key Laboratory of Marine Environmental Science (MEL) and Department of Physical Oceanography, College of Ocean and Earth Sciences, Xiamen University, Xiamen 361102, China

Yannick Donnadieu
Laboratoire des Sciences du Climat et de l'Environnement, LSCE-IPSL, CEA-CNRS-UVSQ, Université Paris-Saclay, 91191 Gif sur Yvette, France

Jean- François Deconinck, Emmanuelle Pucéat and Mathieu Moiroud
Biogéosciences Dijon, Université Bourgogne-Franche-Comté, UMR CNRS 6282, Dijon 21000, France

François Guillocheau
Géosciences Rennes, Université de Rennes, UMR CNRS 6118, Rennes 35042, France

Thomas Felis, Cyril Giry, Jürgen Pätzold and Martin Kölling
MARUM—Center for Marine Environmental Sciences, University of Bremen, 28359 Bremen, Germany

Denis Scholz
Institute for Geosciences, Johannes Gutenberg University Mainz, 55099 Mainz, Germany

Madlene Pfeiffer
Alfred Wegener Institute, Helmholtz Centre for Polar and Marine Research (AWI), 27570 Bremerhaven, Germany

Gerrit Lohmann
MARUM—Center for Marine Environmental Sciences, University of Bremen, 28359 Bremen, Germany
Alfred Wegener Institute, Helmholtz Centre for Polar and Marine Research (AWI), 27570 Bremerhaven, Germany

Sander R. Scheffers
Marine Ecology Research Centre, Southern Cross University, Lismore, New South Wales 2480, Australia

Gerald A. Meehl, Aixue Hu and Haiyan Teng
National Center for Atmospheric Research, 3090 Center Green Drive, Boulder, Colorado 80301, USA

F. Leprieur and F. Guilhaumon
UMR 9190 MARBEC, IRD-CNRS-IFREMER-UM, Université de Montpellier, Montpellier 34095, France

D. Mouillot
UMR 9190 MARBEC, IRD-CNRS-IFREMER-UM, Université de Montpellier, Montpellier 34095, France
Australian Research Council Centre of Excellence for Coral Reef Studies, James Cook University, Townsville, Queensland 4811, Australia

D.R. Bellwood and T.P. Hughes
Australian Research Council Centre of Excellence for Coral Reef Studies, James Cook University, Townsville, Queensland 4811, Australia

V. Parravicini
CRIOBE, USR 3278 CNRS-EPHE-UPVD, Labex 'Corail', University of Perpignan, Perpignan 66860, France

D. Huang
Department of Biological Sciences and Tropical Marine Science Institute, National University of Singapore, Singapore 117543, Singapore

P.F. Cowman
Department of Ecology & Evolutionary Biology, Yale University, 21 Sachem St, New Haven, Connecticut 06511 USA

C. Albouy
Département de biologie, chimie et géographie, Université du Québec à Rimouski, 300 Alée des Ursulines, Rimouski, Canada G5L 3A1

W. Thuiller
Laboratoire d'E´cologie Alpine (LECA), Univ. Grenoble Alpes, Grenoble F-38000, France
Laboratoire d'E´cologie Alpine (LECA), CNRS, Grenoble F-38000, France

Erik van Sebille
ARC Centre of Excellence for Climate System Science and Climate Change Research Centre, School of Biological, Earth and Environmental Sciences, University of New SouthWales, Sydney, New SouthWales 2010, Australia

Chris Turney
ARC Centre of Excellence for Climate System Science and Climate Change Research Centre, School of Biological, Earth and Environmental Sciences, University of New SouthWales, Sydney, New SouthWales 2010, Australia

Paolo Scussolini
Earth and Climate Cluster, Faculty of Earth and Life Sciences, VU University, 1081
HV Amsterdam, The Netherlands
Institute for Environmental Studies (IVM), VU University Amsterdam, The Netherlands (P.S.)

Frank J. C. Peeters
Earth and Climate Cluster, Faculty of Earth and Life Sciences, VU University, 1081
HV Amsterdam, The Netherlands

Jonathan V. Durgadoo and Arne Biastoch
GEOMAR Helmholtz Centre for Ocean Research Kiel 24148 Kiel, Germany

Wilbert Weijer
Los Alamos National Laboratory, Los Alamos, New Mexico 87545

Claire B. Paris
Rosenstiel School for Marine and Atmospheric Science, University of Miami, Miami, FL 33149 Florida, USA

Rainer Zahn
Institució Catalana de Recerca i Estudis Avançats (ICREA), 08010 Barcelona, Spain Departament de Física, Institut de Ciència i Tecnologia Ambientals (ICTA), Universitat Autònoma de Barcelona, 08193 Bellaterra (Cerdanyola), Spain

Mathieu Mongin, Mark E. Baird, Richard J. Matear, Andrew Lenton, Mike Herzfeld, Karen Wild-Allen, Jenny Skerratt, Nugzar Margvelashvili and Andrew D.L. Steven
CSIRO Oceans and Atmosphere, Hobart, Tasmania 7000, Australia

Bronte Tilbrook
CSIRO Oceans and Atmosphere, Hobart, Tasmania 7000, Australia
Antarctic Climate and Ecosystems Co-operative Research Centre, Hobart, Tasmania 7000, Australia

Barbara J. Robson
CSIRO Land and Water, Canberra, Australian Capital Territory 2601, Australia

Carlos M. Duarte
Red Sea Research Center, King Abdullah University of Science and Technology, Thuval 23955-6900, Kingdom of Saudi Arabia

Malin S.M. Gustafsson and Peter J. Ralph
Plant Functional Biology and Climate Change Cluster (C3), Faculty of Science, University of Technology Sydney, Sydney, New SouthWales 2007, Australia

M. Holcomb
Marine Biology Department, Centre Scientifique de Monaco, 8 Quai Antoine 1er, Monaco 98000, Monaco

ARC Centre of Excellence in Coral Reef Studies, School of Earth and Environment & Oceans Institute, The University of Western Australia, Crawley, WA 6009, Australia

E. Tambutté, A.A. Venn, N. Segonds, N. Techer, D. Zoccola, D. Allemand and S. Tambutté
Marine Biology Department, Centre Scientifique de Monaco, 8 Quai Antoine 1er, Monaco 98000, Monaco Laboratoire Européen Associé 647 (BIOSENSIB), Centre Scientifique de Monaco- Centre National de la Recherche Scientifique, 8 Quai Antoine 1er, Monaco 98000, Monaco

Didier Swingedouw
Environnements et Paléoenvironnements Océaniques et Continentaux (EPOC), UMR CNRS 5805 EPOC— OASU—Université de Bordeaux, Allée Geoffroy Saint-Hilaire, Pessac 33615, France

Pablo Ortega and Myriam Khodri
LOCEAN/IPSL Sorbonne Universités (UPMC, Univ Paris 06)-CNRS-IRD-MNHN, 4 place Jussieu, Paris F-75005, France

Juliette Mignot
LOCEAN/IPSL Sorbonne Universités (UPMC, Univ Paris 06)-CNRS-IRD-MNHN, 4 place Jussieu, Paris F-75005, France
Climate and Environmental Physics, Physics Institute, University of Bern, Falkenplatz 16, 3012 Bern, Switzerland
Oeschger Centre of. Climate Change Research, University of Bern, Falkenplatz 16, Bern 3012, Switzerland

Eric Guilyardi
LOCEAN/IPSL Sorbonne Universités (UPMC, Univ Paris 06)-CNRS-IRD-MNHN, 4 place Jussieu, Paris F-75005, France
NCAS-Climate, Univeristy of Reading, Reading RG6 6BB, UK

Valérie Masson-Delmotte
Laboratoire des Sciences du Climat et de l'Environnement (Institut Pierre Simon Laplace, CEA-CNRS-UVSQ, UMR8212), 91191 Gif-sur-Yvette, France

Paul G. Butler
School of Ocean Sciences, Bangor University, Menai Bridge, Anglesey LL59 5AB, UK

Roland Séférian
Centre National de Recherches Météorologiques– Groupe d'Etude de l'Atmosphére Météorologique/Groupe de Météorologie de Grande Echelle et Climat, Toulouse 31100, France

María Aranguren-Gassis
Departamento de Ecología y Biología animal, Universidad de Vigo, E36310 Vigo, Spain
4WK Kellogg Biological Station, Michigan State University, 3700 E. Gull Lake Drive, Hickory Corners MI 49060, USA (M.A.-G.)

Enma Elena García-Martín
Departamento de Ecología y Biología animal, Universidad de Vigo, E36310 Vigo, Spain
School of Environmental Sciences, University of East Anglia, Norwich Research Park, Norwich NR4 7TJ, UK (E.E.G.-M.)

Pablo Serret and José Lozano
Departamento de Ecología y Biología animal, Universidad de Vigo, E36310 Vigo, Spain
Estación de Ciencias Marinas de Toralla, Universidad dc Vigo, Toralla island, E-36331 Vigo, Spain

Carol Robinson
School of Environmental Sciences, University of East Anglia, Norwich Research Park, Norwich NR4 7TJ, UK

John Stephens, Carolyn Harris, Niki Gist and Vassilis Kitidis
Plymouth Marine Laboratory, Prospect Place, Plymouth PL1 3DH, UK

Rob Thomas
British Oceanographic Data Centre, Joseph Proudman Building, 6 Brownlow Street, Liverpool L3 5DA, UK

Xavier Crosta, D. Swingedouw, S. Schmidt, S. Capo and V. Marieu
EPOC, UMR CNRS 5805, Université de Bordeaux, Allée Geoffroy St Hilaire, Pessac 33615, France

P. Campagne
EPOC, UMR CNRS 5805, Université de Bordeaux, Allée Geoffroy St Hilaire, Pessac 33615, France
LOCEAN, UMR CNRS/UPMC/IRD/MNHN 7159, Université Pierre et Marie Curie, 4 Place Jussieu, Paris 75252, France
TAKUVIK, UMI 3376 UL/CNRS, Université Laval, 1045 avenue de la Médecine, Quebec City, Quebec, Canada G1V 0A6

I. Closset, A. Martin and M.N. Houssais
LOCEAN, UMR CNRS/UPMC/IRD/MNHN 7159, Université Pierre et Marie Curie, 4 Place Jussieu, Paris 75252, France

G. Massé
LOCEAN, UMR CNRS/UPMC/IRD/MNHN 7159, Université Pierre et Marie Curie, 4 Place Jussieu, Paris 75252, France
TAKUVIK, UMI 3376 UL/CNRS, Université Laval, 1045 avenue de la Médecine, Quebec City, Quebec, Canada G1V 0A6

E. Devred
TAKUVIK, UMI 3376 UL/CNRS, Université Laval, 1045 avenue de la Médecine, Quebec City, Quebec, Canada G1V 0A6

Samuel L. Jaccard
Institute of Geological Sciences and Oeschger Centre for Climate Change Research, University of Bern, 3012 Bern, Switzerland

Olivier Cartapanis
Institute of Geological Sciences and Oeschger Centre for Climate Change Research, University of Bern, 3012 Bern, Switzerland
Department of Earth and Planetary Sciences, McGill University, Montreal, Canada H3A 2A7

Daniele Bianchi
epartment of Earth and Planetary Sciences, McGill University, Montreal, Canada H3A 2A7
School of Oceanography, University of Washington, Seattle, Washington 98105, USA
Department of Atmospheric and Oceanic Sciences, University of California Los Angeles, Los Angeles, California 90095-1565, USA

Eric D. Galbraith
Department of Earth and Planetary Sciences, McGill University, Montreal, Canada H3A 2A7
Institució Catalana de Recerca i Estudis Avancats (ICREA), 08010 Barcelona, Spain
Institut de Ciència i Tecnologia Ambientals and Department of Mathematics, Universitat
Autonoma de Barcelona, 08193 Barcelona, Spain

Ayako Yamamoto
Department of Atmospheric and Oceanic Sciences, McGill University, 805 Sherbrooke StreetWest, Montreal, Quebec, Canada H3A 2K6

Jaime B. Palter
Department of Atmospheric and Oceanic Sciences, McGill University, 805 Sherbrooke StreetWest, Montreal, Quebec, Canada H3A 2K6
Graduate School of Oceanography, University of Rhode Island, Narragansett Bay Campus, Narragansett, Rhode Island 02882, USA

Tom Flodén
Department of Geological Sciences, Stockholm University, Stockholm 106 91, Sweden

Jan Backman, Matthew O'Regan, Natalia Barrientos Macho, Helen Coxall, Björn Eriksson, Carina Johansson and Rezwan Mohammad
Department of Geological Sciences, Stockholm University, Stockholm 106 91, Sweden
Bolin Centre for Climate Research, Stockholm University, Stockholm 106 91, Sweden

Martin Jakobsson
Department of Geological Sciences, Stockholm University, Stockholm 106 91, Sweden
Bolin Centre for Climate Research, Stockholm University, Stockholm 106 91, Sweden
UNIS - The University Centre in Svalbard, Longyearbyen N-9171, Svalbard

Riko Noormets
UNIS - The University Centre in Svalbard, Longyearbyen N-9171, Svalbard

Johan Nilsson
Bolin Centre for Climate Research, Stockholm University, Stockholm 106 91, Sweden
Department of Meteorology, Stockholm University, Stockholm 106 91, Sweden

Leif Anderson and Göran Björk
Department of Marine Sciences, University of Gothenburg, Gothenburg 405 30, Sweden

Thomas M. Cronin and Laura Gemery
US Geological Survey Reston, 12201 Sunrise Valley Drive, Reston, Virginia 20192, USA

Nina Kirchner
Department of Physical Geography, Stockholm University, Stockholm 106 91, Sweden

Alexey Khortov
National Research Tomsk Polytechnic University, Tomsk 634050, Russia

Andrey Koshurnikov and Roman Ananiev
National Research Tomsk Polytechnic University, Tomsk 634050, Russia
Department of Geocryology, Moscow State University, Moscow 119991, Russia

Larry Mayer and Kevin Jerram
Center for Coastal and Ocean Mapping, University of New Hampshire, 24 Colovos Road, Durham, New Hampshire 03824, USA

Denis Cherniykh and Igor Semiletov
National Research Tomsk Polytechnic University, Tomsk 634050, Russia
Russian Academy of Sciences, Pacific Oceanological Institute, 43 Baltiiskaya Street, Vladivostok 690041, Russia

Christian Stranne
Department of Geological Sciences, Stockholm University, Stockholm 106 91, Sweden
Bolin Centre for Climate Research, Stockholm University, Stockholm 106 91, Sweden
Center for Coastal and Ocean Mapping, University of New Hampshire, 24 Colovos Road, Durham, New Hampshire 03824, USA

örjan Gustafsson
Bolin Centre for Climate Research, Stockholm University, Stockholm 106 91, Sweden
Department of Environmental Science and Analytical Chemistry, Stockholm University, Stockholm 106 91, Sweden

Qian Yang and Timothy H. Dixon
School of Geosciences, University of South Florida, 4202 E Fowler Avenue, Tampa, Florida 33620, USA

Paul G. Myers
Department of Earth and Atmospheric Sciences, University of Alberta, 1-26 ESB, Edmonton, Alta, Canada T6G 2E3

Jennifer Bonin and Don Chambers
College of Marine Science, University of South Florida, St. Petersburg, Florida 33701, USA

M.R. van den Broeke
Institute for Marine and Atmospheric Research Utrecht, Utrecht University, P.O. Box 80.005, 3508 TA, Utrecht, The Netherlands

Wenju Cai
CSIRO Oceans and Atmosphere, Aspendale, Victoria 3195, Australia

Matthew H. England
Climate Change Research Centre, University of New South Wales, Sydney, New
South Wales 2052, Australia
ARC Centre of Excellence for Climate System Science, University of New South Wales, Sydney, New South Wales 2052, Australia

Ariaan Purich
CSIRO Oceans and Atmosphere, Aspendale, Victoria 3195, Australia
Climate Change Research Centre, University of New South Wales, Sydney, New
South Wales 2052, Australia
ARC Centre of Excellence for Climate System Science, University of New South Wales, Sydney, New South Wales 2052, Australia

Tim Cowan
CSIRO Oceans and Atmosphere, Aspendale, Victoria 3195, Australia
School of GeoSciences, University of Edinburgh, Edinburgh EH9 3FE, UK

Zunli Lu, Xiaoli Zhou, Kristina M. Gutchess and Wanyi Lu
Department of Earth Sciences, Syracuse University, Syracuse, New York 13244, USA

Luke Jones, Rosalind E.M. Rickaby and Babette A.A. Hoogakker
Department of Earth Sciences, University of Oxford, Oxford OX1 3AN, UK

Claus-Dieter Hillenbrand
British Antarctic Survey, Cambridge CB3 0ET, UK

Ellen Thomas
Department of Geology and Geophysics, Yale University, New Haven, Connecticut, USA

Index

A

Agung-like Eruption, 141

Aragonite Crystal, 133, 135

Aragonite Saturation, 124, 126-128

Asexually Reproducing Organisms, 10, 13

Atlantic Meridional Overturning, 38-39, 45, 141-142, 169, 176, 178-179, 196, 202

Atlantic Meridional Overturning Circulation, 38-39, 45, 141-142, 151-152, 169, 176, 178, 196

Atoll-reef Ecosystems, 31-35

Autotrophic, 36, 129, 131, 153-156, 160

B

Bathymetric Slope, 30-33, 35-36

Benthic Diatoms, 24-26, 29

Bidecadal, 141-143, 145, 147-151, 200, 202

Biogeochemical Cycling, 24, 35, 160

Biogeophysical Drivers, 31, 33, 35

Biological Processes, 47, 128-129

Biomass Parameters, 139

Biomineralized Cell Walls, 24

C

Catch Reconstruction, 15, 19-20, 22-23

Centennial Timescales, 1-2

Circumnavigation, 2-3, 7-8

Climate Models, 35, 38, 42-43, 45, 78, 90, 99-100, 104, 142, 149, 152, 211

Connectivity Modeling System, 117

Coral Calcification, 125, 129-130, 133, 135, 137, 139

Coral Reef Islands, 30-33, 35

Coral Skeleton, 132-133, 135-137, 139

D

Dark Global Ocean, 1, 3, 5

Declining Discards, 15, 17

Deep-ocean Ventilation, 47, 51

Deep-sea Sediments, 56, 171-172

Deglaciation, 47-48, 51, 53-57, 100, 122, 176, 195, 218

Diatoms, 24-27, 29, 58-64, 66-69, 163-168, 170, 217

Discarded Bycatch, 15, 17

Dissolved Organic Matter, 1, 8

Dissolved Silicic Acid, 24-25

Dsi Fluxes, 24-25, 27

E

Epipelagic Zone, 154-158

Exoskeleton, 124

F

Fisheries Catches, 15-17, 19, 22-23

Fluorescence Intensity, 5, 7

Fluorophores, 1-2

Footprint, 22, 37, 114, 116-118, 129

Foraminifer Coatings, 48, 52, 55

Foraminiferal Habitat, 217

Foraminiferal Shell, 48

G

Gfdl Model Suite, 45

Glacial Period, 47-48, 51, 53-57, 173, 213, 218

Global Marine, 15, 22, 36, 108

Global Ocean, 1-3, 5, 7-10, 12-14, 44, 49, 53, 57, 79, 98-99, 125, 131, 154, 167, 175, 177, 187, 193, 213, 217

Great Barrier Reef, 36, 114, 124-125, 129-131

H

Humic-like Components, 2-3

I

Infeasible Number, 10, 13

Interglacials, 93-94, 98, 171, 173, 217

Island Mass Effect, 30-31, 33, 36

L

Lagrangian Particle Tracking, 9, 12

Last Glacial Cycle, 57, 171, 173, 176, 195

Last Millennium, 141-143, 146, 150, 152

M

Marine Ecosystem, 30

Marine Fisheries, 15-17, 19-20, 22-23

Marine Isotope, 51, 56, 186-187, 213, 218

Marine Microorganisms, 116

Millennial-scale, 47-48, 51-57, 100, 122

Multidecadal Variability, 99, 101, 145, 152, 178-179, 184-185

N

Net Community Production, 154, 156, 160-161

O

Ocean Acidification, 9, 14, 124-126, 128-133, 137, 139

Ocean Currents, 10, 13-14, 31-32, 35, 79, 84, 116, 119, 130

Ocean-atmosphere Co2, 47-48

Oceanographic, 10, 14, 36, 41, 43, 45, 55, 71, 142, 150, 153, 170, 190, 192-193, 197, 219

Oligotrophic Gyres, 153, 158

Oligotrophic Ocean, 153-154, 156-158, 160

Organic Carbon, 7-8, 37, 47-49, 51-56, 158, 160, 171-177

Organic Carbon Pump, 47-48, 52-53

Organic Matrix, 59, 132, 136-137

Organic Matter, 1-2, 8, 48-49, 54, 56, 153-154, 157-158, 160, 171, 173-177, 217

Oscillation Index, 38, 170

Oxygenation, 51-57, 81-82, 171, 174, 176-177, 212-213, 217-219

P

Palaeoclimatological, 116

Particle Seeding, 13

Phytoplankton Biomass, 30-37

Phytoplankton Production, 30-31, 160

Planktic Foraminifera, 116-117, 121-122

Plankton Metabolism, 153-157, 160-161

Planktonic Communities, 9-10

Planktonic Ecosystem, 154

Proxy Signals, 94, 116-117

R

Reduced Carbon, 1-2

Reef-building Corals, 124-125

Reservoirs, 1, 176

Respiration, 8, 48-49, 53, 125-130, 133, 135, 139, 153-154, 156, 158, 160-161, 174

Respiration Rates, 133, 135, 139, 160

S

Sea Ice Concentrations, 163, 170

Sea Surface Temperature, 73, 77, 93-96, 99, 107, 113, 122, 145, 152, 168, 178, 181, 184, 204, 211

Sea-level Rise, 38-39, 45, 94, 99, 193

Silicic Acid, 24-26, 29, 64, 68, 173, 217, 219

Skeletal Porosity, 132-133, 137-138

Skeleton Morphology, 133, 136-137

Slr Rates, 39-40, 43-44

Small-scale Fisheries, 15-16, 20, 22

Species, 10, 13-14, 20, 22, 24-27, 49, 51, 55, 59-63, 66, 68-69, 98, 108-110, 112-117, 120-122, 125, 128, 130, 135-137, 139, 165-170, 213-215, 218-219

Stylophora Pistillata, 132-133, 137, 139-140

T

Tide Gauge, 37-39, 45

Timescales, 1-2, 9-10, 12-13, 26, 53-54, 76, 97, 102, 130, 142, 145, 166, 171-172, 184, 192, 206-207

Trophic Levels, 25, 30-31, 34, 109, 114

U

Unicellular Algae, 24-25